T0192809

International Series in Operations Research & Management Science

Founding Editor
Frederick S. Hillier, Stanford University, Stanford, CA, USA

Volume 333

The book series **International Series in Operations Research and Management Science** encompasses the various areas of operations research and management science. Both theoretical and applied books are included. It describes current advances anywhere in the world that are at the cutting edge of the field. The series is aimed especially at researchers, advanced graduate students, and sophisticated practitioners.

The series features three types of books:

• Advanced expository books that extend and unify our understanding of particular areas.

• Research monographs that make substantial contributions to knowledge.

• Handbooks that define the new state of the art in particular areas. Each handbook will be edited by a leading authority in the area who will organize a team of experts on various aspects of the topic to write individual chapters. A handbook may emphasize expository surveys or completely new advances (either research or applications) or a combination of both.

The series emphasizes the following four areas:

Mathematical Programming : Including linear programming, integer programming, nonlinear programming, interior point methods, game theory, network optimization models, combinatorics, equilibrium programming, complementarity theory, multiobjective optimization, dynamic programming, stochastic programming, complexity theory, etc.

Applied Probability: Including queuing theory, simulation, renewal theory, Brownian motion and diffusion processes, decision analysis, Markov decision processes, reliability theory, forecasting, other stochastic processes motivated by applications, etc.

Production and Operations Management: Including inventory theory, production scheduling, capacity planning, facility location, supply chain management, distribution systems, materials requirements planning, just-in-time systems, flexible manufacturing systems, design of production lines, logistical planning, strategic issues, etc.

Applications of Operations Research and Management Science: Including telecommunications, health care, capital budgeting and finance, economics, marketing, public policy, military operations research, humanitarian relief and disaster mitigation, service operations, transportation systems, etc.

This book series is indexed in Scopus.

Fausto Pedro García Márquez • Benjamin Lev
Editors

Sustainability

Cases and Studies in Using Operations Research and Management Science Methods

Editors
Fausto Pedro García Márquez
Ingenium Research Group
University of Castilla-La Mancha
Ciudad Real, Spain

Benjamin Lev
LeBow College of Business
Drexel University
Philadelphia, PA, USA

ISSN 0884-8289 ISSN 2214-7934 (electronic)
International Series in Operations Research & Management Science
ISBN 978-3-031-16622-8 ISBN 978-3-031-16620-4 (eBook)
https://doi.org/10.1007/978-3-031-16620-4

This Springer imprint is published by the registered company Springer Nature Switzerland AG
The registered company address is: Gewerbestrasse 11, 6330 Cham, Switzerland

Preface

Sustainability is being an important discipline for the society and governments, a research field that requires the use and application in Management Science (MS) and Operations Research (OR). In this book, we introduce the topics of MS and OR together with Sustainability. We start with basic concept and will present cases, applications, theory, and potential future. The authors contribute chapters to the book to cover as wide array of topics as space permits. Examples are from renewable energy, smart industry, city, transportation, home and smart devices. We discuss future applications, trends, and potential future of this new discipline.

This book is intended for professionals in the field of energy, engineering, information science, mathematics, economics, and researchers who wish to develop new skills in sustainability, or who employ the sustainability discipline as part of their work. The authors of this volume describe their original work in the area or provide material for cases and studies successfully applying the sustainability discipline in real-life cases and theoretical approaches.

Ciudad Real, Spain
Philadelphia, PA

Fausto Pedro García Márquez
Benjamin Lev

Contents

About the Editors

Fausto Pedro García Márquez works at UCLM as Full Professor (Accredited as Full Professor from 2013), Spain, Honorary Senior Research Fellow at Birmingham University, UK, Lecturer at the Postgraduate European Institute, and he has been Senior Manager in Accenture (2013–2014). He obtained his European PhD with a maximum distinction. He has been distinguished with the prices: Grand Prize (2021), Runner Prize (2020) and Advancement Prize (2018), Runner (2015), Advancement (2013) and Silver (2012) by the International Society of Management Science and Engineering Management (ICMSEM), First International Business Ideas Competition 2017 Award (2017). He has published more than 200 papers (125 JCR (58Q1; 30Q2; 33Q3; 4Q4), 65 % ISI, 30% JCR and 92% internationals), some recognized as: "Applied Energy" (Q1, IF 9.746, as "Best Paper 2020"), "Renewable Energy" (Q1, IF 8.001, as "Best Paper 2014"); "ICMSEM" (as "excellent"), "Int. J. of Automation and Computing" and "IMechE Part F: J. of Rail and Rapid Transit" (most downloaded), etc. He is author and editor of more than 45 books (Elsevier, Springer, Pearson, Mc-GrawHill, Intech, IGI, Marcombo, AlfaOmega, etc.), >100 international chapters, and 6 patents. He is Editor of 5 Int. Journals, Committee Member of more than 70 Int. Conferences. He has been Principal Investigator in 4 European Projects, 8 National Projects, and more than 150 projects for Universities, Companies, etc. He is being: Expert in the European Union in AI4People (EISMD), and ESF; Director of www.

ingeniumgroup.eu; Senior Member at IEEE, 2021–. . .; Honoured Honourary Member of the Research Council of Indian Institute of Finance, 2021–. . .; Committee Chair of The International Society for Management Science and Engineering Management (ISMSEM), 2020–. . . . His main interests are: Artificial Intelligence, Maintenance, Management, Renewable Energy, Transport, Advanced Analytics, Data Science. More information in https://blog.uclm.es/faustopedrogarcia

Benjamin Lev is the University Trustee Professor at LeBow College of Business, Drexel University, Philadelphia, Pennsylvania, USA. He holds a PhD in Operations Research from Case Western Reserve University. Prior to joining Drexel University, Dr. Lev held academic and administrative positions at Temple University, University of Michigan-Dearborn, and Worcester Polytechnic Institute. He is the Editor-in-Chief of OMEGA-The International journal of Management Science www.OmegaJournal.org; Co-Editor-in-Chief of International Journal of Management Science and Engineering Management; and serves on several other journal editorial boards (INFORMS JAA (formerly Interfaces), IAOR, ORPJ, Financial Innovation, OPSEARCH, IDIM, IIE-Transactions, ERRJ, INFORMS JOR). He currently holds faculty appointments at seven Chinese Universities (Beijing Jiaotong University, Chengdu University, Nanjing University of Aeronautics and Astronautics, Nanjing University of Information Science and Technology, Nanjing Audit University, Xidian University, Tianjin University). Professor Lev is the former TIMS VP-Meetings and INFORMS VP-Meetings. INFORMS Fellow since 2003. He has published sixteen books, numerous articles and organized many national and international conferences including ORSA/TIMS Philadelphia; INFORMS Israel; IFORS Scotland; ALIO-INFORMS Buenos Aires; ICMSEM Philadelphia. https://en.wikipedia.org/wiki/Benjamin_Lev

Sustainability: Cases and Studies in Using Operations Research and Management Science Methods

Fausto Pedro García Márquez and Benjamin Lev

Sustainability is being an important discipline for the society and governments, a research field that require the use and application in Management Science (MS) and Operations Research (OR). In this book, we introduce the topics of MS and OR together with Sustainability. We start with basic concept and present cases, applications, theory, and potential future. Authors contribute chapters to the book to cover as wide array of topics as space permits. Examples are from renewable energy; smart industry; city; transportation; home and smart devices, etc. We discuss future applications, trends, and potential future of this new discipline.

Sustainability, a challenge for mankind for our planet, can be understood as the quality of being able to continue over a period of time, where Management Science (MS) and Operations Research (OR) are being used to that propose. MS and OR could include machine, equipment, human, animal, anything. It could transfer text, signals, media, software, etc. To build and manage sustainability, new disciplines have emerged in MS&OR such as advanced analytics, machine learning, etc. Specifically, this book provides an interface between the main disciplines of engineering/technology and the organizational, economic, administrative, and planning capabilities of managing sustainability.

This book aims to provide relevant theoretical frameworks and the latest empirical research findings in sustainability in Management Science and Operations Research. It is written for professionals who want to improve their understanding of the strategic role of sustainability at various levels of the information and knowledge organization, that is, sustainability at the global economy level, at

F. P. García Márquez (✉)
Ingenium Research Group, University of Castilla-La Mancha, Ciudad Real, Spain
e-mail: FaustoPedro.Garcia@uclm.es

B. Lev
LeBow College of Business, Drexel University, Philadelphia, PA, USA
e-mail: bl355@drexel.edu

F. P. García Márquez, B. Lev (eds.), *Sustainability*, International Series in Operations Research & Management Science 333,
https://doi.org/10.1007/978-3-031-16620-4_1

networks and organizations, at teams and work groups, of information systems and, finally, sustainability at the level of individuals as players in the networked environments.

This book is intended for professionals in the field of energy, engineering, information science, mathematics, economics, and researchers who wish to develop new skills in sustainability, or who employ the sustainability discipline as part of their work. The authors of this volume describe their original work in the area or provide material for cases and studies successfully applying the sustainability discipline in real-life cases and theoretical approaches.

Sustainability and sustainable development have specifically focused on people's well-being and equity levels. The basic objective of the present research study is to elaborate the meaning of sustainability and sustainable development for laymen or beginners. In chapter "What Is Sustainability? A Layman Perspective", the topics of city sustainability, business sustainability, dimensions and pillars of sustainability are discussed in depth. Nowadays, the developing countries are facing severe socio-economic pressure and environmental challenges to achieve sustainable development goals. The challenges faced by developing countries include lack of financial resources, political instability, pollution, emission of smoke from vehicles, factories, or brick kilns, water contamination, food adulteration, social inequalities, poverty, and population growth. Sustainability and sustainable development have specifically addressed the people's well-being and equity. Developing countries should develop and implement social policies to promote social unity, tolerance, and justice, while integrating economic, social, and environmental paradigms of sustainability. Several examples are discussed concerning developing and developed countries. Finally, this study highlights the challenges and opportunities of sustainability in developing countries. Thorough this chapter, a beginner will understand the core concepts of sustainability and sustainable development in the context of developing and developed countries.

With the emergence of the concept of sustainability, sustainable supply chain management (SSCM) has become a challengeable area for both academics and industrialists since sustainable supply chain is the cornerstone of a company to achieve sustainable goals. Sustainable supply chain management has been a hot research topic and many achievements on SSC have been obtained from different angles. To help readers understand the advances of this field and explore future research directions, chapter "Sustainable Supply Chain Management: Definition, Bibliometrics, Applications and Future Directions" firstly distinguishes different statements regarding the sustainability of supply chain and then specify the concept of SSC. Afterwards, a bibliometrics analysis regarding the publications on SSCM is provided. Given that the fuzzy set theory has been widely used for information expression in SSCM, the applications of fuzzy set theory in SSCM are reviewed.

Machine Learning, a branch of artificial intelligence (AI) and computer science, focuses on the usage of data and algorithms to copy the humans learning method, slowly increasing its accurateness. Chapter "Sustainable Supply Chain Management: Definition, Bibliometrics, Applications and Future Directions" aims at discussing the applications of the machine learning algorithms, essential for developing predictive modeling and for carrying out classification and prediction in both supervised and unsupervised scenarios. The Machine Learning techniques have been applied to many application domains as a result of a humongous amount of data being created, processed, and mined from the evolution of the World Wide Web, mobile applications, and the rise of social media applications. Some of these applications are virtual personal assistants, predictions, surveillance, social media services, malware filtering, search engine result refining, and online fraud detections. The chapter includes the introduction, State of the Art, Machine Learning Algorithms, Applications of Machine Learning Algorithms in data sciences, followed by conclusion and future recommendations.

In recent years, social networking sites have gained popularity on the Internet, attracting hundreds of thousands of users who spend billions of minutes utilizing them. Social networking sites have now taken over many people's life. Every day, a large number of people create profiles on social media sites and contact others regardless of their location or time. Users' personal information is at risk on social networking sites posing security risks. It is important to identify the bogus user's social media identities so that the person who is pushing threats can be uncovered. In chapter "The Machine Learning Model for Identifying Bogus Profiles in Social Networking Sites", the Ensemble learner approach helps to increase the accuracy of detecting fake profiles through automated user profile classification. The study investigates the different text, image, audio, and video attributes that can be used to discern the difference between a fake and a legitimate account. The Ensemble learner approach utilizes these attributes and assesses their performance on real-world social media datasets using performance metrics such as accuracy, precision, recall, and F-1 score. The experimental evaluation indicates that the average accuracy achieved by the proposed ensemble learner approach is 97.67%, precision of 92.56%, and a recall of 82%, compared to learning models (Support Vector Machine (SVM), K-Nearest Neighbors (KNN), Logistic regression).

Wind energy is a growing market due to improved monitoring systems, decreased downtime, and optimal predictive maintenance. Classification algorithms based on machine learning efficiently categorize large volumes of complex data. K-Fold cross validation is a competent tool for measuring the success rate of classification algorithms. The purpose of chapter "Support Vector Machine and K-fold Cross Validation to Detect False Alarms in Wind Turbines" is to compare the performance of various kernel functions for the support vector machine algorithm, as well as to analyze the different values of K-fold cross-validation and holdout validation. The approach is applied to a real dataset of wind turbine, with the aim to detect false alarms. The results indicate an accuracy of 99.2%, and the F_1 is 0.996. These values indicate that the methodology proposed is efficient to detect and identify false alarms.

The field of artificial intelligence and, in particular, that which deals with artificial neural networks, is experiencing a great interest in companies for the inspection and verification of images. The current trend in the sector is to implement these novel methodologies in industrial environments so that they can benefit from their advantages over traditional systems. Quality and production managers are increasingly interested in replacing the classic inspection methods with this new approach due to its flexibility and precision. Traditional methods have some weaknesses when it comes to inspecting parts, such as sensitivity to disturbances. In an industrial environment these disturbances can be changes in lighting during the day or during the year, the appearance of external elements such as dust or dirt. The use of new convolutional neural network techniques allows training including disturbance scenarios, teaching the artificial neural network to detect non-verse defects influenced by changes in light or by the appearance of dust. In this way, it is possible to drastically reduce false positives, avoiding costly stops in production and maximizing the precision of detection and classification of each defect. Chapter "A Hybrid Neural Network Model Based on Convolutional Cascade Neural Networks: An Application for Image Inspection in Production" studies the implementation of a hybrid model based on a cascade detection neural network with a classification neural network in an industrial environment.

For urban road networks with hazardous materials (hazmat) transportation, a Multi-Objective Sustainable traffic Control (MOSAC) is proposed in chapter "A Multi-objective Sustainable Traffic Signal Control for Smart Cities Under Uncertainty". A multi-objective stochastic program is proposed to simultaneously minimize total travel delay and reduce maximum risk exposure to all road users. Following Wardrop's principle, traveler's route choice is considered. In order to effectively solve the proposed stochastic program, a novel heuristic is proposed. Numerical experiments using a realistic road network are performed to demonstrate feasibility of proposed approach. For varying traffic conditions, computational comparisons are also made with recently proposed one, indicating that system performance can be greatly improved by MOSAC in all cases.

Traffic flow on highways exhibits a dynamic phenomenon in different operational settings. The concept of sustainability in transportation engineering is elucidated using multi-criteria decision making analysis (MCDA), a discipline of Operation Research (OR) which is in wide range of application and practice in real-time traffic engineering. The principle criteria considered in the study of short-term prediction of traffic flow rate in chapter "Multi Criteria Decision Making for Sustainable Transport: A Case Study on Traffic Flow Prediction Using Spatial-Temporal Traffic Sequence" are spatial and temporal information. The spatial–temporal components of physical traffic flow are the parametric measures. Exploring the intrinsic relationship between these measures helps in realizing the dynamics of traffic flow rate. Hence, the objective of this chapter is to elucidate the significance of MCDA considering the spatial and temporal measures of traffic flow rate. Also, this chapter presents a case study on formulation of algorithm for the prediction of traffic flow rate on highways. Experimental results are reported with estimation of time complexity of algorithms.

As an energy conversion device that converts hydrogen into electrical energy, the fuel cell provides high power density with a compact structure and is thus considered as a good candidate for a wide variety of future power applications. However, its performance is suffered from multiple failure mechanisms, this is the reason why sustainability and reliability become the barriers to mass deployment of the fuel cell.

In order to predict the fuel cell state of health (SOH) and further estimate its remaining useful lifetime, prognostic and health management (PHM) is a powerful technology. As a crucial step of PHM, the prognostics use a set of monitoring fuel cell data to indicate the future aging performance. Chapter "A Robust Prognostic Indicator for Renewable Energy Fuel Cells: A Hybrid Data-Driven Prediction Approach" proposes a novel hybrid prognostic approach for fuel cell aging prediction based on a combination of state space model and machine learning methods. The proposed method can capture both the fade trend and nonlinear features of aging data. Experiment results demonstrate the accuracy and robustness. The major contributions can be summarized as follows: (1) The proposed hybrid prognostic approach considers state space model and machine learning methods, where the state space model method can be efficiently used to forecast the aging data trend in a long-range time, and machine learning method can accurately describe the local non-linear characteristics for the short-range time prediction. Therefore, the proposed prognostic algorithm can achieve a higher prediction accuracy compared with conventional single method; (2) The moving window method is applied to iteratively update the prediction process based on newly measured data. During the iterative process, each dataset in one window is divided into three phases: training, evaluation, and prediction. By using this iterative design, both prediction models can be efficiently retrained in dynamic, and then fused in real time at each moving window step; (3) The prediction accuracy and robustness of the proposed method have been experimentally validated and compared with other algorithms from literature. Experiment results show that the proposed method has strong robustness properties to stabilize the errors for long-term forecasting. This performance robustness is meaningful for fuel cell maintenance management in terms of earlier decision-making, more effective cost minimization, and risk mitigation.

Chapter "An Assessment of Electricity Markets in Turkey: Price Mechanisms, Regulations, and Methods" is focused on the price determination problem of the Turkish day-ahead electricity market. Management science techniques are used to analyze the problem. An optimization objective function is defined and explained. Once the problem formulation is stated, and the price is compared to the market clearing price, there is an acceptance guarantee for the affordable block, flexible, and, specifically, paradoxical bids (offers). The objective function of the economic model has been determined by maximizing the total market surplus as a requirement of a market that operates efficiently. The goal is to find the best solution using the algorithms and procedures that have been developed. The problem of price determination in the day-ahead electricity markets includes many research subjects and development areas. With the ongoing research in this field, it will be useful to academically examine both the suggestions from the sector and the applications in the developed markets. Thereby, some of the applications and simulations can be put into practice.

Chapter "The Role of DevOps in Sustainable Enterprise Development" considers the problem of sustainable enterprise development from the perspective of the organization of software development. The goal is to provide a comprehensive analysis of the characteristics of the DevOps approach and present how it enables organization of continuous software development. By utilizing these DevOps principles, an enterprise can quickly adapt to ever-changing business requirements while fostering an environment capable of sustainable growth. This growth is primarily enabled by a shift in culture, where teams are encouraged to be functionally diverse, which allows them to consider the whole lifecycle of an activity, rather than one single aspect. A case study of a university research and development center is presented to demonstrate the peculiarities arising from the DevOps approach. Conclusions and recommendations for future works are drawn from the literature analysis and own experiences.

Chapter "Practices and Indicators of Waste and Resource Management in Commercial Buildings" examines the indicators and practices used by facilities managers regarding waste and resource management in commercial buildings and whether these current practices and indicators align with standards impacting facility sustainability and performance. A review of existing literature was undertaken to outline relevant information, which addresses the primary objectives of the report. The research adopted a qualitative approach that employs an embedded case study research design to obtain primary research data from interviews with facilities managers of commercial buildings. Furthermore, secondary data relating to energy, water, and waste was collected from company records. The scope of the study was limited to three Green Building Council of South Africa (GBCSA) Green star-rated (rating of green buildings in South Africa) commercial buildings in Cape Town. The key findings of the research indicate that the level of energy and water consumption is within the prescribed standards. The research concludes that indicators (resource consumption per annum/month/m2) and practices (continuous service planning, facility plant, and equipment maintenance, accessing tenant requirements) used in waste and resource management by the facilities managers of the sampled commercial buildings are aligned with international and local standards and are adequate in addressing the needs of commercial buildings regarding the management of energy and water resources. Based on these findings, the study recommends that facilities managers use innovative technologies such as metering and the Building Management System Software, and cradle-to-cradle technology to manage the waste generated in commercial buildings. It also recommends that further research be undertaken using a larger sample size to allow for the generalization of the results.

Chapter "A Descriptive Comprehension Study on Solar Energy, Solar Products, and Solar Product Marketing in Indian Context" focuses in detail on the three main components, i.e., solar energy, solar product, and solar product marketing and tries to connect them to know their applicable part. The research gap of solar product marketing is also well focused. Geographically, India is an ideal country for solar energy. Despite India is having the potential power to tap solar energy at mass level; however, the solar product users are very less which therefore impacts solar product marketing. The literature shows that this gap is due to the quality of solar products, technology of solar products, unready mindset of consumers, and promotions. The present study is the sum of the aspects of the solar energy sector, marketing of solar products, and perspective of Indian consumers; thus, this study will benefit the researchers, academicians, and business practitioners, and this study stands out from the past literature as it covers data from various sources (research papers, government agencies, and business sector) from the year 2000 to 2020. Subsequently going through this research work the researchers, academicians, and business practitioners will get to know the status of solar energy and solar product marketing.

Precision Livestock Farming (PLF) plays a key role in the advancement of animal housing, since it is associated with the improvement of animal's health and welfare status, ensuring sustainability and efficiency of farms. The main objective of researchers is the development of systems for real-time continuous monitoring of the animals' everyday lives (i.e., animal-centric tools). Such systems based on both steady state and dynamic models should have low installation costs, be precise, accurate, easy to use, and environmentally friendly and provide the farmers with valuable information serving as decision support tools for the improvement of management practices. The data could be collected within the unit by simple sensors such as accelerometers, RFID sensors, etc., or more complex computer-based vision or sound and audio analysis systems. Chapter "Precision Livestock Farming (PLF) Systems: Improving Sustainability and Efficiency of Animal Production" presents various PLF systems in basic livestock (i.e., dairy cows, sheep and goats, pigs, and poultry) indicating their benefits upon the production process.

Construction of road pavement consumes significant quantities of raw materials and energy and involves use of heavy machineries. In India, total length of road network is 6.38 million km, out of which highest share (71%) is that of the rural roads. The main aim of chapter "Environmental Assessment of Rural Road Construction in India" is to present the development of detailed life cycle inventory for material phase and construction phase of rural road pavement followed by evaluation of their environmental impacts. The scope of the study included the raw material extraction; the transportation of raw materials to the site; and the construction of rural road using heavy machineries for a typical rural road section in Bihar, India using GaBi software version 10.5. Maintenance and end-of-life phases were beyond the scope. The functional unit adopted was 1 km of road pavement constructed. A total of seven impact categories were considered. The pavement was constructed in three layers, viz., granular sub-base layer, water bound macadam layer, and top bituminous layer. It was found that the granular sub-base layer construction had the highest contribution towards the impact on environment for all impact categories due to the high consumption of aggerates.

Synthetic natural gas (SNG) may be obtained by methanation. Many thermodynamic reaction details involved in this process are not yet fully known. In chapter "Perspectives on the Sustainable Steel Production Process: A Critical Review of the Carbon Dioxide (CO_2) to Methane (CH_4) Conversion Process", a comprehensive thermodynamic analysis of the reactions involved in the methanation of carbon oxides (CO and CO_2) is performed using the Gibbs free energy minimization method. The equilibrium constants of eight reactions involved in the methanation reactions were calculated at different temperatures. The effects of temperature, pressure, ratio of H_2/CO (and H_2/CO_2) and addition of other compounds (H_2O, O_2, and CH_4) in the feed gas on the conversion of CO and CO_2, CH_4 selectivity and yield, and carbon capture were carefully studied. It was found that low temperature, high pressure, and large H_2/CO (and H_2/CO_2) ratio are favorable for methanation reactions. However, the conversion of carbon dioxide (CO_2) into methane may be a solution for a new technology.

Exergy and Anergy are important terms for understanding sustainability in the context of energy consumption. Well established in the scientific world, these terms are nearly unknown in public perception. This is astonishing regarding the man driven climate change and the related energy politics. Simply spoken, Exergy is the usable feature of any form of energy and Anergy is its not usable feature. In this sense, Exergy is the necessary driving feature for all living species and for all economic activities of mankind. Our most important sustainable Exergy source is the sun radiation. Chapter "Exergy, Anergy and Sustainability" explains the essence of Exergy and Anergy with its significance in a wide scope of applications like in daily routine, in education, in statistics as well as in science and technology development. A well-adapted understanding of these notations supports sustainable behavior within all education levels. A generalist reader of this chapter may omit reading the (later) subchapters addressed to thermodynamic specialists. However, the last chapter on statistics is addressed to generalists as well.

Pollutant concentrations in the environment are functions of source variables, environmental variables, and time. Relationships among these variables are complex and unknown; therefore, multiple linear regression may not be applicable. Chapter "Simulating Complex Relationships Between Pollutants and the Environment Using Regression Splines: A Case Study for Landfill Leachate" considers Multivariate Adaptive Regression Splines (MARS). MARS is a flexible statistical method, used for machine learning, but it is not typically appropriate for sustainability applications due to limited data. While most machine learning algorithms require a large amount of data and may not yield interpretable results, by contrast, MARS in conjunction with design of experiments has the ability to find important interpretable relationships in limited data. Laboratory data were collected under a controlled environment to discover how leachate composition changes as functions of ambient temperature, rainfall intensities, and waste composition. The MARS application to landfill leachate demonstrates a promise in capturing complex relationships, illustrated via MARS 3D interaction plots.

Finally, as a researcher and author on sustainability, I (Ben) would be remiss not to acknowledge that I have been sustained through the love of family. Since my last book, my family tree has sprouted new roots in new locations, but it continues to sustain me and lays a foundation for generations to come. This remains a source of pride and is by far the greatest life accomplishment of my beloved wife Debbie and I. Thank you for your sustaining dedication and love to my son Ron (M.D.) and his wife Melitza and my daughter Nurit (J.D., Lieutenant Colonel Retired) and her husband James (Colonel Retired) and my five beautiful grandchildren: Hannah (Carnegie Mellon University, B.S., Neuroscience), Jimmy (Junior, Virginia Polytechnic Institute), Veronica, Sebastian, and Arya.

It is also dedicated to my (Fausto) family, and Gino, my best friend!

What Is Sustainability? A Layman Perspective

Muhammad Hashim, Muhammad Nazam, Sajjad Ahmad Baig, Sadia Samar Ali, and Manzoor Ahmad

Abstract Sustainability and sustainable development have specifically focused on people's well-being and equity levels. The basic objective of the present research study is to elaborate the meaning of sustainability and sustainable development for laymen or beginners. In this research study, the topics of city sustainability, business sustainability, dimensions and pillars of sustainability are discussed in depth. Nowadays, the developing countries are facing severe socio-economic pressure and environmental challenges to achieve sustainable development goals. The challenges faced by developing countries include lack of financial resources, political instability, pollution, emission of smoke from vehicles, factories, or brick kilns, water contamination, food adulteration, social inequalities, poverty, and population growth. Sustainability and sustainable development have specifically addressed the people's well-being and equity. Developing countries should develop and implement social policies to promote social unity, tolerance, and justice, while integrating economic, social, and environmental paradigms of sustainability. Several examples are discussed concerning developing and developed countries. Finally, this study highlights the challenges and opportunities of sustainability in developing countries. Thorough this chapter, a beginner will understand the core concepts of sustainability and sustainable development in the context of developing and developed countries.

This research study is composed of the following sections, Sect. 1 discuss the introduction, Sect. 2 describes the literature, Sect. 3 describes the challenges and opportunities, and last Sect. 4 summarize the study.

M. Hashim · S. A. Baig · M. Ahmad
Faisalabad Business School, National Textile University, Faisalabad, Pakistan

M. Nazam (✉)
Institute of Business Management Sciences, University of Agriculture, Faisalabad, Pakistan

S. S. Ali
Department of Industrial Engineering, Faculty of Engineering, King Abdul-Aziz University, Jeddah, Saudi Arabia

F. P. García Márquez, B. Lev (eds.), *Sustainability*, International Series in Operations Research & Management Science 333,
https://doi.org/10.1007/978-3-031-16620-4_2

Keywords Challenges · Developing countries · Development goals · Opportunities · Sustainability · Sustainable City

Abbreviations

GDP	Gross Domestic Product
ICT	Information and Communication Technologies
KPMG	Klynveld Peat Marwick Goerdeler
MDGs	Millennium Development Goals
SDGs	Sustainable Development Goals
SDSN	Sustainable Development Solutions Network
TBL	Triple Bottom Line
UNDESA	United Nations Department of Economic and Social Affairs

1 Introduction

Sustainability is a new idea that has gained popularity after the movements of social justice, conservationism, internationalism, and global warming. Sustainability requires social responsibility, environmental protection, and a sustainable society's balance between human and natural systems. Nowadays, societies face economic, environmental, and social problems across the world. Due to this reason, the sustainability principles are getting the attention of practitioners and academicians. Usually, sustainability describes the process or actions that ease the way for mitigation of reduction in the availability of natural resources and to maintain ecological balance at the world level. The concept of sustainability is applied to all sectors, e.g., textile, agriculture, infrastructure, urbanization, water availability, energy use, and transportation (Youmatter, n.d.; Emerald Built Environments_Green Building Consulting, n.d.; Allam & Jones, 2021).

The urban population has significantly increased globally since the industrial revolution. Urbanization attracts large numbers of families and skilled or unskilled workers for exploring better employment opportunities, quality of life, higher standard of living, better education, better infrastructure, sustainable transportation, and so on. Growth in urbanization was observed first in America, Europe, Africa, and Asia's larger megacities (Vardoulakis & Kinney, 2019). United Nations statistics indicate that the population in urban areas has been steadily increasing since the last century. Due to economic growth, better urban infrastructure, and adequate employment opportunities, people are migrating from rural areas to urban areas. Around 1.1 billion people will migrate from rural areas to urban areas in Asian countries to explore better opportunities in the coming years (UN Environment Programme, n.d.). A majority of the world's population (approximately 55% or 70 million people) lives in urban areas. As per United Nations estimates, the urban population will rise to 70% of the world's population (approximately 6.3 billion populations in urban areas, two-thirds of the world population) by 2050 (United Nations Water and Urbanizations n.d.). Six hundred big cities contribute about the world's 60% Gross Domestic Product (GDP). Globally, urban areas generate approximately 55% of the national GDP in developing countries and 85% in developed countries. Across the Asia Pacific region, urban areas contribute 80% of GDP, which support economic growth and ultimately improve peoples' lifestyles. The estimated annual investment in urban infrastructure and buildings by 2050 will increase $20 trillion (Unsdsn, n.d.). Globally, urban cities consume more than 75% of the total world's available natural resources across the world like raw materials, food, water, energy, fossil fuels, etc. (Swilling et al., 2018). Sustainable cities, also known as an eco-cities and green cities that considered an essential sustainability pillar (triple bottom line), i.e., social, economic, and environmental areas (Catalano et al., 2021).

The sustainable development city concept was introduced in the last twentieth century. A sustainable city is defined as a city, which fulfills the needs of all people and enhances their well-being without damaging the natural resources, environment, and safe resources for future generations(Girardet, 1999). The world is facing issues in urban areas, like increasing the human population, inadequate urban infrastructure, air, water, and land pollution, traffic congestion, degradation environment, non-availability of green and open spaces due to deforestation, lack of necessities of life related to appropriate system of food, water supply, sanitation, and waste management. Currently, sustaining economic growth and establishing sustainable cities are major challenges for all stakeholders in Asia and Pacific countries. A sustainable city means the provision of necessities of life to all levels of the society along with robust and durable infrastructure, better facilities of health, medical, public transport, education, housing, good governance, and excess employment opportunities without any discrimination. A sustainable city is significant for responding to global climate change and unforeseen circumstances, which becomes more intense day by day and cannot be ignored. Developing and implementing sustainable strategies is crucial for mitigating the problems of non-sustainability in developing countries. Currently, Pakistan's Lahore city is declared the world's most polluted city (World air quality, n.d.). In the current scenario, it is imperative to discuss and highlight the opportunities to adopt the sustainability concept and, at the same time, to discuss the implementation challenges faced by developing countries (Fig. 1).

A sustainable city is one that combines the three main sustainability pillars (economic development, environmental management, and social development) based on the urban government (Unisdr, n.d.). The world urban forum was held at the headquarters of the United Nations Human Settlements Program in Nairobi, Kenya (from 29th April 2002 to 3rd May 2002), in which it was confirmed that economic, environmental, social development as well as urban governance are important for sustainable urbanization (Unhabitat, n.d.). Therefore, urban governance should be the foundation of economic development, environmental development, and social development. In this way, a city can ensure its sustainability.

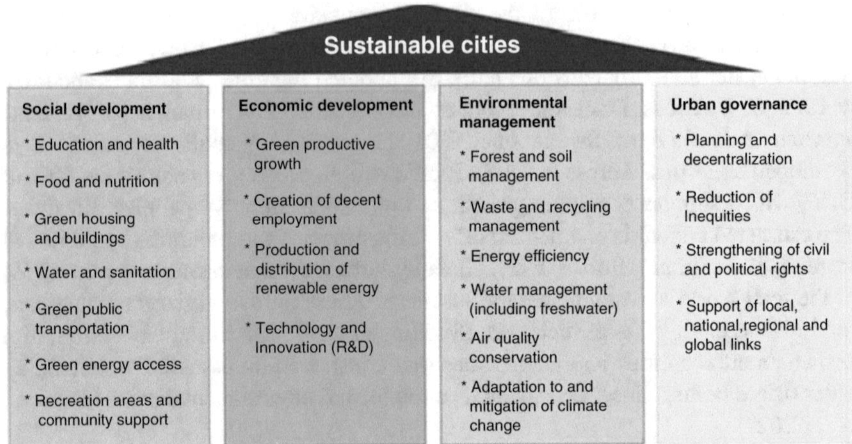

Fig. 1 World urbanization perspectives (UNDESA, 2012)

2 Literature Review

Sustainability means sustaining or maintaining the resources for the next generations and the resources are scarce and finite on the earth. Therefore, resources should be used carefully and conservatively without compromising the quality of life. In short, sustainability relates to the better future of our children and grandchildren for their healthy life (Robertson, 2021). Currently, the world is facing problems of global warming (climate change and environmental issues) along with increasing poverty, socio and economic inequality, and limitation of natural resources, which increase the importance of sustainable development. Social, economic, and environmental sustainability importance has been increased across the globe (Giovannoni & Fabietti, 2013). Due to this reason, efforts are being made at national and international levels as well as at institutional levels. Policymakers are making policies and make initiatives, and academicians conducted research on sustainability and sustainable development, e.g. the United Nations has launched Sustainable Development Solutions Network (SDSN) in 2012 (United Nations Sustainable Development Solutions Network (UNSDN), 2013). Klynveld Peat Marwick Goerdeler (KPMG) issued sustainability report in detail in 2011 (KPMG, 2011) and Joseph conducted research on sustainability in 2011 (Joseph, 2012). Sustainability meaning has been analyzed in-depth, while it has been discussed rarely (Abdel-khalik, 2002). Nowadays, everyone is deemed to recognize the importance of sustainability and sustainable development. Now, we have seen the essence of sustainability and sustainable development in vision (mission) statements, and incoming as well as outgoing reporting systems of organizations (institutions). In this regard, some initiatives have been taken and seen by municipalities to build sustainable cities with less emission of carbon, sustainable building structures, etc., and many countries are adopting and implementing sustainable practices with all respect (Sodiq et al., 2019).

In the twenty-first century, the societies are struggling to provide the best possible facilities and services to their residents like sustainable building and housing, better transportation facilities, better healthcare, and education facilities, proper sanitation, and waste management system, a clean and green environment, public security and safety to increase the well-being and uplift the living standards of their residents (Pardo-Bosch et al., 2019). United Nations promulgated Millennium Development Goals (MDGs) in 2000. It has been estimated that eight goals of (MDGs) were attained by UN member countries in 2015 (UN, 2015). Although the parameter of success varies from country to country, most targets were achieved in 2015. Keeping in view, the MDG's accomplishment and implementation of the remaining MDGs goals, United Nations General Assembly passed a resolution in September 2015 for sustainable development by 2030 to transform our world to become more sustainable and resilient.

The top priority of Sustainable Development Goals (SDGs) is to establish resilient, safe, and sustainable cities. As per UN Sustainable Development Goal 11, sustainable cities are those cities that are dedicated to attaining economic sustainability, social sustainability, and green sustainability (Wiedmann & Allen, 2021; Pradhan et al., 2017). There is a close interconnection among the elements of sustainable city (economy, environment, and preservation of natural resources) which enrich the acceptability or quality of life (Development Alternatives Group, n.d.). At the time of designing, building, and managing cities, the principles of sustainable development should be followed to ensure economic development, reduce inequalities among citizens, and improve the quality of life as well as the environment (World Health Organization, 2016). United Nations Agenda of 2030 for the sustainable development goal 11 (SDG11) defines a sustainable city as a city more sustainable, inclusive, and resilient (Jayasooria, 2016). These SGD 11 goals provide guidelines for researchers, practitioners, decision-makers, and policymakers related to sustainability (economics, environment, and health in urban areas). Sustainable Development Goal 11 is also known as SDG 11 as well as Global Goal 11 (Wiedmann & Allen, 2021; Pradhan et al., 2017). It is about "Sustainable cities and communities" to enhance the housing prices at affordable levels, sustainable urbanization, preservation of natural resources and cultural heritage, sustainable transport system at affordable rates, disaster and calamities risk reduction, resource efficiency, provision of green, safe, and public spaces, devise policies for combating with climate changes and unforeseen circumstances, sustainable building, etc. The United Nations General Assembly established 17 sustainable development goals in 2015. The purpose of these goals is to establish resilient, safe, and sustainable cities.

The following are some important definitions of sustainability provided by different authors (Table 1)

Table 1 Definitions of sustainability

Sustainability	
Author's name	Definition
Brundtland (1987)	"Sustainable development is development that meets the needs of the present without compromising the needs of future generations to meet their own needs"
Pearce and Barbier (1989)	"Sustainable development involves designing a social and economic system, which ensures that these goals are sustained, i.e. that real income rise, that educational standards increase that the health of the nation improves, that the general quality of life is advanced"
Wlaschek et al. (1994)	"No single approach to 'sustainable development' or framework is consistently useful, given the variety of scales inherent in different conservation programs and different types of societies an institutional structure"
Lélé (1991)	"Sustainable development, sustainable growth, and sustainable use have been used interchangeable, as if their meanings were the same. They are not. Sustainable growth is a contradiction in terms: nothing physical can grow indefinitely. Sustainable use is only applicable to renewable resources. Sustainable development is used in this strategy to mean: improving the quality of human life whilst within the carrying capacity of the ecosystems"
Holdgate (1993)	"Development is about realizing resource potential, sustainable development of renewable natural resources implies respecting limits to the development process, even though these limits are adjustable by technology. The sustainability of technology may be judged by whether it increase production, but retains it other environmental and other limits"
Pearce and Atkinson (1993)	"Sustainable development is concerned with the development of a society where the costs of development are not transferred to future generations, or at least an attempt is made to compensate for such costs"
Altman and Bland (1994)	"Most societies want to achieve economic development to secure higher standards of living, now and for future generations. They also seek to protect and enhance their environment, now and for their children. Sustainable development tries to reconcile these two objectives"
Sustainable city	
United Nation SGD Tracker (n.d.)	"The challenges cities face can be overcome in ways that allow them to continue to thrive and grow, while improving resource use and reducing pollution and poverty. The future we want includes cities of opportunities for all, with access to basic services, energy, housing, transportation and more"
City of Dover, New Hampshire (n.d.)	"Balancing the values of environmental steawardship, social responsibility and economic vitality to meet our present needs while ensuring the ability of future generations to meet their needs"
Giffinger et al. (2007)	"A city well performing in a forward-looking way in [economy, people, governance, mobility, environment, and living] built on the smart combination of endowments and activities of self-decisive, independent and aware citizens"

(continued)

Table 1 (continued)

Sustainability	
Author's name	Definition
Landers et al. (2012)	"Smart sustainable cities use information and communication technologies (ICT) to be more intelligent and efficient in the use of resources, resulting in cost and energy savings, improved service delivery and quality of life, and reduced environmental footprint—all supporting innovation and the low-carbon economy"
Meijer and Bolivar (2013)	"We believe a city to be smart when investments in human and social capital and traditional (transport) and modern (ICT) communication infrastructure fuel sustainable economic growth and a high quality of life, with a wise management of natural resources, through participatory governance"

2.1 How Does Sustainability Work?

Sustainability can be measured through different dimensions or pillars of sustainability. Three pillars of sustainability must be applied consistently for achieving sustainable development performance effectively at industrial and society levels (Emerald Built Environments_Green Building Consulting, n.d.; TWI Global, n.d.). The pillars of sustainability included economic, social, and environmental factors which are also known as profit, people, and the planet. It is also called the TBL (Triple Bottom Line) in the corporate world (Tech Target, n.d.). These are also called the foundation of sustainability. John Elkington and Volans are pioneer authors (scholars) who introduce three dimensions (pillars) of sustainability. They emphasize that companies adopt and implement these dimensions (principles) to achieve long-run goals. Economic sustainability refers to business profit, revenue, and increased future growth. Environmental sustainability refers to reducing carbon emission, wastage of water, water treatment, and non-decomposable packaging. Social sustainability refers to control of the impacts of organization process on people, both positive and negative (Investopedia: Sharper insight, better investing, n.d.).

2.2 Why Sustainability Is Important?

Sustainability plays a vital role in improving the quality of mass lives, protecting ecosystems, and saving natural resources for future generations (TWI Global, n.d.). In order to fulfill the necessities of the increasing world population, it is a need of time to build modern, clean, safe, and sustainable cities. The modern era urban planning is very important for the survival and prosperity of future generations that will establish the base of safe, green, and resilient cities. According to the United Nations, urban areas will be approximately 68% of the total population by 2050. The United Nations has included sustainable cities and communities as goal no. 11 in its sustainable development goals and also defines the importance of a sustainable future. A city's system is composed of many parts and for a sustainable city, all parts should work in collaboration with each other in a sustainable way. The existence of a sustainable, affordable, reliable, and free-from fossil fuels transportation system plays a very significant role in making a sustainable city. Sustainability highlight and encourage the stakeholders to focus on sustainable materials that should be used in building construction to resist earthquakes and other natural calamities. Streets should be wide, clean, and walkable, and designate places for rest and social gatherings. Safe and affordable housing should be available for all citizens. Informal settlements and unplanned cities should not grow speedily, which is harmful and destructive to the natural environment (USF Blogs, n.d.).

Business sustainability is also called corporate sustainability; it is the relationship between environmental, social, and economic demands for the sake of ethical and ongoing success (Harvard Business School Online Courses & Learning Platforms, n. d.). In a business context, sustainability means doing business without harming the society or community as a whole. The business stakeholders should be very careful about these two serious concerns: (a) what are the effects of business on the environment? (b) what are the effects of business on society? The first and foremost objective of sustainable business development is to minimize the harmful impacts of business on the environment and society. If companies fail to fulfill their responsibility, this may cause environmental degradation, injustice, and inequality in society. The organizational decision makers must consider the environmental, economic, and social factors while they are designing organizational processes. Many smart organizations have adopted and implemented sustainable business practices to achieve a competitive edge (Harvard Business School Online Courses & Learning Platforms, n.d.). Sustainable business strategies ease the way for organizations to achieve long-term business goals and enhance organizational values. The basic elements of business sustainability are (1) production with sustainable materials, (2) using sustainable supply chain practices to reduce greenhouse gases emission, (3) allocation of funds for education and technical training for local community youth, and (4) emphasizing the use of renewable energy sources, e.g. water, solar, water, etc. (Hashim et al. 2021a, b).

According to the Uswitch report, 2021, the following are the most sustainable cities in the world after meticulously studying of city's energy, CO_2 emission, air quality, affordability, green space percentage, and transportation infrastructure (Sustainability Magazine, n.d.; Uswitch, 2021):

1. *Canberra, Australia (Index Score 427):* To meet the city's energy consumption, Canberra, Australia, relies on wind farms and solar-powered energy. Besides providing a sustainable living to its residents, Canberra ensures Internet access to approximately 94% of residents (Fig. 2).

2. *Madrid, Spain (Index Score 403):* Madrid, Spain, had curbed the deforestation and affirmed to protect the forest at all cost; due to this reason, Madrid, Spain,

Fig. 2 Canberra, Australia

Fig. 3 Madrid, Spain

became a green city. There is using sustainable energy for transportation (Fig. 3).

3. *Brisbane, Australia (Index Score 382):* Keeping Brisbane clean, green, and sustainable; protecting waterways, habitat areas, healthy native plants, and wildlife. Brisbane aims that by 2031, there will be 40% of natural habitat areas to combat the average CO_2 emissions of six tonnes from transport, waste, and energy (Brisbane, n.d.) (Fig. 4).

4. *Dubai, United Arab Emirates (Index Score 375):* Dubai has a sustainable building development system that's why it is called a "Sustainable City."

Fig. 4 Brisbane, Australia

Fig. 5 Dubai, United Arab Emirates

Dubai recycles its waste and water. It will be able to produce 75% of energy from renewable sources by 2050 (UAE Sustainability Initiatives, n.d.) (Fig. 5).

5. *Copenhagen, Denmark (Index Score 369):* In Copenhagen, most people ride more cycles than driving vehicles to avert any emission of CO_2 and there are only 29% of persons having their own vehicles (Fig. 6).

6. *Frankfurt, Germany (Index Score 365):* Frankfurt, Germany, has admitted to mitigate CO_2 emission up to 50% till 2030. In Frankfurt, Germany, the concrete industry and burning fossil fuels cause CO_2 emissions. In 2000, Frankfurt had declared to become an environmentally friendly city in the world (Fig. 7).

Fig. 6 Copenhagen, Denmark

Fig. 7 Frankfurt, Germany

7. *Hamburg, Germany (Index Score 364):* Hamburg, Germany, relies on cycle paths, waste management, electric mobility, and redevelopment of neighborhoods. It is moving towards a "green city" due to the green spaces. Forests and entertaining areas have been increased up to 16.5% in metropolitan areas (Fig. 8).

8. *Prague, Czech Republic (Index Score 359):* For mitigation of CO_2 emissions, Prague promotes electric cars and bikes. Prague has aimed to reduce 45% CO_2 emission by 2030 and eliminating CO_2 emission by 2050 (Fig. 9).

9. *Abu Dhabi, United Arab Emirates (Index Score 357):* Abu Dhabi relies on other renewable energy, e.g., solar energy, the establishment of a free zone,

Fig. 8 Hamburg, Germany

Fig. 9 Prague, Czech Republic

Fig. 10 Abu Dhabi, United Arab Emirates

construction of residential buildings nearest shops, restaurants, and public green spaces (UAE Sustainability Initiatives, n.d.) (Fig. 10).

10. *Zurich, Switzerland (Index Score 355):* Zurich, Switzerland, focuses on increasing the awareness of sustainability among residents, e.g., energy efficiency, sustainable building and usage of renewable energies, recycling and waste management, improving the water quality of lakes and waterways, celebration of annual environment days, and the Zurich Multimobile action day (Sustainability Management School, n.d.) (Fig. 11 and Table 2).

3 Challenges and Opportunities

Most cities are at different levels of development stages, they have specific priorities and responses that are related to policies and procedures at the local as well as national levels. In this context, different cities have different challenges and opportunities (United nations, n.d.). There are many challenges and opportunities faced by developing countries in the adoption of sustainability models. World's population is increasing day by day and more than 50% of people live in urban areas. However, a detail understanding of this filed and its challenges is relatively important in current situation of environmental issues. It can help to create a sustainable and healthy life for all the stakeholders. The cities may face numerous problems and challenges due

Fig. 11 Zurich, Switzerland

Table 2 Most sustainable cities in the world

Rank	City	Index score
1	Canberra, Australia	427
2	Madrid, Spain	403
3	Brisbane, Australia	382
4	Dubai, United Arab Emirates	375
5	Copenhagen, Denmark	369
6	Frankfurt, Germany	365
7	Hamburg, Germany	364
8	Prague, Czech Republic	359
9	Abu Dhabi, United Arab Emirates	357
10	Zurich, Switzerland	355

Source: Uswitch report, 2021

to the increase in population. The following challenges as discussed below (Janik et al., 2020):

- Lack of financial resources and social inequalities.
- Limitation of natural resources.
- Corruption.
- Natural disasters (viruses, flood, and earthquakes) damage the infrastructure and peoples.

- Conflict of interest between current profit and invest in sustainable technologies at industrial levels.
- To develop the sustainable development mindset. It is very important for developing sustainable city, society, industry, and nations.
- Informal settlement by unplanned house construction and expansion may cause deficiencies in administrative services as well as inadequacy in necessities and infrastructure.
- Due to the population density, developing countries are facing infrastructure, population control, scarcity of resources, climate challenges, and environmental challenges.
- Developing countries are primarily dependent on imported latest technology items in engineering and manufacturing.
- Developing countries are facing issues of controlling prices, illiteracy, and energy crises (load-shedding of electricity, gas, water, etc.).
- Long-term (strategic) planning is required to fulfill the demand of public services for the growing urban population, such as healthcare, education, public transportation, sanitation, etc.

Sustainability assists the society in taking concrete actions for less emission of CO_2, better infrastructure, housing at affordable rates, sustainable housing, a clean and green environment, healthier life, increase the literacy rates, fight against climate changes, etc. Numerous challenges faced by humankind such as poverty alleviation, resources scarcity (water, climate change, gender discrimination, and inequality) can be efficiently and effectively addressed with the help of sustainable development. The United Nations has approved the sustainable development 2030 agenda, i.e. protection of the planet and well-being of people. Worldwide active participation of countries, individuals, and business societies is very important for achieving these goals (Acciona, n.d.). A sustainable city has numerous benefits, whether these may be short term or long term.

Nowadays, experts, analysts, and policymakers measure sustainable city practices and their impacts using environmental, social, and governance (ESG) metrics. It focuses on the provision of necessities of life, planned housing, better health care facilities, better infrastructure, sustainable transportation, a clean and green environment, renewable energy, etc. Following are the significant benefits of a sustainable city:

1. *Reduction in energy usage:* Sustainable city and society emphasizes the use of renewable energy, e.g. solar and wind energy. With the adoption of sustainability practices, long-term energy costs can be reduced. The use of energy-efficient solar and wind energy equipment causes a reduction in monthly utility bills.
2. *A healthy life for all:* Sustainable city and society helps the healthier life of people and focuses on purity of food, clean water, recycling of waste management, and reduction in CO_2 emission into the environment.
3. *Better living conditions:* Due to the sustainable city and society, availability of housing at affordable prices, construction of buildings with sustainable materials, planned construction, transport facility at affordable rates, and use of electricity,

living conditions of mass have been improved. The availability and accessibility of resources, opportunities, and institutions for everyone is important for social equality and better life.

4 Concluding Remarks

There is an interconnection between how cities manage natural sources and mass quality of life. The tendency toward urbanization causes immense pressure on the environment, increasing the demand for necessities of life, better infrastructure, affordable transport system, employment opportunities, safe and affordable housing, etc. Sustainable cities and communities are essential to decrease the human, social, and economic losses. It is also very important to protect the culture, environment, mitigate climate change and calamities (disaster risks). A green, safe, and clean environment contributes to a nation's better health. Better infrastructures, affordable transport systems, employment opportunities, safe and affordable housing contributed to residents' economic growth. The level of involvement of members from academia in planning and implementing environmental projects can be fruitful for adopting sustainability practices. The environmental issues should be highlighted by reporting and educating to all stakeholders of the society about the consequences of non-sustainability. Sustainable development is a very important issue which should be considered very seriously as it affects everyone in societies.

References

Abdel-khalik, A. (2002). Reforming corporate governance post Enron: Shareholders Board of Trustees and the auditor. *Journal of Accounting and Public Policy, 21*, 97–103.

Acciona. (n.d.). https://www.acciona.com/sustainable-development/?_adin=11551547647

Allam, Z., & Jones, D. S. (2021). Future (post-COVID) digital, smart and sustainable cities in the wake of 6G: Digital twins, immersive realities and new urban economies. *Land Use Policy, 101*, 105201.

Altman, D. G., & Bland, J. M. (1994). Diagnostic tests 1: Sensitivity and specificity. *BMJ. British Medical Journal, 308*(6943), 1552.

Brisbane. (n.d.). *Clean, green, sustainable 2017–2031*. https://www.brisbane.qld.gov.au/clean-and-green/2017-31

Brundtland. (1987). *Our common future book by Brundtland Commission*. https://sustainabledevelopment.un.org/content/documents/5987our-common-future.pdf

Catalano, C., Meslec, M., Boileau, J., Guarino, R., Aurich, I., Baumann, N., et al. (2021). Smart sustainable cities of the new millennium: Towards design for nature. *Circular Economy and Sustainability, 1*(3), 1053–1086.

City of Dover, New Hampshire. (n.d.). http://www2.dover.nh.gov/government/city-operations/planning/sustainable-dover/sustainability-definition/

Development Alternatives Group. (n.d.). https://www.devalt.org/newsletter/jul00/lead.htm

Emerald Built Environments_Green Building Consulting. (n.d.). http://emeraldbe.com/sustainable-development-important/

Giffinger, R., Fertner, C., Kramar, H., & Meijers, E. (2007). *City-ranking of European medium-sized cities* (pp. 1–12). Centre of Regional Science.

Giovannoni, E., & Fabietti, G. (2013). What is sustainability? A review of the concept and its applications. *Integrated reporting*, 21–40.

Girardet, H. (1999). *Creating sustainable cities* (No. 2). Schumacher Briefings.

Harvard Business School Online Courses & Learning Platforms. (n.d.). https://online.hbs.edu/blog/post/what-is-sustainability-in-business

Hashim, M., Nazam, M., Abrar, M., Hussain, Z., Nazim, M., & Shabbir, R. (2021b). Unlocking the sustainable production indicators: A novel TESCO based fuzzy AHP approach. *Cogent Business & Management, 8*(1), 1870807.

Hashim, M., Nazam, M., Zia-ur-Rehman, M., Abrar, M., Baig, S. A., Nazim, M., & Hussain, Z. (2021a). Modeling supply chain sustainability-related risks and vulnerability: Insights from the textile sector of Pakistan. *Autex Research Journal, 22*(1), 123–134.

Holdgate, M. W. (1993). *The sustainable use of tropical coastal resources-a key conservation issue.* Ambio, 22(7), 481–482.

Investopedia: Sharper insight, better investing. (n.d.). https://www.investopedia.com/terms/s/sustainability.asp

Janik, A., Ryszko, A., & Szafraniec, M. (2020). Scientific landscape of smart and sustainable cities literature: A bibliometric analysis. *Sustainability, 12*(3), 779.

Jayasooria, D. (2016). Sustainable development goals & Malaysia society: Civil society perspectives. *UKM Ethnic Studies, (45*).

Joseph, G. (2012). Ambiguous but tethered: An accounting basis for sustainability reporting. *Critical Perspectives on Accounting, 23*, 93–106.

KPMG. (2011). *KPMG international survey of corporate responsibility reporting 2011.* http://www.kpmg.com/PT/pt/IssuesAndInsights/Documents/corporate-responsibility2011.pdf

Landers, T. F., Cohen, B., Wittum, T. E., & Larson, E. L. (2012). A review of antibiotic use in food animals: Perspective, policy, and potential. *Public Health Reports, 127*(1), 4–22.

Lélé, S. M. (1991). Sustainable development: A critical review. *World Development, 19*(6), 607–621.

Meijer, A., & Bolivar, M. (2013, Governing the smart city: Scaling-up the search for socio-techno synergy. In *2013 EGPA Conference Proceedings*.

Pardo-Bosch, F., Aguado, A., & Pino, M. (2019). Holistic model to analyze and prioritize urban sustainable buildings for public services. *Sustainable Cities and Society, 44*, 227–236. https://doi.org/10.1016/J.SCS.2018.02.028

Pearce & Barbier. (1989). *Environment and society portal.* https://www.environmentandsociety.org/sites/default/files/key_docs/pearce_1_1.pdf

Pearce, D. W., & Atkinson, G. D. (1993). Capital theory and the measurement of sustainable development: An indicator of "weak" sustainability. *Ecological Economics, 8*(2), 103–108.

Pradhan, P., Costa, L., Rybski, D., Lucht, W., & Kropp, J. P. (2017). A systematic study of sustainable development goal (SDG) interactions. *Earth's Future, 5*(11), 1169–1179.

Robertson, M. (2021). *Sustainability principles and practice*. Routledge.

SGD Tracker. (n.d.). *SGD Tracker: Measuring, progress towards the sustainable.* https://sdg-tracker.org/cities

Sodiq, A., Baloch, A. A., Khan, S. A., Sezer, N., Mahmoud, S., Jama, M., & Abdelaal, A. (2019). Towards modern sustainable cities: Review of sustainability principles and trends. *Journal of Cleaner Production, 227*, 972–1001.

Sustainability Magazine. (n.d.). https://sustainabilitymag.com/top10/top-10-most-sustainable-cities-world

Sustainability Management School. (n.d.). https://sumas.ch/5-examples-of-sustainability-in-switzerland/

Swilling, M., Hajer, M., Baynes, T., Bergesen, J., Labbé, F., Musango, J. K., & Tabory, S. (2018). *The weight of cities: Resource requirements of future urbanization.* IRP Reports.

Tech Target. (n.d.). https://whatis.techtarget.com/definition/business-sustainability

TWI Global. (n.d.). https://www.twi-global.com/technical-knowledge/faqs/faq-what-is-sustainability

UAE Sustainability Initiatives. (n.d.). https://emiratesgbc.org/uae-sustainability-initiatives/

UN. (2015). *The Millennium Development Goals Report 2015*. United Nations.

UN Environment Programme. (n.d.). https://www.unep.org/regions/asia-and-pacific/regional-initiatives/supporting-resource-efficiency/sustainable-cities

Unhabitat. (n.d.). https://www.unhabitat.org/pmss/listitemsdetails.aspx?publicationID=1234

Unisdr. (n.d.). https://www.unisdr.org/files/28240 rcreport.pdf

United Nations. (n.d.). https://www.un.org/en/development/desa/policy/wess/wess current/wess2013/WESS2013.PDF

United Nations Sustainable Development Solutions Network (UNSDN). (2013). *An action agenda for sustainable development*. http://www.unsdsn.org/files/2013/06/130613-SDSN-An-Action-Agenda-for-Sustainable-Development-FINAL.pdf

United Nations Water and Urbanizations. (n.d.). https://www.unwater.org/water-facts/urbanization/

Unsdsn. (n.d.). https://resources.unsdsn.org/files/2013/11/an-action-agenda-for-sustainable-development

UNDESA, U. (2012). Population dynamics. United Nations, May.

USF Blogs. (n.d.). *Educational blogs from our community*. https://usfblogs.usfca.edu/sustainability/2020/04/23/the-importance-of-sustainable-cities/

Uswitch. (2021). https://www.uswitch.com/gas-electricity/most-sustainable-cities/

Vardoulakis, S., & Kinney, P. (2019). Grand challenges in sustainable cities and health. *Frontiers in Sustainable Cities, 1*, 7.

Wiedmann, T., & Allen, C. (2021). City footprints and SDGs provide untapped potential for assessing city sustainability. *Nature Communications, 12*(1), 1–10.

Wlaschek, M., Heinen, G., Poswig, A., Schwarz, A., Krieg, T., & Scharffetter-Kochanek, K. (1994). UVA-induced autocrine stimulation of fibroblast-derived collagenase/mmp-1 by interrelated loops of interleukin-1 and interleukin-6. *Photochemistry and Photobiology, 59*(5), 550–556.

World Air Quality. (n.d.). https://www.iqair.com/world-most-polluted-cities

World Health Organization. (2016). *Health as the pulse of the new urban agenda: United Nations conference on housing and sustainable urban development*, Quito.

Youmatter. (n.d.). *Understand today better, take action for our future*. https://youmatter.world/en/definition/definitions-sustainability-definition-examples-principles/

Sustainable Supply Chain Management: Definition, Bibliometrics, Applications, and Future Directions

Yilu Long, Huchang Liao, and Benjamin Lev

Abstract With the emergence of the concept of sustainability, sustainable supply chain management (SSCM) has become a challengeable area for both academics and industrialists since sustainable supply chain is the cornerstone of a company to achieve sustainable goals. Sustainable supply chain management has been a hot research topic and many achievements on SSC have been obtained from different angles. To help readers understand the advances of this field and explore future research directions, this chapter firstly distinguishes different statements regarding the sustainability of supply chain and then specify the concept of SSC. Afterwards, a bibliometrics analysis regarding the publications on SSCM is provided. Given that the fuzzy set theory has been widely used for information expression in SSCM, we then review the applications of fuzzy set theory in SSCM. It is hoped that this chapter could provide suggestions for the future research on SSCM.

Keywords Bibliometrics · Decision-making methods · Fuzzy theory · Sustainability · Sustainable supply chain management

1 Introduction

According to the World Commission on Environment and Development for Sustainable Development (WCED, 1987), sustainability was defined as "*the use of resources to meet needs of the present without comprising the ability of future generations to meet their own needs*." Supply chains are critical driving force behind business competitive advantages. With the emergence of the concept of sustainability, sustainable supply chain management (SSCM) has become a challengeable area for both academics and industrialists.

Y. Long · H. Liao (✉)
Business School, Sichuan University, Chengdu, China

B. Lev
Decision Sciences Department, Drexel University, Philadelphia, PA, USA

F. P. García Márquez, B. Lev (eds.), *Sustainability*, International Series in Operations Research & Management Science 333,
https://doi.org/10.1007/978-3-031-16620-4_3

A sustainable supply chain (SSC) can be described as a complex network system that involves diverse entities that manage products and associated returns from suppliers to customers, accounting for social, environmental, and economic impacts (Barbosa-Póvoa, 2014). Beside the SSC, there are other common terms regarding the sustainability of a supply chain, such as reverse logistics (RL), green supply chain (GSC), closed-loop supply chain (CLSC), and circular supply chain (CSC). Till now, scholars did not make a strict conceptual distinction between these terms. Given that the scopes and emphasis of these narratives are different, it is necessary to distinguish them.

Existing studies on SSCM covered many aspects and different literature reviews on SSCM have been published from various angles, including the triple bottom line-based SSCM (Ashby et al., 2012), quantitative models for SSCM (Brandenburg et al., 2014), and the effects of sustainable practices on firms (Beske-Janssen et al., 2015). In order to facilitate readers to understand current developments of SSCM and explore future research directions, this chapter provides bibliometric analysis regarding the research status of existing literature on SSCM from the following aspects: (1) publication and citation trends, (2) the most productive countries/regions and institutions, (3) the most highly cited papers of reviewed publications, and (4) keywords co-occurrence.

In supply chain management, it is usually necessary to collect relevant information of a supply chain and then use various methods and technologies for decision-making. The fuzzy set theory has been widely used to deal with vague and uncertain information in SSCM. Researchers (Seuring, 2013; Barbosa-Povoa et al., 2018) reviewed various methods and techniques used in SSCM, but did not review different information representation forms. Therefore, we further review the applications of fuzzy set theory in SSCM and then give future research directions.

This chapter is organized as follows: Sect. 2 describes the differences between sustainability narratives. Sect. 3 presents the bibliometric analysis on the research status of SSCM. Sect. 4 reviews the fuzzy theory used in SSCM. Research directions are given in Sect. 5. Section 6 ends the chapter with concluding remarks.

2 Conceptual Distinctions

There are several common sustainable narratives regarding SSCM, and many studies did not make a strict distinction between them. This chapter mainly introduces the differences between SSC and several other related concepts including RL, GSC, CLSC, and CSC.

Firstly, through sorting and analyzing relevant literature on SSCM, definitions of the above-mentioned five common narratives related to the sustainability of a supply chain are shown in Table 1.

In order to more clearly explain several supply chain narratives in Table 1, we refer to the archetype of circular supply chain by (Batista et al., 2018), as shown in Fig. 1. Figure 1 describes each link of the supply chain in detail, which involves RL

Table 1 Definitions of five common narratives related to the sustainability of a supply chain

Category	Definition
Reverse logistics (RL) (Rogers & Tibben-Lembke, 2001)	*"The process of planning, implementing, and controlling the efficient, cost-effective flow of raw materials, in-process inventory, finished goods and related information from the point of consumption to the point of origin for the purpose of recapturing value or proper disposal."*
Closed-loop supply chains (CLSC) (Guide, 2009)	*"Design, control, and operation of a system to maximize value creation over the entire life cycle of a product with dynamic recovery of value from different types and volumes of returns over time."*
Circular supply chain (CSC) (Batista et al., 2018)	*"The coordinated forward and reverse supply chains* via *purposeful business ecosystem integration for value creation from products/services, by-products and useful waste flows through prolonged life cycles that improve the economic, social and environmental sustainability of organisations."*
Sustainable supply chain management (SSCM) (Ahi & Searcy, 2013)	*"Creation of coordinated supply chains through the voluntary integration of economic, environmental, and social considerations with key inter-organizational business systems designed to efficiently and effectively manage the material, information, and capital flows associated with the procurement, production, and distribution of products or services in order to meet stakeholder requirements and improve the profitability, competitiveness, and resilience of the organization over the short- and long-term."*
Green supply chain management (GSCM) (Tseng et al. 2019a, 2019b, 2019c)	*"GSCM can be defined as the integration of environmental management system into the supply chain process including collaboration with customers, suppliers, and logistics service providers to share information and knowledge with an aim to improve environmental performance."*

and CLSC. The SSC emphasizes taking economic, environmental, and social factors (see Fig. 2) into account in all links of supply chain management, while the GSC emphasizes the environmental level.

Next, the differences of the above five concepts are distinguished in detail.

- The differences between RL and CLSC

  RL and CLSC are fundamentally different in coverage. The RL focuses on the reverse flows of materials from the consumption point to the origin point, while the CLSC considers both forward and reverse supply chains. In other words, the CLSC covers the whole product life cycle (Govindan & Soleimani, 2016).

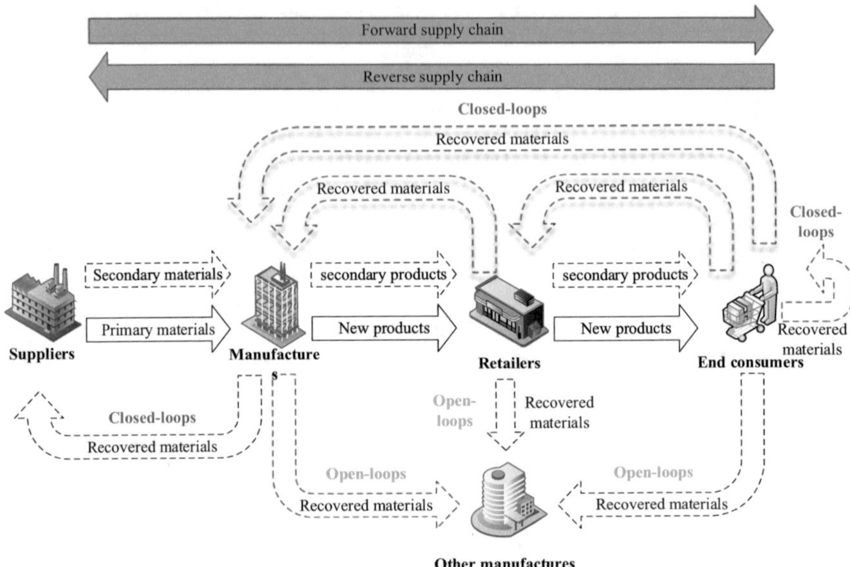

Fig. 1 General framework of a circular supply chain (adapted from Batista et al., 2018)

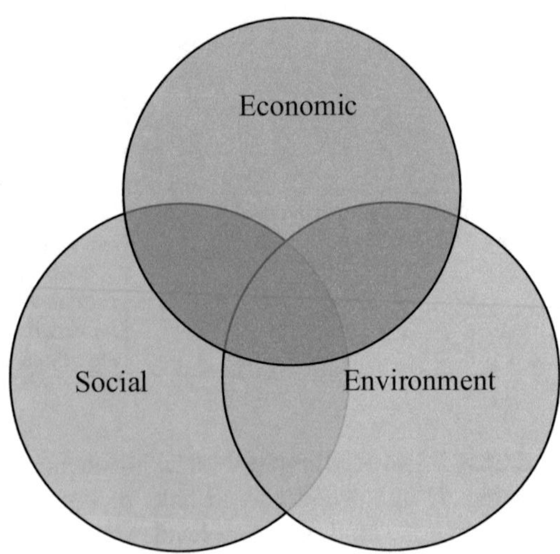

Fig. 2 Triple bottom line principle in sustainable supply chain

- The differences between CLSC and CSC

The fundamental difference between the perspectives of "closed-loop" and "circular" supply chain lies in the scopes and focuses of their related value chain systems. CLSCs tend to focus more on the flow of major products than on the flow of by-products and useful waste, and support value creation systems derived from entire product life cycles and related returns. By comparison, in terms of scope, CSCs extend the boundaries of CLSCs by taking into account post-production stewardship to include forward feeding flows into alternative supply chains. In terms of focus, CSCs support sustainable value chain systems derived not only from products and their end-of-life returns, but also from associated by-product synergies, services, and waste flows (Guide, 2009; Batista et al., 2018).

- The differences between GSC and SSC

  There is a considerable overlap between GSC and SSC (Glover et al., 2014), but the scope of GSC is narrower. The GSC mainly focuses on the environmental dimension of sustainability, while the SSC expands the environmental perspective and includes the social and economic perspectives. In this sense, the SSCM can adopt a comprehensive triple bottom line approach (Fabbe-Costes et al., 2014).

3 Bibliometric Analysis on the Research Status of SSCM

There are many related studies on SSCM. This section analyzes the relevant literature review such that readers can have a general understanding of this field. Here, we only consider the papers written in English and published in peer-reviewed journals. The related papers were retrieved from the core database of Web of Science, and there is no limit on the date range but the final retrieved date for this study was on October 17, 2021. To include as many papers as possible, the search term was set as "sustainable supply chain." A total of 1770 papers were identified, and 153 of them were reviews. It should be noted that there may be some papers related to SSCM but were not included in our retrieved records, especially considering that we use the precise retrieval strategy to search publications. In the following, we analyze these records from five perspectives: (1) publication and citation trends, (2) the most productive countries/regions, (3) the top 10 highly cited papers of reviewed publications, and (4) keywords co-occurrence.

3.1 Publication and Citation Trends

The development status and trend of a research field can be predicted according to the number of publications and citations. The number of publications, number of total citations, and average citations per publication of reviewed papers in each year are presented in Fig. 3.

As can be seen in Fig. 3, the first paper about SSCM was published in 2000. From 2000 to 2020, the number of publications is increasing exponentially, which shows that the SSCM has gradually become a hot research field. Note that there is a slight decrease in 2021 since we only have the number of publications till October 2021. The average citations per article were relatively high from 2007 to 2015, and peaked in 2008. Since 2012, the average citations per paper have been in a downward trend,

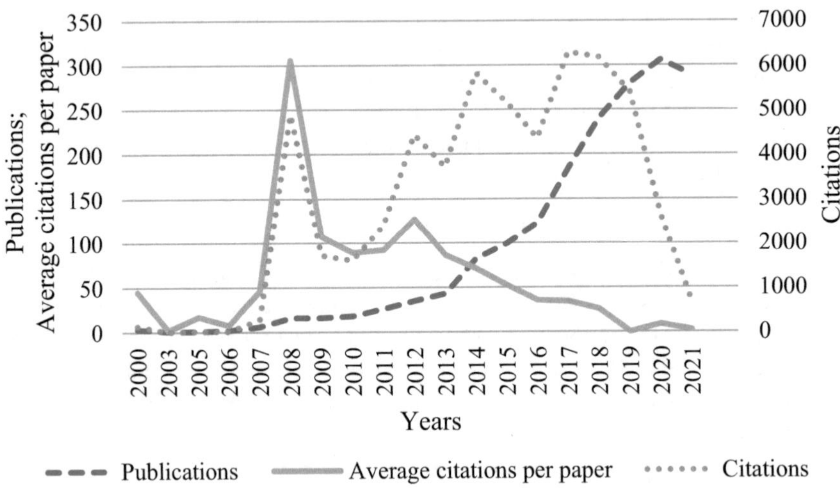

Fig. 3 Publication and citation trends from 2000 to 2021 (till Oct 17, 2021) (Data source: Web of Science)

which may be caused by the sharp increase in the number of publications. However, it is worth noting that scholars should not only increase the number of publications, but also improve the quality of papers, so as to be recognized by more other scholars.

3.2 The Most Productive Countries/Regions and Institutions

By studying publications and citations from different countries/regions and institutions, we can focus on the most creative places in a research field, which can guide future cooperation with world-leading scholars. Table 2 shows the top 15 regions in the number of publications in the field of SSCM.

From Table 2, we can see that Chinese scholars published the most papers related to SSCM (327), followed by the scholars from the USA (251), England (222), and India (190). As for citations, publications from China were not the most highly cited. Papers published by American scholars were cited the most (13,453), followed by the scholars from Germany (9336) and England (8596). With regard to the average citations per paper, the most cited countries/regions are Canada (80.3), Germany (67.17), and the USA (53.6). Although the number of publications of Canada is small, the average citation of Canadian publications is the highest.

Table 3 presents the top ten most productive institutions in the field of SSCM. It can be seen that the Islamic Azad University (70) is the largest publication source, followed by the Universitat Kassel (45) and the Indian Institute of Technology System (IIT System) (43). It can be seen that, among the top ten institutions, three belong to Iran. Combined with the total number of papers by Iranian scholars in Table 2, it can be found that the three Iranian institutions in Table 3 are the main research forces in Iran in SSCM research.

Table 2 The most productive countries/regions regarding the publications on SSCM (till Oct, 2021)

Country/Region	TP	TC	TC/TP
China	327	6109	18.68
USA	251	13,453	53.6
England	222	8596	38.72
India	190	4966	26.14
Iran	168	4178	24.87
Germany	139	9336	67.17
France	112	2325	20.76
Italy	90	2647	29.41
Brazil	81	1748	21.58
Australia	74	2420	32.7
Canada	73	5862	80.3
Taiwan	63	1537	24.4
Spain	56	858	15.32
Malaysia	49	1359	27.73
South Korea	45	687	15.27

Note: TP represents the total number of publications, *TC* represents the total number of citations

Table 3 The most productive institutions regarding the publications on SSCM (till Oct, 2021)

Institution	Country/Region	Publications
Islamic Azad University	Iran	70
Universitat Kassel	Germany	45
Indian Institute of Technology System (IIT System)	India	43
University of Tehran	Iran	37
Indian Institute of Management (IIM System)	India	31
National Institute of Technology (NIT System)	Bangladesh	31
University of Southern Denmark	Denmark	28
Iran University Science & Technology	Iran	27
Worcester Polytechnic Institute	USA	27
Montpellier Business School	France	25

3.3 *The Top 10 Most Highly Cited Papers of Reviewed Publications*

To some extent, we can learn the topics concerned by scholars in a certain field by analyzing highly cited papers since highly cited articles are generally of high quality. Table 4 shows the top ten most highly cited papers in the field of SSCM.

From Table 4, we can find that the most highly cited articles were published between 2008 and 2014, and most of them were literature reviews. These reviews can help scholars quickly understand the development of a certain branch. Among

Table 4 The top 10 most highly cited papers of reviewed publications (till Oct, 2021)

Reference	Author(s)	Journal	Year	Citations
From a literature review to a conceptual framework for sustainable supply chain management	Seuring and Mueller	Journal of Cleaner Production	Seuring and Müller (2008)	2551
A framework of sustainable supply chain management: moving toward new theory	Carter and Rogers	International Journal of Physical Distribution & Logistics Management	Carter and Rogers (2008)	1543
Building a more complete theory of sustainable supply chain management using case studies of 10 examples	Pagell and Wu	Journal of Supply Chain Management	Pagell and Wu (2009)	789
Sustainable supply chain management: evolution and future directions	Carter and Easton	International Journal of Physical Distribution & Logistics Management	Carter and Easton (2011)	693
Quantitative models for sustainable supply chain management: Developments and directions	Brandenburg et al.	European Journal of Operational Research	Brandenburg et al. (2014)	594
A comparative literature analysis of definitions for green and sustainable supply chain management	Ahi and Searcy	Journal of Cleaner Production	Ahi and Searcy (2013)	572
International trade drives biodiversity threats in developing nations	Lenzen et al.	Nature	Lenzen et al. (2012)	567
A literature review and a case study of sustainable supply chains with a focus on metrics	Hassini et al.	International Journal of Production Economics	Hassini et al. (2012)	549
A review of modeling approaches for sustainable supply chain management	Seuring	Decision Support Systems	Seuring (2013)	543
A fuzzy multi criteria approach for measuring sustainability performance of a supplier based on triple bottom line approach	Govindan et al.	Journal of Cleaner Production	Govindan et al. (2013)	508

the most highly cited papers, three of them were published in *Journal of Cleaner Production* and two of them were from *International Journal of Physical Distribution & Logistics Management*. It can be seen that there is no article published in the past 5 years (2017–2021) in Table 4, although many articles have been published in the past 5 years.

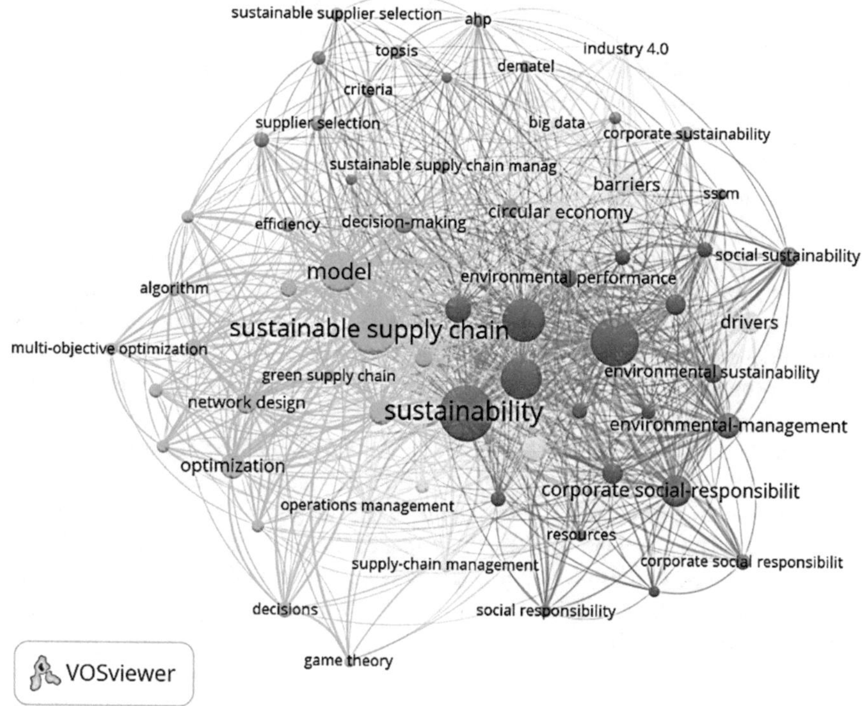

Fig. 4 Keywords analysis of reviewed papers in the field of SSCM

3.4 Keywords Co-occurrence

Through keywords, we can quickly understand the research scale and content of a paper. Keyword analysis is helpful for finding innovative research directions and their links. Figure 4 shows the keyword co-occurrence of reviewed 1770 papers in the field of SSCM.

From Fig. 4, we can find that most of the keywords are related to sustainability, and the emerging keywords include three levels: economy, society, and environment. Some methods, including "AHP," "TOPSIS," "DEA," "DEMATEL" appear in the figure, which shows that applying decision-making methods to SSCM is a hot and important research topic. Scholars can make further exploration in this direction.

4 The Fuzzy Theory Applied in SSCM

SSC is a complex system because it involves many aspects. In the practice of implementing SSCM, it is usually necessary to collect relevant information of a supply chain and then use various methods and technologies to make decisions.

Researchers have reviewed various methods and technologies used in SSCM. For instance, Seuring (2013) summarized researches on quantitative models for forward supply chains. Barbosa-Povoa et al. (2018) reviewed the trends and directions of applying operational research methods towards the achievement of SSCM. Tozanli et al. (2017) reviewed the research status and related methods of environmentally concerned logistics operations in fuzzy environment and found that fuzzy technology is a common method in environmentally concerned logistics operations. However, Tozanli et al. (2017) did not make a detailed introduction to specific fuzzy environment, but only used the concept of fuzzy set theory. Coenen et al. (2018) reviewed fuzzy set theory-based methods used in CLSC management in complex and uncertain environments. In order to explore research directions regarding the applications of fuzzy theories in SSCM, this chapter reviews the fuzzy theory used in SSCM from the aspects of fuzzy information expression, specific methods, and specific applications.

To make our study more specific, we researched the core database of Web of Science with the search strategy being "sustainable supply chain" and "fuzzy or uncertain" and the time slot of publications being "all years." A total of 245 papers were identified on October 20, 2021. We screened these papers manually and limited the journal category division of publications being Q1 and Q2. Finally, we selected 88 papers with high correlation shown in Table 5 for analyzing the application of fuzzy set theory in SSCM.

4.1 Fuzzy Information Representations Applied in SSCM

Collecting real and effective information is the basis for SSCM. In SSCM, quantitative and qualitative information are often involved at the same time. To represent vague or uncertain information, the fuzzy set theory has been widely used in SSCM. In SSCM, fuzzy information expressions were often used to evaluate the importance of the criteria (Kannan, 2018) or the performances of alternatives (Govindan et al., 2013). In the 88 papers we screened and presented in Table 5, fuzzy numbers, interval values, linguistic terms, rough sets, and so on have been applied to SSCM. Among them, triangular fuzzy numbers (35) were used most, followed by interval values (14) and linguistic term sets (7). Fuzzy preference relation and plithogenic set were also used.

Different expressions have different emphases. Fuzzy numbers can be used to express uncertain and imprecise information. Triangular fuzzy number can describe a fuzzy event by the smallest possible value, the most promising value, and the largest possible value. The most promising value of a triangular fuzzy number is represented by a single value, while the most promising value of a trapezoidal fuzzy number is in an interval form. Type-2 fuzzy sets incorporate uncertainty about the membership function in their definition. Traditional fuzzy sets use a membership function to describe information. Intuitionistic fuzzy set is an extension of the traditional fuzzy set, which considers the information of membership degree,

Table 5 Publications related to fuzzy theory applications in SSCM

Fuzzy representation	Decision-making methods	Specific application	Involved sector	Reference
Fuzzy set (67)				
Triangular fuzzy numbers (35)	Fuzzy preference programming +TOPSIS	Supplier selection	Textile manufacturing company	Fallahpour et al. (2017)
	Fuzzy Delphi +ISM + ANP + COPRAS		Textile industry	Kannan (2018)
	MOORA + MEA + multi-objective mathematical model		Home appliance industry	Arabsheybani et al. (2018)
	Approximate fuzzy DEMATEL method		Food	Lin et al. (2018)
	A two-phase decision model using FIS along with Fuzzy Kano philosophy		Iron and steel industry	Jain and Singh (2020)
	BWM + COCOSO		Home appliance manufacturer	Ecer and Pamucar (2020)
	Bayesian network + SWARA		–	Cui et al. (2021)
	BWM + ISM + machine learning		Petrochemical holding company	Alavi et al. (2021)
	Multi-objective optimization	Supplier selection and order allocation	Air-conditioning industry	Liaqait et al. (2021)
	AHP + VIKOR	Third-Party logistics selection	–	Wang et al. (2021)
	Six FMCDM methods	Framework for selecting SSC processes	Manufacturing	Padhi et al. (2018)
	A multi-criteria framework based on FMAUT	Performance evaluation	An alert management system	Erol et al. (2011)
	TOPSIS		–	Govindan et al. (2013)
			Apparel manufacturer	Jakhar (2015)

(continued)

Table 5 (continued)

Fuzzy representation	Decision-making methods	Specific application	Involved sector	Reference
	SEM + ANP+ fuzzy multi-objective linear programming			
	Fuzzy Delphi + ANP		Printed circuit board electronics focal firm	Tseng et al. (2015)
	Audition check-list based FIS		Automotive spare part industry	Ghadimi et al. (2017)
	ANP + TOPSIS		Agri-food value chain	Liu et al. (2019a, 2019b)
	Total ISM + ANP	Supplier evaluation and SSC strategies development	Food industry	Chen et al. (2018)
	Fuzzy double frontier network DEA model	SSC evaluation	Food	Tavassoli et al. (2021)
	Multi-objective fuzzy programming	Supply chain network design	–	Pourjavad and Mayorga (2019)
	ANP + multi-objective optimization model		Solar photovoltaic	Nili et al. (2021)
	Multi-objective fuzzy robust stochastic model		Water heater industry	Nayeri et al. (2021)
	Simulated annealing		Hospitals	Eskandari-Khanghahi et al. (2018)
	QFD	SSC designing	HAVI Logistics	Buyukozkan and Cifci (2013)
	Multi-objective model	SSC framework design	Energy	Ahmed and Sarkar (2019)
	TOPSIS	Identify the decisive aspects of SSC finance	Textile	Tseng et al. (2018)

	Principal component analysis + DEMATEL	Identify major causes	–	Yadav and Singh (2020)
	AHP	Key barriers identification	Food industry	Nazam et al. (2020)
	Total ISM + MICMAC	Investigate the drivers of SSC	Emerging economy	Karmaker et al. (2021)
	ISM + TODIM	Build a successful SSC finance	Textile industry	Tseng et al. (2019a, 2019b, 2019c)
	AHP	Prioritization of SSC practices	Electronics parts/components manufacturing organizations	Laosirihongthong et al. (2020)
	ANP + ISM approach using blockchain	SSC ranking	–	Yadav and Singh (2021)
	AHP + fuzzy multi-objective optimization + clustering methods	Strategy formulation	Electric industry	Gracia and Quezada (2016)
	ISM + AHP	Analyzing the interactions among barriers of SSCM practices	Rubber products manufacturing industry	Narayanan et al. (2019)
	Fuzzy rule + multi-objective	Inventory management	–	Rout et al. (2020)
Trapezoidal fuzzy numbers (6)	Fuzzy logic + FISM	Supplier selection	–	Amindoust et al. (2012)
	DEA	Supplier sustainability assessment	–	Izadikhah et al. (2017)
	TOPSIS+CRITIC	SSC risk evaluation	–	Rostamzadeh et al. (2018)
	Fuzzy multi-objective programming + two-phase stochastics programming	Supply chain network design	–	Tsao et al. (2018)
	Multi-objective robust optimization model		Sugarcane-to-biofuel	Gilani and Sahebi (2020)
	DEA + a robust possibilistic programming model		–	Gilani et al. (2020)

(continued)

Table 5 (continued)

Fuzzy representation	Decision-making methods	Specific application	Involved sector	Reference
Fuzzy numbers (7)	DEA	Supplier selection	Resin production company	Azadi et al. (2015)
	The mathematical model	SSC Designing	Agricultural	Ahmed and Sarkar (2018)
	AHP + TODIM	Investigate the potential risks of the transition from a linear to a circular economy	Logistics company	Kazancoglu et al. (2021)
	SEM + fuzzy set qualitative comparative analysis	Analyzing the mediating role of SSC performance	–	Belhadi et al. (2021)
	Fuzzy logic + multi-objective optimization + genetic algorithm	Supply chain network design	–	Soleimani et al. (2017)
	Modular FIS	Supplier selection	Textile	Amindoust and Saghafinia (2017)
	ISM methodology integrated with fuzzy MICMAC analysis	SSCM practices strategies analysis	Automotive and related companies	Luthra and Mangla (2018)
Intuitionistic fuzzy sets (2)	AHP + TOPSIS	Sustainability assessment	Manufacturing organizations	Choudhary (2021)
	AHP + DEMATEL	Barrier analysis and evaluation	Industry and service sectors	Boutkhoum et al. (2021)
Intuitionistic fuzzy hypersoft set	Aggregation operators	Supplier selection	–	Zulqarnain et al. (2021)
Hesitant fuzzy sets (3)	DEMATEL	Examines the nature of behavioral factors for SSCM practices	Footwear industry	Kumar et al. (2020)
	SWARA-COPRAS	Supplier selection	–	Rani et al. (2020)
	Cumulative prospect theory + VIKOR	Barriers and solutions evaluation	Chemical, agri-food, and textile	Khan et al. (2021)
Picture fuzzy numbers (3)	VIKOR + the grey correlation coefficient	Supplier selection	–	Peng et al. (2020)
	VIKOR			

			Chemical manufacturing enterprise	Bai and Sarkis (2018)
	VIKOR		Food	Meksavang et al. (2019)
Rough set (2)	Pairwise comparison method + DEMATEL + rough set theory	Developing sustainable supplier selection criteria	Solar air-conditioner manufacturer	Song et al. (2017)
	A rough cloud TOPSIS approach	Supplier selection	Photovoltaic modules supplier	Li et al. (2019)
Neutrosophic fuzzy set (2)	MADM framework comprising the SVNHWS and entropy measures	Supplier selection	–	Zeng et al. (2020)
	TODIM+TOPSIS+BWM	SSC finance measurements evaluation	Gas industry	Abdel-Basset et al. (2020)
Type-2 fuzzy sets (2)	Multi-objective DEA model	Supplier evaluation	–	Zhou et al. (2016)
	DEMATEL+QFD	Drivers of SSC evaluation and ranking	Agricultural production systems	Yazdani et al. (2020)
Fuzzy preference relation (2)	ANP	Supplier selection	–	Buyukozkan and Cifci (2011)
	Fuzzy time series with clustering techniques	Analyze demand forecast strategies	–	Ewbank et al. (2020)
Plithogenic set	TOPSIS+CRITIC model	SSC risk management	Telecommunications equipment company	Abdel-Basset and Mohamed (2020)
Pythagorean fuzzy set	AHP	Evaluate the enablers of SSC innovation	Manufacturing industry	Shete et al. (2020)
Interval-value numbers (14)				
Interval-valued fuzzy set (2)	FMEA	Supplier selection	Manufacturing services	Foroozesh et al. (2018)
	PIPRECIA +MABAC		Agricultural production	Puska et al. (2021)
Interval-valued neutrosophic sets	ANP + TOPSIS		Dairy company	Abdel-Basset et al. (2018)

(continued)

Table 5 (continued)

Fuzzy representation	Decision-making methods	Specific application	Involved sector	Reference
Interval-valued intuitionistic uncertain linguistic sets	BWM + alternative queuing method		Watch manufacturer	Liu et al. (2019a, 2019b)
Interval type-2 fuzzy set	AHP sort II		–	Xu et al. (2019)
Interval grey numbers	Grey BWM + TODIM		Manufacturing company	Bai et al. (2019)
Interval-valued Pythagorean fuzzy set	FGRA+TOPSIS+FMEA+ cloud computing-entropy weight method		Chemical industry	Wu et al. (2021)
Interval-valued intuitionistic fuzzy sets	The projection model +entropy method	Third-party reverse logistics provider selection	Manufacturing companies	Chen et al. (2021)
Interval fuzzy number	Robust network goal programming-DEA model	SSC evaluation	Doors and windows manufacturer	Yousefi et al. (2017)
Interval type-2 fuzzy data	Dynamic double frontier network DEA model			Zhou et al. (2019)
Interval-valued triangular fuzzy numbers (2)	GRA	Performance evaluation	Semiconductor manufacturer	Wu et al. (2016)
	TOPSIS	The competitive priorities within SSCM evaluation	Electronic focal manufacturing firms	Lin and Tseng (2016)
The interval number	Robust optimization + two-layer multi-slave programming	The trade-offs of two-tier supply chain enterprises	Urban Construction Group	Jin et al. (2018)
Interval type-2 Pythagorean fuzzy set	Choquet integral + CI-DEMATEL +CI-GRA-MABAC	Risk management	Automotive industry	Mondal and Roy (2021)
Linguistic term sets (7)				
Fuzzy linguistic term sets (4)	Multi-objective optimization model	"resiliently sustainable" supply chain design	Sportswear clothing	Fahimnia and Jabbarzadeh (2016)
	DEMATEL+AEW + VIKOR	Sustainable recycling partner selection	Recycling organizations	Zhou et al. (2018)
	Fuzzy synthetic method + DEMATEL	Performance assessment	Textile industry	

				Tseng et al. (2019a, 2019b, 2019c)
	Fuzzy screening system +DEA	Supplier clustering	–	Izadikhah et al. (2021)
Hesitant fuzzy linguistic term sets (2)	MCDM outranking method	Bioenergy technologies evaluate	Energy	Khishtandar et al. (2017)
	MULTIMOORA	Third-party reverse logistics provider (3PRLP) selection	E-commerce express industry	Luo and Li (2019)
Probabilistic linguistic term sets	DEMATEL	Sustainable recycling partners evaluation and selection	Recycling organizations	Li et al. (2020)

Note: The full names of abbreviations in Table 5 are as follows: *FMAUT* Fuzzy entropy and fuzzy multi-attribute utility, *SEM* Structural equation modeling, *ISM* Interpretive structural model, *FMEA* Failure mode and effects analysis, *GRA* Grey relational analysis, *FIS* Fuzzy Interference System, *VIKOR* VIse Kriterijumska Optimizacija kompromisno Resenje, in Serbian; means multiple criteria optimization compromise solution, *SWARA* Step-wise Weight Assessment Ratio Analysis, *DEMATEL* Decision-making trial and evaluation laboratory, *TOPSIS* Technique for Order Performance by Similarity to Ideal Solution, *AHP* Analytic Hierarchy Process, *ELECTRE* ELimination Et Choix Traduisant la REalité in French, ELimination and Choice Expressing the Reality in English, ANP Analytic Network Process, *TODIM* Portuguese acronym for Interactive Multi-Criteria Decision Making, *COPRAS* Complex proportional assessment, *MABAC* Multi attributive border approximation area comparison, *CoCoSo* Combined Compromise Solution, *DEA* Data envelopment analysis, *MULTIMOORA* Multi-Objective Optimization on the basis of a Ratio Analysis plus the full MULTIplicative form, *MOORA* Multi-Objective Optimization Ratio Analysis, *BWM* Best Worst Method, *QFD* Quality Function Deployment, *COPRAS* The Complex Proportional Assessment, *MICMAC* Matriced Impacts Croisés Multiplication Appliquée á un Classement techniques, *CRITIC* criteria importance through inter-criteria correlation, *AEW* anti-entropy weighting, *PIPRECIA* Pivot pairwise relative criteria importance assessment, *SVNHWS* single-valued neutrosophic hybrid weighted similarity non-membership degree, and hesitancy degree at the same time, so it is more flexible and practical than the traditional fuzzy set in dealing with fuzziness and uncertainty. Hesitant fuzzy set describes the hesitant information of decision-makers in detail. Picture fuzzy numbers, involving membership, neutral membership, and non-membership information simultaneously, can depict uncertainty in a subjective evaluation.

Interval-valued fuzzy numbers can be considered as special forms of generalized fuzzy numbers, which use intervals instead of crisp values to express imprecise information, thus reducing the probability of decision error. The fuzzy linguistic approach is a common approach for modeling linguistic information of linguistic variables. A linguistic value is less precise than a number, but it is closer to human cognitive processes (Zadeh, 1975). Decision problems in the real world are usually complex because of uncertainty and fuzziness. Linguistic terms are direct and natural tools to model complex cognition of experts.

At present, many fuzzy information expressions have not been applied to SSCM, such as the cognitive fuzzy set (CFS) (Jiang & Liao, 2020), which comprehensively analyzed the cognitive characteristic of experts by introducing the joint degree of membership degree and non-membership degree to reflect the cognitive overlap between them. It is worth advocating to choose an appropriate expression according to practical problems.

4.2 *Decision-making Methods Used in SSCM*

As mentioned in Sect. 3.4, the applications of decision-making methods in SSCM is a hot direction. Below we review the applications of decision-making methods used in SSCM.

Since an SSC is a complex system, to make it sustainable, multiple criteria need to be considered to evaluate suppliers, success factors, drivers, and challenges in SSCM. Therefore, multiple criteria decision making (MCDM) methods have been widely used in SSCM. From Table 5, we can find that some publications used individual MCDM methods such as TOPSIS (Govindan et al., 2013), AHP (Shete et al., 2020), ANP (Buyukozkan & Cifci, 2011), while some used integrated MCDM methods, i.e., the combination of two or more MCDM methods, such as BWM + TODIM (Bai et al., 2019), DEMATEL+AEW + VIKOR (Zhou et al., 2018). There were only a few studies which combined MCDM methods with other operational research methods, like AHP + fuzzy multi-objective optimization + clustering methods (Gracia & Quezada, 2016). It is necessary to combine MCDM methods with other operational research methods to improve the accuracy of decision-making results. As pointed out by Paul et al. (2021), combining MCDM method with mathematical modeling and optimization can help to obtain the best decision in SSCM. In the future, scholars can further study from this perspective.

In addition, we find that intelligent optimization algorithms such as simulated annealing (Eskandari-Khanghahi et al., 2018), machine learning (Alavi et al., 2021), genetic algorithm (Soleimani et al., 2017), have also been applied to SSCM. These are popular research directions, which can be further studied in the future.

4.3 *Specific Applications in Different Sectors for SSCM*

SSC has become the cornerstone of a company seeking to achieve sustainable goals. The image of a company is no longer related to the sustainability of its own activities, but to the strong cooperation among all supply chain stakeholders to achieve sustainable activities (Barbosa-Povoa et al., 2018). Therefore, the implementation of SSCM has been the consensus of various industries. As can be seen from Table 5, researches on SSCM mainly include supplier selection, performance evaluation, risk management, identification of successful factors and barriers, logistics management, and supply chain network design.

There were many sectors implementing SSCM, mainly including food industry, textile industry, energy, and agriculture. There were abundant researches on the management of each link in SSC. However, there were few studies on the cooperation and trade-off between various links of a supply chain, such as the trade-offs of two-tier supply chain enterprises (Jin et al., 2018). In addition, the public participation in all links can be considered, fully implementing the triple bottom line principle.

5 Lessons Learned and Future Directions

The implementation of SSCM is important in today's society, which helps to realize the coordination of economic benefits and environmental optimization, and promote the sustainable development of the society. In this section, we summarize the lessons learned and give future research directions.

1. Many studies did not strictly distinguish several supply chain sustainability narratives and some involved concepts such as forward logistics, reverse logistics, open-loop and closed-loop supply chain. It is important to clearly distinguish between different sustainability narratives of the supply chain. In this sense, more detailed researches are needed.
2. The research related to supply chain should be close to practice and provide guidance and suggestions for the operation of supply chain. Additionally, the social level has been less considered in existing studies about SSCM. In this regard, the triple bottom line principle could be implemented in SSCM, taking into account economic, environmental, and social factors simultaneously.
3. With the rapid development of society, people are facing more and more complex problems. Due to the uncertainty of the objective world and the limited cognitive ability of human beings to the real world, decision-making in an uncertain

environment has become a practical problem with strong demand. Fuzzy set theory is a powerful tool to deal with uncertain problems. At present, the fuzzy set theory has been widely used in SSCM. In the future, more fuzzy expressions such as the CFS (Jiang & Liao, 2020) can be applied to SSCM to describe the original information in different forms.

4. Researchers can explore other methods to handle SSCM problems, such as combining MCDM methods with mathematical modeling and intelligent optimization algorithms. The environmental and social factors involved in SSC are difficult to measure and calculate with crisp values given that the environmental and social level may involve unmeasurable indicators. Therefore, the management of SSC needs interdisciplinary knowledge and methods. Scholars in different fields can cooperate in research.
5. Regarding the applications involved in SSCM, we can further study the cooperation and trade-offs between various links of a supply chain. In the context of energy conservation and emission reduction, the impact of carbon tax on supply chain management and related operation decisions have become a hot topic. Zhou et al. (2021) reviewed the research status of supply chain management under carbon tax. In the future, the impact of carbon tax policy can be fully considered in the research of SSCM.

6 Conclusion

SSCM has become a hot research field, and a large number of relevant papers have been published. To help readers understand the advances of this field and explore future research directions, this chapter first clarifies the concept of SSC and analyzes its differences from other supply chain sustainability narratives. Then, a bibliometric analysis is provided to demonstrate the research status of SSCM. Furthermore, we summarize the applications of fuzzy set theory in SSCM and put forward some future research directions. It is acknowledged that, although we try to be inclusive in the methods of searching and selecting papers in bibliometric analysis and literature review, our search strategy may limit the breadth of samples, resulting in some omissions.

Acknowledgments The work was supported by the National Natural Science Foundation of China (71771156, 71971145, 72171158).

References

Abdel-Basset, M., & Mohamed, R. (2020). A novel plithogenic TOPSIS–CRITIC model for sustainable supply chain risk management. *Journal of Cleaner Production, 247*, 119586.

Abdel-Basset, M., Mohamed, R., Sallam, K., & Elhoseny, M. (2020). A novel decision-making model for sustainable supply chain finance under uncertainty environment. *Journal of Cleaner Production, 269*, 122324.

Abdel-Basset, M., Mohamed, M., & Smarandache, F. (2018). A hybrid neutrosophic group ANP–TOPSIS framework for supplier selection problems. *Symmetry, 10*(6), 226.

Ahi, P., & Searcy, C. (2013). A comparative literature analysis of definitions for green and sustainable supply chain management. *Journal of Cleaner Production, 52*, 329–341.

Ahmed, W., & Sarkar, B. (2018). Impact of carbon emissions in a sustainable supply chain management for a second-generation biofuel. *Journal of Cleaner Production, 186*, 807–820.

Ahmed, W., & Sarkar, B. (2019). Management of next-generation energy using a triple bottom line approach under a supply chain framework. *Resources Conservation and Recycling, 150*, 104431.

Alavi, B., Tavana, M., & Mina, H. (2021). A dynamic decision support system for sustainable supplier selection in circular economy. *Sustainable Production and Consumption, 27*, 905–920.

Amindoust, A., Ahmed, S., Saghafinia, A., & Bahreininejad, A. (2012). Sustainable supplier selection: A ranking model based on fuzzy inference system. *Applied Soft Computing, 12*(6), 1668–1677.

Amindoust, A., & Saghafinia, A. (2017). Textile supplier selection in sustainable supply chain using a modular fuzzy inference system model. *Journal of the Textile Institute, 108*(7), 1250–1258.

Arabsheybani, A., Paydar, M. M., & Safaei, A. S. (2018). An integrated fuzzy MOORA method and FMEA technique for sustainable supplier selection considering quantity discounts and supplier's risk. *Journal of Cleaner Production, 190*, 577–591.

Ashby, A., Leat, M., & Hudson-Smith, M. (2012). Making connections: A review of supply chain management and sustainability literature. *Supply Chain Management: An International Journal, 17*(5), 497–516.

Azadi, M., Jafarian, M., Saen, R. F., & Mirhedayatian, S. M. (2015). A new fuzzy DEA model for evaluation of efficiency and effectiveness of suppliers in sustainable supply chain management context. *Computers & Operations Research, 54*, 274–285.

Bai, C. G., Kusi-Sarpong, S., Ahmadi, H. B., & Sarkis, J. (2019). Social sustainable supplier evaluation and selection: A group decision-support approach. *International Journal of Production Research, 57*(22), 7046–7067.

Bai, C. G., & Sarkis, J. (2018). Integrating sustainability into supplier selection: A grey-based topsis analysis. *Technological and Economic Development of Economy, 24*(6), 2202–2224.

Barbosa-Póvoa, A. P. (2014). Process supply chains management–where are we? Where to go next? *Frontiers in Energy Research, 2*, 1–13.

Barbosa-Povoa, A. P., da Silva, C., & Carvalho, A. (2018). Opportunities and challenges in sustainable supply chain: An operations research perspective. *European Journal of Operational Research, 268*(2), 399–431.

Batista, L., Bourlakis, M., Smart, P., & Maull, R. (2018). In search of a circular supply chain archetype–a content-analysis-based literature review. *Production Planning & Control, 29*(6), 438–451.

Belhadi, A., Kamble, S., Gunasekaran, A., & Mani, V. (2021). Analyzing the mediating role of organizational ambidexterity and digital business transformation on industry 4.0 capabilities and sustainable supply chain performance. *Supply Chain Management: An International Journal.* https://doi.org/10.1108/SCM-04-2021-0152

Beske-Janssen, P., Johnson, M. P., & Schaltegger, S. (2015). 20 Years of performance measurement in sustainable supply chain management-what has been achieved? *Supply Chain Management: An International Journal, 20*(6), 664–680.

Boutkhoum, O., Hanine, M., Nabil, M., El Barakaz, F., Lee, E., Rustam, F., & Ashraf, I. (2021). Analysis and evaluation of barriers influencing blockchain implementation in Moroccan sustainable supply chain management: An integrated IFAHP-DEMATEL framework. *Mathematics, 9*(14), 1601.

Brandenburg, M., Govindan, K., Sarkis, J., & Seuring, S. (2014). Quantitative models for sustainable supply chain management: Developments and directions. *European Journal of Operational Research, 233*(2), 299–312.

Buyukozkan, G., & Cifci, G. (2011). A novel fuzzy multi-criteria decision framework for sustainable supplier selection with incomplete information. *Computers in Industry, 62*(2), 164–174.

Buyukozkan, G., & Cifci, G. (2013). An integrated QFD framework with multiple formatted and incomplete preferences: A sustainable supply chain application. *Applied Soft Computing, 13*(9), 3931–3941.

Carter, C. R., & Easton, P. L. (2011). Sustainable supply chain management: Evolution and future directions. *International Journal of Physical Distribution & Logistics Management, 41*(1), 46–62.

Carter, C. R., & Rogers, D. S. (2008). A framework of sustainable supply chain management: Moving toward new theory. *International Journal of Physical Distribution & Logistics Management, 38*(5), 360–387.

Chen, L. J., Duan, D. T., Mishra, A. R., & Alrasheedi, M. (2021). Sustainable third-party reverse logistics provider selection to promote circular economy using new uncertain interval-valued intuitionistic fuzzy-projection model. *Journal of Enterprise Information Management*. https://doi.org/10.1108/JEIM-02-2021-0066

Chen, Y. G., Wang, S., Yao, J. R., Li, Y. X., & Yang, S. Q. (2018). Socially responsible supplier selection and sustainable supply chain development: A combined approach of total interpretive structural modeling and fuzzy analytic network process. *Business Strategy and the Environment, 27*(8), 1708–1719.

Choudhary, A., De, A., Ahmed, K., & Shankar, R. (2021). An integrated fuzzy intuitionistic sustainability assessment framework for manufacturing supply chain: A study of UK based firms. *Annals of Operations Research*. https://doi.org/10.1007/s10479-019-03452-3

Coenen, J., Van Der Heijden, Rob, E. C. M., Riel, V., & Allard, C. R. (2018). Understanding approaches to complexity and uncertainty in closed-loop supply chain management: Past findings and future directions. *Journal of Cleaner Production, 201*, 1–13.

Cui, L., Wu, H., & Dai, J. (2021). Modelling flexible decisions about sustainable supplier selection in multitier sustainable supply chain management. *International Journal of Production Research*. https://doi.org/10.1080/00207543.2021.1924412

Ecer, F., & Pamucar, D. (2020). Sustainable supplier selection: A novel integrated fuzzy best worst method (F-BWM) and fuzzy CoCoSo with Bonferroni (CoCoSo'B) multi-criteria model. *Journal of Cleaner Production, 266*, 121981.

Erol, I., Sencer, S., & Sari, R. (2011). A new fuzzy multi-criteria framework for measuring sustainability performance of a supply chain. *Ecological Economics, 70*(6), 1088–1100.

Eskandari-Khanghahi, M., Tavakkoli-Moghaddam, R., Taleizadeh, A. A., & Amin, S. H. (2018). Designing and optimizing a sustainable supply chain network for a blood platelet bank under uncertainty. *Engineering Applications of Artificial Intelligence, 71*, 236–250.

Ewbank, H., Roveda, J. A. F., Roveda, S. R. M. M., Ribeiro, A. I., Bressane, A., Hadi-Vencheh, A., & Wanke, P. (2020). Sustainable resource management in a supply chain: A methodological proposal combining zero-inflated fuzzy time series and clustering techniques. *Journal of Enterprise Information Management, 33*(5), 1059–1076.

Fabbe-Costes, N., Christine, R., Margaret, T., & Andrew, T. (2014). Sustainable supply chains: a framework for environmental scanning practices. *International Journal of Operations & Production Management, 34*(5), 664–694.

Fahimnia, B., & Jabbarzadeh, A. (2016). Marrying supply chain sustainability and resilience: A match made in heaven. *Transportation Research Part E-Logistics and Transportation Review, 91*, 306–324.

Fallahpour, A., Olugu, E. U., Musa, S. N., Wong, K. Y., & Noori, S. (2017). A decision support model for sustainable supplier selection in sustainable supply chain management. *Computers & Industrial Engineering, 105*, 391–410.

Foroozesh, N., Tavakkoli-Moghaddam, R., & Mousavi, S. M. (2018). Sustainable-supplier selection for manufacturing services: A failure mode and effects analysis model based on interval-valued fuzzy group decision-making. *International Journal of Advanced Manufacturing Technology, 95*(9–12), 3609–3629.

Ghadimi, P., Dargi, A., & Heavey, C. (2017). Sustainable supplier performance scoring using audition check-list based fuzzy inference system: A case application in automotive spare part industry. *Computers & Industrial Engineering, 105*, 12–27.

Gilani, H., & Sahebi, H. (2020). A multi-objective robust optimization model to design sustainable sugarcane-to-biofuel supply network: The case of study. *Biomass Conversion and Biorefinery*. https://doi.org/10.1007/s13399-020-00639-8

Gilani, H., Sahebi, H., & Oliveira, F. (2020). Sustainable sugarcane-to-bioethanol supply chain network design: A robust possibilistic programming model. *Applied Energy, 278*, 115653.

Glover, J. L., Champion, D., Daniels, K. J., & Dainty, A. J. D. (2014). An institutional theory perspective on sustainable practices across the dairy supply chain. *International Journal of Production Economics, 152*, 102–111.

Govindan, K., Khodaverdi, R., & Jafarian, A. (2013). A fuzzy multi criteria approach for measuring sustainability performance of a supplier based on triple bottom line approach. *Journal of Cleaner Production, 47*, 345–354.

Govindan, K., & Soleimani, H. (2016). A review of reverse logistics and closed-loop supply chains: A journal of cleaner production focus. *Journal of Cleaner Production, 142*, 371–384.

Gracia, M. D., & Quezada, L. E. (2016). A framework for strategy formulation in sustainable supply chains: A case study in the electric industry. *Netnomics, 17*(1), 3–27.

Guide, V. D. R., & Van Wassenhove, L. N. (2009). OR FORUM—The evolution of closed-loop supply chain research. *Operations Research, 57*(1), 10–18.

Hassini, E., Surti, C., & Searcy, C. (2012). A literature review and a case study of sustainable supply chains with a focus on metrics. *International Journal of Production Economics, 140*(1), 69–82.

Izadikhah, M., Saen, R. F., & Ahmadi, K. (2017). How to assess sustainability of suppliers in volume discount context? A new data envelopment analysis approach. *Transportation Research Part D-Transport and Environment, 51*, 102–121.

Izadikhah, M., Saen, R. F., Ahmadi, K., & Shamsi, M. (2021). How to use fuzzy screening system and data envelopment analysis for clustering sustainable suppliers? A case study in Iran. *Journal of Enterprise Information Management, 34*(1), 199–229.

Jain, N., & Singh, A. R. (2020). Sustainable supplier selection under must-be criteria through fuzzy inference system. *Journal of Cleaner Production, 248*, 119275.

Jakhar, S. K. (2015). Performance evaluation and a flow allocation decision model for a sustainable supply chain of an apparel industry. *Journal of Cleaner Production, 87*, 391–413.

Jiang, L. S., & Liao, H. C. (2020). Cognitive fuzzy sets for decision making. *Applied Soft Computing, 93*, 106374.

Jin, M. Z., Song, L. J., Wang, Y. A., & Zeng, Y. C. (2018). Longitudinal cooperative robust optimization model for sustainable supply chain management. *Chaos Solitons & Fractals, 116*, 95–105.

Kannan, D. (2018). Role of multiple stakeholders and the critical success factor theory for the sustainable supplier selection process. *International Journal of Production Economics, 195*, 391–418.

Karmaker, C. L., Ahmed, T., Ahmed, S., Ali, S. M., Moktadir, M. A., & Kabir, G. (2021). Improving supply chain sustainability in the context of COVID-19 pandemic in an emerging economy: Exploring drivers using an integrated model. *Sustainable Production and Consumption, 26*, 411–427.

Kazancoglu, Y., Ozkan-Ozen, Y. D., Sagnak, M., Kazancoglu, I., & Dora, M. (2021). Framework for a sustainable supply chain to overcome risks in transition to a circular economy through Industry 4.0. *Production Planning & Control*. https://doi.org/10.1080/09537287.2021.1980910

Khan, S. A. R., Zkik, K., Belhadi, A., & Kamble, S. S. (2021). Evaluating barriers and solutions for social sustainability adoption in multi-tier supply chains. *International Journal of Production Research, 59*(11), 3378–3397.

Khishtandar, S., Zandieh, M., & Dorri, B. (2017). A multi criteria decision making framework for sustainability assessment of bioenergy production technologies with hesitant fuzzy linguistic term sets: The case of Iran. *Renewable & Sustainable Energy Reviews, 77*, 1130–1145.

Kumar, A., Moktadir, M. A., Khan, S. A. R., Garza-Reyes, J. A., Tyagi, M., & Kazancoglu, Y. (2020). Behavioral factors on the adoption of sustainable supply chain practices. *Resources Conservation and Recycling, 158*, 104818.

Laosirihongthong, T., Samaranayake, P., Nagalingam, S. V., & Adebanjo, D. (2020). Prioritization of sustainable supply chain practices with triple bottom line and organizational theories: Industry and academic perspectives. *Production Planning & Control, 31*(14), 1207–1221.

Lenzen, M., Moran, D., Kanemoto, K., Foran, B., Lobefaro, L., & Geschke, A. (2012). International trade drives biodiversity threats in developing nations. *Nature, 486*(7401), 109–112.

Li, J., Fang, H., & Song, W. Y. (2019). Sustainable supplier selection based on SSCM practices: A rough cloud TOPSIS approach. *Journal of Cleaner Production, 222*, 606–621.

Li, P., Liu, J., & Wei, C. P. (2020). Factor relation analysis for sustainable recycling partner evaluation using probabilistic linguistic DEMATEL. *Fuzzy Optimization and Decision Making, 19*(4), 471–497.

Liaqait, R. A., Warsi, S. S., Agha, M. H., Zahid, T., & Becker, T. (2021). A multi-criteria decision framework for sustainable supplier selection and order allocation using multi-objective optimization and fuzzy approach. *Engineering Optimization*. https://doi.org/10.1080/0305215X.2021.1901898

Lin, Y. H., & Tseng, M. L. (2016). Assessing the competitive priorities within sustainable supply chain management under uncertainty. *Journal of Cleaner Production, 112*, 2133–2144.

Lin, K. P., Tseng, M. L., & Pai, P. F. (2018). Sustainable supply chain management using approximate fuzzy DEMATEL method. *Resources Conservation and Recycling, 128*, 134–142.

Liu, Y., Eckert, C., Yannou-Le Bris, G., & Petit, G. (2019a). A fuzzy decision tool to evaluate the sustainable performance of suppliers in an agrifood value chain. *Computers & Industrial Engineering, 127*, 196–212.

Liu, H. C., Quan, M. Y., Li, Z. W., & Wang, Z. L. (2019b). A new integrated MCDM model for sustainable supplier selection under interval-valued intuitionistic uncertain linguistic environment. *Information Sciences, 486*, 254–270.

Luo, Z. H., & Li, Z. X. (2019). A MAGDM method based on possibility distribution hesitant fuzzy linguistic term set and its application. *Mathematics, 7*(11), 1063.

Luthra, S., & Mangla, S. K. (2018). When strategies matter: Adoption of sustainable supply chain management practices in an emerging economy's context. *Resources Conservation and Recycling, 138*, 194–206.

Meksavang, P., Shi, H., Lin, S. M., & Liu, H. C. (2019). An extended picture fuzzy VIKOR approach for sustainable supplier management and its application in the beef industry. *Symmetry, 11*(4), 468.

Mondal, A., & Roy, S. K. (2021). Application of Choquet integral in interval type-2 Pythagorean fuzzy sustainable supply chain management under risk. *International Journal of Intelligent Systems*. https://doi.org/10.1002/int.22623

Narayanan, A. E., Sridharan, R., & Kumar, P. N. R. (2019). Analyzing the interactions among barriers of sustainable supply chain management practices A case study. *Journal of Manufacturing Technology Management, 30*(6), 937–971.

Nayeri, S., Torabi, S. A., Tavakoli, M., & Sazvar, Z. (2021). A multi-objective fuzzy robust stochastic model for designing a sustainable-resilient-responsive supply chain network. *Journal of Cleaner Production, 311*, 127691.

Nazam, M., Hashim, M., Baig, S. A., Abrar, M., & Shabbir, R. (2020). Modeling the key barriers of knowledge management adoption in sustainable supply chain. *Journal of Enterprise Information Management, 33*(5), 1077–1109.

Nili, M., Seyedhosseini, S. M., Jabalameli, M. S., & Dehghani, E. (2021). A multi-objective optimization model to sustainable closed-loop solar photovoltaic supply chain network design: A case study in Iran. *Renewable & Sustainable Energy Reviews, 150*, 111428.

Padhi, S. S., Pati, R. K., & Rajeev, A. (2018). Framework for selecting sustainable supply chain processes and industries using an integrated approach. *Journal of Cleaner Production, 184*, 969–984.

Pagell, M., & Wu, Z. H. (2009). Building a more complete theory of sustainable supply chain management using case studies of 10 examples. *The Journal of Supply Chain Management, 45*(2), 37–56.

Paul, A., Shukla, N., Paul, S. K., & Trianni, A. (2021). Sustainable supply chain management and multi-criteria decision-making methods: A systematic review. *Sustainability, 13*(13), 7104.

Peng, J. J., Tian, C., Zhang, W. Y., Zhang, S., & Wang, J. Q. (2020). An integrated multi-criteria decision-making framework for sustainable supplier selection under picture fuzzy environment. *Technological and Economic Development of Economy, 26*(3), 573–598.

Pourjavad, E., & Mayorga, R. V. (2019). Multi-objective fuzzy programming of closed-loop supply chain considering sustainable measures. *International Journal of Fuzzy Systems, 21*(2), 655–673.

Puska, A., Nedeljkovic, M., Zolfani, S. H., & Pamucar, D. (2021). Application of interval fuzzy logic in selecting a sustainable supplier on the example of agricultural production. *Symmetry, 13*(5), 774.

Rani, P., Mishra, A. R., Krishankumar, R., Mardani, A., Cavallaro, F., Ravichandran, K. S., & Balasubramanian, K. (2020). Hesitant fuzzy SWARA-complex proportional assessment approach for sustainable supplier selection (HF-SWARA-COPRAS). *Symmetry, 12*(7), 1152.

Rogers, D. S., & Tibben-Lembke, R. (2001). An examination of reverse logistics practices. *Journal of Business Logistics, 22*(2), 129–148.

Rostamzadeh, R., Keshavarz Ghorabaee, M., Govindan, K., Esmaeili, A., & Nobar, H. B. K. (2018). Evaluation of sustainable supply chain risk management using an integrated fuzzy TOPSIS- CRITIC approach. *Journal of Cleaner Production, 175*, 651–669.

Rout, C., Paul, A., Kumar, R. S., Chakraborty, D., & Goswami, A. (2020). Cooperative sustainable supply chain for deteriorating item and imperfect production under different carbon emission regulations. *Journal of Cleaner Production, 272*, 122170.

Seuring, S. (2013). A review of modeling approaches for sustainable supply chain management. *Decision Support Systems, 54*(4), 1513–1520.

Seuring, S., & Müller, M. (2008). From a literature review to a conceptual framework for sustainable supply chain management. *Journal of Cleaner Production, 16*(15), 1699–1710.

Shete, P. C., Ansari, Z. N., & Kant, R. (2020). A Pythagorean fuzzy AHP approach and its application to evaluate the enablers of sustainable supply chain innovation. *Sustainable Production and Consumption, 23*, 77–93.

Soleimani, H., Govindan, K., Saghafi, H., & Jafari, H. (2017). Fuzzy multi-objective sustainable and green closed-loop supply chain network design. *Computers & Industrial Engineering, 109*, 191–203.

Song, W. Y., Xu, Z. T., & Liu, H. C. (2017). Developing sustainable supplier selection criteria for solar air-conditioner manufacturer: An integrated approach. *Renewable & Sustainable Energy Reviews, 79*, 1461–1471.

Tavassoli, M., Fathi, A., & Saen, R. F. (2021). Assessing the sustainable supply chains of tomato paste by fuzzy double frontier network DEA model. *Annals of Operations Research*. https://doi.org/10.1007/s10479-021-04139-4

Tozanli, O., Duman, G. M., Kongar, E., & Gupta, S. M. (2017). Environmentally concerned logistics operations in fuzzy environment: A literature survey. *Logistics, 1*(1), 4. https://doi.org/10.3390/logistics1010004

Tsao, Y. C., Thanh, V. V., Lu, J. C., & Yu, V. (2018). Designing sustainable supply chain networks under uncertain environments: Fuzzy multi-objective programming. *Journal of Cleaner Production, 174*, 1550–1565.

Tseng, M. L., Islam, M. S., Karia, N., Firdaus Ahmad Fauzi, F. A., & Samina Afrin, S. (2019a). A literature review on green supply chain management: Trends and future challenges. *Resources, Conservation and Recycling, 141*, 145–162.

Tseng, M. L., Lim, M., & Wong, W. P. (2015). Sustainable supply chain management A closed-loop network hierarchical approach. *Industrial Management & Data Systems, 115*(3), 436–461.

Tseng, M. L., Lim, M. K., & Wu, K. J. (2019c). Improving the benefits and costs on sustainable supply chain finance under uncertainty. *International Journal of Production Economics, 218*, 308–321.

Tseng, M. L., Wu, K. J., Hu, J. Y., & Wang, C. H. (2018). Decision-making model for sustainable supply chain finance under uncertainties. *International Journal of Production Economics, 205*, 30–36.

Tseng, M. L., Wu, K. J., Lim, M. K., & Wong, W. P. (2019b). Data-driven sustainable supply chain management performance: A hierarchical structure assessment under uncertainties. *Journal of Cleaner Production, 227*, 760–771.

Wang, C. N., Nguyen, N. A. T., Dang, T. T., & Lu, C. M. (2021). A compromised decision-making approach to third-party logistics selection in sustainable supply chain using fuzzy AHP and fuzzy VIKOR methods. *Mathematics, 9*(8), 886.

WCED. (1987). *WCED-World Commission on Environment and Development: Our common future. Environment: Science and Policy for Sustainable Development.*

Wu, K. J., Liao, C. J., Tseng, M. L., & Chiu, K. K. S. (2016). Multi-attribute approach to sustainable supply chain management under uncertainty. *Industrial Management & Data Systems, 116*(4), 777–800.

Wu, C., Lin, Y., & Barnes, D. (2021). An integrated decision-making approach for sustainable supplier selection in the chemical industry. *Expert Systems with Applications, 184*, 115553.

Xu, Z., Qin, J. D., Liu, J., & Martinez, L. (2019). Sustainable supplier selection based on AHPSort II in interval type-2 fuzzy environment. *Information Sciences, 483*, 273–293.

Yadav, S., & Singh, S. P. (2020). Blockchain critical success factors for sustainable supply chain. *Resources Conservation and Recycling, 152*, 104505.

Yadav, S., & Singh, S. P. (2021). An integrated fuzzy-ANP and fuzzy-ISM approach using blockchain for sustainable supply chain. *Journal of Enterprise Information Management, 34*(1), 54–78.

Yazdani, M, Wang, ZX, Chan, FTS: A decision support model based on the combined structure of DEMATEL, QFD and fuzzy values. *Soft Computing* 24(16), 12449–12468 (2020).

Yousefi, S., Soltani, R., Saen, R. F., & Pishvaee, M. S. (2017). A robust fuzzy possibilistic programming for a new network GP-DEA model to evaluate sustainable supply chains. *Journal of Cleaner Production, 166*, 537–549.

Zadeh, L. A. (1975). The concept of a linguistic variable and its applications to approximate reasoning. *Information Sciences, 8*, 199–249, II, III8,9: 301–357,43–80.

Zeng, S. Z., Hu, Y. J., Balezentis, T., & Streimikiene, D. (2020). A multi-criteria sustainable supplier selection framework based on neutrosophic fuzzy data and entropy weighting. *Sustainable Development, 28*(5), 1431–1440.

Zhou, X. Y., Pedrycz, W., Kuang, Y. X., & Zhang, Z. (2016). Type-2 fuzzy multi-objective DEA model: An application to sustainable supplier evaluation. *Applied Soft Computing, 46*, 424–440.

Zhou, X. Y., Wang, Y., Chai, J., Wang, L. Q., Wang, S. Y., & Lev, B. (2019). Sustainable supply chain evaluation: A dynamic double frontier network DEA model with interval type-2 fuzzy data. *Information Sciences, 504*, 394–421.

Zhou, F. L., Wang, X., Lim, M. K., He, Y. D., & Li, L. X. (2018). Sustainable recycling partner selection using fuzzy DEMATEL-AEW-FVIKOR: A case study in small-and-medium enterprises (SMEs). *Journal of Cleaner Production, 196*, 489–504.

Zhou, X. Y., Wei, X. Y., Lin, J., Tian, X., Lev, B., & Wang, S. Y. (2021). Supply chain management under carbon taxes: A review and bibliometric analysis. *Omega, 98*(102295).

Zulqarnain, R. M., Siddique, I., Ali, R., Pamucar, D., Marinkovic, D., & Bozanic, D. (2021). Robust aggregation operators for intuitionistic fuzzy hypersoft set with their application to solve MCDM problem. *Entropy, 23*(6), 688.

Applications of Machine Learning Algorithms in Data Sciences

Adeel Ansari, Seema Ansari, Fatima Maqbool, Rabia Zaman, and Kubra Bashir

Abstract Machine Learning, a branch of artificial intelligence (AI) and computer science, focuses on the usage of data and algorithms to copy the humans learning method, slowly increasing its accurateness. The chapter aims at discussing the applications of the machine learning algorithms, essential for developing predictive modeling and for carrying out classification and prediction in both supervised and unsupervised scenarios. The Machine Learning techniques have been applied to many application domains as a result of a humongous amount of data being created, processed, and mined from the evolution of the World Wide Web, mobile applications, and the rise of social media applications. Some of these applications are virtual personal assistants, predictions, surveillance, social media services, malware filtering, search engine result refining, and online fraud detections. The chapter includes the introduction, State of the Art, Machine Learning Algorithms, Applications of Machine Learning Algorithms in data sciences, followed by conclusion and future recommendations.

Keywords Artificial intelligence · Big data · Data mining · Data science · Machine learning · Neural networks

---

A. Ansari
Department of Computer Sciences, Shaheed Zulfikar Ali Bhutto Institute of Science & Technology, Karachi, Pakistan
e-mail: adeel.ansari@szabist.edu.pk

S. Ansari (✉) · R. Zaman · K. Bashir
Electrical Engineering and Engineering Management Department, Institute of Business Management, Karachi, Pakistan
e-mail: seema.ansari@iobm.edu.pk; rabia.zaman@iobm.edu.pk; kubra.bashir@iobm.edu.pk

F. Maqbool
Department of Computer Sciences, Shaheed Zulfikar Ali Bhutto Institute of Science & Technology, Karachi, Pakistan

Electrical Engineering and Engineering Management Department, Institute of Business Management, Karachi, Pakistan
e-mail: fatima.maqbool@iobm.edu.pk

F. P. García Márquez, B. Lev (eds.), *Sustainability*, International Series in Operations Research & Management Science 333,
https://doi.org/10.1007/978-3-031-16620-4_4

1 Introduction

Machine learning plays a vital role in the developing field of data sciences. By using statistical techniques, algorithms are designed to classify or project, the main insights in projects of data mining which later enable decision-making in applications and commerce, impacting vital growth metrics. With the continuous expansion of big data, the demand for data scientist will also continue to grow as they will be required to help in identifying the most related business queries and the data to answer the queries. Machine Learning may be defined as making a model through inductive learning and patterns using a dataset by a cognitive system (Tzanis et al., 2006). Machine Learning tasks can be categorized into three groups: supervised learning, unsupervised learning, and reinforcement learning.

In supervised learning, input data is labeled. Supervised learning is used for classification and prediction. In unsupervised learning, input data is unlabeled and unstructured. Unsupervised learning is used for clustering (Supervised learning vs unsupervised learning, n.d.). Reinforcement learning is different from supervised and unsupervised learning. Reinforcement learning is used for the purpose of reward and punishments based on the performance (Reinforcement learning, n.d.). Reinforcement learning is applicable for policies and standards. Supervised learning consists of multiple algorithms such as support vector machine (SVM), decision trees, and neural networks. Unsupervised learning includes: K mean clustering and Gaussian mixture, whereas reinforcement learning involves Q-Learning, R-Learning, and TD-Learning.

2 State of the Art

In this section, the researcher's contribution in the field of Machine Learning and data sciences is presented.

In Baraneetharan (2020), authors have discussed the role of machine learning algorithms in Wireless Sensor Network (WSN) environment. Machine learning algorithms can be used to increase the ability of WSN environment by eliminating the need of redesigning the network. With the help of machine learning algorithms, security can be enhanced by securing the data from hackers.

In Asthana and Hazela (2020), authors have discussed the role of machine learning algorithms in academics. Machine Learning based systems are capable of experiential learning. It can learn from experience and can update itself for better performance. It has different algorithms for classification and prediction. Some of them are Artificial Neural Network, KNN, and Principle Component Analysis.

In Fabisch et al. (2019), authors have highlighted the role of machine learning algorithms in robotics. Machine learning technology is playing a remarkable role in the field of Robotics. Mostly, robots are designed on hard computing, they are programmed for specific tasks. Whereas, machine learning is based on soft computing, it can train the machines to take intelligent decisions. Humanoid Robots are the most remarkable contribution of machine learning in robotics, currently the intelligent Humanoid robot named as Erica and Sophia have been designed, which can behave like human and efficient to take decision.

In Sarke et al. (2020), authors have discussed the role of machine learning algorithms in data security. The data security is the main requirement of all industries and cyber-attacks are growing very rapidly in the digital world. Machine learning is playing a vital role in making data secure and safe. Many firms have adopted machine learning technologies such as AEG bot, AI2 Platform to predict cyber-attacks and bugs.

In Schrage and Kiron (2018), the role of machine learning has been discussed in automotive industry. Many automotive industries are using AI technology to provide better performance and virtual assistance to their customers. Tesla has introduced an intelligent virtual assistant named Tesla Bot. Tesla Bot is used to find the magnitude of electricity and it is also known as Geiger Counter. Multiple industries are using machine learning algorithms for creating driver free or self-driven vehicles.

Recent studies in the field of machine learning have revealed that computers are in self-learning mode, working without any programming. Data scientist have found machine learning very much useful in the field of data sciences. Machine learning algorithms can be applied for data collection, preparation of data, training the model, evaluating the model, tuning the parameters, and making predictions (Role of Machine Learning in Data Science Applications, 2019).

From the research (Kononenko, 2001), it is quite observed that the machine learning can play a significant role in medical diagnostic era and treatments due to the transparency of data examination. For intelligent data diagnosis, the author emphasized on Naive Bayesian classifier, neural networks, and decision trees algorithms of machine learning.

According to the study (Sanou, 2015; Sarker et al., 2019) currently, smartphones are considered as the most basic and beneficial need of our life. In today's world, 96.8% of the people are recorded to use mobile, and this amount can be increased up to 100% in many advanced countries. In this research, smartphone usage using machine learning algorithms has been predicted that can be helpful for both the researchers and application designers to design and plan smart and intellectual context-aware systems for smartphone operators.

Current developments in Machine Learning (ML) proved huge potential in manufacturing domain through advanced systematic tools for processing the vast amount of manufacturing data. It is achieving the goal of refining operations from conceptualization, dropping error rates, improving maintenance, and expressively growing inventory turns to the final delivery as elaborated in Kang et al. (2020).

In Sarker (2021), author elaborated the role of ML algorithm in the real-world applications. In this era of data, everything is interrelated to a data source and digitally recorded. In present times there are treasure of data in several real-world *application* areas, such as cybersecurity data, the Internet of Things (IoT), smart city data, smartphone data, business data, social media data, COVID-19 data, and many more. An effective machine learning model is determined by data and performance of the learning algorithms. It has capability of *extracting valuable knowledge* or *insights from* the data in an intelligent and an appropriate way on which the real-world applications are based.

In Qu et al. (2019), author enlightened the role of machine learning in the field of microbiology. As microorganisms have great impact on human heath, crop growth, environmental management, livestock farming, industrial chemical manufacture, and food production, ML can be used in the field of microbiology, specifically identifying problems and for discovering the interaction between bacteria/viruses and the surrounding situation.

In Jabbar et al. (2018), authors expanded the role of machine learning algorithm in the area of health care. ML assists in resolving issues of both the patients and the physicians, by dealing and analyzing the medical data, which includes X-ray results, blood samples, DNA sequences, vaccination, vital signs, etc., in most efficient way. With the help of Machine Learning, there will be a reduction in patient-handling time period and will provide quicker healthcare results, which can have a decremental impact over the incurred expenses in healthcare sector.

3 Machine Learning Algorithms

Machine Learning (ML) algorithms are a critical part of Data Science. ML is a technique to analyze data that automates analytical model building. ML is a domain of AI built on the concept that systems can be trained from data, recognize patterns, and yield decisions with minimum human involvement. ML teaches a machine how to gather knowledge and be trained (Machine Learning, n.d.). They can be defined as processes in sets of rules to solve various problems, depicted in Fig. 1.

Machine Learning Algorithms and their classifications are shown in Fig. 2. There are many algorithms that organizations develop to serve their unique needs.

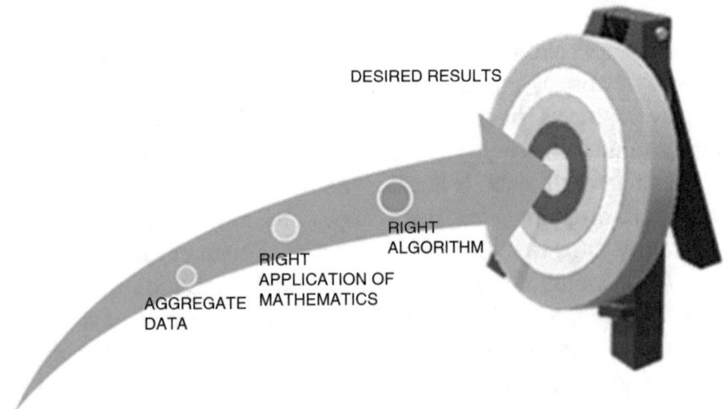

Fig. 1 Machine learning process (Machine Learning Process, n.d.)

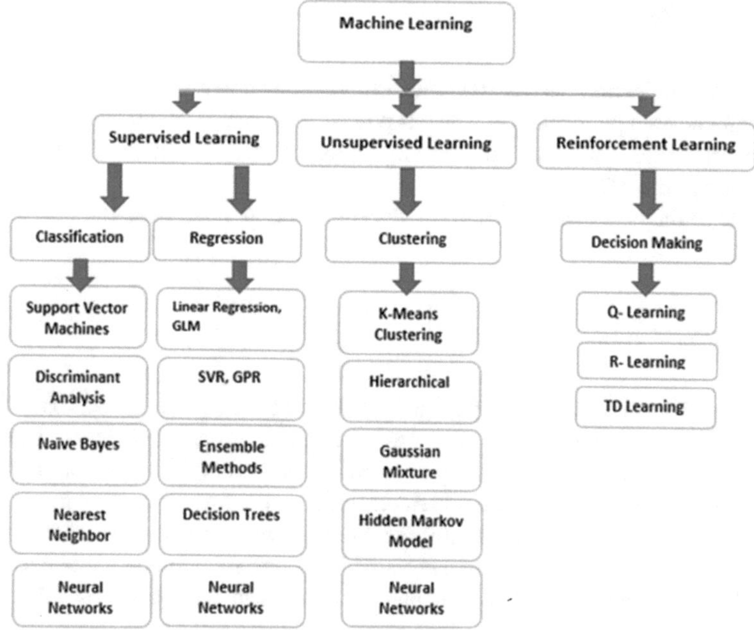

Fig. 2: Classification of machine learning algorithm (Machine Learning Algorithms, n.d.)

3.1 Supervised Learning

It is used for the structured dataset. It analyzes the training data and generates a function that will be used for other datasets. Supervised learning is further divided into classification and regression. Classification is used to classify the data into classes, whereas regression is used for prediction.

3.2 Unsupervised Learning

It is used for raw datasets. Its main task is to convert raw data to structured data. In today's world, there is a huge amount of raw data in every field. Even the computer generates log files which are in the form of raw data. Therefore it is the most important part of machine learning.

3.3 Reinforcement Learning

Reinforcement learning is based on rewards and punishments. Rewards are given on achieving desired performance, whereas punishments are given on undesired behavior/performance. This type of learning involves decision-making, Q-Learning, R-Learning, and TD Learning.

Some of the best machine learning algorithms are defined below:

3.4 Linear Regression

Linear regression is a statistical machine learning algorithm which is used to estimate linear relationship between one or more than one predicators. There are two types of linear regression: simple regression and multiple regression. In simple regression we have one dependent variable which is predicted from one independent variable. Whereas, in multiple regression we have an dependent variable which is predicted from the set of independent variables. Regression modeling is performed to find linear relationship between one or more than one variables (Maulud & Abdulazeez, 2020).

3.5 Logistic Regression

Logistic regression algorithm is considered as one of the most popular machine learning algorithms which is based on statistics. Logistic regression is used for binary classification of multiple data points. It consists of two components: sigmoid curve and hypothesis. Sigmoid curve represents cost function, whereas hypothesis function is used to limit cost function between 0 and 1. There are three types of Logistic Regression: Binary Logistic Regression, Ordinal Logistic Regression, and Multinomial Logistic Regression (Rymarczyk et al., 2019).

3.6 Decision Trees

Decision trees are used for making decision with provided set of input and also help in classification and prediction. There are two approaches to build decision trees: pruning and induction. Pruning technique is used to reduce size of decision tree by eliminating non-critical data. Induction is used for the purpose of data mining (Jijo & Abdulazeez, 2021).

3.7 Naive Bayes

Naïve Bayes is a classification algorithm which is based on Bayes Theorem. Accuracy of this algorithm is very high and small amount of training data is required to obtain true class. This algorithm is famous for data mining and probabilistic classification. Bayes theorem formula is given below:

$$P(A|B) = \frac{P(B|A)P(A)}{P(B)}$$

where $P(A|B)$ is called prior probability and $P(B|A)$ is known as marginal probability (Chen et al., 2020).

3.8 Artificial Neural Networks

Artificial Neural Networks (ANN) are based on human nervous system. It has neuron to pass information from input layer to the output layer. It comprises input layer, hidden layer, and output layer. There are two types of ANN: Single-Layer Perceptron (SLP) and Multi-Layer Perceptron (MLP). Single-Layer Perceptron is designed for one-dimension data. It consists of Input Layer (i) and neuron or output layer (j). In order to understand the concept of Single-Layer Perceptron it is important to know: how many input attributes are there? How many output attributes are there? and which network is being used? Multi-Layer Perceptron is designed for two-dimension and diffused data. It consists of Input Layer (i), hidden layer (p), and neuron or output layer (j). In order to understand the concept of multi-layer perceptron it is important to know: how many input attributes are there? How many output attributes are there? how many hidden layers are there? and which network is being used? (Yang & Wang, 2020).

3.9 K-Means Clustering

K-Means Clustering is a machine learning algorithm which is used to partition data containing n values into subsets. This algorithm is mainly utilized to narrow Euclidean distance to lowest value. This comes under the category of supervised machine learning algorithms. This algorithm consists of three steps: (i) specification of cluster (k), (ii) initialization of centroids by rearranging, and (iii) keep on iterating until all the centroids eliminate (Chakraborty et al., 2020).

3.10 Anomaly Detection

Anomaly detection plays a vital role in enhancement of communication and reduction of threats to the software system. Anomaly detection technique is used to identify unusual datasets. With the help of anomaly detection, we can remove unnecessary information. Machine learning based anomaly detection system works accurately, timely and handles complex and large datasets (Johnson, 2020).

3.11 Gaussian Mixture Model

Gaussian Mixture Model comes under the category of unsupervised machine learning algorithms. This model is used to represent normally distributed subpopulation of the main population. This algorithm facilitates the process of data mining and feature extraction. The procedure of this algorithm is similar to K-Mean Clustering. Model can be used to cluster unlabeled classes/datasets (Xing & Cao, 2020).

3.12 Principal Component Analysis

Principle Component Analysis (PCA) is used to reduce the dimension of large dataset into small units. These small units have main information of the large dataset. This algorithm helps in exploring and understanding the complex datasets. This method consists of six steps: (i) getting data, (ii) normalization of data, (iii) calculation of covariance matrix, (iv) calculation of Eigen values and Eigen matrices, (v) formation of feature vector, and (vi) deriving new dataset (Ramirez-Figueroa et al., 2021).

3.13 K-Nearest Neighbor (KNN)

K-Nearest Neighbor (KNN) is a supervised machine learning algorithm. This algorithm is used for prediction regression and classification. This facilitates in finding missing values and sampling the data. In KNN, K represents number of the nearest neighbors. In this algorithm, data is classified on the basis of neighbor. Euclidean distance is calculated to classify the datasets (Peng et al., 2020).

3.14 *Support Vector Machines*

Support Vector Machines (SVM) lies under the category of supervised machine learning algorithm. This algorithm is used to analyze data through classification and regression. With the help of SVM, user can label training data into different categories. This tool differentiates two classification problems with the help of classification algorithms (Wadkar et al., 2020).

4 Applications of Machine Learning Algorithms in Data Sciences

Machine learning algorithms are a critical part of data science that has been applied extensively in various application domains. Some of the most popular applications include medical diagnosis, credit risk analysis, customer profiling, market segmentation, targeted marketing, retail management, and fraud detection (Tzanis et al., 2006). The machine learning techniques have been applied to many application domains as a result of a humongous amount of data being created, processed, and mined from the evolution of the World Wide Web, mobile applications, and the rise of social media applications.

Some of these applications are as follows:

4.1 *Virtual Personal Assistants*

Virtual assistants are also known as digital assistants which can understand natural language. Machine learning plays a vital role in creating virtual personal assistants with the help of algorithm, known as "Natural Language Processing." Many Higher Educational Institutions have adopted Virtual Personal Assistants to impart knowledge. These are capable of solving queries raised by the students and making decisions (Gupta & Harsh, 2020).

4.2 *Predictions*

Machine learning algorithms are widely used to predict future outcomes. There are various algorithms such as pattern recognition, clustering, image recognition and classification can be used for prediction. Many healthcare centers have implemented machine learning to predict the diseases and systems. MATLAB based tool, regression learner can be used to do predictions. Many business industries have adopted machine learning to predict sales and to make accurate guesses. (Data Science: Machine Learning and Predictions, n.d.).

4.3 Video Surveillance

Video surveillance possesses a significant role in managing unstructured big data. Machine learning algorithms such as pattern recognition and classification are used for the purpose of video surveillance. With the help of anomaly detection machine learning algorithm, an intelligent video surveillance can be made to ensure surveillance and security (Durai & Saleem, 2019).

4.4 Social Media Services

Social media is mainly based on information. Machine learning is a buzzword for collecting, analyzing, processing, and evaluating data to convert it into useful information. It is easy to market products by knowing about social media audiences. In order to know when people are marketing your products on social media, image recognition is needed. By using ML algorithms systems can easily recognized faces, logos, and objects, in both images and video. Also, machine learning algorithms are a powerful tool to help our customers to provide them reliability and authenticity on their products, ambassadors, and brands (AI Basics: How AI & Machine Learning Supercharge Your Social Media Marketing, n.d.).

4.5 Email Spam and Malware Filtering

Machine learning algorithms have been widely adopted by Google and other platforms for the purpose of malware filtering and to spamming emails which can be harmful for the system. It can be used to identify and filter out unwanted email called spam emails. In this era of Internet, the number of useless emails is increasing day by day. So the filtering of such emails in Yahoo, Gmail, and Outlook is much needed. By applying deep learning techniques, spam emails can be effectively dealt with and filtered out (Dada et al., 2019).

4.6 Online Customer Care

Machine learning facilitates online customer service by responding customers, resolving their queries, and giving decisions. ML is useful to work machines more efficiently and also in elevating customer support through channels such as on calls, emails, chats or through documentation and much more. ML helps to know what customer exactly want to hear by responding to their frequently asked questions. (How Machine Learning is Optimizing Customer Support, n.d.).

4.7 *Search Engine Result Refining*

Machine Learning has revolutionized search engines. Machine learning algorithms have refined the search engine by filtering out irrelevant sources and information. Tensor Flow is a platform for machine learning for both beginners and experts. It helps the researchers to insert their creativity in the field of machine learning and developer can easily made their own ML application (How Machine Learning in Search Works: Everything You Need to Know, n.d.).

4.8 *Product Recommendations*

Product recommendations refers to the suggestions about the products. With the help of machine learning algorithms, software tools are designed which are capable of recommending products to the stakeholders. There are three major types of product recommendation system which gives the relationship between user–product, user–user, and product–product. It relates these connections and recommends products accordingly in online business, which helps to boost up revenue, sales and create a positive impact on customer by given them loyalty, reliability, and brand affinity (Mrukwa & Grzegorz, 2019).

4.9 *Online Fraud Detection*

Various machine learning algorithms can be applied to detect fraud especially when physical loss of credit card or loss of sensitive credit card information. To figure out any irregularity, many algorithms including Naïve Bayes, Random Forest, Multi-layer Perceptron, and Logistic Regression have been used to identify transaction as fraud or genuine with high accuracy (Varmedja et al., 2019).

5 Conclusion

New technologies provide the prospects for innovative and better approaches for analysis of data. Some evolving developments of machine learning have been discussed. In this chapter, some most popular machine learning algorithms and their implication in different application domains have been discussed in detail. Machine learning provides a better interpretation of the human cognitive intelligence and offers better facilitation towards services like image and audio recognition, feature extraction, strategic and marketing decision-making and plays now a significant role in consumer buying behavior and for customer segmentations for new start-ups. The demand for Machine Learning Engineers is increasing and this is due to evolving technology and generation of huge amounts of data. The machine learning techniques have been applied to many application domains as a result of a massive amount of data being produced, processed, and mined from the evolution of the World Wide Web, mobile applications, and the rise of social media applications. With the features of the new kind of available data, and variety of new problems, the importance of machine learning research and the critical issues are still open and active collaborative research by academic researchers and industry is recommended.

References

AI Basics. (n.d.). How AI & machine learning supercharge your social media marketing. *Artificial Intelligence*. Accessed August 17, 2021, from https://www.linkfluence.com/blog/ai-basics-how-ai-machine-learning-supercharge-social-media-marketing

Asthana, P., & Hazela, B. 2020. Applications of machine learning in improving learning environment. In *Multimedia big data computing for IoT applications* (pp. 417-433). Springer, .

Baraneetharan, E. (2020). Role of machine learning algorithms intrusion detection in WSNs: A survey. *Information Technology and Digital World, 2*, 161–173.

Chakraborty, S., Paul, D., Das, S. and Xu, J. (2020). Entropy Weighted power k-means clustering. In *Twenty Third International Conference on Artificial Intelligence and Statistics*.

Chen, S., Webb, G. I., Liu, L., & Ma, X. (2020). A novel selective naïve Bayes algorithm. *Knowledge-Based System, 192*, 105361.

Dada, E. G., Bassi, J. S., Chiroma, H., Abdulhamid, S. M., Adetunmbi, A. O., & Ajibuwae, O. E. (2019). Machine learning for email spam filtering: Review, approaches and open research problems. *Heliyon, 5*(6), e01802.

Data Science: Machine Learning and Predictions. (n.d.). Accessed August 17, 2021, from https://www.edx.org/course/foundations-of-data-science-prediction-and-machine

Durai, G. S., & Saleem, M. A. (2019). Intelligent video surveillance: A review through deep learning techniques for crowd analysis. *Journal of Big Data, 6(1), 1–27.*

Fabisch, A., Petzoldt, C., Otto, M., & Kirchner, F. (2019). A survey of behavior learning applications in robotics—state of the art and perspectives. *International Journal of Robotics Research*.

Gupta & Harsh. (2020). *Machine learning by virtual assistants*. Accessed August 17, 2021, from https://whataftercollege.com/machine-learning/machine-learning-virtual-assistants/

How Machine Learning in Search Works: Everything You Need to Know. (n.d.). Accessed August 17, 2021, from https://www.searchenginejournal.com/search-engines/machine-learning/#close

How Machine Learning is Optimizing Customer Support. (n.d.). Accessed August 17, 2021, from https://freshdesk.com/customer-support/machine-learning-optimizing-customer-support-blog/

Jabbar, A., Samreen, S., & Aluvalu, R. (2018). The future of health care: Machine learning. *International Journal of Engineering & Technology, 7*, 23–25. ISSN 2227-524X. Accessed November 20, 2021, from https://doi.org/10.14419/ijet.v7i4.6.20226

Jijo, B. T., & Abdulazeez, A. M. (2021). Classification based on decision tree algorithm for machine learning. *Applied Science and Technology Trends, 2*, 20–28.

Johnson, J. 2020. Anomaly detection with machine learning: An introduction. In *Machine learning & big data blog*.

Kang, Z., Catal, C., & Tekinerdogan, B. (2020). Machine learning applications in production lines: A systematic literature review. *Journal of Computers & Industrial Engineering, 149*, 106773.

Kononenko, I. (2001). Machine learning for medical diagnosis: History, state of the art and perspective. *Artificial Intelligence in Medicine, 23*(1), 89–109. http://citeseerx.ist.psu.edu/viewdoc/download?doi=10.1.1.96.184&rep=rep1&type=pdf

Machine Learning. (n.d.). Accessed August 18, 2021, from https://www.sas.com/en_us/insights/analytics/machine-learning.html

Machine Learning Algorithms. (n.d.). Accessed August 17, 2021, from https://intellipaat.com/mediaFiles/2015/11/Picture7.jpg

Machine Learning Process. (n.d.). Accessed August 17, 2021, from https://intellipaat.com/mediaFiles/2015/11/Picture6.png

Maulud, D. H., & Abdulazeez, A. M. (2020). A review on linear regression comprehensive in machine. *Applied Science and Technology Trends, 1*, 140–147.

Mrukwa & Grzegorz. (2019). *How to build a product recommendation system using machine learning*.

Peng, X., Chen, R., Yu, K., Ye, F., & Xue, W. (2020). An improved weighted k-nearest neighbor algorithm for indoor localization. *Electronics, 9*(12), 2117.

Qu, K., Guo, F., Liu, X., Lin, Y., & Zou, Q. (2019). Application of machine learning in microbiology. *Frontiers in Microbiology, 10*, 827. https://doi.org/10.3389/fmicb.2019.00827

Ramirez-Figueroa, J. A., Martin-Barreiro, C., Nieto-Librero, A. B., Leiva, V., & Galindo-Villardón, M. P. (2021). A new principal component analysis by particle swarm optimization with an environmental application for data science. *Stochastic Environmental Research and Risk Assessment, 35(10), 1969–1984*.

Reinforcement Learning. (n.d.). Accessed November 23, 2021, from https://www.geeksforgeeks.org/what-is-reinforcement-learning/

Role of Machine Learning in Data Science Applications. (2019). *Technology trsends*. Accessed August 17, 2021, from https://akki-greatlearning.medium.com/role-of-machine-learning-in-data-science-applications-c92bf2b695f1

Rymarczyk, T., Kozłowski, E., Kłosowski, G., & Niderla, K. (2019). Logistic regression and artificial neural network classification models: A methodology review. *Intelligent Sensor Signal in Machine Learning, 19*.

Sanou, B. (2015). *Measuring the information society report*. https://www.itu.int/en/ITU-D/Statistics/Documents/publications/misr2015/MISR2015-w5.pdf

Sarke, I. H., Kayes, A. S. M., Badsha, S., Alqahtani, H., Watters, P., & Ng, A. (2020). Cybersecurity data science: An overview from machine learning perspective. *Journal of Big data, 7*(1), 1–29.

Sarker, I. H. (2021). Machine learning: Algorithms, real-world applications and research directions. *SN Computer Science, 2*(3), 1–21.

Sarker, I. H., Kayes, A. S. M., & Watters, P. (2019). Effectiveness analysis of machine learning classification models for predicting personalized context-aware smartphone usage. *Journal of Big Data, 6*(1), 1–28.

Schrage, M., & Kiron, D. (2018). Machine learning in the automotive industry: Aligning investments and incentives. In *Big idea: Strategic measurement*.

Supervised Learning Vs Unsupervised Learning. (n.d.). Accessed November 23, 2021, from https://www.javatpoint.com/difference-between-supervised-and-unsupervised-learning

Tzanis, G., Katakis, I., Partalas, I., & Vlahavas, I. (2006, July). Modern applications of machine learning. *In Proceedings of the 1st Annual SEERC Doctoral Student Conference–DSC*, 1(1), pp. 1-10.

Varmedja, D., Karanovic, M., Sladojevic, S., Arsenovic, M. & Anderla, A. (2019). Credit card fraud detection–machine learning methods. In *International Symposium on Infoteh-Jahorina (INFOTEH)*, East Sarajevo, Bosnia and Herzegovina.

Wadkar, M., Troia, F. D., & Stamp, M. (2020). Detecting malware evolution using support vector machines. *Expert System with Applications, 143*, 113022.

Xing, Y., & Cao, D. (2020). Design of integrated road perception and lane detection system for driver intention inference. In *Advanced driver intention inference*.

Yang, G. R., & Wang, X.-J. (2020). Artificial neural networks for neuroscientists: A primer. *Neuron, 109*(6), 1048–1070.

The Machine Learning Model for Identifying Bogus Profiles in Social Networking Sites

C. Hema

Abstract In recent years, social networking sites have gained popularity on the internet, attracting hundreds of thousands of users who spend billions of minutes utilising them. Social networking sites have now taken over many people's life. Every day, a large number of people create profiles on social media sites and contact others regardless of their location or time. Users' personal information is at risk on social networking sites posing security risks. It is important to identify the bogus user's social media identities so that the person who is pushing threats can be uncovered. In this proposed work, the Ensemble learner approach helps to increase the accuracy of detecting fake profiles through automated user profile classification. The study investigates the different text, image, audio, and video attributes that can be used to discern the difference between a fake and a legitimate account. The Ensemble learner approach utilises these attributes and assesses their performance on real-world social media datasets using performance metrics such as accuracy, precision, recall, and F-1 score. The experimental evaluation indicates that the average accuracy achieved by the proposed ensemble learner approach is 97.67%, precision of 92.56% and a recall of 82%, compared to learning models (Support Vector Machine (SVM), K-Nearest Neighbors (KNN), Logistic regression).

Keywords Ensemble learner · Bogus profile · Social media · Machine learning

1 Introduction

A social networking service, or SNS (Romanov et al., 2017) (sometimes known as a social networking site), is an online platform that allows people to develop social networks or relationships with others who have similar personal or professional interests, hobbies, backgrounds, or real-life connections. Examples of a few social networking sites are Facebook, WhatsApp, Twitter and Instagram. The users of

C. Hema (✉)
Department of Computer Science and Engineering, B.S. Abdur Rahman Crescent Institute of Science and Technology, Chennai, Tamil Nadu, India

F. P. García Márquez, B. Lev (eds.), *Sustainability*, International Series in Operations Research & Management Science 333,
https://doi.org/10.1007/978-3-031-16620-4_5

these social networking sites build profiles with personalised information, images and interact with other user profiles. However, increased security concerns and the protection of OSN (Online Social Networks) privacy are key considerations.

Two persons exchange certain private information using social networking sites (SNS). Private information getting exposed to the public in whole or in part makes us prime targets for various forms of attacks, the most serious of which may be identity theft. When a person uses a character's knowledge for personal gain or intent, this is known as identity theft. Online identity fraud is a major issue in the current decade, affecting millions of people all over the world. Identification theft victims may face a variety of consequences, such as losing time or money, being sent to prison, their public image getting destroyed, or harming their relationships with associates and loved ones. Currently, the vast majority of SNs do not verify ordinary user's profiles and hence impacting privacy and security policies. Actually, most SNS applications have their privacy settings set to the bare minimum and as a result, SNs have become a prime forum for fraud and scams. Identity theft and impersonation attacks have been made easier by social networking services for both experienced and naive attackers. Moreover, users must provide accurate information to create an account on social networking websites. However, if such profile data are compromised, easy tracking of what users post on the internet would result in disastrous losses.

In online networks, profile information is mainly static. Static knowledge refers to the information that can be provided by an individual at the time of profile development, while dynamic knowledge refers to the post that is spread by the device inside the network. Static knowledge includes a person's demographics and interests, while dynamic knowledge includes a person's routine behaviours and network location.

The majority of today's research relies on both static and dynamic data. However, this is not true for many social networks, where only a few static profiles are visible and dynamic profiles are rarely visible to the user network. Few researchers have suggested several procedures for detecting false identities and malicious content material in online social networks. Each procedure had its own merits and drawbacks.

Privacy issues, online abuses, misuses, trolls, and other issues surround social networking. Bogus profiles on social networking sites have been used in many illegitimate cases. Bogus profiles are those that are not specific, i.e., profiles of men and women with forged credentials. Bogus Facebook, Instagram, LinkedIn, WhatsApp accounts are more often involved in malicious and undesirable practices, creating problems for users of the social networking site. People create fake profiles for a variety of reasons, including social engineering, online impersonation to shame a person, supporting and campaigning for a character or a group of individuals. Social media sites have their security scheme in place to protect user credentials from spam, phishing, and other forms of fraud. But most of the schemes have not been able to track fake profiles generated by users on social media sites to a large extent.

2 Existing Approaches

In their work, Adikari and Dutta (2020) focused on the minimum collection of profile data required for detecting fake profiles on LinkedIn and suggested a data mining approach for detecting fake profiles. They proved that their approach can detect fake profiles with 87% accuracy and 94% True-Negative Rate using minimal profile data, which is comparable to results obtained using larger data sets and more extensive profile information. Furthermore, when compared to other methods that use similar quantities and types of data, their system showed improved accuracy by about 14%.

To protect legitimate users from nefarious intent, it is crucial to detect these false identities on social media. To combat this issue, researchers intended to use a feature-based approach (Elovici et al., 2017) to identify fake social media profiles. They used 24 features to quickly recognise fake accounts. Three classification algorithms were used to validate the classification performance. Their model was able to achieve 97.9% accuracy using the Random Forest algorithm, according to experimental results. As a result, the proposed method is effective at detecting fake profiles.

Romanov et al. (2017) have suggested that the methods for identifying fake social media accounts can be divided into two types: those that analyse individual accounts and those that capture orchestrated activities through a large number of accounts. The article discusses false identities involved in network intrusions and the methods for identifying fake social media accounts have also been listed.

Chu et al. (2010) introduced COMPA, a method for detecting compromised social media accounts. This scheme is focused on user activity on social media sites. Standard user behaviour is consistent, but COMPA detects compromised accounts with more erratic behaviour. In COMPA, a behavioural profile is generated based on the account's previous post. When a new message is generated, the behavioural profile is compared to it. COMPA flags a message as a compromise if it differs from a behavioural profile. This method is used on both Twitter and Facebook and yields positive results. Twitter and Facebook have a false-positive rate of 4% and 3.6%, respectively.

In the research work of Viswanath et al. (2014), a taxonomy is provided for the classification of malicious information content at various stages and prevalent technologies to deal with the issue, including origin, dissemination, and detection. Authors also identified a research gap and suggested potential research directions to make web information material more reliable and secure for decision-making and knowledge sharing.

In the research work by Farooqi et al. (2017), the authors addressed the problem of identifying and establishing phantom profiles in social popular games on the internet. The research paper looks at a Facebook programme called "Fighters Club," which is popular for offering prizes and users who ask their friends to play to have a gameplay advantage. According to the authors, the game forces players to construct phoney profiles by offering such incentives. The users will get motivated for himself/herself by presenting those fake profiles into the game.

Stringhini et al. (2013) investigated Twitter fan markets. They group the clients of the business sectors and identify the characteristics of Twitter devotee advertisers. The authors argue that there are two types of accounts that pursue the "client": false accounts (also known as "sybils") and compromised accounts, whose owners are unaware that their follower list is growing. Clients of adherent markets may be celebrities or politicians who want to appear as having a larger fan base or cybercriminals who want to make their profile look more credible so they can distribute malware and spam more quickly.

3 Materials and Methods

We have first suggested the framework, followed by ensemble learning models, and performance metrics.

3.1 Proposed Framework

Following an exhaustive analysis of the literature, it was discovered that there are a variety of ways to identify bogus profiles. We should first examine the user's social media profiles to discover who is spreading threats on social media platforms. Based on the classification algorithms, we can identify the difference between genuine and bogus user profiles on social media sites. Nevertheless, the accuracy of bogus profile detection on social media sites should be improved.

In this study, we describe how to spot untrustworthy people on social media sites using an ensemble learner technique. Ensemble learning is a generic machine learning metamethod that aims to improve predictive performance by integrating predictions from many models. Ensemble learning approaches are divided into three categories: bagging, stacking and boosting. The technique of matching numerous decision trees to different samples of the same dataset and merging the results is known as bagging. The technique of relating many types of models to the same data is known as stacking and then comparing the other model to discover the optimal approach to incorporate the forecasts. The technique of adding ensemble members successively to correct earlier model predictions and obtaining a weighted average of the outcomes is known as boosting.

As shown in Fig. 1, our proposed approach is based on the existing literature by incorporating an ensemble learning model with diverse sets to categorise the profiles in social media networks as genuine or bogus. Data collection, data visualisation, training, fine-tuning and evaluation parameters are all included in the proposed framework's five modules.

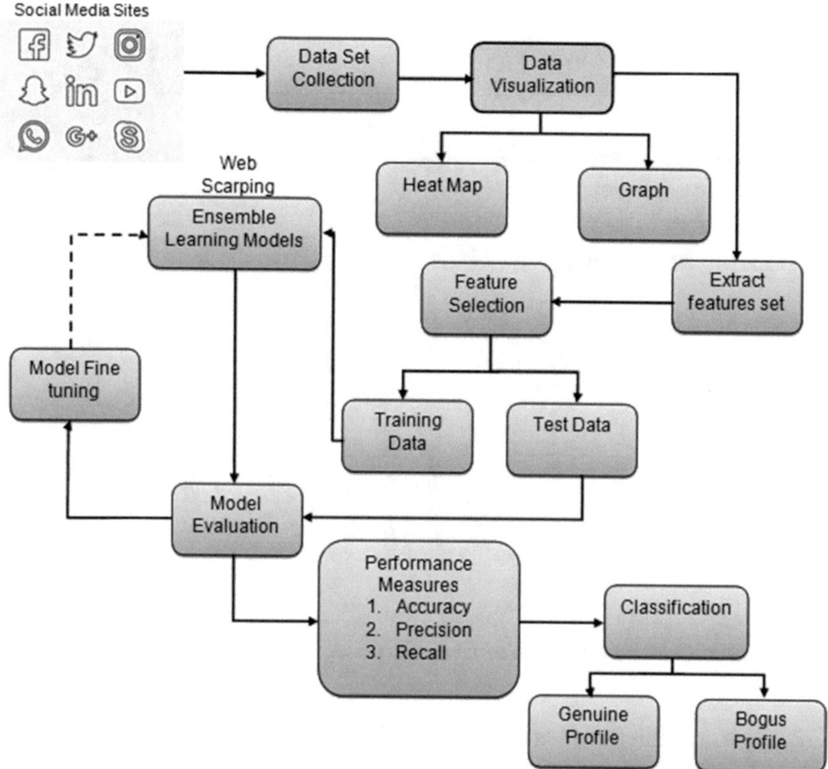

Fig. 1 Process flow for identifying bogus profiles

The data collecting module is used to gather information for the machine learning model's training. We acquire datasets from social networking websites using web scraping technology, and each dataset contains all types of information about each user, resulting in a total of more than 3000–4000 datasets getting collected and saved in excel as CSV format. Data collection allows us to keep track of prior events so that we can utilise data analysis to look for recurring trends. Predictive models are built utilising machine learning algorithms that seek patterns and predict future changes based on those patterns. Since predictive models can only be as effective as the data they are dependent on, effective data collection techniques are critical for building high-performing models.

The data visualisation module converts the data into a graph, which is useful since it makes trends and patterns more visible. We can visualise how the data appears and can identify the relationship between the properties of available data with the help of data visualisation. It is the quickest way to understand if the features correspond to the output. Heat maps and histograms are effective ways to visualise data.

A heat map is a graphical representation of user behaviour data that uses a warm-to-cool colour scheme to depict hot and cold regions which are presented in Fig. 2.

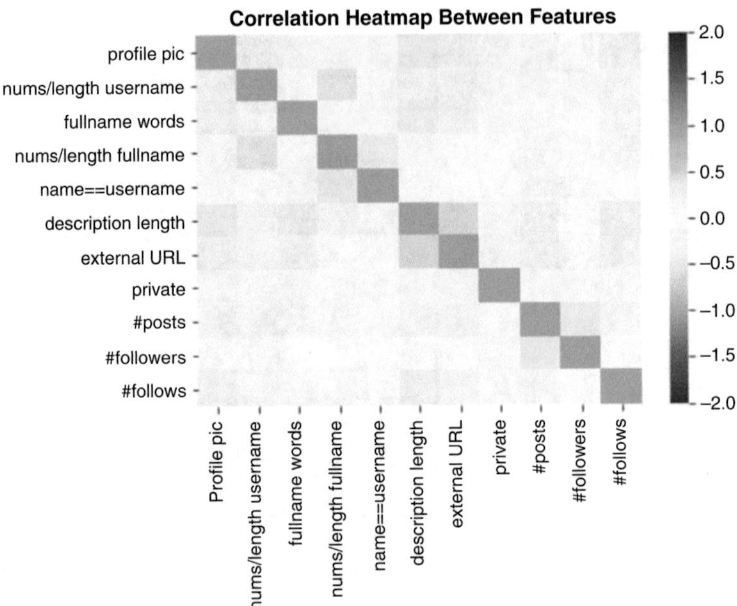

Fig. 2 Heat map for visualising the social media dataset

Colours denoted areas with the most visitor interaction, with red denoting the highest level of interaction, and cool colours denoting areas with the least interaction. In Histograms data is divided into bins and it is the quickest way to present how each attribute in a dataset is distributed.

Gradient boost, random forest, Naïve Bayes classifier, support vector machine (SVM), logistic regression and KNN are the six machine learning techniques used in the training module. Using these six techniques, we were able to train the system and assess the training score. Learning good values for all the weights and the bias from labelled instances is a prerequisite for the optimal training of the model. A machine learning algorithm generates a model by studying numerous instances and attempts to discover a model that minimises loss in supervised learning. This method is known as practical risk aversion.

Model training is the process of feeding data into a machine-learning algorithm to help in the discovery and learning of the appropriate values for all of the included variables. Supervised and unsupervised learning are the two most frequent forms of machine learning models. Supervised learning is feasible once the training data comprises both the input and output values. The training effectiveness is determined by the difference between the completed output and the specified outcome after the inputs are inserted into the model.

Unsupervised learning helps identify the patterns in data. Following that, additional data is being used to match patterns or clusters. By evaluating the observed patterns to the predicted patterns, we confirm that the accuracy getting enhanced.

Fine-tuning is the procedure of excellent-tuning or changing a model that has already been trained for one mission to make it execute a second related mission. Although fine-tuning is useful for training new machine learning models, it can only be applied whenever the datasets of an existing model and the new machine learning model are identical. A machine learning network that can identify a tiger, for example, may be fine-tuned to identify cats. Machine learning can distinguish features of a cat, such as legs, eyes, tail, and so on, since it can recognise the features of a tiger, such as legs, eyes, tail, and so on. Due to the fact, tiger and cat have similar characteristics; the deep gaining knowledge of the model does not need to be retrained to comprehend them. The same rationale may be applied to develop machine learning approaches for detecting fake social media profiles. As a result, data from a previously trained machine learning method can be used to save a lot of time and money.

To classify the bogus and genuine profiles on social media sites, the accuracy score is calculated through a confusion matrix in the evaluation module. A Confusion matrix is a N x N matrix used to evaluate the accuracy of machine learning models, where N specifies the number of target classes. The matrix involves comparing target values to the predictions made by the machine learning model. The confusion matrix shows us a complete view of how effectively our classification model performs and the types of issues it generates.

3.2 Ensemble Learning Models

To measure the efficiency of bogus profile detection classifiers, by our proposed methodology, we have used the ensemble learning models listed below.

Random Forest Models

Random Forest Model (Akar & Güngör, 2012) is similar to Bagging, although there are a few differences. When it comes to deciding where to split and how to decide things, Bagged Decision Trees provides a variety of opportunities. As an effect, the developed and tested samples may differ substantially; the data in each model will often split in at the same points. Random Forest models use a random selection of features to determine where to split. Because each tree separates focusing on various attributes rather than dividing at similar attributes at each node, Random Forest models include a degree of variation all across the model as represented in Fig. 3. This degree of differentiation allows a larger ensemble to be aggregated, resulting in a more accurate predictor.

There are several strategies for identifying where to split a decision tree depending on a regression problem. We used the Gini index as a cost function to predict a split in the dataset for the classification problem. By removing the total of each class's squared probability from one, the Gini index is computed. The Gini index () is calculated by Eq. 1.

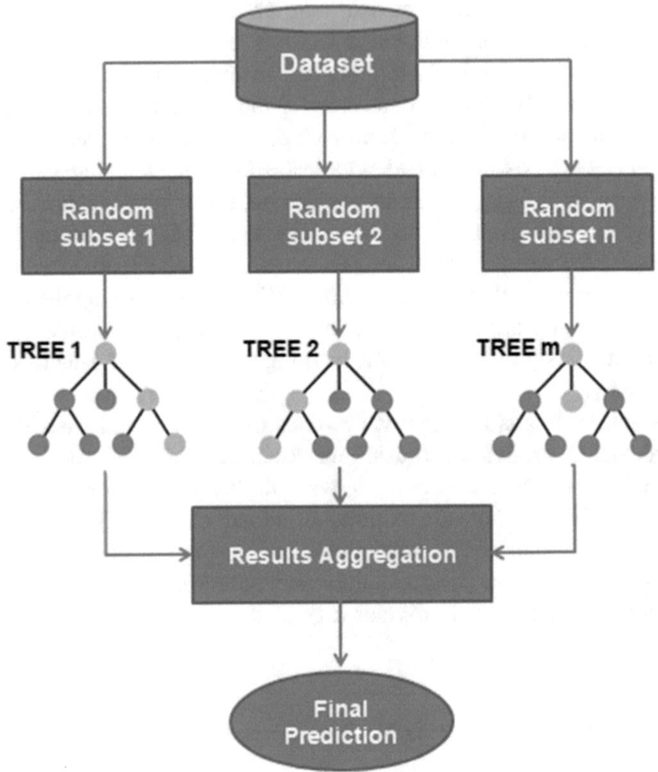

Fig. 3 Random Forest model

$$\text{Gini} = 1 - \sum_{i=1}^{C} (p_i)^2 \tag{1}$$

while P_i denotes the relative frequency of the class.

Gradient Boosting

Boosting is an ensemble strategy that learns from previous predictor errors to improve future predictions. The strategy incorporates a set of weak base classifiers into a single strong learner, resulting in significantly improved prediction dependability. Boosting operates by sequentially inserting weak learners so that they can train from the next learner in the pipeline, resulting in more accurate forecasting models. Boosting techniques include gradient boosting, adaptive boosting, and XGBoost. Gradient boosting (Felix & Sasipraba, 2019) is a method of building predictive models that are most commonly employed in regression and classification operations as depicted in Fig. 4. Decision trees are frequently used to provide prediction models so that the best prediction can be selected. Gradient boosting,

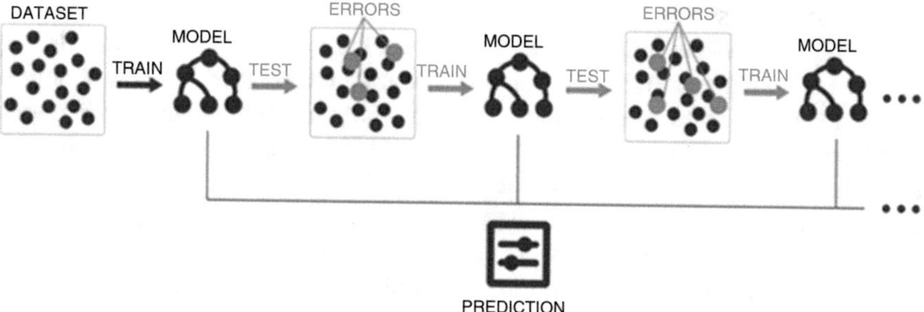

Fig. 4 Gradient Boost algorithm

like other boosting approaches, allows the generalisation and optimisation of differentiable loss functions where presenting models are created in phases.

Algorithm:

1. Set the model's input values to a constant

$$F_0(x) = \arg\min_{\gamma} \sum_{i=1}^{n} L(y_i, \gamma). \tag{2}$$

2. For m ranging from 1 to M:

   (a) Determine the pseudo-residuals:

$$r_{im} = -\left[\frac{\partial L(y_i,\ F(x_i))}{\partial F(x_i)}\right]_{F(x)=F_{m-1}(x)} \quad for\ i = 1, \ldots, n. \tag{3}$$

   (b) Fitting a base learner to $h_m(x)$ pseudo-residuals, i.e., train it with the training set. $\{(x_i,\ r_{im})\}_{i=1}^{n}$.

   (c) Solve the one-dimensional optimisation problem below to find the multiplier. γ_m

$$\gamma_m = \arg\min_{\gamma} \sum_{i=1}^{n} L(y_i, F_{m-1}(x_i) + \gamma h_m(x_i)). \tag{4}$$

   (d) Revise the model:

$$F_m(x) = F_{m-1}(x) + \gamma_m h_m(x) \tag{5}$$

3. Output $F_M(x)$.

Naïve Bayes Classifier

The ensemble method is also referred to as stacking generalisation. This strategy involves allowing a training algorithm to integrate the predictions of a variety of relevant machine learning algorithms. Stacking has been used effectively in regression, density estimation, distance learning, and classification. It could also be used to calculate the error rate associated with bagging.

Bayes' theorem (Mukherjee & Sharma, 2012) is used to create a naive Bayesian classifier and it assumes that predictors are independent. A Naive Bayesian model is easy to build and does not need time-consuming repeated calculation, making it appropriate for large datasets. Due to its simple structure, the Naive Bayesian classifier is widely used because it surpasses more sophisticated classification methods.

The posterior likelihood, P(c|x) is calculated from P(c), P(x), and P(x|c) using the Naive Bayes classifier.

$$P(c|x) = P(x|c)\ P(c)/P(x) \tag{6}$$

P(c|x) is the posterior likelihood of class (target) given predictor (attribute).
P(c) is the prior likelihood of class.
P(x|c) is the likelihood of a predictor in a particular class.
P(x) denotes the predictor's likelihood function.

3.3 *Performance Metrics*

We employed a variety of criteria to assess the performance of ensemble learning models. A confusion matrix is a tabular depiction of machine learning models performance on a test data set that includes four values: True-Positive, False-Positive, True-Negative, and False-Negative as represented in the Table 1.

Accuracy

Accuracy is a widely employed measure that calculates the proportion of correctly anticipated true or false observations. Use the equation given below to estimate the accuracy of the machine learning model

$$\text{Accuracy} = \frac{TP + TN}{TP + TN + FP + FN} \tag{7}$$

Table 1 Confusion matrix

	Predicted True	Predicted False
Actual True	True-Positive (TP)	False-Negative (FN)
Actual False	False-Positive (FP)	True-Negative (TN)

In the majority of cases, a high accuracy value implies a successful model; however, in this case, we are training a classification model, a user profile that was anticipated as true but was untrue (false-positive) can have negative effects.; Likewise, if a user profile was anticipated to be incorrect yet contained true data, trust concerns may arise. Consequently, we have applied three performance metrics including precision, recall, and F1-score, to account for the wrongly classified observation.

Recall

The number of positive categorisations that occur outside of the true class is known as the recall. It represents the percentage of user profiles estimated to be real in our example, from a total of user profiles.

$$\text{Recall} = \frac{\text{TP}}{\text{TP} + \text{FN}} \tag{8}$$

Precision

The precision value is defined as the proportion of true-positives to all real-world instances anticipated. In our example, precision is used to describe the percentage of favourably anticipated (true) profiles that are designated as true.

$$\text{Precision} = \frac{\text{TP}}{\text{TP} + \text{FP}} \tag{9}$$

F1-Score

The exchange of precision and recall is indicated by the F1-score. It determines the Pythagorean mean of the two values. It considers both false-positive and false-negative results. The following formula can be used to compute the F1-score:

$$\text{F1-Score} = 2\frac{\text{Precision} \times \text{Recall}}{\text{Precision} + \text{Recall}} \tag{10}$$

4 Result and Discussion

We analysed a social media dataset. The confusion matrix has been used to show the results of the analysis of social media datasets by applying ensemble learning models. Gradient boost, Random Forest Model, and Naive Bayes classifier are the ensemble learning models used for detection. Python code uses the cognitive learning package to generate the confusion matrix automatically. On all performance criteria, the Gradient Boost algorithm outperformed other models. The operating principle that efficiently finds faults and eliminates them in each iteration is the major element resulting in Gradient boosting's outstanding performance. The core idea underlying Gradient Boosting is to use multiple categorisations and regression trees

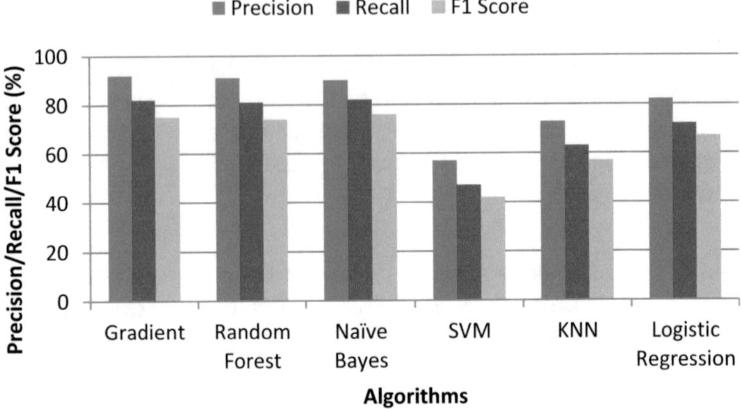

Fig. 5 Performance of Ensemble learning models

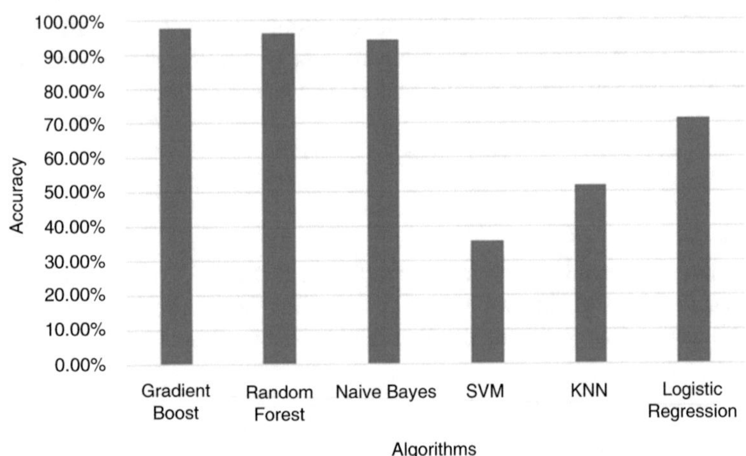

Fig. 6 The overall accuracy of ensemble learning algorithms

that combine many weak learners to provide higher weights for incorrectly classified data sets. As a result, the model is capable of reliably predicting outcomes during each successive iteration, while regularisation parameters are employed to alleviate the overfitting problem.

Figure 5 depicts the performance of learning models on all datasets using precision, recall, and F1-score. Except for SVM, KNN, and logistic regression, there is little difference in the performance of ensemble learning algorithms employing multiple performance criteria.

Figure 6 illustrates how effectively our ensemble learning model can distinguish between "bogus" and "genuine" user profiles. Gradient boost, Random Forest, and Naive Bayes algorithms were more effective in detecting bogus accounts, with

97.6%, 96.5% and 94.14% accuracy, respectively, whereas KNN had 51.52% accuracy. Gradient boost, Random Forest, and Naive Bayes algorithms also outperformed SVM and KNN in terms of precision and F-score.

5 Conclusion

Social networking services such as Facebook, LinkedIn, Instagram, Pinterest, and Twitter, among others, are rapidly expanding. As social media has grown in popularity, a large number of fraudulent user profiles have been created for nefarious purposes. Sybils or social Bots are other names for fake profiles. Many of these profiles attempt to follow innocent people in the hopes of acquiring access to confidential information. Social engineering tactics and fake profiles have posed a serious threat to information theft and deliberate manipulation for ulterior objectives. Fake profiles have become a huge issue on social media sites, and they are hard to spot. As a result, this research offers a method for recognising bogus profiles based on their attributes. The primary goal of the ensemble learning model is to efficiently detect bogus profiles by utilising specific traits. In this study, we used machine learning models and ensemble approaches to solve the challenge of detecting bogus profiles on social media sites. The information we used in our research is obtained from the internet and included user profiles from a variety of social media sites. In this proposed work Ensemble learning approach is used to improve the accuracy of recognising bogus profiles using automatic user profile classification. The research explored many texts, images, audio and video attributes that can be used to distinguish between a bogus and genuine account. These attributes are used by the Ensemble learner technique, which evaluates their performance on real-world social media datasets using performance metrics including as accuracy, precision, recall and F-1 score. In comparison to existing learning models (Support Vector Machine, KNN, Logistic regression), the suggested ensemble learner approach achieves an average accuracy of 97.67%, precision of 92.56%, and a recall of 82%, according to the experimental evaluation.

References

Adikari, S., & Dutta, K. (2020). Identifying fake profiles in LinkedIn arXiv preprint arXiv: 2006.01381.

Akar, Ö., & Güngör, O. (2012). Classification of multispectral images using Random Forest algorithm. *Journal of Geodesy and Geo information, 1*(2), 105–112.

Chu, Z., Gianvecchio, S., Wang, H., & Jajodia, S. (2010, December). Who is tweeting on Twitter: human, bot, or cycyborg. In *Proceedings of the 26th annual computer security applications conference* (pp. 21–30).

Elovici, Y., Michael, F. I. R. E., & Katz, G. (2017). *U.S. patent no. 9,659,185*. U.S. Patent and Trademark Office.

Farooqi, S., Jourjon, G., Ikram, M., Kaafar, M. A., De Cris-tofaro, E., Shafiq, Z., … & Zaffar, F. (2017, April). Characterizing key stakeholders in an online black-hat marketplace. In *2017 APWG Symposium on Electronic Crime Research (eCrime)*. (pp. 17–27). IEEE.

Felix, A. Y., & Sasipraba, T. (2019, December). Flood detection using gradient boost machine learning approach. In *2019 International conference on computational intelligence and knowledge economy (ICCIKE)* (pp. 779–783). IEEE.

Hema, C., & Sankar, S. (2021). *Collision mitigation algorithm for tracking of RFID based assets in defence* (p. e27). EAI Endorsed Transactions on Energy Web.

Mukherjee, S., & Sharma, N. (2012). Intrusion detection using naive Bayes classifier with feature reduction. *Procedia Technology, 4*, 119–128.

Romanov, A., Semenov, A., Mazhelis, O., & Veijalainen, J. (2017). Detection of fake profiles in social media-literature review. *International Conference on Web Information Systems and Technologies, 2*, 363–369. SCITEPRESS.

Stringhini, G., Wang, G., Egele, M., Kruegel, C., Vigna, G., Zheng, H., & Zhao, B. Y. (2013, October). Follow the green: growth and dynamics in Twitter follower markets. In *Proceedings of the 2013 conference on Internet measurement conference* (pp. 163–176).

Viswanath, B., Bashir, M. A., Crovella, M., Guha, S., Gum-madi, K. P., Krishnamurthy, B., & Mislove, A. (2014). To-wards detecting anomalous user behaviour in online social networks. In *23rd {USENIX} Security Symposium ({USENIX} Security 14)*. (pp. 223–238).

Support Vector Machine and K-fold Cross-validation to Detect False Alarms in Wind Turbines

Ana Maria Peco Chacon and Fausto Pedro García Márquez

Abstract Wind energy is a growing market, due to improved monitoring systems, decreased downtime, and optimal predictive maintenance. Classification algorithms based on machine learning efficiently categorize large volumes of complex data. K-Fold cross-validation is a competent tool for measuring the success rate of classification algorithms. The purpose of this research is to compare the performance of various kernel functions for the support vector machine algorithm, as well as to analyze the different values of K-fold cross-validation and holdout validation. The approach is applied to a real dataset of wind turbine, with the aim to detect false alarms. The results indicate an accuracy of 99.2%, and the F_1 is 0.996. These values indicate that the methodology proposed is efficient to detect and identify false alarms.

Keywords Wind turbine · Maintenance management · SCADA · False alarm · Support vector machine · Accuracy · Prediction model · Cross-validation

1 Introduction

Wind energy is a major competitor to fossil fuel energy. In 2020, the total installed wind power exceeded 743 GW (GWEC, 2022). Wind Turbines (WTs) are complicated electromechanical systems with a rotor that converts wind energy into mechanical energy and, subsequently, into electrical energy. In the last decade, the size and number of WTs have increased rapidly this means that maintenance is becoming more important and complex. Maintenance and repair costs represents between 25% and 30% of the total cost of wind power generation (Márquez et al., 2012), and 80% of this amount is spent on unplanned maintenance problems due to unexpected failures in WTs (Ruiz et al., 2018). Maintenance management aims to reduce potential component failures and ensure optimal performance of WTs. This is

A. M. Peco Chacon (✉) · F. P. García Márquez
Ingenium Research Group, Universidad Castilla-La Mancha, Ciudad Real, Spain
e-mail: anamaria.peco@uclm.es; faustopedro.garcia@uclm.es

F. P. García Márquez, B. Lev (eds.), *Sustainability*, International Series in Operations Research & Management Science 333,
https://doi.org/10.1007/978-3-031-16620-4_6

done to reduce resources and costs and to avoid economic losses due to downtime (Márquez and Chacón, 2020).

There are three types of maintenance: reactive, preventive, and predictive. Reactive maintenance is activated when the breakdown has occurred, it has high costs (Márquez et al., 2012). Preventive maintenance is activated before failure occurs. And predictive maintenance is based on Control Monitoring (CM) system; this type of maintenance predicts failures before they occur using a robust and efficient CM; therefore, the cost of operation and maintenance can be considerably reduced (Helsen et al., 2015). The CM system in wind farms can detect and collect information indicating the status of WTs. Improved sensor, big data management, machine learning, signal processing, and computational capabilities make CM system essential to identify failures before they occur (Stetco et al., 2019). The CM system and Supervisory Control and Data Acquisition (SCADA) system improve the productivity and profitability of wind farms (Gómez Muñoz et al., 2014). SCADA data is sampled at a speed of 1 s; to save storage space it is very common to use an average of this data (Velandia-Cardenas et al., 2021). In this case, a time window of 10 min is used; this SCADA data is called low frequency or slow speed (Wang et al., 2014).

The three main disadvantages of sensor-based monitoring methods are as follows:

1. The installation or replacement of sensors in operating WTs requires a high cost of resources.
2. Sensors require maintenance, in some cases complex.
3. The sensor signal must have high transmission reliability in order not to impair the reliability of the power supply.

Most studies have focused on SCADA signals rather than SCADA alarms (Qiu et al., 2020). The SCADA database collects alarm information. When signals from essential components surpass threshold limits, alarms are generated and recorded (Qiu et al., 2012). SCADA alarms indicate to operators possible emergency events and that decrease the risk of catastrophic failure (Qiu et al., 2020). Models use SCADA or simulated data, with classification accounting for 67% of the methods and regression accounting for the remainder 33% (Stetco et al., 2019).

There are large data sets that provide information on the status of WTs due to the low cost of sensors. Machine learning algorithms can process large amounts of data. The use of efficient methods for the prevention and early detection of failures in WTs is a subject of intense study during the last years, with the implementation of new combined intelligent methods or machine learning algorithms (Yang et al., 2021). Machine learning can be defined as the computational ability to solve problems based on prior knowledge, i.e., data (Xu et al., 2021). Moreno et al. (2020) obtained an average accuracy of 98.6% in the classifiers for the detection of WT anomalies.

The Support Vector Machine (SVM) method created by Vapnik (1998) aims to categorize the set of features into two classes. The SVM is a supervised learning model, that uses its learning algorithms to perform data analysis for classification and regression. SVM is a widely used algorithm, and Fig. 1 shows how publications on this method have increased. The SVM is used for intelligent fault identification and diagnosis in WTs (Wang et al., 2020).

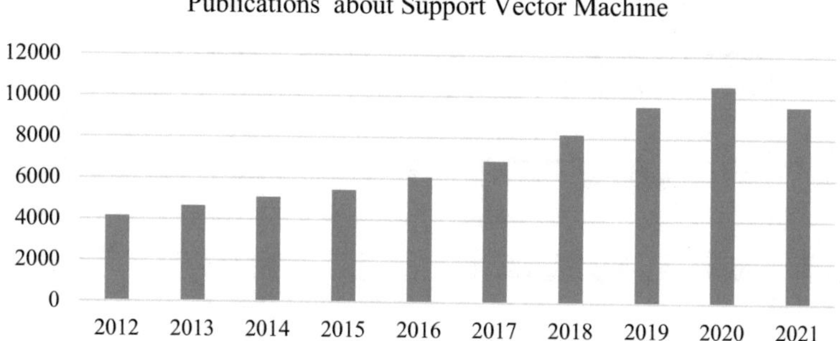

Fig. 1 Evolution of SVM publications in recent years based on (2021)

In Hübner et al. (2021), SVM is used to detect mass imbalance in the rotor of WTs, resulting in an accuracy higher than 84%. Yang et al. (2021) applied SVM to identify damage to WT blades, and they obtained an average accuracy of 82.93%. Turnbull et al. (2019) used SVM algorithms to predict the failure of the WT generator bearing. They used 5 K-fold cross-validation (CV) and obtained an accuracy beyond 70%. Carroll et al. (2019) also applied SVM to detect WT gearbox faults, and in this research, they used holdout validation; 70% of the data was used for training and 30% for testing the machine learning algorithms. The results showed an accuracy of 60% and a false-positive rate of 4%. Santos et al. (2015) obtained an accuracy of 98.26% for the detection of certain faults in WTs, using a linear kernel SVM. Dhiman et al. (2021) applied SVM and adaptive threshold to detect anomalies in WTs gearbox. The method achieved an accuracy of 91.11%, and false-positives were 3.5% with a tenfold CV; however, with twin SVM the results improved; the accuracy was 95.75%, and the false-positives were 2.11%. The K-fold CV is used to check the success rate of prediction models, but there are very few studies on the results as a function of the k-value (Jahangiri et al., 2020). Table 1 shows a summary of the latest research where SVM is applied for fault detection in WTs.

Majority of the works in literature focuses on fault diagnosis and detection. However, it does not analyze those alarms that are false and cause unnecessary stoppages in WTs. False alarms cause significant economic losses due to downtime of WTs (Pliego Marugán et al., 2019). This research analyzes the accuracy of the classification algorithm and the causes that may have led to incorrect predictions.

2 Approach

The proposed methodology relates the alarm log with the SCADA variables. This methodology for this work is shown in Fig. 2. The alarm log records information about alarms and their activation periods, including the following information: alarm code; description; timestamp of data collection for activation; and timestamp of data collection for deactivation.

Table 1 Summary of research applying SVM in WTs

Ref	Method	Component	Accuracy	Validation
(Yang et al., 2021)	SVM gaussian radial	Blades	82.93%	No data
(Hübner et al., 2021)	SVM kernel lineal	Rotor	84%	No data
(Turnbull et al., 2019)	SVM with polynomial kernel	Generator	64%	5-fold CV
(Carroll et al., 2019)	SVM kernel lineal	Gearbox	60%	Holdout: 70% training, 30% testing
(Santos et al., 2015)	SVM kernel lineal	WT	98.26%	5 × 2 CV
(Dhiman et al., 2021)	Adaptive threshold and twin support vector machines	Gearbox	95.75%	Tenfold CV
(Dhiman et al., 2021)	SVM with class weight	Generator heating fault	Precision = 0.24	Holdout: 80% training, 20% testing
(Tang et al., 2014)	Shannon wavelet support vector machine	Transmission system	92%	No data
(Zhang et al., 2020)	Combination of one-dimensional convolutional neural network, SVM and particle swarm optimization	Bearing	98.2%	No data

In the first step, the data from the SCADA system and the alarm log are obtained. The alarm log records information about alarms and their activation periods, including the following information: alarm code; description; timestamp of data collection for activation; and timestamp of data collection for deactivation. The data collected from alarm log is synchronized with the SCADA variables on a similar time scale. In this phase also, the SCADA dataset is required to eliminate any values that are erroneous, i.e., not a number (Nan), this is the filtering of the data.

The next step is the classification with different SVM models and several values of holdout and K-fold validation. The SVM algorithm is selected due to the better accuracy and low computing time. The confusion matrix is used in the third phase to categorize the results. Finally, misclassifications are analyzed with information from the alarm log and the maintenance log to determine whether or not they are false alarms.

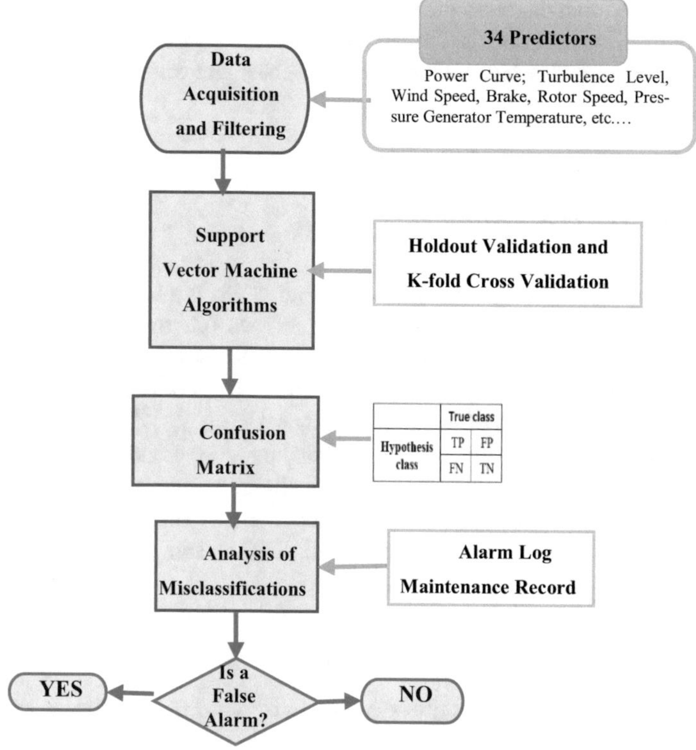

Fig. 2 Flowchart of the approach

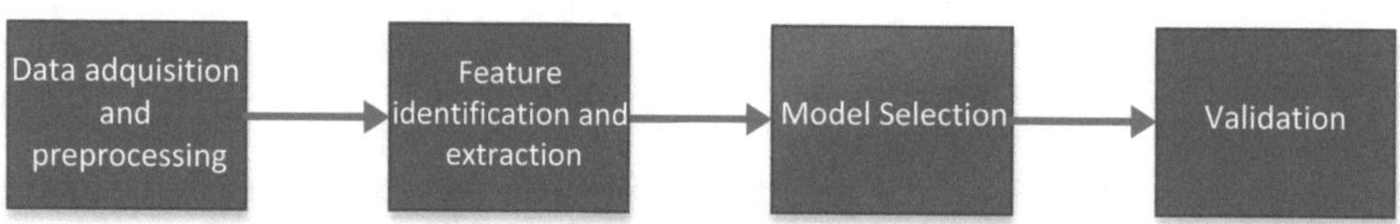

Fig. 3 Machine learning flowchart

2.1 Machine Learning

The machine learning methods comprise the main steps shown in Fig. 3. During the data acquisition and preprocessing phase, outliers must be removed, and the different classes must be brought in the same format. In the second step, only the most important features should be retained. Depending on the problem to be solved, one classification model or another should be chosen, for which it is important to clearly establish the parameters of the model. Finally, validation serves to measure the performance of the model used.

Based on the input data into the model, supervised learning can predict an output variable. Supervised learning is divided into models that predict a numerical variable and those that predict a categorical variable. These are known as regression and classification, respectively.

2.2 *Support Vector Machine*

SVM is a supervised machine learning algorithm. SVM is a widely used method for fault detection and for analyzing complex data sets (Géron, 2019). SVM solves classification and regression problems. The aim of SVM is to find boundaries with the maximum margin of separation between different classes of data.

For the case of linear SVM, given a training set $\{(x_i,\ y_i)\}_{i=1}^{M}$. The input vectors have dimensions D and x$\in R^D$, and their labels are (Abreu and Ribeiro, 2003). For the linear case, the hyperplane is found by maximizing the margin between classes by Eq. (1), c is the vector containing the coefficients that determine the vector orthogonal to the hyperplane (Chacón et al., 2021) and A in this case is one observation of SCADA data.

$$h(x) = A'x + c \tag{1}$$

The algorithm separates the observations into two different classes divided by two parallel hyperplanes (Santos et al., 2015). In this case, the different classes are alarm activation or no alarm activation.

$$h(x) \geq 1, if\ y = 1$$

$$h(x) \leq -1, if\ y = -1$$

The separation between the parallel hyperplanes corresponds to $2/_{\|A\|}$, and to maximize this margin $1/_{2\|A\|}$ must be minimized. Thus, it becomes an optimization problem, Eq. (2)

$$\min_{A,c} y\frac{1}{2}\ \|A\|^2 \tag{2}$$

Subject to $y_i(A^{'}x_i + c) \geq 1,\ i = 1, \ldots, M.$

For non-linearly separable cases, a slack variable, α, and a regulation parameter, P, must be included. Eq. (3) shows the general formulation for the linear kernel.

$$min_{A,c}\, y\frac{1}{2}\ \|A\|^2 + P\sum\nolimits_{i=1}^{M}\alpha_i \tag{3}$$

Subject to $y_i(A^{'}x_i + c) \geq 1 - \alpha_i,\ \alpha_i > 0,\ i = 1, \ldots, M$

The slack variable indicates the allowed error, and the regularization parameter indicates the weight of the second term for minimization. Equation (4) shows the general case of SVM, and in this case, the hyperplanes cannot be formulated directly in the original feature space. For this reason, the "kernel trick" is used for the non-linear transformation of the data (Santos et al., 2015). The inner product and a function β are used to obtain the hyperplanes with the kernel transformation.

$$min_{A,c,\alpha}\, y\frac{1}{2}\ \langle A, A\rangle^2 + P\sum\nolimits_{i=1}^{M}\alpha_i \tag{4}$$

Subject to $y_i(A^{'}, \beta_x > \ + c) \geq 1 - \alpha_i,\ \alpha_i > 0,\ i = 1, \ldots, M$

The classifier, in this example, is established as follows:

$$h(x) = sign\ (\langle A, \beta_x\rangle + c \tag{5}$$

SVM is used in dual form to avoid the infinite dimensions that inner products can have. K is the kernel function and is defined for vectors of different characteristics x_i, x_j as $K(x_j, x_j) = \left\langle \beta_{x_j}, \beta_{x_j} \right\rangle$.

$$min_\gamma \sum\nolimits_{i=1}^{M}\sum\nolimits_{j=1}^{M}\gamma_i\gamma_j y_i y_j K(x_i, x_j) - \sum\nolimits_{i=1}^{M}\gamma_i \tag{6}$$

Subject to $\sum_{i=1}^{M} y_i\gamma_i = 0{,}0 \leq \gamma_i \leq P$

Applying the theorem of Aronszajn (Suárez, 2014), Eq. (7) is obtained

$$K(x_i, x_j) = \langle \beta(x_{i)}, \beta(x_j) \rangle \forall x_i, x_j \in X \tag{7}$$

To solve the dual problem, it is not necessary to know the set of transformation basis functions; it is only necessary to know the corresponding kernel functional form.

The most used kernel functions are the following:

Lineal:

$$K(x_i, x_j) = \langle x_i, x_j \rangle \tag{8}$$

Polynomial:

$$K(x_i, x_j) = \langle 1 + x_i, x_j \rangle^P \tag{9}$$

Gaussian:

$$K(x_i, x_j) = \exp\left(\frac{\|x_i - x_j\|^2}{2\tau^2}\right) \tag{10}$$

The parameter p indicates the degree of the polynomial and σ is the standard deviation parameter.

In this case study, the following models are analyzed: Linear, Quadratic, Cubic, and Gaussian. In Gaussian Kernel cases, it is divided into the following cases:

- Fine Gaussian: fine distinctions are made between classes. The kernel scale in this case is equal to $\sqrt{N}/4$, N being the number of predictors.
- Medium Gaussian: as its name suggests it makes fewer distinctions than the previous case, so its kernel scale is $\sqrt{N}$.
- Coarse Gaussian: in this case the kernel scale is $4 \cdot \sqrt{N}$, and the distinctions between classes are less precise.

2.3 Holdout Validation

In holdout validation, the data is first divided into training and validation sets, and then the model is trained and validated (Zhou et al., 2020). The holdout part is then used for testing. The remaining part of the data learns from the model. The holdout validation can have different ratios of data being held out for testing. The election of this ratio is significant to avoid overfitting or the possibility of the learning model not being correctly distributed. The time of holdout validation for learning the model is relatively lesser than the time of k-fold CV. The disadvantage of holdout is that it does not take into account the variance with respect to the training data set, so it does not use the data efficiently (Dietterich, 1998) (Fig. 4).

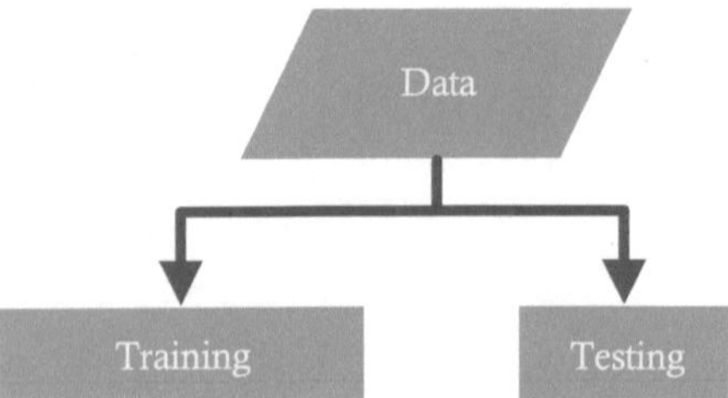

Fig. 4 Holdout validation

2.4 K-Fold Cross-validation

Cross-validation results in an unbiased model, while maintaining a high accuracy value (Choe et al., 2021). The data is divided into K segments or folds that are all equal or nearly equal. Training and testing of these partitioned folds are done in K iterations, leaving with each iteration onefold for testing and training the model on the remaining K -1 folds. The model accuracy is the average of the accuracy obtained in each iteration. The data is frequently stratified before being divided into K segments (Yadav and Shukla, 2016). Stratification consists of readjusting data, so each fold is a good representation of the whole. Its main advantages are the repeated use of random sub-samples for training and validation, as well as the verification of results of each iteration (Yuanyuan et al., 2017). The main disadvantages of K-fold CV are that its training and validation processes require more time than holdout validation, and that it also needs more computer memory than holdout validation. This is due to the different cycle calculations (Fig. 5).

In 50% holdout validation, data is divided into two equal parts, one for training and the other for testing. In twofold CV, the data is split into two equal parts again; the model is trained on each part in two different iterations, and the testing is carried out on the other part, respectively. The accuracy obtained in both iterations is calculated to achieve the average accuracy of the model. This is the main difference between the two methods, so it cannot be assumed that 50% holdout validation is equal to 2 K-fold CV. Analogously, fivefold CV and 20% hold validation are similar in that they train 80% of the data while testing the remaining 20%. The data is divided into 5 equal folds in a 5 K-fold CV, with 20% of the data available in each fold. Onefold is reserved for testing, while the other four are used for training. In

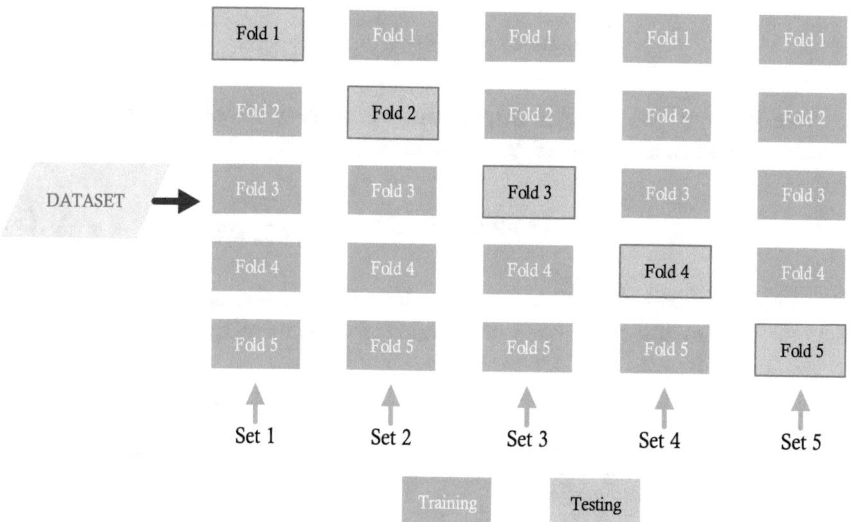

Fig. 5 Cross-validation

each fold 20% of the data is available, and in 4 folds correspond to 80% of the data. The fivefold CV differs from the other in that the model is trained and tested five times. As a point of comparison, the percentage of data withheld for testing can be used. The data being held out is different in each iteration, and because the K-fold CV is random, the accuracy may vary.

2.5 Confusion Matrix and False Alarms Indicators

The confusion matrix is a widely used tool in statistical classification problems. The confusion matrix (Table 2) shows the performance of classification models. In this case, it is a binary model, thus the dimensions of the matrix are 2 × 2. The confusion matrix displays the results of a classification that distinguishes between real classes and predicted outcomes. The cases on the diagonal show the cases that have been correctly classified. The data are classified into the following categories:

True-positive (TP): an alarm has been triggered and the classification model categorizes it correctly.
False-positive (FP): indicates the positive sample prediction error number, the classification model determines an alarm triggered when it is not triggered.
True-negative (TN): the alarm is deactivated, and the classification model detects it correctly.
False-negative (FN): The alarm is activated, and the classification model indicates deactivation, also called missed alarms.

The ratio of the total number of correct predictions is the accuracy, see Eq. (11).

$$\text{Accuracy} = \frac{\text{TP} + \text{TN}}{\text{TP} + \text{TN} + \text{FP} + \text{FN}} \tag{11}$$

True-Positive Rate (TPR) is the percentage of successfully classified observations in each true class, also referred to as sensitivity, Eq. (12).

$$\text{True} - \text{Positive Rate (TPR)} = \frac{\text{TP}}{\text{TP} + \text{FN}} \tag{12}$$

Equation 13 shows the True-Negative Rate (TNR), also called to specificity, is the ratio of correctly categorized observations for each negative class.

Table 2 Confusion matrix

True Class	Alarm no activation	TP	FP
	Alarm activation	FN	TN
		Alarm no activation	Alarm activation
		Predicted Class	

$$\text{True} - \text{Negative Rate (TNR)} = \frac{\text{TN}}{\text{TN} + \text{FP}} \tag{13}$$

There are several metrics for calculating the performance of classification methods. In this case, the F_1 value is used which combines the measures of precision and recall. This makes it simpler to compare the combined performance of the precision and completeness of the classification model. F_1, Eq. (14), is very useful when the distribution of classes is uneven and considers both false-positives and false-negatives.

$$F_1 = \frac{2 \cdot \text{TP}}{2\text{TP} + \text{FP} + \text{FN}} \tag{14}$$

3 Case Study and Results

Classification models search for the relationship between independent variables, called predictors, and the response variable. In this case, the predictor variables are the 34 SCADA variables shown in Table 3, and the response variable is the alarm log, this has two class labels: alarm activation or deactivation. This research employs SCADA data from the European project OPTIMUS. The SCADA signals were measured every 10 minutes for 2 months.

Table 3 Variables SCADA

1	General accumulator 1	18	General accumulator pressure
2	General accumulator 2	19	Transformer 1 temperature
3	General accumulator 3	20	Transformer 2 temperature
4	Phi cosine	21	Transformer 3 temperature
5	Turbulence level	22	Brake pressure
6	Oscillation level	23	Hydraulic group pressure
7	Vibration level	24	SP pitch angle
8	Pitch 1 angle	25	Hydraulic group oil temperature
9	Pitch 2 angle	26	Gearbox oil temperature
10	Pitch 3 angle	27	Environmental temperature
11	Active power	28	Nacelle temperature
12	Non-drive end side generator bearing	29	Drive end side generator bearing
13	Generator winding temperature	30	Total reactive power
14	Nacelle temperature	31	Generator speed
15	Lower gearbox radiator	32	Rotor speed
16	Upper radiator	33	Wind speed
17	Grid voltage	34	Yaw

Holdout validation is one type of validation that is used. The dataset is split between the training and validation sets. The training set establishes the most important characteristics of the classifier, and the validation set assesses the classifier (Pomares et al., 2018). Table 4 shows the accuracy of holdout validation and CV for different values.

The accuracy values are very similar; therefore, another important factor is the validation computation time. Table 5 shows the time required for the different classification models with different validation values. In this case, the quadratic model obtains a high accuracy with a reduced computational time.

The power curve is a graph that relates the power of WT to the wind speed. Power curve modeling, which provides a basic representation of the data, is often used to detect faults in WTs (Pelletier et al., 2016). Figure 6 shows the historical data of the power curve. The orange dots are SCADA data when the alarm is activated. The blue dots are SCADA data when the alarm is deactivated.

Figure 7 shows the points misclassified by the quadratic SVM model with fivefold CV. In this scenario, the black crosses indicate that the model has been classified as a non-activation alarm when in fact there was an alarm; this is known as a false-negative. The opposite case is false-positives, where the classification model predicts that the alarm is activated when it is not.

The confusion matrix is used to analyze the fit of the prediction to each class. Table 6 shows the confusion matrix of quadratic SVM with fivefold CV. The classification model correctly estimates 8575 data SCADA, although there are 57 FN and 11 FP. The confusion matrix reveals where the classification failed more than any other. The results indicate a TPR of 99.31%, the TNR is 97% and F_1 is considerably high at 0.996. These indicators suggest that the prediction model is appropriate for alarm prediction in WTs.

The misclassified points are analyzed below to examine the causes of this prediction error. Firstly, the 57 FN cases are analyzed.

The alarm for high turbulence has been activated 14 times; however, the approach has not been able to detect it.

Six maintenance activities were done during the study period, with four missed cases by using classification algorithm.

In 15 cases, several alarms were activated for a few seconds, and these alarm activations are deemed as false alarms generated by the SCADA system.

Finally, in 24 cases, there is a desynchronization between the start and finish of the alarm with respect to the intervals predicted by the classification. In this type of misclassification, the model fails to detect the alarm period accurately.

The classification model generates false alarms, which are represented by FP points in the confusion table. There are 11 points in this scenario, and the alarms occurred just before or after the alarm forecast. Maintenance activities were generated at the same time as expected points in two situations; as a result, it is common for the model to notice anomalies in the WT and classify them as an alarm.

Table 4 Accuracy of the methods

Model	Holdout 50%	2 K	Holdout 20%	5 K	10 K	15 K	20 K	30 K	50 K
Linear	98%	98.3%	98.4%	98.4%	98.4%	98.4%	98.4%	98.4%	98.4%
Quadratic	98.9%	99.1%	99%	**99.2%**	**99.2%**	99.2%	99.2%	99.2%	99.2%
Cubic	99%	99.2%	99%	99.1%	99.1%	99.2%	99.2%	99.2%	99.2%
Fine gaussian	97.9%	97.7%	98.1%	98.2%	98.3%	98.2%	98.3%	98.3%	98.2%
Medium gaussian	98.3%	98.6%	98.6%	98.6%	98.7%	98.8%	98.8%	98.8%	98.8%
Coarse gaussian	96.8%	97%	97.3%	97.4%	97.6%	97.7%	97.6%	97.7%	97.7%

Table 5 Computational time (units seconds)

Model	Holdout 50%	2 K	Holdout 20%	5 K	10 K	15 K	20 K	30 K	50 K
Linear	2.59	2.39	3.71	5.45	11.49	14.52	26.19	30.59	75.19
Quadratic	3.3	3.12	4.55	**8.99**	**22.33**	33.43	67.25	67.84	145.53
Cubic	3.5	4.02	6.13	11.96	26.69	39.81	60.81	82.56	158.29
Fine gaussian	4.54	5.33	5.86	11.59	26.57	38.8	55.95	68.92	133.88
Medium gaussian	1.01	1.44	1.40	2.86	6.56	10.13	14.96	17.48	34.99
Coarse gaussian	1.02	1.98	1.84	3.36	7.62	11.41	17.37	19.58	38.96

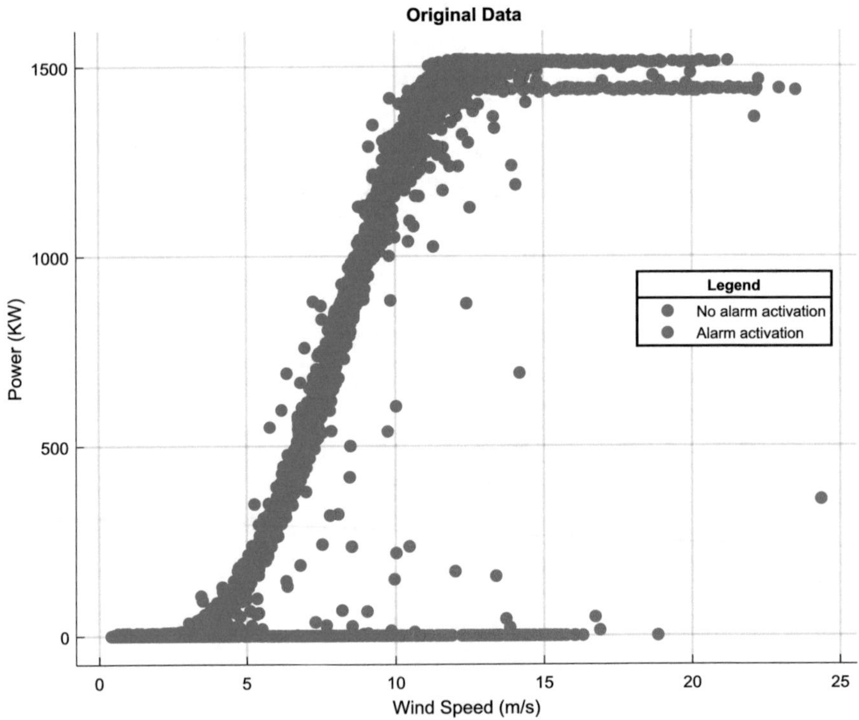

Fig. 6 Power curve

4 Conclusions

The reliability of a wind farm is dependent on operations and maintenance tasks. False alarm detection is a fundamental factor of wind turbine maintenance management. Machine learning algorithms process a large volume of data with a high degree of accuracy and in a short period of time. This chapter presents a novel approach based on support vector machine with the comparison of holdout validation and

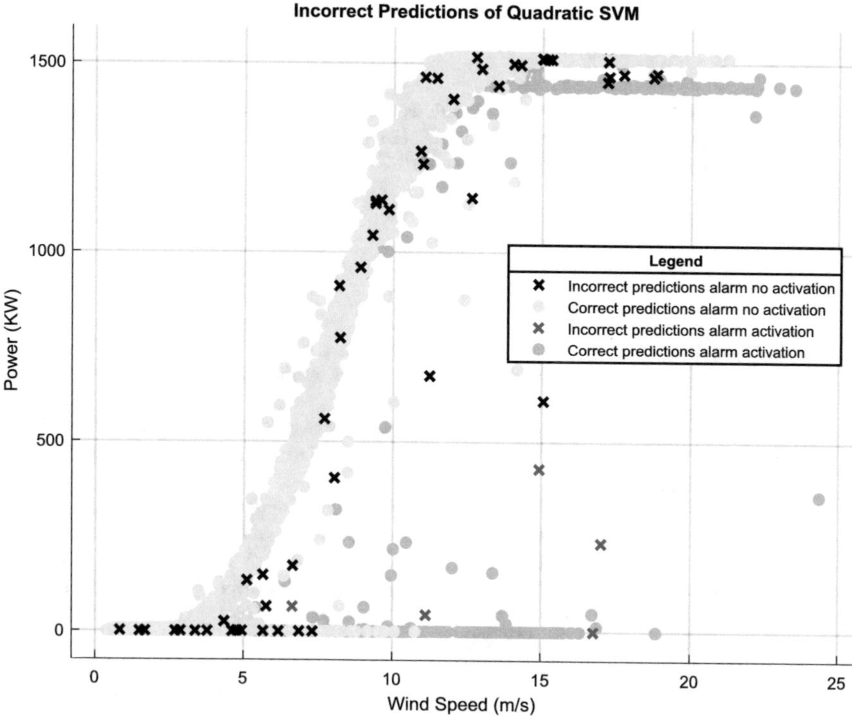

Fig. 7 Misclassifications of quadratic SVM

Table 6 Confusion matrix of quadratic SVM model

True Class	Alarm no activation	8220	11
	Alarm activation	57	355
		Alarm no activation	Alarm activation
		Predicted Class	

k-fold cross-validation for different values. The methodology is used for the detection of false alarms. The approach is applied to a real dataset from a supervisory control and data acquisition system along with the alarm log as response.

The quadratic support vector machine model with fivefold CV presents an accuracy of 99.2%. The true-positive rate, or sensitivity, is 99.31% and the true-negative rate is 97%. The classification model has an F_1-value of 0.996. These very high values indicate that the model is a good predictor for the alarms in wind turbines. In addition, the points categorized as misclassifications, in particular false-negatives, indicate that false alarms represent 26.32%.

Future research could be expanded to include the classification of alarm types and the use of other artificial intelligence models. It would be significant to find the correlation between the type of alarm and the different variables of the data acquisition system, as well as the use of other classification methods for to improve the detection of false alarms.

Acknowledgments The work reported herewith has been financially by the Dirección General de Universidades, Investigación e Innovación of Castilla-La Mancha, under Research Grant ProSeaWind project (Ref.: SBPLY/19/180501/000102).

References

Abreu, G. L., & Ribeiro, J. F. (2003). On-line control of a flexible beam using adaptive fuzzy controller and piezoelectric actuators. *Sba: Controle & Automação Sociedade Brasileira de Automatica, 14*, 377–383.

Carroll, J., Koukoura, S., Mcdonald, A., Charalambous, A., Weiss, S., & Mcarthur, S. (2019). Wind turbine gearbox failure and remaining useful life prediction using machine learning techniques. *Wind Energy, 22*, 360–375.

Chacón, A. M. P., Ramirez, I. S. & Márquez, F. P. G. (2021). Support vector machine for false alarm detection in wind turbine management. In *2021 7th International Conference on Control, Instrumentation and Automation (ICCIA)*, 23–24 Feb. 2021 (pp. 1–5).

Choe, D.-E., Kim, H.-C., & Kim, M.-H. (2021). Sequence-based modeling of deep learning with LSTM and GRU networks for structural damage detection of floating offshore wind turbine blades. *Renewable Energy, 174*, 218–235.

Dhiman, H. S., Deb, D., Muyeen, S., & Kamwa, I. (2021). Wind turbine gearbox anomaly detection based on adaptive threshold and twin support vector machines. *IEEE Transactions on Energy Conversion, 36*, 3462.

Dietterich, T. G. (1998). Approximate statistical tests for comparing supervised classification learning algorithms. *Neural Computation, 10*, 1895–1923.

Digital Science & Research Solutions [Online]. (2021). Accessed November 29, 2021, from https://app.dimensions.ai/analytics/publication/overview/timeline?search_mode=content&search_text=SVM&search_type=kws&search_field=text_search

Géron, A. (2019). *Hands-on machine learning with Scikit-Learn, Keras, and TensorFlow: Concepts, tools, and techniques to build intelligent systems*. O'Reilly Media.

Global Wind Energy Council. Accessed July 14, 2022, from https://gwec.net/global-wind-report-2022/

Gómez Muñoz, C. Q., et al. (2014). A novel approach to fault detection and diagnosis on wind turbines. *Global NEST Journal, 16*(6), 1029–1037.

Helsen, J et al. (2015). Condition monitoring by means of scada analysis. En *Proceedings of European wind energy association international conference Paris*.

Hübner, G. R., Pinheiro, H., de Souza, C. E., Franchi, C. M., da Rosa, L. D., & Dias, J. P. (2021). Detection of mass imbalance in the rotor of wind turbines using support vector machine. *Renewable Energy, 170*, 49–59.

Jahangiri, M., Jahangiri, M., & Najafgholipour, M. (2020). The sensitivity and specificity analyses of ambient temperature and population size on the transmission rate of the novel coronavirus (COVID-19) in different provinces of Iran. *Science of the Total Environment, 728*, 138872.

Márquez, F. P. G., & Chacón, A. M. P. (2020). A review of non-destructive testing on wind turbines blades. *Renewable Energy, 161*, 998–1010.

Márquez, F. P. G., Tobias, A. M., Pérez, J. M. P., & Papaelias, M. (2012). Condition monitoring of wind turbines: Techniques and methods. *Renewable Energy, 46*, 169–178.

Moreno, S. R., Coelho, L. D. S., Ayala, H. V., & Mariani, V. C. (2020). Wind turbines anomaly detection based on power curves and ensemble learning. *IET Renewable Power Generation, 14* (19), 4086–4093.

Pliego Marugán, A., Peco Chacón, A. M., & García Márquez, F. P. (2019). Reliability analysis of detecting false alarms that employ neural networks: A real case study on wind turbines. *Reliability Engineering and System Safety, 191*, 106574. https://doi.org/10.1016/j.ress.2019.106574

Pelletier, F., Masson, C., & Tahan, A. (2016). Wind turbine power curve modelling using artificial neural network. *Renewable Energy, 89*, 207–214.

Pomares, A., Martínez, J. L., Mandow, A., Martínez, M. A., Morán, M. & Morales, J. (2018). Ground extraction from 3D lidar point clouds with the classification learner App. In *2018 26th Mediterranean conference on control and automation (MED)* (pp. 1–9). IEEE.

Qiu, Y., Feng, Y., Tavner, P., Richardson, P., Erdos, G., & Chen, B. (2012). Wind turbine SCADA alarm analysis for improving reliability. *Wind Energy, 15*(8), 951–966.

Qiu, Y., Feng, Y., & Infield, D. (2020). Fault diagnosis of wind turbine with SCADA alarms based multidimensional information processing method. *Renewable Energy, 145*, 1923–1931.

Ruiz, M., Mujica, L. E., Alferez, S., Acho, L., Tutiven, C., Vidal, Y., et al. (2018). Wind turbine fault detection and classification by means of image texture analysis. *Mechanical Systems and Signal Processing, 107*, 149–167.

Santos, P., Villa, L. F., Reñones, A., Bustillo, A., & Maudes, J. (2015). An SVM-based solution for fault detection in wind turbines. *Sensors, 15*, 5627–5648.

Stetco, A., Dinmohammadi, F., Zhao, X., Robu, V., Flynn, D., Barnes, M., Keane, J., & Nenadic, G. (2019). Machine learning methods for wind turbine condition monitoring: A review. *Renewable Energy, 133*, 620–635.

Suárez, E. J. C. (2014). Tutorial sobre máquinas de vectores soporte (sVM). *Tutorial sobre Máquinas de Vectores Soporte (SVM), 1*, 1–12.

Tang, B., Song, T., Li, F., & Deng, L. (2014). Fault diagnosis for a wind turbine transmission system based on manifold learning and Shannon wavelet support vector machine. *Renewable Energy, 62*, 1–9.

Turnbull, A., Carroll, J., Koukoura, S., & Mcdonald, A. (2019). Prediction of wind turbine generator bearing failure through analysis of high-frequency vibration data and the application of support vector machine algorithms. *The Journal of Engineering, 2019*, 4965–4969.

Vapnik, V. (1998). The support vector method of function estimation. En Nonlinear modeling. Springer, p. 55–85.

Velandia-Cardenas, C., Vidal, Y., & Pozo, F. (2021). Wind turbine fault detection using highly imbalanced real SCADA data. *Energies, 14*, 1728.

Wang, K.-S., Sharma, V. S., & Zhang, Z.-Y. (2014). SCADA data based condition monitoring of wind turbines. *Advances in Manufacturing, 2*, 61–69.

Wang, J., et al. (2020). An integrated fault diagnosis and prognosis approach for predictive maintenance of wind turbine bearing with limited samples. *Renewable Energy, 145*, 642–650.

Xu, Z., & Saleh, J. H. (2021). Machine learning for reliability engineering and safety applications: Review of current status and future opportunities. *Reliability Engineering and System Safety, 211*, 107530.

Yadav, S. & Shukla, S. (2016). Analysis of k-fold cross-validation over hold-out validation on colossal datasets for quality classification. In *2016 IEEE 6th International Conference on Advanced Computing (IACC)*, 27–28 Feb. 2016 (pp. 78–83).

Yang, X., Zhang, Y., Lv, W., & Wang, D. (2021). Image recognition of wind turbine blade damage based on a deep learning model with transfer learning and an ensemble learning classifier. *Renewable Energy, 163*, 386–397.

Yuanyuan, S., Yongming, W., Lili, G., Zhongsong, M. & Shan, J. (2017). The comparison of optimizing SVM by GA and grid search. In *2017 13th IEEE International Conference on Electronic Measurement & Instruments (ICEMI)* (pp. 354–360). IEEE.

Zhang, X., Han, P., Xu, L., Zhang, F., Wang, Y., & Gao, L. (2020). Research on bearing fault diagnosis of wind turbine gearbox based on 1DCNN-PSO-SVM. *IEEE Access, 8*, 192248–192258.

Zhou, S., Chu, X., Cao, S., Liu, X., & Zhou, Y. (2020). Prediction of the ground temperature with ANN, LS-SVM and fuzzy LS-SVM for GSHP application. *Geothermics, 84*, 101757.

A Hybrid Neural Network Model Based on Convolutional Cascade Neural Networks: An Application for Image Inspection in Production

Diego Ortega Sanz, Carlos Quiterio Gómez Muñoz, Guillermo Benéitez, and Fausto Pedro García Márquez

Abstract The field of artificial intelligence and, in particular, that which deals with artificial neural networks, is experiencing a great interest in companies for the inspection and verification of images. The current trend in the sector is to implement these novel methodologies in industrial environments so that they can benefit from their advantages over traditional systems. Quality and production managers are increasingly interested in replacing the classic inspection methods with this new approach due to its flexibility and precision. Traditional methods have some weaknesses when it comes to inspecting parts, such as sensitivity to disturbances. In an industrial environment, these disturbances can be changes in lighting during the day or during the year, the appearance of external elements such as dust or dirt. The use of new convolutional neural network techniques allows training including disturbance scenarios, teaching artificial neural network to detect non-verse defects influenced by changes in light or by the appearance of dust. In this way, it is possible to drastically reduce false-positives, avoiding costly stops in production and maximizing the precision of detection and classification of each defect. This work studies the implementation of a hybrid model based on a cascade detection neural network with a classification neural network in an industrial environment.

Keywords Convolutional neural networks · Fault detection · Classification · Hybrid model · Deep learning

---

D. Ortega Sanz · C. Q. Gómez Muñoz (✉)
HCTLab Research Group, Electronics and Communications Technology Department, Universidad Autonoma de Madrid, Madrid, Spain
e-mail: carlosq.gomez@uam.es

G. Benéitez
Universidad Europea de Madrid, Villaviciosa de Odón, Spain

F. P. García Márquez
Ingenium Research Group, University of Castilla-La Mancha, Ciudad Real, Spain
e-mail: FaustoPedro.Garcia@uclm.es

F. P. García Márquez, B. Lev (eds.), *Sustainability*, International Series in Operations Research & Management Science 333,
https://doi.org/10.1007/978-3-031-16620-4_7

1 Introduction

Neural networks are a computational model that has evolved from various scientific contributions over time and covers many areas. These models consist of a set of units, called neurons, connected to transmit signals to each other. The input information passes through the neural network (where it is subjected to various operations, depending on its architecture and complexity) producing output values (Hartmann & Van der Auweraer, 2021; López & Fernandez, 2008; Priddy & Keller, 2005; Zhou et al., 2019).

There are multiple types of neural networks that are optimized for specific applications. One of the areas that uses artificial neural networks is medicine, (Dalton, 1995; Papik et al., 1998), forecasting (Fadlalla & Lin, 2001; Huang et al., 2007), autonomous driving (Arcos-Garcia et al., 2018; Gómez et al., 2015; Yi et al., 2017), or military area (Garcia Marquez & Gomez Munoz, 2020; Ramirez et al., 2017; Speri et al., 1998).

Another interesting area is related to manufacturing and production (Akyol & Bayhan, 2007; Garetti & Taisch, 1999). The main lines of research are focused on automated robots and control systems (artificial vision and sensors for pressure, temperature, gas, etc.) (Huang & Zhang, 1994; Zhang & Huang, 1995). Another important aspect is the image analysis for detection and classification in real time, acquiring data for production control in process lines, quality inspection, etc. (Melin & Castillo, 2007; Niaki & Davoodi, 2009). Digitals twins are a growing field with many opportunities for neural networks and predictive models (El Saddik, 2018). Most of these applications consist of performing pattern recognition, such as: searching for a pattern in a series of examples, classifying patterns, completing a signal from partial values, or reconstructing the correct pattern from a distorted one. Regarding the analysis of patterns in images, there is a complete architecture of neural networks, created specifically for image analysis: Convolutional Neural Networks (CNNs) (Matich, 2001; Widrow et al., 1994).

Convolutional Neural Networks (CNNs) are based on the *Neocognitron*, introduced by Kunihiko Fukushima (Fukushima et al., 1983). Most of this type of neural networks are focused on edge detection, shape segmentation, and motion detection (Juan & Mario, 2011). In 2006, a CNN architecture was developed to obtain a segmentation of shape and edges (Fujita et al., 2008; Nishizono & Nishio, 2006). CNNs are specifically designed to delimit contours with a little amount of noise (Babatunde et al., 2010).

As neural networks evolve, their architectures become increasingly complex and their computational weight increases. This meant that neural networks were slow and not supported by all hardware. A new field of research was opened up in hardware to run neural networks and achieve near real-time speeds (García Márquez et al., 2019). The first significant breakthrough in this respect was achieved when was implemented in hardware specialized analog-digital chips that have been designed for character recognition and image preprocessing applications. Speeds of more than

1000 characters per second were obtained with network architectures with around 100,000 connections (LeCun & Bengio, 1995; Louati et al., 2021).

A hybrid artificial intelligence system is formed by the integration of several intelligent subsystems, which maintains its own language of representation and different mechanisms for inferring solutions, i.e., condition monitoring (Gómez Muñoz et al., 2019; Ramirez et al., 2017). The aim of hybrid systems is to improve the efficiency and performance of the overall system (Márquez, 2010). The efficiency of a model can be improved with the use of control mechanisms that determine those subsystems that should be used at each moment (Jiménez et al., 2020). Hybrid systems have the potential to solve some problems that are difficult to address using a single method and combine several solutions of different approaches to increase the accuracy (Gómez Muñoz et al., 2018; Riverola & Corchado, 2000). The main approach in this type of networks has been the creation of hybrid models in which different types of base units are used in the hidden layer (Martínez et al., 2007).

As a result, hybrid models have been defined where the hidden layer nodes have different activation/transfer functions. Among these works, it is worth highlighting, on the one hand, the work of Duch and Jankowski (2001), in which works different transfer functions are proposed in the hidden layer nodes, and on the other hand, the proposal of Cohen and Intrator, (2002a, 2002b), which contemplate the duality between functions using projection-based function approximations and projection-based functions and radial basis functions). To design the system is necessary to define the approaches necessary. The CNN was chosen to classify the input images and to detect components within it (Herraiz et al., 2020). Then, according to the information obtained, a series of operations is carried out to determine the condition of the component (Garcia Marquez et al., 2017; Gómez Muñoz et al., 2015; Muñoz & Márquez, 2018a, 2018b).

This paper studies the integration of different neural network models with each other to generate a new hybrid model whose overall robustness and efficiency is greater than the sum of its parts, in order to satisfy the needs of image detection and classification in manufacturing environments.

The present work is structured as follows: modeling and approach, where the workflow of the future neural network model is established, case study, in which a real case and problems on which the hybrid model will be based are explained, and results and conclusions.

2 Modeling and Approach

The following approach arises as a need to reduce the failures that occur in the packaging process. This company manufactures office furniture, so it uses hundreds of different parts to assemble the final product. The parts to be analyzed are parts used to assemble office desks, furniture, chairs, etc. One of the biggest problems that occurs in the packaging process is that some of the parts are missing or defective,

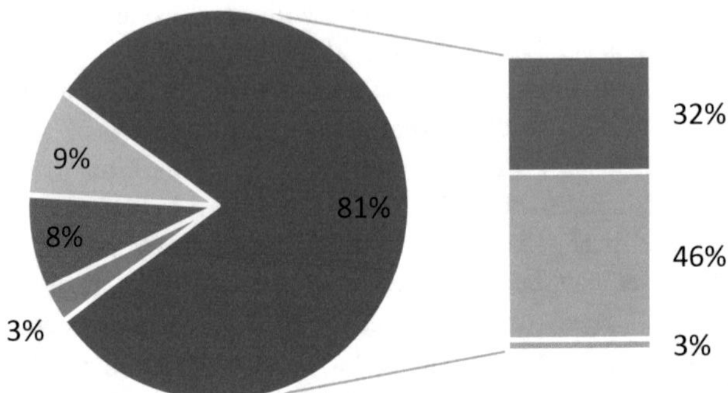

Fig. 1 Quality failures per year

which generates many after-sales service expenses and reduces customer satisfaction.

The main goal is to find a solution that allows verifying the content of all the pieces of each product in its packaging process in real time to reduce costs for the company. This aspect represents one of its biggest problems in terms of quality and related costs. Figure 1 shows a summary of the production quality failures extracted from an internal company quality report. It can be seen that the impact of quality failures due to lack of components is very large and is accentuated in products with a higher production volume and which require more specific parts and components.

The impact of non-quality forced the company until now to solve the problem by increasing the applied force on the production line. As shown in Fig. 2, an increase in personnel, and incentives to increase attention manage to reduce quality failures but significantly increase the cost, which is not feasible.

The presented hybrid system is tested to solve this problem. A part of the hybrid system will oversee detecting the pieces with respect to the background. The second cascade network will be in charge of classifying the different pieces, counting the number of each of them, and detecting the defective parts. Figure 3 shows the flowchart of the presented approach for image inspection.

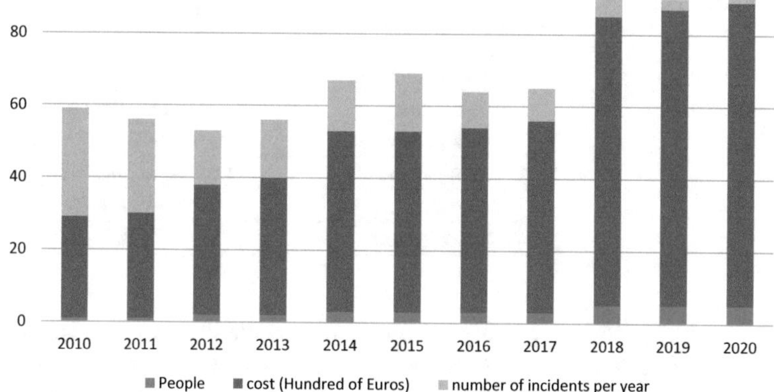

Fig. 2 Ratio of costs and personnel to minimize quality failures

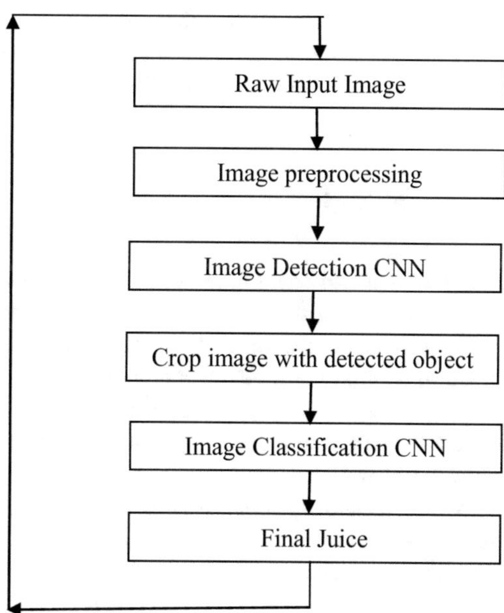

Fig. 3 Chart flow for CNN deployment

The images are obtained through a camera, to obtain a raw image, which needs to be processed, normalizing the image, its size, equalizing the histogram and morphological operations. Thus, the input to the first neural network would be generated.

This image is processed by the first neural network, a classification network, trained with the models of parts that may appear. The network classifies the objects found and generates a bounding box over them. The coordinates of the bounding box over the whole image are used to crop and generate a new image containing only the object.

Table 1 Parameters

Variable	Status
Image input	RGB
Images dimensions	640 x 640
Datasets folders	Defined in .mat file
Datasets parameters	20% validation 5% test
Training parameters	Defined in .mat file
Training epochs	50–100–500–1000
CNN architecture	SSD-MOBILENET; RESNET;

The new image containing only the object will be the input to the second neural network, this time, a classification neural network, which will determine whether the part contained in the image has defects or is a uniform part, according to its meshing. The first network is able to classify the type of object in the image, this information is transmitted to the second network. That is the network which corresponding to the type of object can be loaded, thus improving the robustness of the network.

Table 1 lists the main variable parameters during training and their value determined to achieve comparable results.

3 Case Study

Different real references have been selected from a production environment, with the aim of classifying them properly and determining if the object has any defect. The problem focuses on a production line and final assembly of the product, in which an operator manually inserts the components of the product into a bag and proceeds to pack the entire batch. A camera takes a picture of the packaging of the components to have information in case of a possible claim or quality failure. Figure 4 shows the process followed.

A new camera system has been installed to obtain the images, then the images have been processed to stabilize their input to the network. The first time identifies the different references and cuts them to minimize the adverse effects of the background, and the second network proceeds to classify the images, identifying whether the component has defects.

Figure 5 shows different models of parts, which must be properly classified and subsequently validated for defects. This input image shall be pre-processed in order to ensure a stable and homogeneous input to the network (Ding et al., 2021; Jiménez et al., 2019).

Image preprocessing involves several sequential steps and processes, as shown in Fig. 6:

The first step is the conversion of the image to greyscale. The image is in RGB mode, it consists of 3 channels. Each channel is a matrix of rows and columns equivalent to the pixels in width and length of the image, each pixel having a value between 0 and 255 (Alomoush et al., 2021; Vincent, 1992).

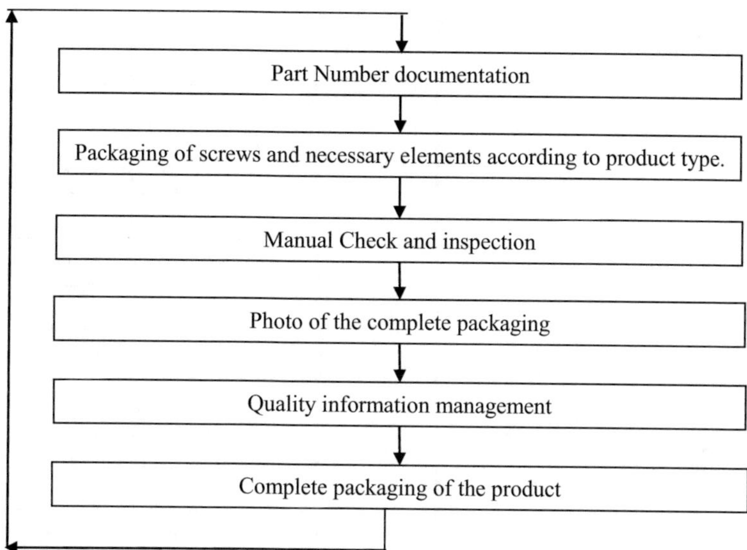

Fig. 4 Manual process flowchart

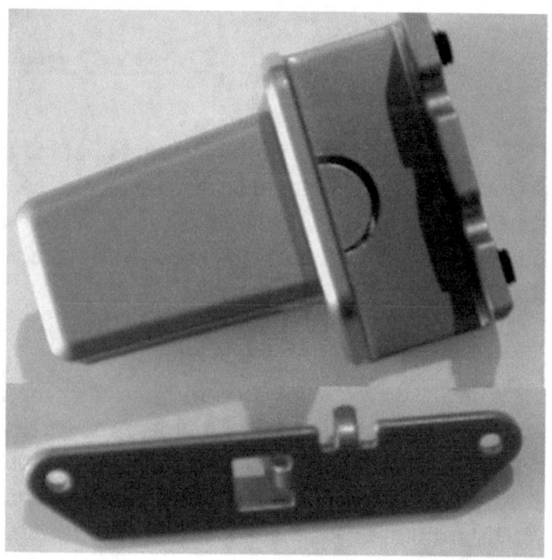

Fig. 5 Different models of parts

There are different methods to convert an image to greyscale, in this case the computational cost of the operations will be considered, for that reason it will be done in the simplest way possible (Kanan & Cottrell, 2012).

An optimal method would be the average of the values of the 3 channels, according to the Grayscale average method (Eq. 1).

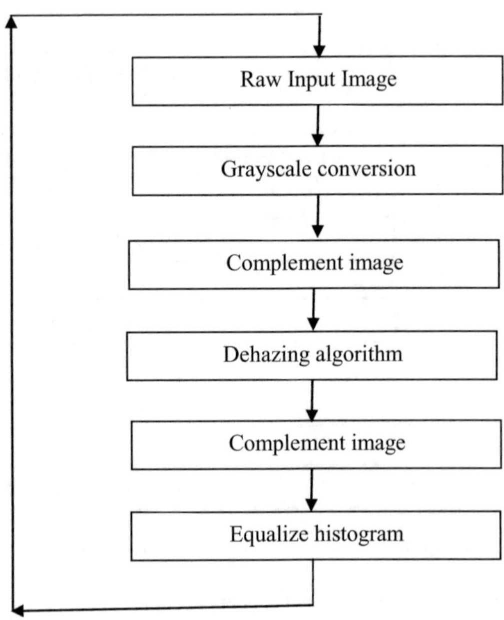

Fig. 6 Pre-processing image chart-flow

$$Y_{\text{average}} = \frac{R_{\text{average}} + G_{\text{average}} + B_{\text{average}}}{3} \quad (1)$$

This method, although the fastest, is not the most correct, as it does not consider factors such as the luminance dependence of each channel. In fact, the perception of the human eye to each channel is different. In order to obtain an adequate greyscale image, each channel must be balanced by a factor (Eq. 2) (Saravanan, 2010).

$$Y_{\text{Linear}} = 0.2126 * R_{\text{Linear}} + 0.7152 * G_{\text{Linear}} + 0.0722 * B_{\text{Linear}} \quad (2)$$

With the image converted to greyscale, the next step is to do the complement image. In the complement of a binary image, zeros become ones and zeros become zeros. Black and white are inverted. In the complement of a greyscale or color image, each pixel value is subtracted from the maximum pixel value supported by the class (or 1.0 for double precision images). The difference is used as the pixel value in the output image. In the output image, dark areas become lighter and light areas become darker. For color images, reds become cyan, greens become magenta, blues become yellow and vice versa (Deng et al., 2000).

Due to the transformations to which the image has been subjected, it will have an excess of illumination, which may cause the loss of details, modify lighting effects on the image and generate, in short, a bad training of the network. This effect is known as an atmospheric haze on the image (Abdulkareem et al., 2021; Hautière et al., 2007; Narasimhan & Nayar, 2003; Qu & Zou, 2017).

Scheneder proposed the following algorithm for dehazing (Eq. 3) (Schechner et al., 2003):

$$I = L^{\text{object}} * t + A \tag{3}$$

I is the irradiance of the image, L^{object} is the irradiance of the object, t is the transmittance of the incoherent light, and A is the light in the air (Tan, 2008).

If other parameters are considered, as A_∞ which is the air light radiance at an infinite distance, β, the coefficient of extinction due to scattering and absorption, and d as the distance from the camera to the scene, the expression could be redefined as follow (Eq. 4) (Tarel & Hautiere, 2009):

$$I = L^{\text{object}} * e^{-\beta d} + A_\infty * \left(1 - e^{-\beta d}\right) \tag{4}$$

The image is uncomplemented again to return it to its original state. The last step consists of equalizing the histogram of the image (Nithyananda & Ramachandra, 2016).

Histogram equalization is a very useful transformation Histogram equalization refers to flattening the grayscale histogram of an image so that the probability of distribution of each grayscale value in the transformed image is the same in the map (Dhal et al., 2021; Kim, 1997).

Before further processing, histogram equalization is usually a very good way to normalize the gray value of the image and can improve the contrast of the image. As a result, processes such as thresholding are substantially improved (Glasbey, 1993).

Histogram equalization is a method in the field of image processing that uses image histograms to adjust contrast. This method is generally used to increase the local contrast of many images, especially when the contrast of the useful image data is quite close. In this way, the brightness can be better distributed in the histogram. This can be used to enhance the local contrast without affecting the overall contrast. Histogram equalization achieves this function by effectively expanding the commonly used brightness. It is a very common and highly used method because of its robustness and efficiency (Luft et al., 2006).

The histogram is a graphical representation of the intensity distribution of pixels in an image. It counts the number of pixels that each intensity value has. First, the number of pixels of each gray level of the original image is counted (Sim & Tso, 2007). The probability of occurrence of pixels with grayscale "i," in the image, where n is the number of all pixels in the image (Eq. 5).

$$p_{x\,(i)} = \frac{n_j}{n} \tag{5}$$

The cumulative distribution function is the cumulative normalized histogram of the image (Eq. 6):

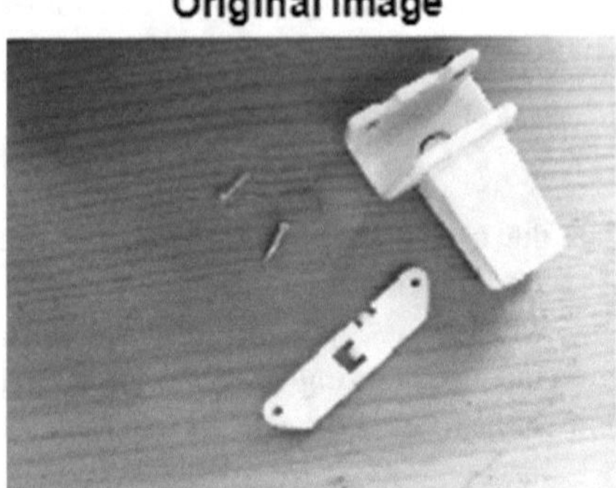

Fig. 7 Results of pre-processing

$$cdf_x(i) = \sum_{j=0}^{i} p_x(j) \tag{6}$$

With these parameters, the histogram equalization calculation formula, represented by h is formed (Eq. 7). For the minimum value of the cumulative distribution function, M and N represent respectively the number of long and wide pixels in the image, and L is the number of gray levels (if the image has a depth of 8 bits, the total number of gray levels is $2^8 = 256$, this is also the most common number of gray levels), v is the pixel value of v in the original image (Dhal et al., 2021).

$$h(v) = \mathrm{round}\left(\frac{\mathrm{cdf}(v) - \mathrm{cdf}_{\mathrm{min}}}{(M * N) - \mathrm{cdf}_{\mathrm{min}}} * (L - 1)\right) \tag{7}$$

Figure 7 shows the results after the transformations were applied to the original image. This result will be the input to the first neural network.

For CNN, it is necessary to analyze different neural network architectures in order to decide on the most suitable one before carrying out the definitive training of the dataset. Different architectures have been considered for achieving real-time video processing and to be implemented on industrial equipment. The tested architectures were:

- *AlexNet* contained eight layers; the first five were convolutional layers, some of them followed by max-pooling layers, and the last three were fully connected layers. It used the non-saturating ReLU activation function, which results in a lower computational cost than the sigmoid function (Krizhevsky et al., 2017).
- *SqueezeNet* was another CNN used in compassion experiments (Howard et al., 2017). This CNN is similar to *AlexNet*, and both have achieved similar accuracy in the tests performed (Howard et al., 2017).

- *MobileNet* is an architecture designed mainly for mobile vision applications. Therefore, it is mainly focused on reducing as much as possible the computational power required for the processing. It is a simpler architecture that sacrifices some precision and performance. This is largely achieved through the use of separable convolutions by depth (Howard et al., 2017).
- *Darknet* is an open-source neural network framework written in C language and *CUDA*. It is fast, easy to install, and supports CPU and GPU computation. *Darknet* is mainly designed for Object Detection, and they have different architecture and features than other deep learning frameworks. It is faster than many other CNN architectures (Vasavi et al., 2020).
- The *Googlenet* network is a CNN inspired by *LeNet* which implements a novel element called inception module. It used batch normalization, image distortions, and RMSprop. This module is based on several very small convolutions to drastically reduce the number of parameters. Their architecture consisted of a 22-layer deep CNN, capable of reducing the number of parameters from 60 million to four million (Anand et al., 2020).
- Residual Network (*ResNet*) is a CNN architecture that was designed to enable hundreds or thousands of convolutional layers. While previous CNN architectures had a drop-off in the effectiveness of additional layers, *ResNet* can add a large number of layers with strong performance (Targ et al., 2016). ResNet was an innovative solution to the "vanishing gradient" problem (Muñoz et al., 2017). Neural networks train via the backpropagation process, which relies on (Gómez et al., 2016; Limonova et al., 2020).
- The You Only Look Once (YOLO) algorithm is a state-of-the-art open-source real-time object detection system that makes use of a unique convolutional neural network to detect objects in images. For its operation, the neural network divides the image into regions, predicting identification frames and probabilities for each region; the boxes are weighted from the predicted probabilities.
- R-CNN is a type of neural network which used an algorithm called selective search to extract candidate regions of interest and used a standard volume neural network product to classify and fit these areas. One of the technologies called region of interest clustering allows the network to share the results of calculations and speed up the model.

4 Results

Figure 8 shows the accuracy of the different neural net architectures employed for the first neural net of the system. AlexNet and MobileNet have the lowest level of accuracy. This is because for the internal architecture because these networks are not suitable for detection images.

RCNN and *YOLO* architectures offer better results. In these cases, the algorithm has satisfactorily found the desired characteristics of the image and has been stable to

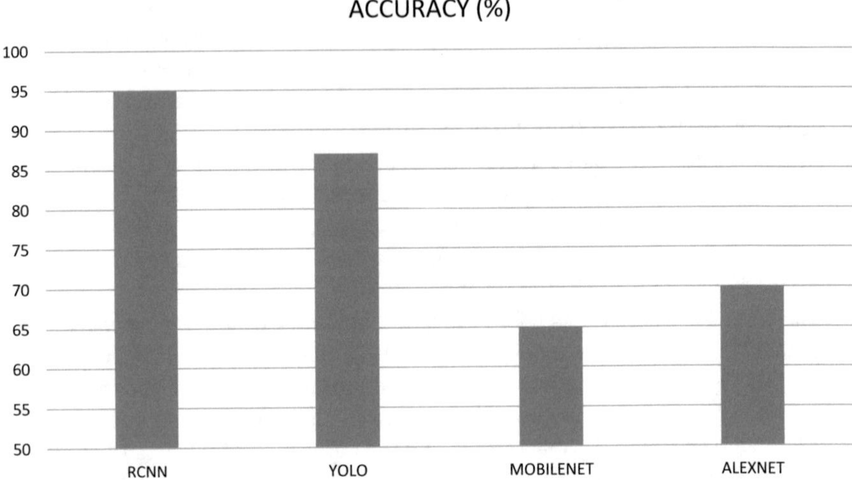

Fig. 8 CNN comparisons

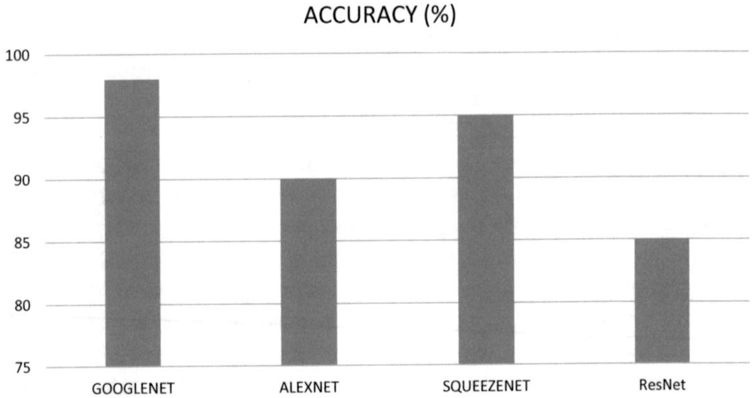

Fig. 9 Second neural network results

changes in the same as brightness or changes in light. This is because both architectures are designed for detection and classification.

CNN-based RCNN shows a very smart approach that greatly improved the performance of the training model.

Figure 9 shows the results obtained by the second neural network. Due to the simplicity of the input data and that is a classification work the performances of CNN are very high. CNN such as GOOGLENET and SQUEEZENET has reached the higher 100% accuracy in the tests, while ALEXNET and ResNet architectures have given the lowest accuracy.

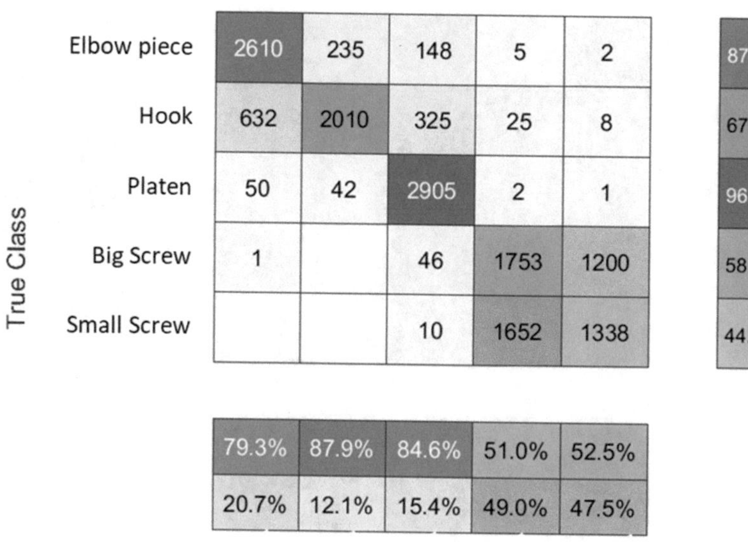

Fig. 10 Resnet confusion matrix

The combination for this hybrid model that has yielded the best results are: For the first network was the *Yolo* architecture, while for the second model *GoogleNet* architecture.

The details of each training are shown below using their respective confusion matrices. Figure 10 shows the confusion matrix for Resnet architecture. It is observed that there is a large number of false positives in all classes, with this problem being greater in the most similar classes such as screws.

The confusion matrix shown in Fig. 11, relating to the Alexnet architecture, shows an improvement in the results, but still retains significant faults. The network still shows failures relative to all classes although to a lesser extent than with the previous architecture.

The results of the confusion matrix shown in Fig. 12 improve significantly. With the Squeezenet architecture, it is possible to correctly identify the classes with the greatest differences, and the classification failures in the most similar classes are substantially reduced.

Finally, Fig. 13 shows the most optimal results achieved for all architectures. The GoogleNet architecture shows excellent performance with all classes, minimizing errors and confusion in the most similar classes.

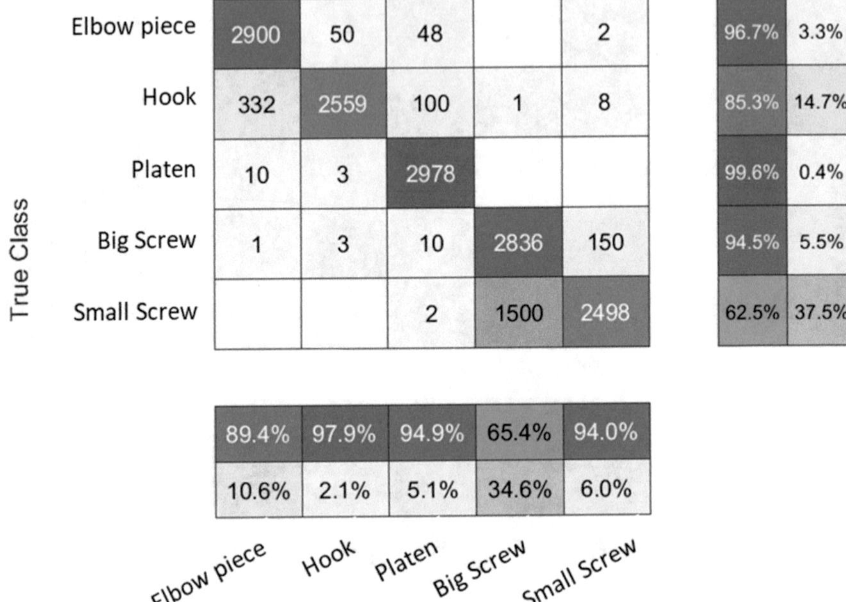

Fig. 11 Alexnet confusion matrix

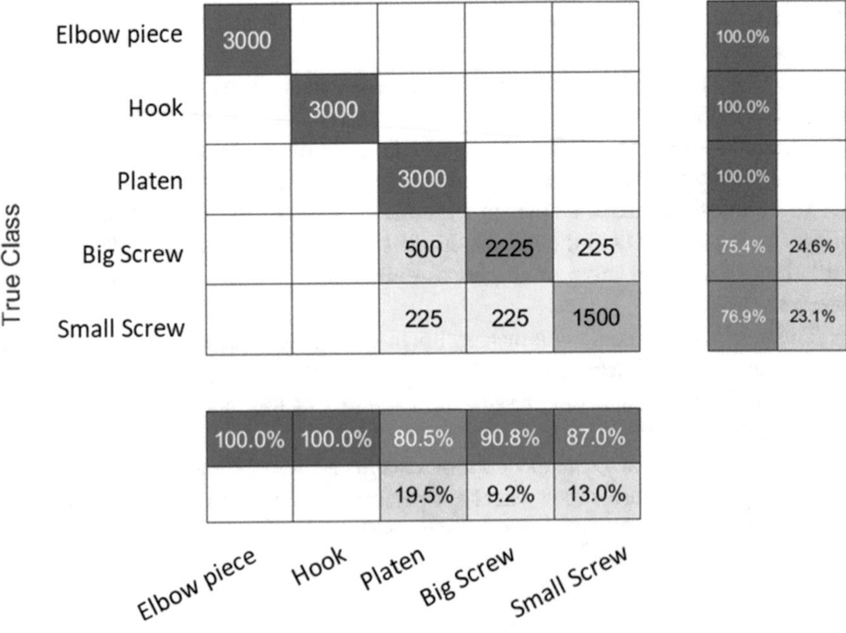

Fig. 12 Squeezenet confusion matrix

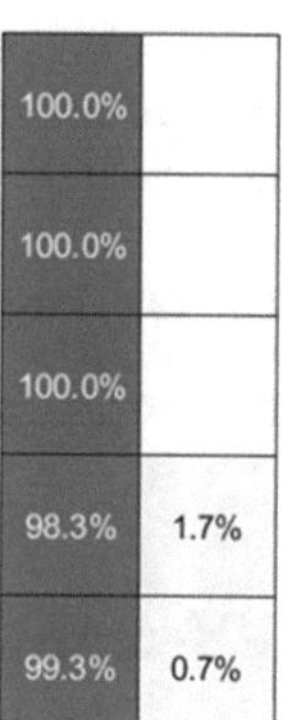

Fig. 13 GoogleNet confusion matrix

5 Conclusions

A solution has been proposed for the detection and classification of defects in an image and its subsequent classification based on the defects found.

Hybrid models could be effective systems for solving problems related with pattern recognition and classification with neural networks. In this work, a hybrid model was employed by combining two different neural networks.

This work offers the possibility of strengthening CNN-based algorithms by combining an image classifier and a linear classifier. This hybrid network system proposes a more robust and efficient algorithm than a simple CNN classifier.

The detection network chosen, RCNN architecture, as the first network offers an accuracy rate of 95%, higher than the other tested architectures.

Googlenet, the classification network chosen as the second network offers an accuracy rate of 98%, much higher than the other tested architectures, except for the *Squeezenet* architecture, which has a good accuracy rate of 95%.

The combination of the *RCNN* and *Googlenet* architecture together yielded results with high accuracy for this hybrid model.

It is demonstrated that the creation of a hybrid model formed by the combination of two or more networks acts in a more robust and efficient way than the same networks separately.

This hybrid model of neural networks offers unprecedented detection for metal components with high surface variability. It allows to drastically reduce false rejects generated by other detection and sorting models.

Funding The work reported herewith has been financially by the Dirección General de Universidades, Investigación e Innovación of Castilla-La Mancha, under Research Grant ProSeaWind project (Ref.: SBPLY/19/180501/000102).

References

Abdulkareem, K. H., Arbaiy, N., Zaidan, A., Zaidan, B., Albahri, O. S., Alsalem, M., & Salih, M. M. (2021). A new standardisation and selection framework for real-time image dehazing algorithms from multi-foggy scenes based on fuzzy delphi and hybrid multi-criteria decision analysis methods. *Neural Computing and Applications, 33*, 1029–1054.

Akyol, D. E., & Bayhan, G. M. (2007). A review on evolution of production scheduling with neural networks. *Computers & Industrial Engineering, 53*, 95–122.

Alomoush, W., Alrosan, A., Alomari, Y. M., Alomoush, A. A., Almomani, A., & Alamri, H. S. (2021). Fully automatic grayscale image segmentation based fuzzy c-means with firefly mate algorithm. *Humanized Computing, 13*, 1–23.

Anand, R., Shanthi, T., Nithish, M., & Lakshman, S. (2020). Face recognition and classification using googlenet architecture. In K. Das, J. Bansal, K. Deep, A. Nagar, P. Pathipooranam, & R. Naidu (Eds.), *Soft computing for problem solving* (pp. 261–269). Springer.

Arcos-Garcia, A., Alvarez-Garcia, J. A., & Soria-Morillo, L. M. (2018). Evaluation of deep neural networks for traffic sign detection systems. *Neurocomputing, 316*, 332–344.

Babatunde, H., Folorunso, O., & Akinwale, A. (2010). A cellular neural network-based model for edge detection. *Journal of Computing and Information Science in Engineering, 5*, 003–010.

Cohen, S., & Intrator, N. (2002a). Forward and backward selection in regression hybrid network. In *International workshop on multiple classifier systems* (pp. 98–107). Springer.

Cohen, S., & Intrator, N. (2002b). A hybrid projection-based and radial basis function architecture: Initial values and global optimisation. *Pattern Analysis & Applications, 5*, 113–120.

Forsström, J. J., & Dalton, K. J. (1995). Artificial neural networks for decision support in clinical medicine. *Annals of Medicine, 27*, 509–517.

Deng, W., Iyengar, S. S., & Brener, N. E. (2000). A fast parallel thinning algorithm for the binary image skeletonization. *The International Journal of High Performance Computing Applications, 14*, 65–81.

Dhal, K. G., Das, A., Ray, S., Gálvez, J., & Das, S. (2021). Histogram equalization variants as optimization problems: A review. *Archives of Computational Methods in Engineering, 28*, 1471–1496.

Ding, K., Ma, K., Wang, S., & Simoncelli, E. P. (2021). Comparison of full-reference image quality models for optimization of image processing systems. *International Journal of Computer Vision, 129*, 1258–1281.

Duch, W., & Jankowski, N. (2001.; Citeseer). *Transfer functions: Hidden possibilities for better neural networks* (pp. 81–94). ESANN.

El Saddik, A. (2018). Digital twins: The convergence of multimedia technologies. *IEEE Multimedia, 25*, 87–92.

Fadlalla, A., & Lin, C.-H. (2001). An analysis of the applications of neural networks in finance. *Interfaces, 31*, 112–122.

Fujita, T., Okamura, T., Nakanishi, M., & Ogura, T. (2008). Cam 2-universal machine: A dtcnn implementation for real-time image processing. In *2008 11th International workshop on cellular neural networks and their applications* (pp. 219–223). IEEE.

Fukushima, K., Miyake, S., & Ito, T. (1983). Neocognitron: A neural network model for a mechanism of visual pattern recognition. *IEEE Transactions on Systems, Man, and Cybernetics, 13*, 826–834.

Garcia Marquez, F. P., & Gomez Munoz, C. Q. (2020). A new approach for fault detection, location and diagnosis by ultrasonic testing. *Energies, 13*, 1192.

Garcia Marquez, F. P., Pliego Marugan, A., Pinar Pérez, J. M., Hillmansen, S., & Papaelias, M. (2017). Optimal dynamic analysis of electrical/electronic components in wind turbines. *Energies, 10*, 1111.

García Márquez, F. P., Segovia Ramírez, I., & Pliego Marugán, A. (2019). Decision making using logical decision tree and binary decision diagrams: A real case study of wind turbine manufacturing. *Energies, 12*, 1753.

Garetti, M., & Taisch, M. (1999). Neural networks in production planning and control. *Production Planning & Control, 10*, 324–339.

Glasbey, C. (1993). An analysis of histogram-based thresholding algorithms. *Graphical Models and Image Processing, 55*, 532–537.

Gómez, C. Q., García Márquez, F. P., Arcos, A., Cheng, L., Kogia, M., & Papaelias, M. (2016). Calculus of the defect severity with emats by analyzing the attenuation curves of the guided waves. *Smart Structures and Systems, 19*(2), 195–202.

Gómez Muñoz, C. Q., Arcos Jiménez, A., García Márquez, F. P., Kogia, M., Cheng, L., Mohimi, A., & Papaelias, M. (2018). Cracks and welds detection approach in solar receiver tubes employing electromagnetic acoustic transducers. *Structural Health Monitoring, 17*, 1046–1055.

Gómez Muñoz, C. Q., García Márquez, F. P., Arcos Jiménez, A., & Papelias, M. (2015). A heuristic method for detecting and locating faults employing electromagnetic acoustic transducers. *Eksploatacja i Niezawodność, 19*, 493.

Gómez Muñoz, C. Q., García Marquez, F. P., Hernandez Crespo, B., & Makaya, K. (2019). Structural health monitoring for delamination detection and location in wind turbine blades employing guided waves. *Wind Energy, 22*, 698–711.

Gómez, C. Q., Villegas, M. A., García, F. P., & Pedregal, D. J. (2015). Big data and web intelligence for condition monitoring: A case study on wind turbines. In N. Zaman, M. E. Seliaman, M. F. Hassan, & F. P. García Márquez (Eds.), *Handbook of research on trends and future directions in big data and web intelligence* (pp. 149–163). IGI Global.

Hartmann, D., & Van der Auweraer, H. (2021). Digital twins. In M. Cruz, C. Parés, & P. Quintela (Eds.), *Progress in industrial mathematics: Success stories* (pp. 3–17). Springer.

Hautière, N., Tarel, J.-P., & Aubert, D. (2007). Towards fog-free in-vehicle vision systems through contrast restoration. In *2007 IEEE conference on computer vision and pattern recognition* (pp. 1–8). IEEE.

Herraiz, Á. H., Marugán, A. P., & Márquez, F. P. G. (2020). Photovoltaic plant condition monitoring using thermal images analysis by convolutional neural network-based structure. *Renewable Energy, 153*, 334–348.

Howard, A. G., Zhu, M., Chen, B., Kalenichenko, D., Wang, W., Weyand, T., Andreetto, M., & Adam, H. (2017). *Mobilenets: Efficient convolutional neural networks for mobile vision applications*. Cornell University Library.

Huang, W., Lai, K. K., Nakamori, Y., Wang, S., & Yu, L. (2007). Neural networks in finance and economics forecasting. *International Journal of Information Technology & Decision Making, 6*, 113–140.

Huang, S. H., & Zhang, H.-C. (1994). Artificial neural networks in manufacturing: Concepts, applications, and perspectives. *IEEE Transactions on Components, Packaging, and Manufacturing Technology: Part A, 17*, 212–228.

Jiménez, A. A., Muñoz, C. Q. G., & Márquez, F. P. G. (2019). Dirt and mud detection and diagnosis on a wind turbine blade employing guided waves and supervised learning classifiers. *Reliability Engineering & System Safety, 184*, 2–12.

Jiménez, A. A., Zhang, L., Muñoz, C. Q. G., & Márquez, F. P. G. (2020). Maintenance management based on machine learning and nonlinear features in wind turbines. *Renewable Energy, 146*, 316–328.

Juan, R. Q., & Mario, C. M. (2011). Redes neuronales artificiales para el procesamiento de imágenes, una revisión de la última década. *RIEE & C, Revista de Ingeniería Eléctrica, Electrónica y Computación, 9*, 7–16.

Kanan, C., & Cottrell, G. W. (2012). Color-to-grayscale: Does the method matter in image recognition? *PLoS One, 7*, e29740.

Kim, Y.-T. (1997). Contrast enhancement using brightness preserving bi-histogram equalization. *IEEE Transactions on Consumer Electronics, 43*, 1–8.

Krizhevsky, A., Sutskever, I., & Hinton, G. E. (2017). Imagenet classification with deep convolutional neural networks. *Communications of the ACM, 60*, 84–90.

LeCun, Y., & Bengio, Y. (1995). Convolutional networks for images, speech, and time series. In J. W. Schwieter (Ed.), *The handbook of brain theory and neural networks* (p. 3361). Wiley.

Limonova, E., Alfonso, D., Nikolaev, D., & Arlazarov, V. V. (2020). *Resnet-like architecture with low hardware requirements*. Cornell University Library.

López, R. F., & Fernandez, J. M. F. (2008). *Las redes neuronales artificiales*. Netbiblo.

Louati, H., Bechikh, S., Louati, A., Hung, C.-C., & Said, L. B. (2021). Deep convolutional neural network architecture design as a bi-level optimization problem. *Neurocomputing, 439*, 44–62.

Luft, T., Colditz, C., & Deussen, O. (2006). Image enhancement by unsharp masking the depth buffer. *ACM Transactions on Graphics, 25*, 1206–1213.

Márquez, F. P. G. (2010). A new method for maintenance management employing principal component analysis. *Structural Durability & Health Monitoring, 6*, 89.

Martínez, C. H., García, F. J. M., Peña, P. A. G., Fernández, J. C., & Ballesteros, A. J. T. (2007). Clasificación mediante la evolución de modelos híbridos de redes neuronales. In F. Almeida Rodríguez (Ed.), *Actas del V Congreso Español sobre Metaheurísticas, Algoritmos Evolutivos y Bioinspirados* (pp. 77–84). Dialnet.

Matich, D. J. (2001). *Redes neuronales: Conceptos básicos y aplicaciones* (p. 41). Universidad Tecnológica Nacional—Facultad Regional Rosario, Departamento de Ingeniería Química.

Melin, P., & Castillo, O. (2007). An intelligent hybrid approach for industrial quality control combining neural networks, fuzzy logic and fractal theory. *Information Sciences, 177*, 1543–1557.

Muñoz, C. Q. G., & Márquez, F. P. G. (2018a). Renewable energies. In C. Q. Gómez Muñoz & F. P. García Márquez (Eds.), *Wind energy power prospective* (pp. 83–95). Springer.

Muñoz, C. Q. G., & Márquez, F. P. G. (2018b). Future maintenance management in renewable energies. In F. P. García Márquez, A. Karyotakis, & M. Papaelias (Eds.), *Renewable energies* (pp. 149–159). Springer.

Muñoz, C. Q. G., Marquez, F. P. G., Lev, B., & Arcos, A. (2017). New pipe notch detection and location method for short distances employing ultrasonic guided waves. *Acta Acustica united with Acustica, 103*, 772–781.

Narasimhan, S. G.; Nayar, S. K. (2003). Interactive (de) weathering of an image using physical models. In *IEEE workshop on color and photometric methods in computer vision* (p. 1). France.

Niaki, S. T. A., & Davoodi, M. (2009). Designing a multivariate–multistage quality control system using artificial neural networks. *International Journal of Production Research, 47*, 251–271.

Nishizono, K., & Nishio, Y. (2006). Image processing of gray scale images by fuzzy cellular neural network. In *RISP International Workshop nonlinear circuits, Honolulu Hawaii, USA* (pp. 90–93).

Nithyananda, C. & Ramachandra, A. (2016). Survey on histogram equalization method based image enhancement techniques. In *2016 International conference on data mining and advanced computing (SAPIENCE)* (pp. 150–158). IEEE.

Papik, K., Molnar, B., Schaefer, R., Dombovari, Z., Tulassay, Z., & Feher, J. J. (1998). Application of neural networks in medicine-a review. *Medical Science Monitor, 4*, MT538–MT546.

Priddy, K. L., & Keller, P. E. (2005). *Artificial neural networks: An introduction* (Vol. 68). SPIE Press.

Qu, Y., & Zou, Z. (2017). Non-sky polarization-based dehazing algorithm for non-specular objects using polarization difference and global scene feature. *Optics Express, 25*, 25004–25022.

Ramirez, I. S., Muñoz, C. Q. G. & Marquez, F. P. G. (2017). A condition monitoring system for blades of wind turbine maintenance management. *Proceedings of the tenth international conference on management science and engineering management* (pp. 3–11). Springer.

Riverola, F. F., & Corchado, J. M. (2000). Sistemas híbridos neuro-simbólicos: Una revisión. *Inteligencia Artificial. Revista Iberoamericana de Inteligencia Artificial, 4*, 12–26.

Saravanan, C. (2010). Color image to grayscale image conversion. In *2010 second international conference on computer engineering and applications* (pp. 196–199). IEEE.

Schechner, Y. Y., Narasimhan, S. G., & Nayar, S. K. (2003). Polarization-based vision through haze. *Applied Optics, 42*, 511–525.

Sim, K. S., Tso, C. P., & Tan, Y. Y. (2007). Recursive sub-image histogram equalization applied to gray scale images. *Pattern Recognition Letters, 28*, 1209–1221.

Speri, L., Schiliro, G., Bezzetto, A., Cifelli, G., De Battisti, L., Marchi, S., Modenese, M., Varalta, F., & Consigliere, F. (1998). The use of artificial neural networks methodology in the assessment of "vulnerability" to heroin use among army corps soldiers: A preliminary study of 170 cases inside the military hospital of legal medicine of verona. *Substance Use & Misuse, 33*, 555–586.

Tan, R. T. (2008). Visibility in bad weather from a single image. In *2008 IEEE conference on computer vision and pattern recognition* (pp. 1–8). IEEE.

Tarel, J.-P., & Hautiere, N. (2009). Fast visibility restoration from a single color or gray level image. In *2009 IEEE 12th international conference on computer vision* (pp. 2201–2208). IEEE.

Targ, S., Almeida, D., & Lyman, K. (2016). *Resnet in resnet: Generalizing residual architectures*. Cornell University Library.

Vasavi, S., Priyadarshini, N. K., & Vardhan, K. H. (2020). Invariant feature based darknet architecture for moving object classification. *IEEE Sensors Journal, 21*, 11417.

Vincent, L. (1992). *Morphological grayscale reconstruction: Definition, efficient algorithm and applications in image analysis* (pp. 633–635). CVPR.

Widrow, B., Rumelhart, D. E., & Lehr, M. A. (1994). Neural networks: Applications in industry, business and science. *Communications of the ACM, 37*, 93–106.

Yi, H., Jung, H. & Bae, S. (2017). *Deep neural networks for traffic flow prediction*. In 2017 IEEE international conference on big data and smart computing (BigComp) (pp. 328–331). IEEE.

Zhang, H.-C., & Huang, S. (1995). Applications of neural networks in manufacturing: A state-of-the-art survey. *International Journal of Production Research, 33*, 705–728.

Zhou, X., Wang, L., Qin, J., Chai, J., & Gómez Muñoz, C. Q. (2019). Emergency rescue planning under probabilistic linguistic information: An integrated fta-anp method. *International Journal of Disaster Risk, 37*, 101170.

A Multi-Objective Sustainable Traffic Signal Control for Smart Cities Under Uncertainty

Suh-Wen Chiou

Abstract For urban road networks with hazardous materials (hazmat) transportation, a Multi-Objective Sustainable trAffic Control (MOSAC) is proposed in this chapter. A multi-objective stochastic program is proposed to simultaneously minimize total travel delay and reduce maximum risk exposure to all road users. Following Wardrop's principle, travelers' route choice is considered. In order to effectively solve the proposed stochastic program, a novel heuristic is proposed. Numerical experiments using a realistic road network are performed to demonstrate the feasibility of proposed approach. For varying traffic conditions, computational comparisons are also made with recently proposed one, indicating that system performance can be greatly improved by MOSAC in all cases.

Keywords Multi-objective sustainable traffic control · Smart city · Stochastic bi-level program · Uncertainty · Hazmat transportation

1 Introduction

For smart cities with hazmat transportation, it becomes increasingly critical to simultaneously minimize the total delay of travelers and mitigate random risk exposure to all road users. In order to effectively compromise conflicting objectives in smart cities while taking route choice of road users into account, a Hazmat Transportation Network Design (HTND) (Bianco et al., 2013) can often be formulated as a bi-level programming problem (BLPP) (Dempe, 2002; Lu et al., 2016; Zhang et al., 2015). For instance, (Kara & Verter, 2004) formulate a hazmat network design problem as a bi-level program with the government as a leader at the upper level and the hazmat carriers as followers in the lower level. To promote equity in the spatial distribution of risk effectively, a bi-level hazmat network design problem has been proposed (Bianco et al., 2009) to reduce the maximum amount of hazmat

S.-W. Chiou (✉)
Department of Information Management, National Dong Hwa University, Shoufeng, Taiwan
e-mail: chiou@mail.ndhu.edu.tw

F. P. García Márquez, B. Lev (eds.), *Sustainability*, International Series in Operations Research & Management Science 333,
https://doi.org/10.1007/978-3-031-16620-4_8

carriers over network links. Among these a system optimum for hazmat carriers is generally considered to minimize total risk in road networks and mitigate the occurrence of accidental release of hazmat in a deterministic bi-level program. In order to compare trade-offs between travel delay and deterministic risk, a bi-objective bi-level program was also presented in (Chiou, 2016). In applications relating to a bi-level program, and multilevel decision-making theory and model, a number of the problem inputs are often subject to uncertainty. In particular, this is true with respect to risk, travel demand, and traffic delay, which are subject to time-dependent fluctuations in hazmat network design for sustainable signal control in most smart cities. For the hierarchical design model of hazmat network design problem, external conditions or measurement errors often introduce uncertainty into the problem. In either case, the uncertainty can be included explicitly by generalizing some of the problem parameters to random variables. However, this generalization might complicate the model substantially. As a result, solution approaches in many cases require some approximation methods to solve the resulting stochastic programs.

To effectively deal with uncertain demand in a hierarchical decentralized organization, (Zhang et al., 2007) presented a fuzzy multi-objective bi-level program. A branch-and-bound algorithm was proposed to solve a multi-objective bi-level program with fuzzy demands where good numerical results can be found in case-based examples. For a general logistics planning problem, (Zhang & Lu, 2007) proposed a fuzzy bi-level program to optimize the objectives of both suppliers and distributors. In order to compare the trade-off between objective function values at upper level and feasibility degree of the constraints, (Ren et al., 2016) proposed an interactive programming approach for a fuzzy bi-level problem. Due to uncertain risk existing in the hazmat network design problem, (Sun et al., 2016) presented a robust bi-level program considering uncertainty on each link for each shipment and uncertainty across all shipments. A heuristic was proposed to solve a robust bi-level problem with good numerical results in real world networks. For smart cities with signal control, the author (Chiou, 2017) proposed a risk-averse bi-level program for hazmat network design. However, from empirical results of past studies (Chiou, 2017, 2018, 2019, 2020), a risk-averse bi-level problem taking a worst-case scenario design appears excessively conservative for some realization of uncertain problem in practice. For the state-of-the-art work in the HTND, readers can refer to (Huang et al., 2021; Mohri et al., 2022) for more details.

In order to improve mobility for smart city and reduce potential risk due to hazmat transportation, in this chapter, a Multi-Objective Sustainable trAffic Control (MOSAC) is proposed. A Stochastic Bi-Level Programming Problem (SBLPP) is presented to simultaneously minimize total delay and risk to all road users in the presence of uncertain travel demand and random risk over links. A multi-objective stochastic program, at the upper level, can be accordingly proposed to minimize total travel delay at links downstream and to simultaneously reduce the maximum risk over links. Following Wardrop's principle, a user equilibrium traffic assignment can be defined for regular traffic at the lower level. Such a user equilibrium traffic assignment can be generalized as a variational inequality, indicating that travelers' route choices can be determined and resulting link flows can be obtained at equilibrium. For hazmat carriers, a system optimum traffic assignment is considered from practical perspective, indicating that a shortest risk path can be identified between specified origin-destination (OD) pairs. In order to effectively solve the stochastic bi-level problem, a novel heuristic using bundle projection approach is proposed. In order to investigate the robustness of the proposed MOSAC for smart cities in practice, numerical experiments using a realistic road network are performed. Moreover, in order to demonstrate the feasibility and computational effectiveness of proposed approach, further computational comparisons are also made with the bi-level program in deterministic conditions, indicating that the performance improvement can be substantially achieved with reasonable computational overhead in all cases. The contribution made from this chapter in the literature for sustainable transportation in smart cities can be summarized as follows:

1. For sustainable traffic signal control, a link delay model at downstream can be presented to calculate a scenario-driven departure cost with random travel demand at each node along a shortest path.
2. In smart cities with random regular OD demands, a user equilibrium traffic assignment can be formulated as a variational inequality problem according to Wardrop's first principle.
3. In smart cities with hazmat traffic, a system optimum traffic assignment with random risk can be formulated as a variational inequality problem according to Wardrop's second principle.
4. For sustainable traffic signal control, a maximum link risk ratio can be particularly considered in constraints.
5. In order to simultaneously minimize travel delay and mitigate traffic risk, a Multi-objective Performance Index (MPI) is considered. A stochastic bi-level program is accordingly proposed to minimize the maximum link risk in a sustainable traffic signal control problem.
6. In order to effectively solve the proposed stochastic bi-level program in practice, sensitivity analysis employing stochastic derivatives for MPI can be carried out. A single-level stochastic program is proposed and approximated with sample averages respectively for random demand and risk exposure.
7. A bundle projection approach is proposed to efficiently minimize travel cost for all road users and maximum risk over all links.
8. To illustrate the effectiveness of MOSAC in practice, numerical experiments are carried out using a real-world Sioux Falls city road network. Numerical comparisons are also made with nominal bi-level problem in the deterministic case under varying traffic conditions.

The rest of this chapter is organized as follows. Section 2 introduces user equilibrium and system optimum respectively for regular traffic assignment and hazmat traffic assignment in the presence of signal delay at links downstream. Section 3 presents a MOSAC with MPI in scenarios. In order to efficiently solve MOSAC, a stochastic bi-level program is proposed. Section 4 conducts sensitivity analysis for the bi-level program by stochastic derivatives. In order to effectively solve the stochastic bi-level problem, a bundle projection approach is proposed in Section 5. Numerical experiments are performed in Section 6. In Section 7, conclusions and discussions of this contribution are summarized.

2 Equilibrium Traffic Assignment for Traffic Flows

In order to minimize travel delay at links downstream incurred by regular travelers, a minimum departure cost at each node along a shortest path can be decided. According to Wardrop's first principle, a user equilibrium traffic assignment for regular traffic can be formulated and solved by a variational inequality (Aashtiani & Magnanti, 1981). Notation used throughout this paper is presented first.

2.1 Notations

Sets

N	set of signal-controlled junctions.
A	set of links, and $\|A\|$ denotes cardinality of links.
$H(a)$	head node set of link a, $\forall a \in A$.
$T(a)$	tail node set of link a,$\forall a \in A$.
Π	set of signal settings.
C	set of OD trips, and $\|C\|$ denotes cardinality of OD.
C^H	set of OD pairs for hazmat carriers.

Parameters

Λ	link-path incidence matrix.
Υ	OD-path incidence matrix.
Γ^1, Γ^2	uncertainty budgets of demand and risk.
k	scenarios for random OD demand, $k = 1, 2, \ldots, \|C\|$.
s	scenarios for random risk over links, $s = 1, 2, \ldots, \|A\|$.
$q = [q_i]$	matrix of hazmat OD demand.
$Q = [Q_i]$	matrix of nominal OD demand.
$\widehat{Q} = \left[\widehat{Q}_i\right]$	matrix of OD demand deviation.
$\widetilde{Q} = \left[\widetilde{Q}_i^k\right]$	matrix of random OD demand in scenario k such that $\widetilde{Q}_i^k \in \left[Q_i, Q_i + \widehat{Q}_i\right]$.
$e = [e_a]$	vector of nominal risk over links.

$\widehat{e} = [\widehat{e}_a]$	vector of link risk deviation.
$\widetilde{e} = [\widetilde{e}^s_a]$	vector of random risk in scenario s such that $\widetilde{e}^s_a \in [e_a, e_a + \widehat{e}_a]$.
g_0	minimum green.
$\Omega_m(j, l)$	collection of numbers 0 and 1 for signal groups at junction m.
υ_{jlm}	clearance time between signal groups j and l at junction m.
s_a	saturation flow over links.
s^H_a	saturation flow for hazmat carriers over links.
c^0_a	link free flow travel time.

Variables

$\Psi = (\zeta, \theta, \phi)$	signal setting variables.
$\theta = [\theta_{jm}]$	start of greens, as proportions of cycle time where θ_{jm} is start of next green for group j at m.
$\phi = [\phi_{jm}]$	duration of greens, as proportions of cycle time where ϕ_{jm} is duration of green for group j at m.
ζ	reciprocal of common cycle time.
ζ_1 and ζ_2	minimum and maximum reciprocal of common cycle time.
λ_a	effective green split on links.
μ	maximum risk ratio.
f^k_a	regular traffic flow in scenario k over links.
h_p	path flow.
x^s_a	hazmat traffic flow in scenario s over links.

Performance Measures

Z_S, Z_N	Multi-objective Performance Index (MPI) for SBLPP and BLPP.
μ_S, μ_N	maximum risk ratio for SBLPP and BLPP
D_a	link delay rate.
σ_D, σ_e	monetary factors associated with delay rate and risk.
d_a	link signal delay.
c_a	link travel cost.
$\tau_{H(a)}$	departure travel cost at head node of link a.
$\tau_{T(a)}$	departure travel cost at tail node of link a.

2.2 A Link Traffic Delay Model

A traffic delay model at links downstream due to signal settings and stochastic flow in scenario k can be formulated as follows. Let

$$f_a = \frac{1}{|C|} \sum_{k=1}^{|C|} f_a^k \tag{1}$$

By definition,

$$D_a(\Psi, f_a) = f_a d_a(\Psi, f_a) = \frac{1}{|C|} \sum_{k=1}^{|C|} f_a^k d_a(\Psi, f_a) \tag{2}$$

The traffic signal delay at links downstream can be determined as

$$d_a(\Psi, f_a) = \frac{D_a(\Psi, f_a)}{f_a} \tag{3}$$

In a signal-controlled road network, the travel cost incurred by travelers at links downstream can be calculated as a sum of free flow travel time and signal delay at links downstream.

$$c_a = c_a^0 + d_a \tag{4}$$

Thus the travel departure cost at tail node of link can be determined as follows.

$$\tau_{T(a)} = Min\{\tau_{T(a)}, c_a + \tau_{H(a)}\} \tag{5}$$

2.3 A User Equilibrium Traffic Assignment with Random Demand

Consider random OD demand $\widetilde{Q}_i^k$ taking a value in $\left[Q_i, Q_i + \widehat{Q}_i\right]$ with scenario k. Let z_i^k denote a normalized scale deviation for random demand in scenario k with OD trip i such that

$$\widetilde{Q}_i^k = Q_i + z_i^k \widehat{Q}_i \tag{6}$$

with

$$\sum_{k=1}^{|C|} z_i^k \leq \Gamma^1; z_i^k \in [0, 1] \tag{7}$$

Let Δ^k denote random flow distribution for OD demand $\widetilde{Q}^k$ in scenario k such that

$$\Delta^k = \left\{ f^k : f^k = \Lambda h^k, \Upsilon h^k = \widetilde{Q}^k, h^k \geq 0 \right\} \tag{8}$$

According to Wardrop's first principle, a user equilibrium can be represented as a following complementarity problem and solved by a variational inequality.

$$f_a^k \left(c_a + \tau_{H(a)} - \tau_{T(a)} \right) = 0 \tag{9}$$

$$c_a + \tau_{H(a)} \geq \tau_{T(a)} \tag{10}$$

$$f_a^k \geq 0, k = 1, \ldots, |C| \tag{11}$$

In (9)–(11), it implies that for every used link a with random positive flow in scenario k, i.e. $f_a^k > 0$, the minimum travel departure cost, $\tau_{T(a)}$ at tail node equals those minimum travel departure cost, $\tau_{H(a)}$ at head node of link a, plus average travel cost at links downstream. A variational equality for (9)–(11) can be presented to find f^k such that

$$c(\Psi, f)\left(g - f^k\right) \geq 0, \forall g \in \Delta^k, k = 1, \ldots, |C| \tag{12}$$

2.4 A System Optimum Traffic Assignment with Random Risk

According to (Chiou, 2018), a data-driven weighted sum risk model is used in this section. For hazmat carriers in a signal-controlled road network, the corresponding risk exposure over links can be represented. Let ξ_a^s denote a normalized scale deviation for random risk e in scenario s on link a such that

$$\widetilde{e}_a^s = e_a + \xi_a^s \widehat{e}_a \tag{13}$$

with

$$\sum_{s=1}^{|A|} \xi_a^s \leq \Gamma^2; \xi_a^s \in [0, 1] \tag{14}$$

By definition, it implies

$$\widetilde{e}_a = \frac{1}{|A|} \sum_{s=1}^{|A|} \widetilde{e}_a^s \tag{15}$$

Similar to (8), let Δ^H denote hazmat traffic distribution such that

$$\Delta^H = \{x : x = \Lambda h,\ \Upsilon h = q,\ h \geq 0\} \tag{16}$$

Suppose regular traffic is a majority at links downstream such that signal delay (3) at links downstream mainly relies on total amount of regular traffic flow in (8). Following the recently proposed data-driven weighted sum risk model in (Chiou, 2018), a link linear delay model for hazmat carriers can be defined as

$$\widetilde{c}_a^s(\Psi, f_a) = c_a(\Psi, f_a)\widetilde{e}_a^s, \forall s = 1, \ldots, |A| \tag{17}$$

According to Wardrop's principle, a shortest path of minimum risk for hazmat carriers can be determined as to find $x_a^s, \forall s = 1, \ldots, |A|$ such that

$$\underset{x_a \in \Delta^H}{Min} \sum_{a=1}^{|A|} \int_0^{x_a^s} \widetilde{c}_a^s(\Psi, f_a)dy = \sum_{a=1}^{|A|} \widetilde{c}_a^s(\Psi, f_a)x_a^s \tag{18}$$

Similar to (12), a system optimum traffic assignment with random risk for hazmat carriers in (18) can be formulated as a following variational inequality to find $x^s \in \Delta^H$, $s = 1, \ldots, |A|$ such that

$$\widetilde{c}^s(\Psi, f)(y - x^s) \geq 0, \forall y \in \Delta^H, s = 1, \ldots, |A| \tag{19}$$

3 A Multi-Objective Sustainable trAffic Control (MOSAC)

For urban road networks, a Multi-Objective Sustainable trAffic Control (MOSAC) is proposed to simultaneously minimize signal delay incurred by travelers at links downstream and mitigate hazmat risk exposure over all links. A delay-minimizing model is firstly considered. The performance measure of sustainable road networks can be evaluated using a well-known traffic model TRANSYT (Vincent et al., 1980) where calculations of indicator of traffic conditions are mathematically derived in (Chiou, 2003). To mitigate risk associated with hazmat transportation and promote equitable risk distribution over the whole network, a risk-reducing model is proposed. In order to simultaneously minimize the total cost for all road users and risk exposure over links, a MOSAC is proposed to find Pareto-optimal solutions in a signal-controlled road network. According to (Miettinen, 1998), a Pareto-optimal solution for MOSAC can be obtained from a weighted optimization program when attached weights to different objectives are determined.

3.1 Signal Control Constraints

A feasible cycle time can take value in the following way:

$$\zeta_1 \leq \zeta \leq \zeta_2 \tag{20}$$

For each signal controlled junction m, the phase j green time for all signal groups at junction m can be expressed as

$$g_0 \zeta \leq \phi_{jm} \leq 1, \forall j, m \tag{21}$$

Road capacity at links downstream in a signal-controlled road network for which all links leading to junction m can be expressed as

$$f_a \leq \lambda_a s_a, \forall a \in A \tag{22}$$

For the sake of road safety, the risk of links can be constrained by the maximum number of hazmat carriers along links leading to junction. That is, for any signal-controlled link a, find a maximum risk ratio μ such that

$$\frac{1}{|A|}\sum_{s=1}^{|A|} \widetilde{e}_a^s x_a^s \leq \mu \frac{1}{|A|}\sum_{s=1}^{|A|} \lambda_a s_a^H \widetilde{e}_a^s, \forall a \in A \tag{23}$$

By definition from (15), it implies

$$\frac{1}{|A|}\sum_{s=1}^{|A|} \widetilde{e}_a^s x_a^s \leq \mu \lambda_a s_a^H \widetilde{e}_a, \forall a \in A \tag{24}$$

A clearance time for incompatible signal groups j and l at junction m can be also expressed as

$$\theta_{jm} + \phi_{jm} + \upsilon_{jlm}\zeta \leq \theta_{lm} + \Omega_m(j, l), \forall j, l, m \tag{25}$$

3.2 A Multi-Objective Performance Index (MPI)

According to (Chiou, 2018), a Multi-objective Performance Index (MPI) considering delay-minimizing and risk-reducing can be expressed in the following way:

$$Z_1(\Psi, f) = \sum_{a \in A} \sigma_D D_a(\Psi, f_a) = \sum_{a \in A} \frac{1}{|C|} \sum_{k=1}^{|C|} \sigma_D f_a^k d_a(\Psi, f_a) \tag{26}$$

and

$$Z_2(\Psi, f, x) = \sigma_e \sum_{a \in A} \frac{1}{|A|} \sum_{s=1}^{|A|} c_a(\Psi, f_a) x_a^s \widetilde{e}_a^s \tag{27}$$

3.3 A Stochastic Bi-Level Programming Problem (SBLPP)

A stochastic bi-level program for MPI in (26) and (27) with equilibrium constraints respectively for regular traffic and hazmat carriers in (12) and (19) can be formulated as follows.

$$\underset{\{\Psi,\ \mu,\ f,\ x\}}{\text{Min}} \quad Z_S(\Psi, f, x) = \begin{pmatrix} Z_1(\Psi,\ f) \\ Z_2(\Psi,\ f,\ x) \end{pmatrix} \tag{28}$$

subject to $\zeta_1 \le \zeta \le \zeta_2$

$$\begin{aligned}
& g_0 \zeta \le \phi_{jm} \le 1, \forall j, m \\
& f_a \le \lambda_a s_a, \forall a \in A \\
& \frac{1}{|A|} \sum_{s=1}^{|A|} \widetilde{e}_a^s x_a^s \le \mu \lambda_a s_a^H \widetilde{e}_a, \forall a \in A \\
& \theta_{jm} + \phi_{jm} + \upsilon_{jlm} \zeta \le \theta_{lm} + \Omega_m(j,\ l), \forall j, l, m \\
& c(\Psi, f)\left(g - f^k\right) \ge 0, \forall g \in \Delta^k, k = 1, \ldots, |C| \\
& \widetilde{c}^s(\Psi, f)(y - x^s) \ge 0, \forall y \in \Delta^H, s = 1, \ldots, |A|
\end{aligned}$$

The Pareto-optimal solution in (28) can be obtained by a weighting method from (Miettinen, 1998). Let α denote a weighting parameter for multi-objective of travel costs in (26), and of risk exposure in (27). The multiple objectives in (28) can be simplified as a following weighted optimization problem through a pre-determined parameter α, $0 \le \alpha \le 1$.

$$Z_S = (1-\alpha)Z_1 + \alpha Z_2 \tag{29}$$

Therefore, a stochastic bi-level program for (28) can be re-expressed.

$$\begin{aligned}\underset{\{\Psi,\ \mu,\ f,\ x\}}{Min}\quad Z_S(\Psi, f, x) = (1-\alpha)\left(\sum_{a\in A}\frac{1}{|C|}\sum_{k=1}^{|C|}\sigma_D f_a^k d_a(\Psi,\ f_a)\right)+\\ \alpha\left(\sigma_e\sum_{a\in A}\frac{1}{|A|}\sum_{s=1}^{|A|}c_a(\Psi,\ f_a)x_a^s\widetilde{e}_a^s\right)\end{aligned} \tag{30}$$

subject to $\zeta_1 \leq \zeta \leq \zeta_2$

$$\begin{aligned}&g_0\zeta \leq \phi_{jm} \leq 1, \forall j,m\\ &f_a \leq \lambda_a s_a, \forall a \in A\\ &\frac{1}{|A|}\sum_{s=1}^{|A|}\widetilde{e}_a^s x_a^s \leq \mu\lambda_a s_a^H \widetilde{e}_a, \forall a \in A\\ &\theta_{jm} + \phi_{jm} + \upsilon_{jlm}\zeta \leq \theta_{lm} + \Omega_m(j,\ l), \forall j,l,m\\ &c(\Psi, f)\left(g - f^k\right) \geq 0, \forall g \in \Delta^k, k = 1, \ldots, |C|\\ &\widetilde{c}^s(\Psi, f)(y - x^s) \geq 0, \forall y \in \Delta^H, s = 1, \ldots, |A|\end{aligned}$$

4 Sensitivity Analysis by Stochastic Derivatives

As notably mentioned in the literature (Luo et al., 1996; Scheel & Scholtes, 2000) that solutions found in the SBLPP (30) may fail at every feasible point to satisfy some constraint qualification in nonlinear programming problem. In order to effectively solve (30) with equilibrium constraints, a sensitivity analysis can be carried out by stochastic derivatives. The stochastic derivatives f'^k of user equilibrium (12) for regular traffic in scenario k, and the stochastic derivatives x'^s of system optimum flow (19) for hazmat carriers in scenario s with respect to the change Ψ' can be determined.

$$\left(\nabla_\Psi c(f,\ \Psi)\Psi' + \nabla_f c(f,\ \Psi)f'^k\right)\left(g - f'^k\right) \geq 0, \forall k = 1, \ldots, |C|, \forall g \in \Delta'^k \tag{31}$$

and

$$\left(\nabla_\Psi \widetilde{c}^s(f,\ \Psi)\Psi' + \nabla_{x^s}\widetilde{c}^s(f,\ \Psi)x'^s\right)\left(y - x'^s\right) \geq 0, \forall s = 1, \ldots, |A|, \forall y \in \Delta'^H \tag{32}$$

In (31) and (32), we introduce

$$\Delta'^k = \left\{ f'^k : f'^k = \Lambda h'^k,\ \Upsilon h'^k = 0 \right\}, \forall k = 1, \ldots, |C| \tag{33}$$

and

$$\Delta'^H = \{x' : x' = \Lambda h',\ \Upsilon h' = 0\} \tag{34}$$

By definition in (17), the first-order stochastic derivatives of risk in (32) with respect to hazmat traffic are zero. The variational inequality (32) can be re-expressed.

$$\left(\nabla_\Psi \tilde{c}^s(f,\ \Psi)\Psi'\right)\left(y - x'^s\right) \geq 0, \forall s = 1, \ldots, |A|, \forall y \in \Delta'^H \tag{35}$$

A single-level for the stochastic bi-level program (30) can be expressed to

$$\begin{aligned} \underset{\{\Psi,\ \mu\}}{Min} \quad Z_S(\Psi) = (1-\alpha)\left(\sum_{a\in A}\frac{1}{|C|}\sum_{k=1}^{|C|}\sigma_D f_a^k d_a(\Psi)\right) + \\ \alpha\left(\sigma_e \sum_{a\in A}\frac{1}{|A|}\sum_{s=1}^{|A|} c_a(\Psi) x_a^s \tilde{e}_a^s\right) \end{aligned} \tag{36}$$

subject to $\zeta_1 \leq \zeta \leq \zeta_2$

$$\begin{aligned} & g_0\zeta \leq \phi_{jm} \leq 1, \forall j,m \\ & f_a \leq \lambda_a s_a, \forall a \in A \\ & \frac{1}{|A|}\sum_{s=1}^{|A|} \tilde{e}_a^s x_a^s \leq \mu \lambda_a s_a^H \tilde{e}_a, \forall a \in A \\ & \theta_{jm} + \phi_{jm} + \upsilon_{jlm}\zeta \leq \theta_{lm} + \Omega_m(j,\ l), \forall j,l,m \end{aligned}$$

5 A Bundle Projection Method

Due to equilibrium constraints in (30), a stochastic bi-level program is usually not convex as mentioned in the literature (Colson et al., 2007; Dempe, 2002). In order to effectively solve the SBLPP via stochastic derivatives, a new Bundle Projection (BP) method is proposed. For the single-level SBLPP (36), a linear approximation $\overline{Z}_S^{(j)}$ for the objective function $Z_S(\Psi)$ at $\Psi^{(j)}$ from below can be constructed first. The stochastic derivatives for $Z_S(\Psi^{(j)})$ in (36) can be obtained.

$$\nabla Z_S^{(j)} = (1-\alpha)\left(\sum_{a\in A}\frac{1}{|C|}\sum_{k=1}^{|C|}\sigma_D\left({f'}_a^k d_a^{(j)} + f_a^k\left({d'}_a^{(j)} + \nabla_f d_a^{(j)} {f'}_a^k\right)\right)\right)+ \\ \alpha\left(\sigma_e\sum_{a\in A}\frac{1}{|A|}\sum_{s=1}^{|A|}\left({c'}_a^{(j)} + \nabla_f c_a {f'}_a^{(j)}\right)x_a^s\tilde{e}_a^s\right) \tag{37}$$

Let

$$\overline{Z}_S^{(j)} = \underset{1\le i\le j}{Max}\left\{\nabla Z_S^{(i)}\left(\Psi - \Psi^{(i)}\right) + \overline{Z}_S^{(i)}\right\} \tag{38}$$

and

$$\varepsilon_{i,j} = \overline{Z}_S^{(j)} - \left(\overline{Z}_S^{(i)} + \nabla Z_S^{(i)}\left(\Psi^{(j)} - \Psi^{(i)}\right)\right) \tag{39}$$

A linear approximation of $\overline{Z}_S^{(j)}$ in (38) can be expressed in terms of the following bundle gradients

$$\overline{Z}_S^{(j)} = \underset{1\le i\le j}{Max}\left\{\nabla Z_S^{(i)}\left(\Psi - \Psi^{(j)}\right) - \varepsilon_{i,j}\right\} + \overline{Z}_S^{(j)} \tag{40}$$

In (40), the minimization of $\overline{Z}_S^{(j)}$ corresponds to one step of cutting plane for the minimization of (36). Since that would give rise to very slow convergence from empirical studies, a bundle projection method can be presented to reliably solve the minimization (36) for $Z_S(\Psi)$ from iteration to iteration. With an approximate Hessian matrix $B^{(j)}$ for $\overline{Z}_S^{(j)}$, a quadratic model for (40) can be given as to

$$\underset{\{\Psi,\ \mu\}}{Min} \quad \nu + \frac{1}{2}\left(\Psi - \Psi^{(j)}\right)B^{(j)}\left(\Psi - \Psi^{(j)}\right) \tag{41}$$

subject to $\nabla Z_S^{(i)}\left(\Psi - \Psi^{(j)}\right) - \varepsilon_{i,j} + \overline{Z}_S^{(j)} \le \nu, 1\le i\le j.$

Let $\Pr_\Pi(\Psi)$ denote the projection of Ψ on a domain set Π for (41) such that

$$\|\Psi - \Pr_\Pi(\Psi)\| = \inf_{x\in\Pi}\|\Psi - x\| \tag{42}$$

A sequence of iterates $\{\Psi^{(j)}\}$ for (41) can be determined in accordance with

$$\Psi^{(j+1)} = \Pr_\Pi\left(\Psi^{(j)} + \left(\Psi - \Psi^{(j)}\right)\right), \quad j = 1,2,\ldots \tag{43}$$

6 Numerical Experiments and Results

To understand feasibility and computational efficiency of MOSAC for sustainable traffic road networks in practice, numerical experiments using a Sioux Falls city road network (Suwansirikul et al., 1987) and grid-sized example network, which are respectively shown in Figs. 1 and 2, are performed in this section. Numerical comparisons are also made with BLPP in deterministic cases for varying random travel demands and risk in a grid-sized signal-controlled road network. Initial data for Sioux Falls city road network with 6 signal-controlled junctions are given in Table 1. In the following computations, the minimum green time for each signal-controlled group is 7 seconds using typical values found in practice, and the clearance times are 5 seconds between incompatible signal groups. The maximum cycle time is set at 180 seconds. The stopping criterion used in BP is set when the relative difference of (36) between iterations is less than 0.15%. Implementations for carrying out computations are made on DELL T7610, Intel Xeon 2.5 GHz processor with 32 GB RAM under Windows 10 using C++ compiler.

6.1 Computational Results on Sioux Falls City Road Network

Computational results for BP are summarized in Tables 2 and 3. As it was observable in Table 2, BP successfully solves MOSAC in the monetary value of \$3318 with significant improvement rates of 36.8% and 57.83% by two random initial data of \$5250 and \$7870 respectively. In particular, the maximum risk ration can be greatly reduced to a value of 0.67 approximately from a fairly high value of 0.95 and 0.97 initially. Computational results for BP over iterations can be summarized in Table 3 for details. As can be seen in Table 3, for data set 1, BP takes 16 iterations to find the solutions with nearly vanishing derivatives of −0.06. For data set 2, BP takes 17 iterations to find the solutions of \$3319 with nearly vanishing derivatives of −0.08. Computational results for MPI (30) in delay-minimizing (26) and risk-reducing (27) can be summarized in Figs. 3 and 4 for two data sets with varying weights from 0 to 1. As shown in Fig. 3 and similarly in Fig. 4, by definition (29), the delay-minimizing objectives in (26) gradually increase as attached weights decrease. On the other hand, the risk-reducing objectives in (27), as expected, gradually decrease as attached weights increase.

In order to understand computational robustness of SBLPP (represented by Z_S) under random data input, numerical comparisons with BLPP in deterministic cases (represented by Z_N) are firstly made. Computational results are summarized in Figs. 5, 6, 7, 8, 9, 10, 11, and 12, respectively, for two data sets. As is observed in Fig. 5, SBLPP employing a stochastic program and utilizing a traffic delay model at links downstream can significantly outperform BLPP nearly by 2% as random demand budget grows up to 6. Similarly, as seen in Fig. 7, SBLPP can outperform BLPP nearly by 2.7% as random risk budget grows up to 8. Moreover, the maximum

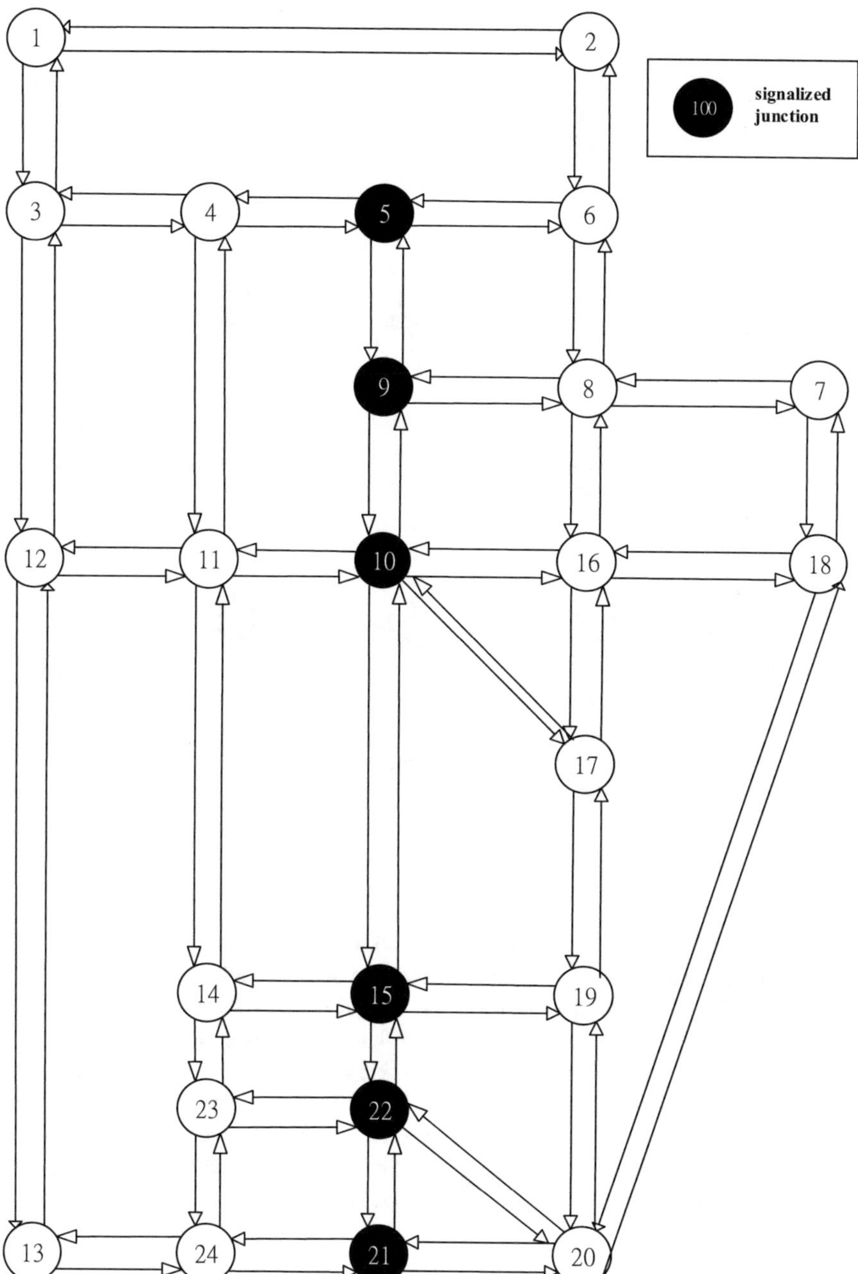

Fig. 1 Sioux Falls network with signalized junctions

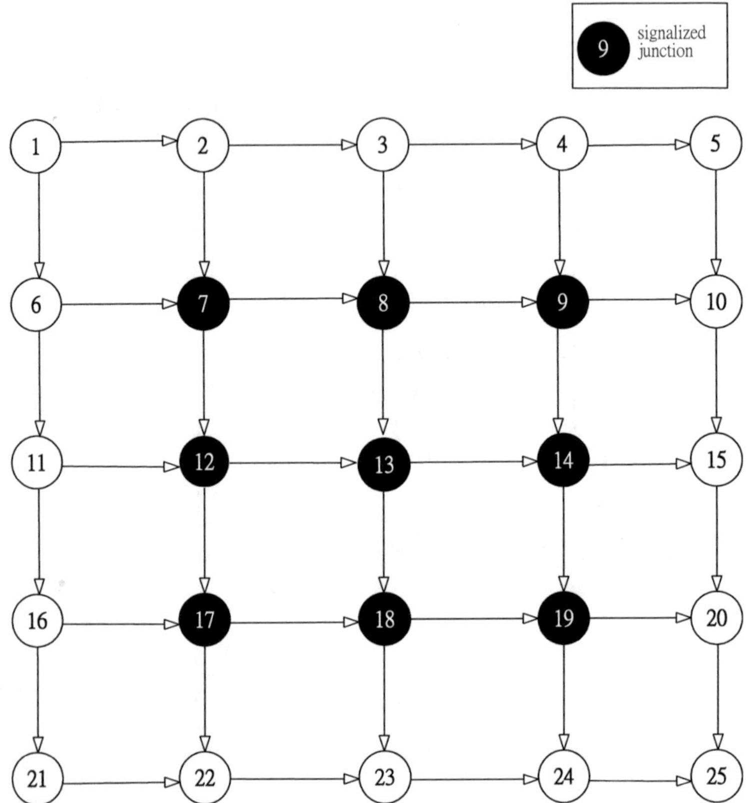

Fig. 2 25-node grid-sized network

risk ratio over links can be greatly reduced with SBLPP (represented by μ_S) under both random demand and risk, as can be observed in Figs. 6 and 8 when compared to those of BLPP (represented by μ_N). For the second data set, as similarly observed in Figs. 9 and 11, SBLPP can again achieve even higher advantage of nearly 2.35% over those did by BLPP as random demand budget grows up to 6. For random risk exposure, as it can be seen in Fig. 11, again SBLPP outperforms BLPP over 3.2% as random risk budget grows up to 8. The maximum risk ratio over links can be again greatly reduced by SBLPP as compared to those did by BLPP under both random demand and risk, as shown in Figs. 10 and 12, respectively.

6.2 *Computational Results on Grid-sized Example Network*

In order to investigate practical advantage of SBLPP, numerical experiments are also made using a grid-sized example network with BLPP under both random demand

Table 1 Initial data for Sioux Falls city network

Signal control	Data set 1	Data set 2
$1/\zeta$	70	90
ϕ_{15}/ζ	30	40
ϕ_{25}/ζ	30	40
ϕ_{19}/ζ	30	40
ϕ_{29}/ζ	30	40
$\phi_{1,\ 10}/\zeta$	30	40
$\phi_{2,\ 10}/\zeta$	30	40
$\phi_{1,\ 15}/\zeta$	30	40
$\phi_{2,\ 15}/\zeta$	30	40
$\phi_{1,\ 21}/\zeta$	30	40
$\phi_{2,\ 21}/\zeta$	30	40
$\phi_{1,\ 22}/\zeta$	30	40
$\phi_{2,\ 22}/\zeta$	30	40
MPI (in \$)	5250	7870
μ	0.95	0.97

where $1/\zeta$ denotes the common cycle time and φ_{jm}/ζ denotes the green durations in seconds for signal group j at junction m

Table 2 MOSAC signals for Sioux Falls city network

Signal control	Data set 1	Data set 2
$1/\zeta$	120	135
ϕ_{15}/ζ	57	62
ϕ_{25}/ζ	53	63
ϕ_{19}/ζ	58	60
ϕ_{29}/ζ	52	65
$\phi_{1,\ 10}/\zeta$	55	65
$\phi_{2,\ 10}/\zeta$	55	60
$\phi_{1,\ 15}/\zeta$	56	61
$\phi_{2,\ 15}/\zeta$	54	64
$\phi_{1,\ 21}/\zeta$	61	58
$\phi_{2,\ 21}/\zeta$	49	67
$\phi_{1,\ 22}/\zeta$	55	60
$\phi_{2,\ 22}/\zeta$	55	65
MPI (in \$)	3318	3319
μ	0.67	0.67

and risk. Numerical results are summarized in Figs. 13 and 14. As can be seen in Fig. 13 and in Fig. 14, as expected, SBLPP with round mark achieves relatively lower MPI values than did those BLPP as random budget varies from zero to 50. As can be seen in Fig. 13, for varying random demand from 30 to 50, SBLPP can outperform BLPP by 4.4% to 10.6% approximately. Moreover, for varying random risk from 30 to 50, as can be seen in Fig. 14, SBLPP can outperform BLPP even more by 5.1–11.2%.

Table 3 MOSAC signal control results over iterations

	Data set 1			Data set 2		
Iteration	$Z_S^{(i)}$ (in \$)	μ	$\nabla Z_S^{(i)}$	$Z_S^{(i)}$(in \$)	μ	$\nabla Z_S^{(i)}$
1	5250	0.95	−227.86	7870	0.97	−312.34
2	4980	0.94	−190.82	7560	0.95	−254.31
3	4680	0.93	−151.53	6840	0.93	−198.77
4	4490	0.89	−101.76	6770	0.89	−143.28
5	4400	0.87	−81.39	6510	0.87	−101.18
6	4190	0.85	−61.58	6190	0.85	−92.36
7	3850	0.84	−42.21	5760	0.84	−87.47
8	3780	0.82	−23.81	5330	0.82	−65.14
9	3660	0.78	−13.00	5120	0.78	−52.60
10	3520	0.76	−8.15	4830	0.76	−42.12
11	3420	0.73	−4.91	4670	0.73	−23.49
12	3370	0.70	−2.96	4330	0.71	−17.76
13	3360	0.70	−1.65	3920	0.70	−9.02
14	3330	0.68	−0.47	3790	0.69	−4.57
15	3320	0.67	−0.19	3560	0.68	−1.42
16	3318	0.67	−0.06	3321	0.68	−0.49
17				3319	0.67	−0.08

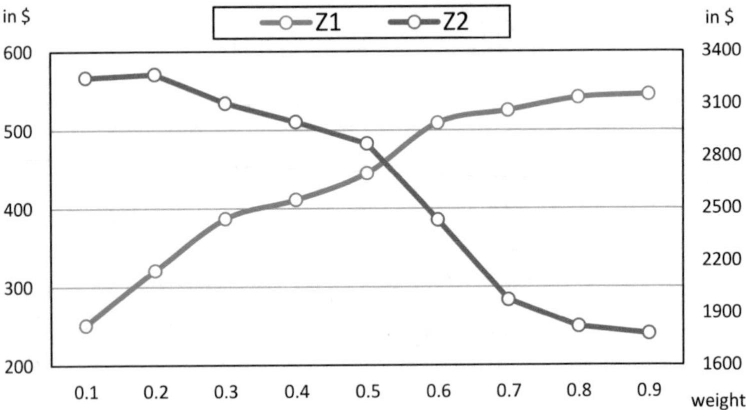

Fig. 3 MPI for two objectives with varying weight for data set 1

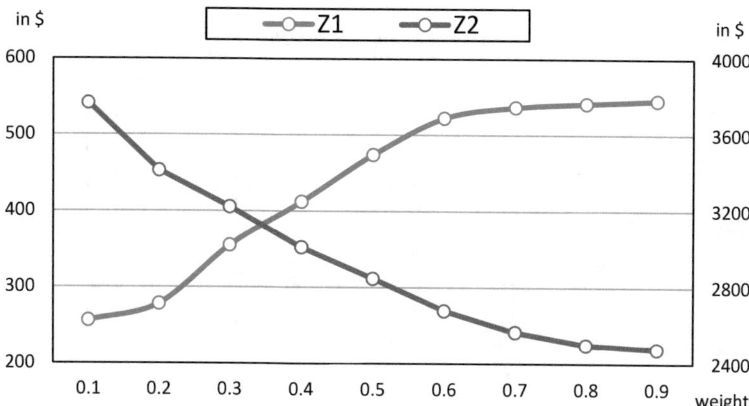

Fig. 4 MPI for two objectives with varying weight for data set 2

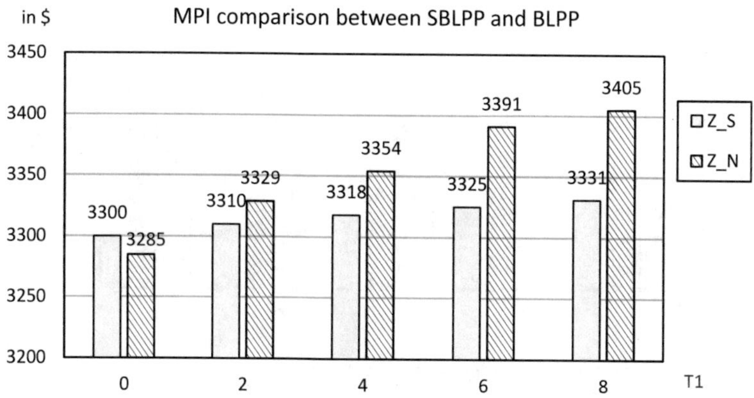

Fig. 5 MPI comparison with varying demand for data set 1

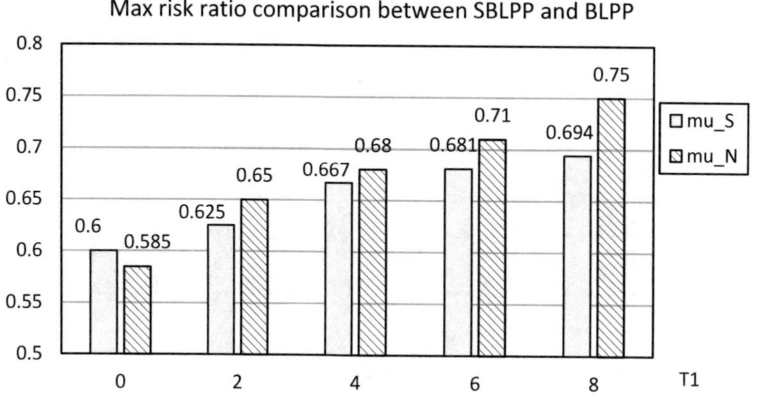

Fig. 6 Max risk comparison with varying demand for data set 1

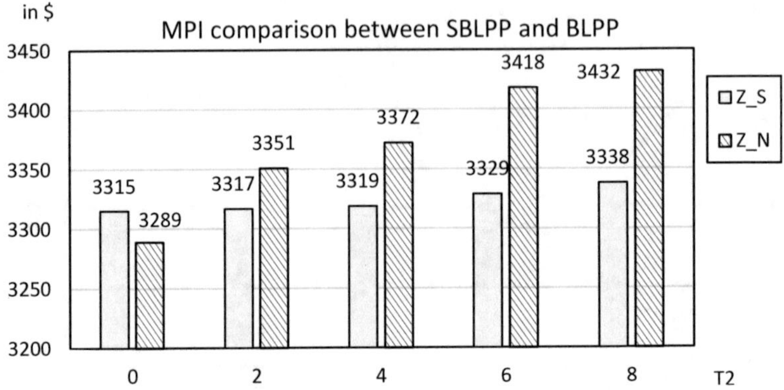

Fig. 7 MPI comparison with varying risk for data set 1

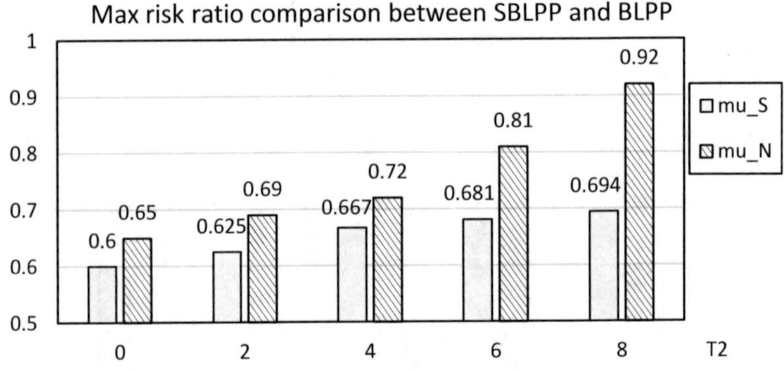

Fig. 8 Max risk comparison with varying risk for data set 1

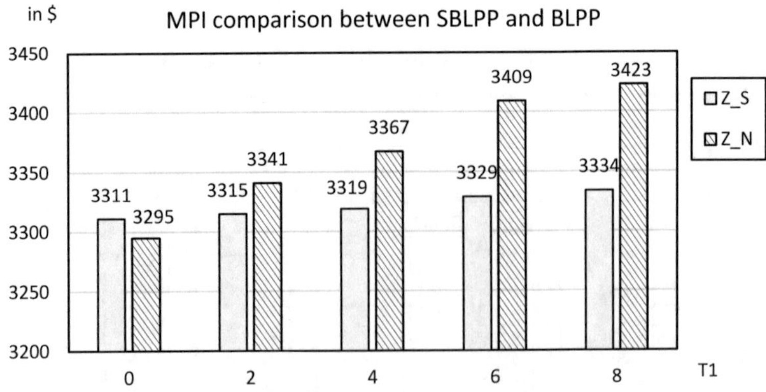

Fig. 9 MPI comparison with varying demand for data set 2

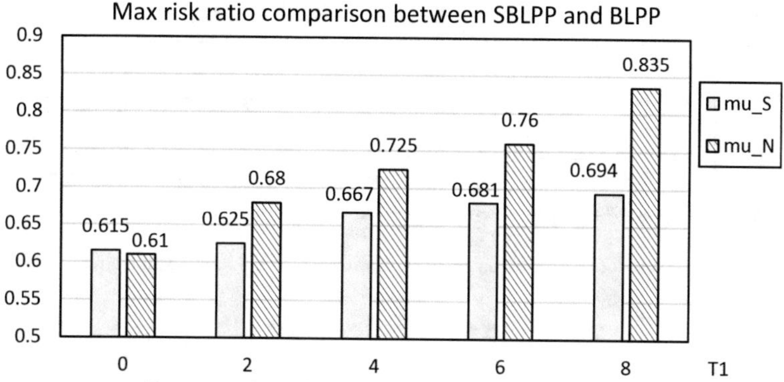

Fig. 10 Max risk comparison with varying demand for data set 2

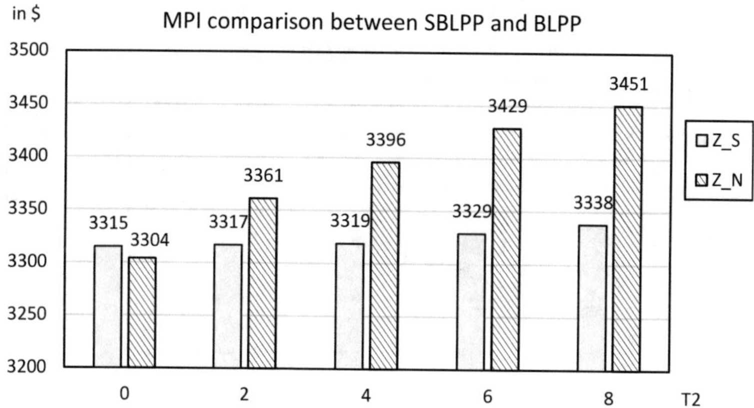

Fig. 11 MPI comparison with varying risk for data set 2

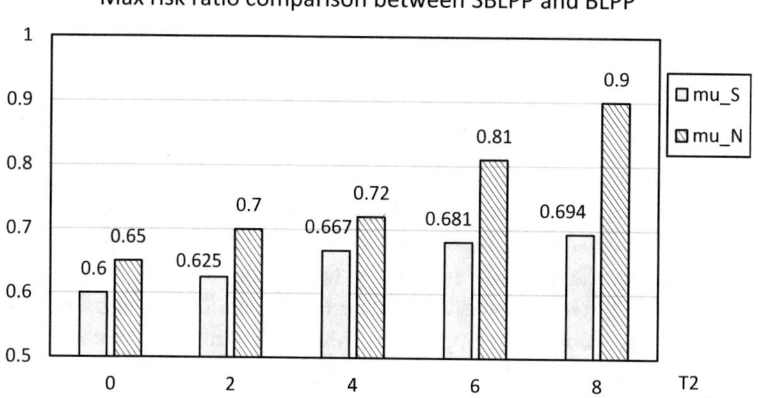

Fig. 12 Max risk comparison with varying risk for data set 2

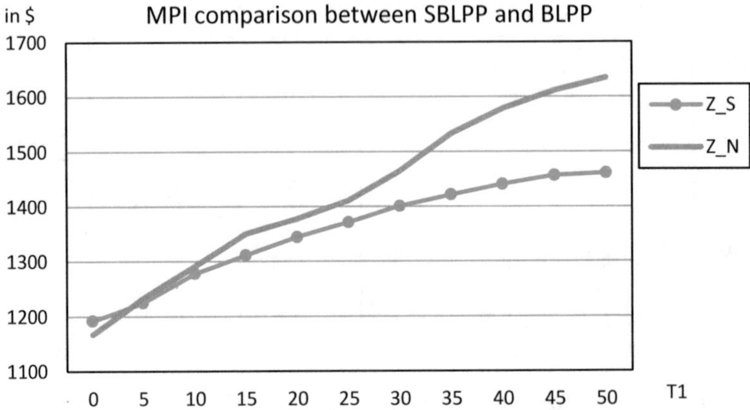

Fig. 13 MPI comparison with varying demand for grid-sized road network

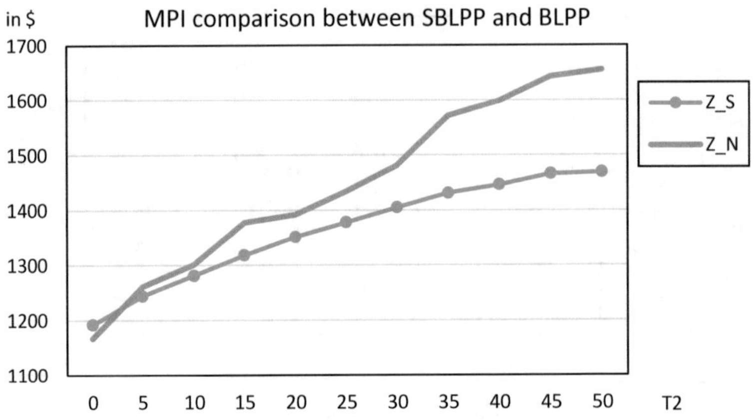

Fig. 14 MPI comparison with varying risk for grid-sized road network

7 Conclusions and Discussions

A MOSAC was proposed for a sustainable road traffic network with hazardous materials transportation to minimize total travel delay and mitigate random risk exposure to all road users. In a signal-controlled road network, an SBLPP was presented to minimize random traffic delay and risk. A multi-objective stochastic program was presented to minimize total delay at links downstream and to reduce maximum risk over links while accounting for travelers' route choices. A bundle projection approach was proposed to solve SBLPP and illustrated by a real-world city road network. Numerical comparisons were made with BLPP where the performance for MOSAC has been greatly improved. The maximum risk ratio over links has been appreciably reduced by SBLPP particularly in the presence of growing random risk over links.

For smart cities with time-varying traffic flow, this chapter utilized a static traffic delay model at links downstream to measure uncertain travel cost in signal-controlled road networks. Based on MOSAC, we have taken a first step in this direction toward reducing the maximum risk ratio over links with a multi-objective signal control for sustainable traffic road networks with random OD trip rates and risk exposure. Since traffic conditions in road networks can be negatively affected by time-varying congestion at downstream links, the traffic delay model used in this chapter, however, would be inadequately able to cope with time-varying random delay incurred both for regular traffic and hazmat carriers. Moreover, for smart cities with hazmat transportation under risk uncertainty, severe traffic congestion and risk of hazard cascades can give rise to catastrophic impact on time-varying performance of road networks such as devastating capacity loss. In order to investigate the sustainability and reliability of dynamic road network systems, the proposed MOSAC can be served as a lower bound for system optimum when comparing with user equilibrium in (Chiou, 2018). In order to leverage the overall robustness of traffic signal control for time-varying travel demand, a stochastic time-varying sustainable signal design can shed some light on evaluating the effectiveness of decisions based on what is achieved in this chapter. In the near future, a dynamic sustainable traffic signal control to quantify redundant capacity in smart cities will be investigated continuously based on what was performed for MOSAC. Further numerical comparison using large-sized real-data example road networks with data-driven BLPP (Chiou, 2018) will be taken into account, and we will discuss these points of interest in subsequent papers.

Acknowledgments The author is grateful to the Editors in Chief for their kindness and anonymous reviewers' constructive comments. Thanks also go to MOST at Taiwan for financial support via grant 110-2221-E-259-017-MY2.

References

Aashtiani, H. Z., & Magnanti, T. L. (1981). Equilibria on a congested transportation network. *SIAM Journal on Matrix Analysis and Applications, 2*, 213–226.

Bianco, L., Caramia, M., & Giordani, S. (2009). A bilevel flow model for hazmat transportation network design. *Transportation Research: Part C, 17*, 175–196.

Bianco, L., Caramia, M., Giordani, S., & Piccialli, V. (2013). Operations research models for global route planning in hazardous material transportation. In R. Batta & C. Kwon (Eds.), *Handbook of OR/MS models in hazardous materials transportation, international series in operations research & management science* (Vol. 193, pp. 49–101). Springer Science + Business Media.

Chiou, S.-W. (2003). TRANSYT derivatives for area traffic control optimisation with network equilibrium flows. *Transportation Research: Part B, 37*, 263–390.

Chiou, S.-W. (2016). A bi-objective bi-level signal control policy for transport of hazardous materials in urban road networks. *Transportation Research Part D, 42*, 16–44.

Chiou, S.-W. (2017). A risk-averse signal setting policy for regulating hazardous material transportation under uncertain travel demand. *Transportation Research Part D, 50*, 446–472.
Chiou, S.-W. (2018). A data-driven bi-level program for knowledge-based signal control system under uncertainty. *Knowledge-Based Systems, 160*, 210–227.
Chiou, S.-W. (2019). An efficient bundle-like algorithm for data-driven multi-objective bi-level signal design for traffic networks with hazardous material transportation. In F. P. García Márquez & B. Lev (Eds.), *Data science and digital business* (pp. 191–220). Springer Nature Switzerland AG.
Chiou, S.-W. (2020). A resilience-based signal control for a time-dependent road network with hazmat transportation. *Reliability Engineering and System Safety, 193*, 106570.
Colson, B., Marcotte, P., & Savard, G. (2007). An overview of bilevel optimization. *Annals of Operations Research, 153*, 235–256.
Dempe, S. (2002). *Foundations of bilevel programming, nonconvex optimization and its applications series*. Kluwer Academic.
Huang, W., Zhou, B., Yu, Y., & Yin, D. (2021). Vulnerability analysis of road network for dangerous goods transportation considering intentional attack: Based on cellular automata. *European Journal of Operational Research, 214*, 107779.
Kara, B. Y., & Verter, V. (2004). Designing a road network for hazardous materials transportation. *Transportation Science, 38*, 188–196.
Lu, J., Han, J., Hu, Y., & Zhang, G. (2016). Multilevel decision-making: A survey. *Information Sciences, 346–347*, 463–487.
Luo, Z. Q., Pang, J. S., & Ralph, D. (1996). *Mathematical programs with equilibrium constraints*. Cambridge University Press.
Miettinen, K. (1998). *Nonlinear multiobjective optimization*. Springer.
Mohri, S., Mohammadi, M., Gendreau, M., Pirayesh, A., Ghasemaghaei, A., & Salehi, V. (2022). Hazardous material transportation problems: A comprehensive overview of models and solution approaches. *European Journal of Operational Research, 214*, 1. (in press). https://doi.org/10.1016/j.ejor.2021.11.045
Ren, A., Wang, Y., & Xue, X. (2016). Interactive programming approach for solving the fully fuzzy bilevel linear programming problem. *Knowledge-Based Systems, 99*, 103–111.
Scheel, H., & Scholtes, S. (2000). Mathematical programs with complementarity constraints: Stationarity, optimality, and sensitivity. *Mathematics of Operations Research, 25*, 1–22.
Sun, L., Karwan, M. H., & Kwon, C. (2016). Robust hazmat network design problems considering risk uncertainty. *Transportation Science, 50*, 1188–1203.
Suwansirikul, C., Friesz, T. L., & Tobin, R. L. (1987). Equilibrium decomposed optimization: A heuristic for continuous equilibrium network design problem. *Transportation Science, 21*, 254–263.
Vincent, R. A., Mitchell, A. I., & Robertson, D. I. (1980). *User guide to TRANSYT, TRRL report, LR888*. Transport and Road Research Laboratory.
Zhang, G., & Lu, J. (2007). Model and approach of fuzzy bilevel decision making for logistics planning problem. *Journal of Enterprise Information Management, 20*, 178–197.
Zhang, G., Lu, J., & Dillon, T. (2007). Decentralized multi-objective bilevel decision making with fuzzy demands. *Knowledge-Based Systems, 20*, 495–507.
Zhang, G., Lu, J., & Gao, Y. (2015). *Multi-level decision making: Models, methods and applications*. Springer.

Multi-Criteria Decision-Making for Sustainable Transport: A Case Study on Traffic Flow Prediction Using Spatial–Temporal Traffic Sequence

Jayanthi Ganapathy

Abstract Traffic flow on highways exhibits a dynamic phenomenon in different operational settings. The concept of sustainability in transportation engineering is elucidated using multi-criteria decision-making analysis (MCDA), a discipline of Operation Research (OR) which is in a wide range of applications and practices in real-time traffic engineering. The principle criteria considered in the study of short-term prediction of traffic flow rate are spatial and temporal information. The spatial–temporal components of physical traffic flow are the parametric measures. Exploring the intrinsic relationship between these measures helps in realizing the dynamics of traffic flow rate. Hence, the objective of this chapter is to elucidate the significance of MCDA considering the spatial and temporal measures of traffic flow rate. Also, this chapter presents a case study on the formulation of algorithm for the prediction of traffic flow rate on highways. Experimental results are reported with an estimation of time complexity of algorithms.

Keywords Transport Data Science · Transportation · Traffic · Temporal · Spatial · Multi-criteria decision analysis · Operation Research

1 Introduction

Transport and mobility is a dynamic activity in real life. Commuters prefer to have multiple choices in the mode of transport in their journey based on vehicular movement and traffic flow. Hence, decision-making is highly demanded in many situations where traffic is unprecedented. Moreover, parametric criterion would better quantify the factors affecting public mobility. These parameters would eventually evaluate indicators for sustainable mobility by various modes of transport. Therefore, the problem of multi-criteria decision analysis would devise better

J. Ganapathy (✉)
Faculty of Engineering and Technology, Sri Ramachandra Institute of Higher Education and Research, Porur, Chennai, Tamil Nadu, India
e-mail: jayanthig@sret.edu.in

F. P. García Márquez, B. Lev (eds.), *Sustainability*, International Series in Operations Research & Management Science 333,
https://doi.org/10.1007/978-3-031-16620-4_9

solutions when commuters have multiple choices in the mode of transport considering the daily traffic profile. Temporal characteristics of traffic flow would enable decision-makers to identify the span of a traffic condition. Spatial characteristics of traffic flow would enable decision makers to identify the road segments too. Ultimately, spatial–temporal traffic behavior is a key factor indicating public mobility in various modes of transportation. The temporal and spatial indicators of traffic flow would devise better decisions thereby enabling sustainable transport and mobility. In preview, researchers across the globe have well-acknowledged road network as spatial–temporal network. The research findings have reported the significance of spatial and temporal measures of traffic flow rate in different operational settings. This has highly motivated the author to present the research prospects on operation research—multi-criteria decision analysis in this problem study. In addition, to manage decisions in uncertain conditions in real-time fuzzy set theory concepts are well appreciated in supply chain management (Seuring, 2013; Barbosa-Povoa et al., 2018). Initially, this chapter would elaborate spatial–temporal traffic information analysis as multi-criteria decision analysis in transportation. Further, algorithms and problem-solving methods involved in spatial–temporal traffic flow analysis will be presented with measures and metrics on indicators for sustainable transport and mobility. Further, the space complexity of algorithm in different operational settings is left to work in the future.

Urban transport system is a time-varying network. The variation in travel time and delay in travel faced by commuters is the adverse effect of traffic congestion. Traffic information in preceding time instances contributes to analyzing traffic in succeeding instances and spatial information of traffic is required for traffic flow assessment on highways. Having proposed travel time-based traffic information sequence on traffic sequence mining framework in Jayanthi and Jothilakshmi (2019, 2021) and Jayanthi and García Márquez (2021a, b) and further evaluation of fieldwork in the prediction of traffic volume, spatial–temporal characteristics of physical traffic flow can be very well observed when traffic sequence mining is employed in upstream and downstream of road segments.

This chapter is organized as follows: Section 2 describes path computation. Section 3 presents MCDA algorithm for the prediction of traffic flow on highways. Section 4 presents data collection and experimental results. Finally, Sect. 5 concludes the chapter with summary and directions for future research.

2 Reliable Path Computation Based on Traffic Flow

Until today over 40 years, transportation research is active and dynamic in all nations across the world. The variation in travel time and delay in travel faced by commuters is the adverse effect of traffic congestion. Therefore, it is essential to manage congestion as it cannot be avoided but can be mitigated. The common problem that was addressed by various researchers in transportation is finding the shortest path between desired source and destination (Dijkstra, 1959). Although this problem has been solved by many researches from a theoretical view point, accurate and realistic shortest path that could overcome travel delay in a time-varying road network is yet to be focused (Nejad et al., 2017). Travel to a location is represented by path traversed between source and destination with the cost of travel represented in terms of speed, distance, time delay, etc. Thus, path computation on time-varying network is a function of space and time. The following section explains the dynamics of traffic flow in spatial–temporal perspective and presents a review on models and methods used in spatial–temporal traffic forecasting.

2.1 Dynamics of Traffic Flow: Spatial–Temporal Perspective

In real road networks, temporal variation in traffic volume has a strong influence on travel delay. Variation in travel time is the effect of congestion in time-varying network. The computation of shortest path considering travel time information like earliest arrival time and latest departure time (Bauer & Delling, 2009) alone is insufficient. Hence, temporal instances of traffic information are highly required in analyzing traffic at preceding instances. Edge weight augmented with temporal information would help in realizing real traffic conditions more effectively.

Road network is a spatial graph $G = (S, R)$ where S is set of vertices representing arterial junctions and R is set of edges representing road segment connecting the junctions. Identification of critical node (arterial junction) in the spatial network based on temporal variation of traffic with analysis on congestion patterns is required in real-time scenarios (Nejad et al., 2017). Traffic congestion incurred due to time-varying travel time needs much consideration in solving path computation problem. Traffic information at different time instances carry useful information about road network. Moreover, traffic information has to be analyzed at different time instances rather than arrival time and departure time alone.

Computation on static network does not yield accurate results as traffic congestion varies with time. Therefore, time-dependent travel time computation over such network is demanded (Brunel et al., 2010). When road network is congested, static information like distance, pre-computed travel time alone are not enough. Rather, traffic conditions in previous time instances are required to assess congestion. In this view, logical analysis of temporal traffic information is required for congestion management. Moreover, logical analysis of temporal traffic information is useful in analyzing traffic in previous instances or intervals. Logical analysis of temporal information is demanded as traffic information in preceding time instances is more significant in estimating the traffic in succeeding instances (Al-Deek et al., 1999). Sequencing spatial and temporal traffic information (Ermagun & Levinson, 2018) is useful in analyzing traffic in previous instances or intervals (Wangyang et al., 2019).

Generalized pre-processing and speed-up technique for Dijkstra's algorithm on dense graph was reported by Bauer and Delling (2009) in which speed-up performance is achieved by introducing goal-directed ALT to hierarchical search. Real road network is time-varying which means traffic information keeps changing. In this view, time-varying network has two ways of defining SPP: (1) Fastest path problem and (2) Minimum cost path problem, where the fastest path is based on travel time while minimum cost path is based on distance, arc length, etc., In both the cases, traffic information or edge cost is a function of time. Although static shortest path algorithms paved solution, they stand far apart when applied to dynamic or time-dependent scenarios in real road networks. Static SPP has proven solution but they cannot be applied to real road network unless it has efficient pre-processing on dynamic update of edge cost. In view of computation overhead and storage efficiency in time-dependent scenarios, speed-up technique was proposed for time-dependent spatial graphs in which pre-processing is performed offline and the fastest path computation is done online (Demiryurek et al., 2011). The offline process involves a non-overlapping partition of graph while online process utilizes heuristic function. This technique has significantly reduced both storage and computational complexity. Alternate path algorithms were formulated to overcome the computational overhead in updating the edge cost between all pairs of vertices. In most of the recent works reviewed here, efforts were made to extend the speed-up technique by pre-processing edge cost based on travel time is yet to devise solution for mitigating traffic congestion, especially for metropolitan transport systems (Goldberg & Harrelson, 2003). In this aspect, the use of temporal and spatial information of traffic has its significance in mitigating transport congestion which was not considered in early achievements as in Bauer et al. (2010). Unlike several speed-up techniques proposed for pre-processing the edge cost in time-dependent shortest path algorithms, this work focuses on the use of temporal and spatial information in effect to overcome travel delay and re-routing the traffic thereby resolving traffic congestion under dynamic scenarios. The logical analysis of temporal and spatial relations on time-varying network was not being considered in the early works of time-dependent shortest path.

Recent advancements in ITS have revolutionized the transport industry nationwide across the globe to serve the public in better way. Traffic flow congestion estimation and management on highways is always in demand worldwide across all nations for safe and hassle-free travel. Traffic congestion problem is inherent in travel time decisions and solving such problem is essential for travel guidance, especially during peak hours of a journey. The two broad spectra of research in transportation that is everlasting in infrastructure planning are: (i) Shortest path computation problem and (ii) Time series traffic forecasting. In spite of technology-driven traffic management using IoT, monitoring and control of vehicular traffic remains a serious issue in real time; thus a fully automated traffic management system is not feasible. Therefore, it is essential to manage traffic flow congestion systematically as it cannot be avoided but can be mitigated. In this view, a speed-up technique is necessary to bridge the gap between traffic flow estimation and path routing of vehicular traffic between origin and destination (OD). The existing speedup techniques that are used in path computation on a dynamic transport network are shown in Table 1.

Table 1 Speed-up technique used in path computations

Author	Algorithm
Hart et al. (1968)	A*, a goal-directed informed search algorithm
Goldberg and Harrelson (2003)	A* + landmark + triangle inequality (ALT), augmented goal-directed search with landmarks satisfying triangle inequality
Bauer and Delling (2009)	Short cut + arc flags (SHARC), augmented hierarchies with edge flag replace
Bauer et al. (2010)	Goal-directed ALT, hierarchical search on ALT
Geisberger et al. (2012)	Hierarchical bi-directional search, augmenting least important vertex in highway hierarchy

2.2 *Spatial–Temporal Traffic Forecasting: A Review on Models and Methods*

In recent years, advancements in data-driven approaches toward the prediction of traffic flow have attracted the attention of researchers worldwide. Short-term traffic forecasting using data-driven approach involves computation-rich problem-solving techniques considering volume, dimension, and characteristic features of the traffic such as vehicle speed, volume, density, and flow (Shi et al., 2019; Wang & Ye, 2015; Wangyang et al., 2019). Park and Rilett (1999) reported that neural network architectures with non-linear activation functions were not found to outperform statistical models in time series traffic forecasting. Recent achievements in transportation research have emphasized deep learning architectures to traffic forecast studies. These architectures extend the feed-forward neural network with more than single hidden layer, thereby increasing parallelism in computation (Xu et al., 2015; Lv et al., 2015) (Table 2).

The efficiency of long short-term memory (LSTM) in learning long-term dependence has replaced the inherent complexity in data computation due to overfit and underfit in data explosion with gradient descent activation in Recurrent Neural Network (RNN) (Huang et al., 2014). Researches in the past have achieved forecast results with variant of deep learning approaches such as unsupervised multitask regression approach, and auto-encoder model with stacked layers, Ma et al. (2015) but they fail to capture spatial and temporal correlation. Origin destination matrix integrated with LSTM was proposed to achieve spatial and temporal correlation of adjacent links (Zhao et al., 2017). Auto-encoder-based LSTM (AE-LSTM) model was formalized considering upstream and downstream traffic flow at different temporal scales.

These two specific works have reported the significance of spatial and temporal information in traffic forecast studies. Dynamism in traffic condition is best anticipated when both spatial and temporal traffic information are considered (Ermagun & Levinson, 2018; Habtemichael & Cetin, 2016). Image-based pattern analysis was reported using evolutionary algorithmic technique (Kartikay & Niladri, 2019). Estimation of traffic congestion at a target location using any one of the traffic variables such as volume, speed, and travel time alone is insufficient. In this view,

Table 2 Models and methods in spatio-temporal traffic forecasting

Author	Model and method	Applications
Min and Wynter (2011), Kamarianakis and Prastacos (2003)	STARIMA	Multivariate modeling for evaluation of spatial dependence using spatial correlation measure
	VARMA	Modeling the restricted and unrestricted traffic flow on upstream
Chandra and Al-Deek (2008)	VAR	Cross-correlation measure of traffic flow on upstream and downstream
Cheng et al. (2012)	Autocorrelation	Autocorrelation measure of spatial links at different temporal lags
Wu et al. (2014), Cai et al. (2016)	k-nn	State vector representation of upstream and downstream traffic flow
Zhang and Zhang (2016)	Recurrent NN	Travel time models using spatial–temporal inputs from freeways
Ma et al. (2017)	Convolution NN	Traffic speed prediction based on learning spatial correlation on images
Wangyang et al. (2019)	Auto-encoder LSTM	Auto-encoded spatial–temporal dependence for prediction of traffic in sequence
Jayanthi (2021)	Sequential pattern mining	IoT-based traffic re-routing based on spatial–temporal traffic information.
Jayanthi and García Márquez (2021a)	TT-PrefixSpan	Re-routing vehicles by augmenting traffic network based on spatial–temporal traffic dependency

traffic information at each temporal instance is required to capture inherent dynamism which is obvious with today's infrastructure. Further, transportation planning using such dynamic traffic information enables reliable congestion management, which is part of ITS.

2.3 *Research Challenges and Problem Statement*

Temporal variation in traffic flow essentially captures the recurring and non-recurring congestion in the dynamics of physical traffic flow. However, temporal traffic information alone is not sufficient in travel decisions when there is a need for reliable path on a spatially connected road network. In route guidance, a path in a travel is said to be reliable when flow rate in successive time instances is made known. This has motivated researchers to focus on the influence of spatial characteristics in dynamics of traffic flow as flow rate from neighboring links contributes significant amount of traffic at current location. From this perspective, a fully automated traffic management system is not feasible. Therefore, it is essential to manage traffic flow congestion systematically as it cannot be avoided but can be mitigated. In this view, a speedup technique is necessary to bridge the gap between traffic flow estimation and path routing of vehicular traffic between origin and destination (OD). These challenges have motivated me in devising the research problem statement as discussed below:

Problem 1: Given travel time based traffic information sequence of uplink and downlink connecting the origin to destination, estimate the traffic flow at spatial intersection on highway.

In this work, both spatial and temporal information are essentially considered as they contribute significant information about traffic. In real road network, temporal variation in traffic volume has a strong influence on travel delay. Variation in travel time is the effect of congestion in time-varying network. Traffic information at different time instances carry useful information about road network. Moreover, traffic information has to be analyzed at different time instances rather than arrival time and departure time alone. Edge weight augmented with temporal information would help in realizing real traffic conditions more effectively.

Identification of critical node (arterial junction) in the spatial network based on temporal variation of traffic with analysis on congestion patterns is required in real-time scenarios. Hence, temporal instances of traffic information are highly required in analyzing traffic at preceding instances. Traffic congestion incurred due to time-varying travel time needs much consideration in solving path computation problems. Sequencing spatial and temporal traffic information in preceding time instances helps in estimating traffic flow in sequence in successive time instances by formalizing sequence convolution-based auto-encoder Long Short-term Memory (SCAE-LSTM) network. Thus, spatial–temporal traffic characteristics extracted using the convolution of upstream and downstream traffic information sequence is fed externally to auto-encoder LSTM. LSTM predicts the traffic flow at target location in successive time instances.

Problem 2: Given origin and destination pair in a travel, augment the spatially connected network to reach destination considering the spatial–temporal dependence of traffic flow.

Given source and destination with start time of journey, path connecting the arterial intersections is checked for congestion in which speed of vehicle captured by sensors in upstream and downstream is sequenced in each time instance using sequence convolution. Auto-encoder is formalized to extract the characteristics of vehicle speed in adjacent links. Auto-encoder is integrated with LSTM to estimate traffic flow rate at an arterial intersection. Path to destination is re-connected considering both distance and spatial dependence of adjacent links at each time instance until the destination is reached.

In a spatially connected network, it is essential to analyze the spatial dependency of road segment with respect to upstream and downstream traffic flow thereby, road segments (a path) that are connecting the source and destination in preceding time instances are analyzed in successive time instances for re-establishing connectivity. In this way, when traffic congestion is detected the path is reconnected based on spatial–temporal traffic information. In the interest of reducing travel delay, Spatial–TemporAl Re-connect (STAR) algorithm is proposed in which traffic flow is estimated by sequencing spatial and temporal traffic information. The traffic information in upstream and downstream road segments represents spatial traffic information, while traffic flow at each time instance forms temporal information. The interest of this work is to formulate speed-up technique by estimating traffic flow at each of arterial junction between OD pair and re-establish path connectivity of road network based on spatial and temporal traffic sequences.

In this work, traffic flow on highways is estimated for different origin-destination (OD) pairs based on spatial–temporal traffic sequences. Hence, Spatial–TemporAl Reconnect (STAR) algorithm is proposed. The performance of STAR is investigated by conducting extensive experimentation on real traffic network of Chennai Metropolitan City. The computational complexity of the algorithm is empirically analyzed. The proposed STAR algorithm is found to estimate traffic flow during peak hour traffic with reduced complexity in computation compared to other baseline methods in short-term traffic flow predictions like LSTM, ConvLSTM, and GRNN. In view of reducing travel delay, traffic flow at the target link has to be predicted, considering spatial and temporal characteristics of the physical traffic flow. Thus, spatially connected road networks can be augmented considering travel time based on frequent traffic sequence pattern generation using the TT-PrefixSpan algorithm on both upstream and downstream of road segments connecting the arterial junctions (Jayanthi & Jothilakshmi, 2019, 2021; Jayanthi & García Márquez, 2021a).

Given an origin and destination pair, procedures are formulated in this chapter for routing vehicles through specific arterial junctions that are identified based on spatial–temporal traffic characteristics to reach the destination. Hence, this chapter presents a review of the augmentation of the road network based on traffic flow from a spatial and temporal perspective. The problem statements identified from research challenges are presented. The formulation of methodology and algorithm are presented. The performance of the proposed algorithms is reported with evaluation measures.

3 Prediction of Traffic Flow Based on Spatial–Temporal Dependence

The complexity in computation with time series representation of traffic flow rate is reduced when time series of traffic flow rate is converted to symbolic sequences (Lin et al., 2007). Convolution of upstream and downstream traffic flow rate derives characteristic sequence containing spatial traffic information. The dependence between the neighboring links is measured using correlation metric in which spatial traffic information sequence of one neighboring link is shifted with respect to the neighbor link which does not change over time. The potential of neural network in modeling the physical traffic flow is widely acknowledged. The ability of neural network (NN) to process heterogeneous input in different data dimension has gained popularity in spatial–temporal traffic forecasting (Karlaftis & Vlahogianni, 2011).

The problem with training and validation in NN is considerably reduced when auto-encoder were used. The layers of perceptron are increased to capture the reality of dynamic traffic behavior. The following section explains the formulation of spatial–temporal traffic sequence and prediction of traffic flow on highways.

3.1 Formulation of Spatial–Temporal Traffic Sequence

Travel time-based traffic information sequences of upstream and downstream segments are used to extract characteristics traffic sequence using sequence convolution method. Consider the sequences $\langle \mathcal{SU}_i \rangle$ and $\langle \mathcal{SD}_i \rangle$ where $i = 1, 2, 3, \ldots, n$. The convolution of upstream and downstream sequence is given by Eq. (1),

$$\langle \mathscr{S}\mathscr{U}_i \rangle_{i=1}^{n} \circ \langle \mathscr{S}\mathscr{D}_i \rangle_{i=1}^{n} = \left\{ \sum_{k=1}^{i} SU_k SD_{i-k} \right\}. \tag{1}$$

The upstream and downstream flow rate in sequence is obtained using sequence convolution. These sequences are fed as input to the auto-encoder. The training dataset for the estimation of vehicle speed at target location is represented as $X[t] = [x_1, x_2, x_3, \ldots, x_m]$, where $x_1, x_2, x_3, \ldots, x_m$ denotes flow rate at time $t_1, t_2, t_3, \ldots, t_{i-1}, t_i, t_{i+1}, \ldots, t_m$. The encoding and decoding process involved in auto-encoder is given by Eqs. (2) and (3), respectively.

$$E_i = f(w_e * S_i + b_e), \tag{2}$$

$$D_i = g(w_d * E_i + b_d), \tag{3}$$

where $w_e, b_e, w_d,$ *and* b_d denotes weight, bias of encoder an decoder respectively.

$$f(*) = g(*) = \left(\frac{e^x}{e^x + 1} \right). \tag{4}$$

The activation function $f(*)$ and $g(*)$is given in Eq. (4). The characteristics of vehicle speed in upstream and downstream road segment are extracted using encode (E_i) and decode (D_i) as given in Eqs. (5) and (6), respectively. The speed of vehicle in sequence at the target location is represented as $S_t = [s_{t1}, s_{t2}, s_{t3}, \ldots\ldots, s_{tm}]$. The upstream and downstream traffic sequence is represented as $S_u = [s_{u1}, s_{u2}, s_{u3}, \ldots\ldots, s_{um}]$ and $S_d = [s_{d1}, s_{d2}, s_{d3}, \ldots\ldots, s_{dm}]$ respectively, where $S_{ui}, S_{di} \in R$, the upstream and downstream vehicle speed of a road segment R is captured using sequence convolution.

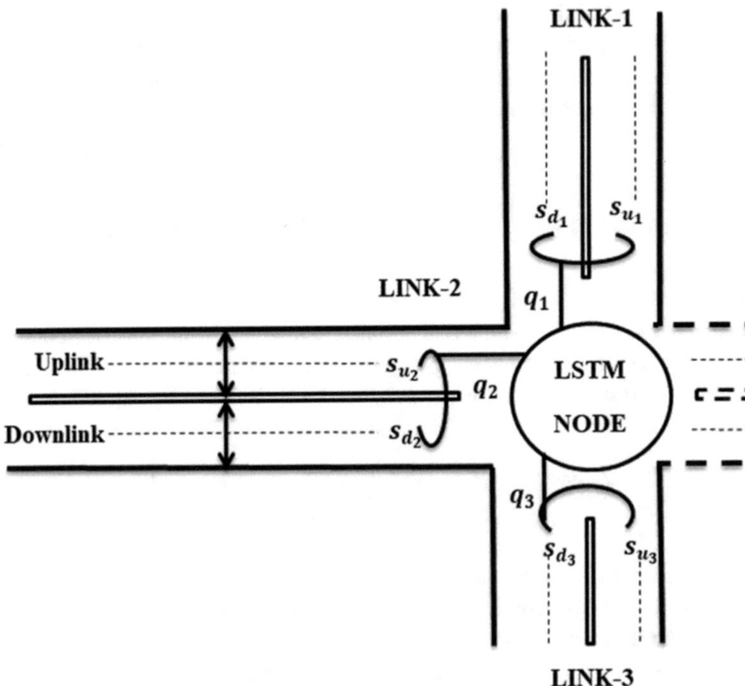

Fig. 1 Spatial–temporal traffic information in a spatially connected road network

3.2 *Sequence Convolution Auto-Encoder Long Short-Term Memory (SCAE-LSTM)*

The spatial and temporal traffic information in a spatially connected road network is shown in Fig. 1. Let $Q_t = [q_1, q_2, q_3, \ldots, q_m]$ denote characteristic sequence extracted from upstream and downstream traffic sequence is obtained using encoding and decoding process. The auto-encoder thus defined is trained for minimal error such that characteristic sequence is valid. The decoder recreates the original data and characteristic sequence is thus obtained when the difference between the original traffic sequence and recreated traffic sequence is minimal. LSTM predicts traffic flow considering the traffic flow at uplink and downlink segments at each intersection as shown in Fig. 1.

$$\text{Encode}(q_t) = f\left(w_q * (S_{ui} + S_{di}) + b_q\right). \tag{5}$$

$$\text{Decode}(z_t) = g(w_z * q_t + b_z). \tag{6}$$

$$f_t = \lambda\left(wf_1 * x_t + wf_2 * q_t + wf_3 * h_{t-1} + b_f\right). \tag{7}$$

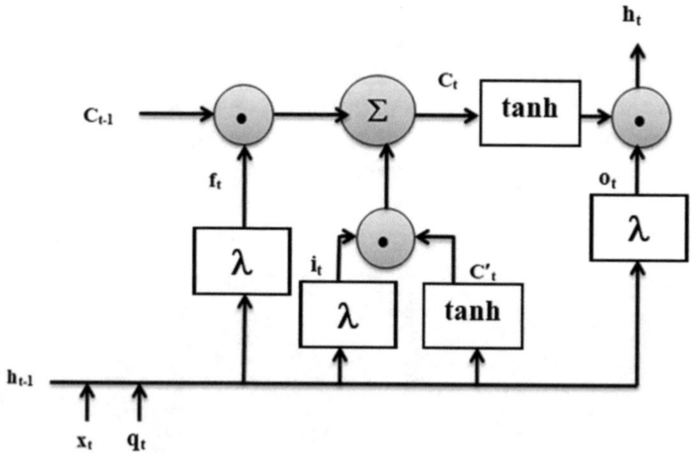

Fig. 2 Structure of LSTM

$$i_t = \lambda(wi_1 * x_t + wi_2 * q_t + wi_3 * h_{t-1} + b_i). \tag{8}$$

$$C'_t = \tan h(wc_1 * x_t + wc_2 * q_t + wc_3 * h_{t-1} + b_c). \tag{9}$$

$$C_t = f_t * C_{t-1} + i_t * C'_t t. \tag{10}$$

$$o_t = \lambda\,(wo_1 * x_t + wo_2 * q_t + wo_3 * h_{t-1} + b_o). \tag{11}$$

$$h_t = o_t * \tan h(C_t). \tag{12}$$

$$\lambda(x) = \frac{1}{(1 + e^{-x})}. \tag{13}$$

$$\tan h(x) = \left(\frac{e^x - e^{-x}}{e^x + e^{-x}}\right). \tag{14}$$

Auto-Encoder LSTM is formalized using Eqs. (7) through (14). The input of the forget gate f_t includes three parts: (i) traffic flow x_t at current location, (ii) characteristic of vehicle speed in sequence q_t extracted from the upstream and downstream traffic sequences, and (iii) previous state of LSTM. The activation function is defined in Eqs. (13) and (14).

The structure of LSTM is shown in Fig. 2 and it is built using Eqs. (7) through (14). Training data sets S_{ui}, S_{di}, and traffic volume at target link Z_t are input parameters to AE-LSTM. The auto-encoder is trained for minimum error using back propagation. The characteristic traffic sequence Q_t obtained using sequence convolution and Z_t are scanned into LSTM network. $Z_{(t+1)}$ is estimated using forward propagation. The state of h_t is updated by forward propagation. The training parameters of AE-LSTM are adjusted and test data is scanned to LSTM to predict Z'. Prediction of traffic flow based on spatial–temporal traffic sequence is given in Procedure 1.

Sequence convolution-based auto-encoder LSTM network (SCAE-LSTM) models the spatial dependence of upstream and downstream vehicle speed captured at every time instances by the sensors and extracts the characteristic traffic sequence. Convolution of sequences is performed to map the sequence $S^{N \times P}$ with P features to Q dimension output. Convolution of sequences of length N has complexity of $O(N\log N)$ which is less expensive compared to convolution of vectors $O(N^2)$. For reducing travel delay and mange peak hour traffic conditions Spatial–TemporAl Reconnect (STAR) algorithm is proposed and explained in the following section.

Procedure 1 Prediction of Traffic Flow Using SCAE-LSTM
Prediction (S_{ui}, S_{di}, Z_t)
Input: the training set S_{ui}, S_{di}, Z_t.
Output: Prediction result Z'.
Method
01: Use the sequence S_{ui}, and S_{di} to create training dataset Z_{ud} for auto-encoder.
02: Assign weight matrices of auto-encoder randomly.
03: Train auto-encoder using Z_{ud}
04: Calculate error between training and test data set.
05: Train the auto-encoder by back propagation until error is minimum.
04: Generate the characteristics traffic sequence Q_t.
05: **FOR** $t = 0$ to epoch do
06: Scan Q_t and Z_t into the LSTM and build forward propagation network using Equation (7) through (14).
07: Estimate $Z_{t+1} = g(w_z * q_t + b_z)$
08: Evaluate error.
09: Update parameters using feedback by back propagation.
10: Update state of h_t by forward propagation using Equation (7) through (14)
11: **ENDFOR**
12: Build LSTM after auto-encoder.
13: Adjust the training initialization parameters and tune the LSTM.
14: Scan test data in AE-LSTM to generate the predicted value Z'.
15: **Return** Z'.

3.3 *Augmentation of Road Network Based on Traffic Flow*

Sequential pattern mining can be used in achieving analysis of temporal traffic information as traffic information in preceding time instances is more significant in estimating the traffic in succeeding instances. Travel to a location is represented by path traversed between source and destination with cost of travel represented in terms of speed, distance, time delay, etc. In this way, when flow rate (Vehicle/h) is high and speed of vehicle is less than free flow speed (60 km/h) congestion is detected, the arterial junctions connecting the destination is augmented on the path leading to destination. In a spatially connected network, the path is traversed between origin and destination based on traffic flow rate of the current location (arterial junction) and its neighboring links.

3.3.1 Path Connectivity Based on Spatial–Temporal Traffic Information

Urban transport network is modeled as a spatial network with time varying traffic flow consisting of N sensors that capture traffic information on upstream and downstream is represented as a weighted graph $G = (J, R)$, where J is set of sensor access points representing intersection of road segments such that $|J| = N$, and R is set of road segments connecting the junction J. W is the adjacency matrix of G representing the proximity of the junction as function of distance between sensors. The traffic information observed by the sensor at each time instance is converted into traffic sequences in P dimension, where P is number of features of each sensor such as traffic speed, volume etc.

Traffic congestion is in effect due to time varying traffic flow through the link. During peak hour the traffic flow at arterial junction is estimated and the path to destination is reconnected based on spatial-temporal characteristic sequence evaluated using SCAE-LSTM. The upstream and downstream traffic flow in preceding time instance $t_1, t_2, t_3, \ldots, t_{i-1}$ is analyzed in estimating traffic flow at successive time instance $t_i, t_{i+1}, \ldots, t_n$ where $[t_1, t_n]$ is the estimated duration of peak hour traffic. The following definitions are considered in formulating ***Spatial-Temporal Reconnect (STAR)***

Definition 1
The indicator of traffic congestion is given by congestion index (CI) measure and defined according to the guideline of HCM as given in Eq. (15)

$$\mathrm{CI} = \left(\frac{\left(\frac{(\mathrm{FreewayVMT})^2}{\mathrm{FreewayLaneMiles}}\right) + \left(\frac{(\mathrm{PrArterialVMT})^2}{\mathrm{PrArterialLaneMiles}}\right)}{(13{,}000 \times \mathrm{FreewayVMT}) + (5{,}500 \times \mathrm{PrArtrialVMT})} \right), \tag{15}$$

where Vehicle Miles Travelled (VMT) is the measure of sum of vehicles travelled; a lane mile is the product of mileage covered by road segment and lane count in highway. FreewayVMT and Pr ArterialVMT are measures of VMT in freeways and principal arterial road segment respectively. The value of CI ranges between 0 and 1. CI > 1 indicates delayed travel condition.

Definition 2
A candidate path is a road segment traversing through arterial junctions $J_1, J_2, \ldots, J_m$ such that speed of vehicle is captured by sensor in upstream and downstream at every

time instance forms traffic sequence of the path. Encode and decode function given in Eqs. (5) and (6) extracts the characteristic traffic sequence of the path from convoluted traffic sequences of upstream and downstream.

Definition 3
Peak hour traffic of a road segment is defined as time varying vehicle speed at every time instance t_i, where $1 \leq i \leq n$ such that $t_1, t_2, t_3, \ldots, t_n \in [t_1, t_n]$ where $[t_1, t_n]$ is the peak hour interval. The vehicle miles travelled during peak hour decreases with respect to time such that travel time increases at each time instance $t_1 < t_2 < t_3 < \ldots < t_{i-1} < t_i < t_{i+1} < \ldots < t_n$

Definition 4
During peak hours, the traffic flow rate on the candidate path that connects origin "*u*" and destination "*v*" is estimated using AE-LSTM formalized using Eqs. (7) through (14).

Definition 5
During peak hour traffic, a candidate path with time-varying traffic speed is said to re-connect two arterial intersections (junctions) J_1 and J_2 between origin and destination in a spatial network if congestion index (CI) of the path is less than 1. The augmentation of road network is done using STAR algorithm explained in Procedure 2. The flow of work is explained in Fig. 3. Traffic congestion on time-varying network is managed using the routines *Spatial_Connect*(*G*, $\boldsymbol{J_i}$, $\boldsymbol{t_i}$) *and Congestion_Detect*($\boldsymbol{R_j}$, $\boldsymbol{Z_t}$, $\boldsymbol{t_i}$) while resolving the traffic congestion by re-establishing connectivity. The *Spatial_Connect*(*G*, $\boldsymbol{J_i}$, t_i) routine is called to identify the non-congested roads connecting the arterial junctions between the origin "*u*" and destination "*v*" as shown in Fig. 3. Spatial connectivity is augmented using this subroutine and given in Procedure 3.

Procedure 2 Spatial–Temporal Reconnect (STAR)
Declare Congestion_Detect(R_j, Z_t, t_i) as Boolean
Declare Spatial_Connect(G, J_i, t_i)
OD: Origin and Destination are spatial location from Google Map.
J_x: Vector consisting of arterial junctions $\{j_1, j_2, j_3, j_4, \ldots, j_n\} \in J_x$ between OD.
R_m: Vector consisting of road segments $\{r_1, r_2, r_3, \ldots, r_n\} \in R_m$, connecting arterial junctions between OD pair
Q_t: The characteristic of traffic flow rate in sequence representing upstream and downstream traffic flow.
$CI(R_i)$: Congestion index of road segment $r_i \in R_m$
G: Spatially connected road network consisting of road segments R_m connecting arterial junctions J_x
W: Adjacency matrix of G representing distance between junction $\{j_1, j_2, j_3, \ldots j_i, j_{i+1}, \ldots, j_n\} \in J_x$

(continued)

Input: G is the weighted graph extracted from Google Map representing the spatially connected road network initialized with distance between arterial junctions using cost adjacency matrix for given origin and destination.
Output: Reliable path to reach destination

Method
01: Get origin and destination locations.
02: Search neighboring arterial junction in close proximity.
03: Call Spatial connect (G, J_i, t_i)
04: $G' = G$
05: $J_C \in G'$, $J_{\text{cost}} =$ old distance + distance(J_c, J_{i+1})
06: Repeat line 02 until destination has reached.

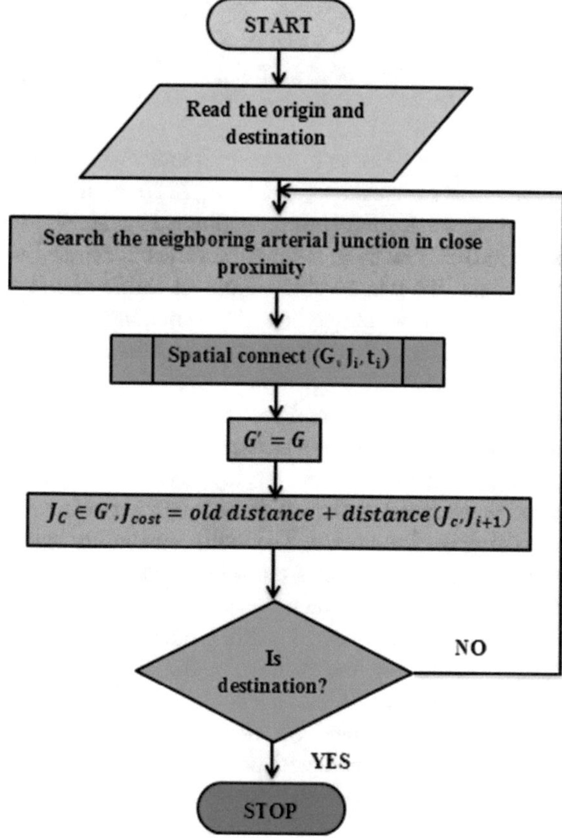

Fig. 3 Flow of work in STAR between origin and destination

Procedure 3 Augmenting Spatial Connectivity
Subroutine Spatial _ Connect(G, J_i, t_i)
Input:
G: The spatially connected road network
R_j, J_i: The list of road segments R_j connecting arterial junctions J_i in G
t_i: 15 minutes flow rate
Output: Augmented road network G'
Method:
01: For $\forall \boldsymbol{R_j} \in \boldsymbol{J_i} \wedge \boldsymbol{t_i}, \boldsymbol{i} = \mathbf{1,2, \ldots, 15}$ ***minutes***
02: **IF** ($Congestion\_Detect(R_j, Z_t, t_i)$)
03: set $R_j \in \mathbb{W}, R_j = \infty$;
04: **ELSE**
05: set $\boldsymbol{R_j} \in \mathbb{W}, \boldsymbol{R_j} = \boldsymbol{d}$
06: **ENDIF**
07: **ENDFOR**
08: $\boldsymbol{G} = \boldsymbol{G} - (\boldsymbol{R_j} \wedge \boldsymbol{J_i})_{\boldsymbol{cost} = \infty}$
09: Return G.

The flow of work in $Spatial\_Connect(G, J_i, t_i)$ routine is shown in Fig. 4. This routine calls $Congestion\_Detect(R_j, Z_t, t_i)$ routine to detect the state of congestion of each road. Peak hour traffic condition is evaluated using congestion index. Compared to free flow time, travel time varies dynamically due to various environmental conditions like when vehicle stops at traffic signal or the path is blocked due to incidents such as temporary non-availability of service, laying and pavement of roads, etc. Thus, the difference in travel time increases the congestion index.

The subroutine $Congestion\_Detect(R_j, Z_t, t_i)$ is given in Procedure 4. When traffic congestion is identified, road segments that are spatially connected to the congested arterial junction are not considered in finding the path between OD pair as the edge cost of the time-varying network becomes infinity. Instead, the vehicle speed is analyzed in sequence, and path between OD pair is re-connected with the remaining road segments. The path connectivity is re-establishing based on spatial–temporal traffic sequences considering upstream and downstream traffic sequences using the proposed sequence convolution-based auto-encoder LSTM. STAR considers temporal and spatial traffic sequence in re-connecting path between OD pair, thereby modifying the path computation held in the conventional shortest path algorithms such as Dijkstra, TDALT, and other time-dependent path computation problems. The proposed STAR re-establishes path connectivity in the spatial network based on the characteristic of vehicle speed in sequence representing the speed of vehicle in upstream and downstream of road segment. Thus, STAR algorithm devises solution by estimating traffic flow at each arterial junction between OD pair and re-connects the path based on spatial–temporal traffic information sequence.

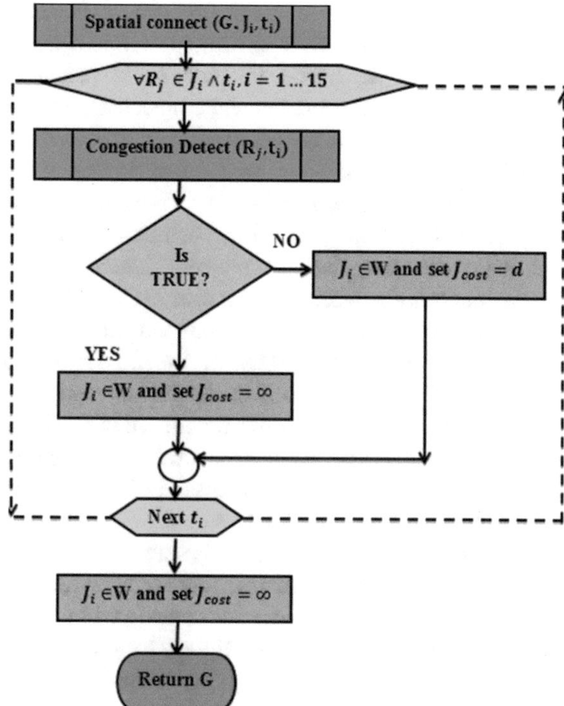

Fig. 4 Flow of work in augmenting path connectivity

Procedure 4 Congestion Detection
Subroutine Congestion _ Detect(R_j, Z_t, t_i)
Input:
R_j, J_i: The list of road segments R_j connecting arterial junctions $J_i$in G
t_i: 15 minutes flow rate
Output: True when congestion is detected.
01: At time instance $t_{i-1} < t_i < t_{i+1} \in [t_1, t_n]$
02: Predict traffic volume Z'using Equation (5) through (14)
02: find Congestion Index (CI) using Equation (15)
03: **IF** ($CI(Z', R_j) > 1$)
04: **Return** TRUE
05: **ENDIF**

3.3.2 Computational Complexity Analysis

The complexity in managing traffic congestion by re-establishing connectivity of the road network based on spatial–temporal traffic sequence is analyzed as follows: *Spatial _ Connect*(G, J_i, t_i) routine identifies the roads without congestion at every time instance. The roads connecting the arterial junction between the origin and destination are analyzed for traffic congestion. Hence, the complexity of this routine is O(cE). *Congestion _ Detect*(R_j, Z_t, t_i) routine detects the state of congestion at any given time instance; hence, this routine is executed in constant time. Temporal instances of traffic information and connectivity of path are analyzed by sequencing traffic information. When congestion is detected, cost adjacency matrix of the spatially connected time-varying network is recreated by eliminating road segments based on characteristic of vehicle speed using auto-encoders. Thus, spatial regions that are involved in traffic flow estimation are considerably reduced, thereby decreasing the computation on several arterial junctions. Hence, augmentation of arterial junctions between OD pair runs in O(logE logV) which is less than O(ElogV) which is the complexity of well-known single shortest path problem algorithm Dijkstra. The time complexity of the STAR in traffic flow assessment is (O (cE) + O(logElogV)) which is linear time execution.

Complexity analysis signifies the importance of spatial and temporal information of time-varying traffic network in traffic management. Unlike several speed-up techniques proposed for pre-processing the edge cost in time-dependent shortest path algorithms, this work focuses on the use of temporal and spatial information in effect to manage congestion and re-establishes path connectivity, thereby resolving traffic congestion under dynamic scenarios. Sequencing temporal and spatial information of traffic network were not considered in the early works of time-dependent shortest path, whereas STAR algorithm shows the significance of such traffic sequences and its analysis using deep neural networks in real-time traffic management on time-varying network.

The training phase of LSTM requires computationally elaborate operational settings. The traffic data sets representing traffic speed, flow rate, and occupancy are voluminous and require intensive computation methods. The potential issues concerned with the implementation of SCAE-LSTM in STAR are detailed as follows:

Given an origin and destination pair a matrix of LSTM nodes is formulated in which each LSTM node represents an arterial junction between origin and destination. During the training phase, the source of input to LSTM is the pre-trained auto-encoders whose primary function is to extract spatial–temporal traffic characteristics from upstream and downstream traffic flow in each hour of the day. Thus, extensive operational settings are required in initial training phase of LSTM nodes.

The core components operated in Procedure 1 are extraction of spatial–temporal characteristics of physical traffic flow using frequent traffic sequences on upstream, downstream, and training auto-encoders using the extracted data sets. The training phase of LSTM requires computationally elaborate operational settings. The traffic data sets representing traffic speed, flow rate collected for 52 weeks is used in training auto-encoders with computationally intensive operational settings on arterial junctions. However, training and validation error does not vary in different operational settings considering the morning and evening peak hours of the day.

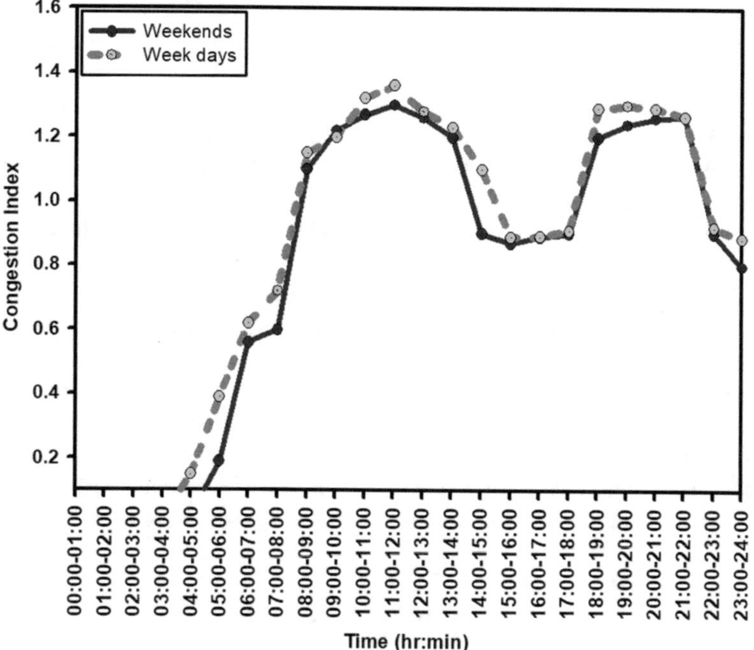

Fig. 5 Congestion Index of SH 49

4 Data Collection and Experimental Results

Estimation of traffic speed using SCAE-LSTM and dynamic path connectivity using STAR algorithm is experimented on road network covering southern region of Chennai city located in Tamil Nadu state of India. The state highways SH 49 and SH 49A are chosen for this study as these highways experience significant traffic on all working days and weekends.

Initially, the length of study area considered for operational setting on state highway SH 49 (East Coast Road) is 50 km and extended till 440 km during testing phase for different origin, destination pairs and SH 49A is 43.7 km. This metropolitan transport network is a highly connected road network covering major arterial junctions connecting urban streets, motorways, and expressways. The state highway of this southern region is facilitated with a centralized toll collection center. Motor vehicle passing through this state highway connects three toll plazas, namely (1) Perungudi Toll Plaza, which is entry to centralized toll collection center from the northern region of Chennai city, (2) East Coast Road (ECR), and (3) Egattur toll plaza, which is the exit of the toll collection center.

In this work, congestion index of the state highway SH 49A and SH 49 used in the experimentation of the STAR is shown in Fig. 5. Travel time during free flow is computed in early morning between 04:00 and 06:00 h, afternoon between 12:00 and 14:00 h, and post-evening between 22:00 and 23:30 h. The actual travel time is computed in morning peak hours between 08:00 and 11:30 h, evening peak hours between 17:00 and 21:00 h during weekdays, while peak hour changes during weekends.

The historical data set describing the vehicle speed in mile/h is used as training data set. Speed of vehicle in state highway SH 49 is captured by the sensor and stored in centralized server which is high end server located in Perungudi Toll Plaza. Estimation of vehicle speed is done by training auto-encoder LSTM with historical data set and validated by testing the model with actual speed of vehicle. Vehicle miles traveled (VMT) recorded for 31 months between January 2017 and July 2019 is used for training the Auto-Encoder LSTM network, while VMT recorded for 20 months between June 2018 and December 2019 is used as test data set in evaluating the system.

Consider a linear time-invariant (LTI) system in which input to the sensor is speed of vehicle. The speed of vehicle at 15 min time interval is captured by the sensor in time sequence $t_1, t_2, t_3, \ldots, t_{i-1}, t_i, t_{i+1}, \ldots, t_m$ by sensors at each arterial junction. This finite length sequence is convoluted with impulse response of LTI sensor system. Let the speed of vehicle be represented as sequence vector given in Eq. (16).

$$x[t] = [x_1, x_2, x_3, \ldots, x_n]. \tag{16}$$

The impulse response of the sensor system for each time instance is converted to sequence $h[t]$. The sequence convolution of two finite-length sequences representing speed of vehicle is given by Eq. (17)

$$y[t] = \sum_{i=0}^{M} x[i]h[t-i]. \tag{17}$$

The output of the system $h[t]$ is a positive non-zero value between

$$\begin{gathered} t_1 + t_3 \leq t \leq t_2 + t_4 \\ T_y = T_x + T_{h-1}. \end{gathered} \tag{18}$$

Convolution of two sequences defined in finite duration is given by $y[t]$ will be at most non-zero in interval $[0, 2(T-1)]$ as given in Eq. (19).

$$x[t] = h[t] = z[t] - z[t-T], T \geq 1, \tag{19}$$

there occur four cases

$$Case\text{-}1 : t < 0, y[t] = 0. \tag{20}$$

Convolution of traffic sequence is not defined for non-positive duration of time. Hence the output is zero as given in Eq. (20).

$$Case\text{-}2: 0 \leq t \leq T-1, y[t] = t+1. \tag{21}$$

Convolution of traffic sequence is defined in Eq. (21) for positive duration of time and output of the system is convoluted sequence of length $t + 1$ that remains non-zero in the interval $[0, (T - 1)]$ when there is overlap with input sequence and impulse response of the LTI system.

$$Case\text{-}3: T-1 \leq t \leq 2T-2, y[t] = 2T-t+1. \tag{22}$$

Convolution of traffic sequence is defined in Eq. (22) when duration overlaps in the interval $[T - 1, 2(T - 1)]$ and the system remains non-zero. The output of the system is convoluted sequence of length $2T - t + 1$ and case-4 is given in Eq. (23)

$$Case\text{-}4: t > 2T-2, when\, there\, is\, no\, overlap, y[t] = 0. \tag{23}$$

Thus, to summarize all four cases, the linear time-invariant system remains positive and non-zero in the interval $[0, 2(T - 1)]$, as given in Eq. (24). Thus,

$$y[t] = \left\{ \begin{array}{ll} t+1 & 0 \leq t \leq T-1 \\ 2T-1-t| & T-1 \leq t \leq 2T-2 \\ 0 & t > 2T-2 \end{array} \right\}. \tag{24}$$

Traffic flow information in upstream and downstream of road link contributes significant traffic information in estimating traffic flow at target link. Vehicle speed on upstream and downstream is recorded by the sensor. The speed data of vehicle is pre-processed and used in estimating congestion index of the road segment. The stability of the system is ensured even with traffic data collection at a temporal resolution of 15 min as illustrated in Fig. 6.

The issues concerned with implementing STAR are space complexity. Augmenting spatially connected road networks as described in procedures 2, 3, and 4 requires the storage of spatial coordinates of the location on a Google Map. The location information and retrieval of spatial coordinates during the inclusion and exclusion of LSTM nodes incurs high space and memory utilization. The experimental setup was made for a maximum distance of 350 miles on state highways SH 49 and SH 49A for testing the performance of the STAR algorithm. However, other practical constraints have to be considered in the further investigation of routing vehicles, which is left to work in the future.

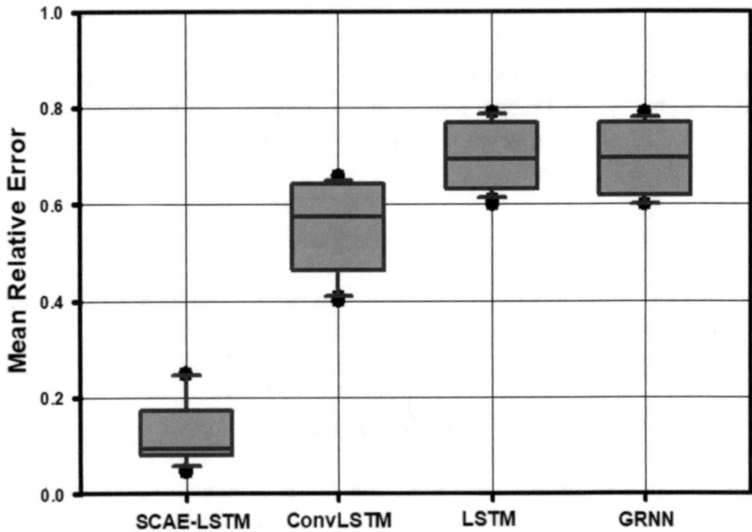

Fig. 6 Mean relative error measure with forecast in 15 min resolution

5 Conclusion

Spatial and temporal traffic information is the principle criteria considered in monitoring transport and mobility on highways. Moreover, physical traffic flow exhibits a dynamic phenomenon in different operational settings. MCDA is a useful tool to elucidate sustainable transport and mobility, a discipline of Operation Research (OR) which is in a wide range of applications and practices in real-time traffic engineering. This chapter introduces reliable path computation methods based on traffic flow. A review of spatial–temporal traffic forecasting models and methods is explored. Research challenges are identified and problem statement is presented.

Formulation of spatial–temporal traffic information sequence is explained. Sequence convolution-based auto-encoder LSTM is formalized for the prediction of traffic flow at road intersection (target link) considering spatial–temporal dependence of neighboring links.

Formulation of Spatial–Temporal Reconnect (STAR) algorithm is presented with an analysis on computational complexity. Training and validation performance of SCAE-LSTM is evaluated and compared with baseline methods. Time complexity of STAR algorithm is evaluated considering estimation of traffic flow with respect to distance measure and augmentation of road network based on origin and destination pair.

This chapter reported the significance of MCDA with an illustrative case study on the prediction of traffic flow on highways considering the spatial and temporal measures of physical traffic flow. Algorithms are formulated to explore the intrinsic relationship between these measures. Experimental results are reported with estimation of time complexity of algorithms

References

Al-Deek, H. M., Mohamed, A. A., & Radwan, E. A. (1999). New model for evaluation of traffic operations at electronic toll collection plazas. *Transportation Research Record, 1710*(1519), 1–10.

Barbosa-Povoa, A. P., da Silva, C., & Carvalho, A. (2018). Opportunities and challenges in sustainable supply chain: An operations research perspective. *European Journal of Operational Research, 268*(2), 399–431.

Bauer, R., & Delling, D. (2009). SHARC: Fast and robust unidirectional routing. *ACM Journal of Experimental Algorithmics, 14*(4), 4–29.

Bauer, R., Delling, D., Sanders, P., Schieferdecker, D., Schultes, D., & Wagner, D. (2010). Combining hierarchical and goal-directed speed-up techniques for Dijkstra's algorithm. *Journal of Experimental Algorithmics, Lecture notes in Computer Science, 5038*, 303–318.

Brunel, E., Delling, D., Gemsa, A., & Wagner, D. (2010). *Space-efficient SHARC-routing, proceedings of ninth international symposium on experimental algorithm* (pp. 47–58).

Cai, P., Wang, Y., Lu, G., Chen, P., Ding, C., & Sun, J. (2016). A spatiotemporal correlative k-nearest neighbor model for short-term traffic multistep forecasting. *Transportation Research - Part C, 62*(1), 21–34.

Chandra, S. R., & Al-Deek, H. (2008). Cross-correlation analysis and multivariate prediction of spatial time series of freeway traffic speeds. *Transportation Research Record Journal of Transportation Research Board, 2061*(1), 64–76.

Cheng, T., Haworth, J., & Wang, J. (2012). Spatio-temporal autocorrelation of road network data. *Journal of Geographical Systems, 14*(4), 389–413.

Demiryurek, U., Banaei-Kashani, E., & Shahabi, C. (2011). Online computation of fastest path in time dependent spatial networks. In *International Symposium on Spatial and Temporal Databases*, pp. 92–111.

Dijkstra, E. W. (1959). A note on two problems in connexion with graphs. *Numerical Mathematics, 1*(1), 269–271.

Ermagun, A., & Levinson, D. (2018). Spatiotemporal traffic forecasting: Review and proposed directions. *Transport Reviews, 38*(6), 786–814.

Geisberger, R., Sanders, P., Schultes, D., Delling, D., & Vetter, C. (2012). Exact routing in large road networks using contraction hierarchies. *Transportation Science, 46*(3), 388–404.

Goldberg, A. V., & Harrelson, C. (2003). Computing the shortest path: A* search meets graph theory. *Proceedings of ACM Symposium on Discrete Algorithms, 2003*, 156–165.

Habtemichael, F. G., & Cetin, M. (2016). Short-term traffic flow rate forecasting based on identifying similar traffic patterns. *Transportation Research Part C, 66*(5), 61–78.

Hart, P. E., Nilsson, N. J., & Raphael, B. (1968). A formal basis for the heuristic determination of minimum cost paths. *IEEE Transactions on Systems Science and Cybernetics, 4*(2), 100–107.

Huang, W., Song, G., & Hong, H. (2014). Deep architecture for traffic flow prediction: Deep belief networks with multitask learning. *IEEE Transaction on Intelligent Transport Systems, 15*(5), 2191–2201.

Jayanthi, G. (2021). Design of Algorithm for IoT-based application: Case study on intelligent transport systems. International series in operations research & management science. In F. P. G. Márquez & B. Lev (Eds.), *Internet of Things, chapter 0* (pp. 227–249). Springer.

Jayanthi, G., & García Márquez, F. P. (2021a). *Travel time based traffic rerouting by augmenting traffic flow network with temporal and spatial relations for congestion management*. Springer. https://doi.org/10.1007/978-3-030-79203-9_43

Jayanthi, G., & García Márquez, F. P. (2021b). Data mining and information technology in transportation—A review. In J. Xu, F. P. García Márquez, M. H. Ali Hassan, G. Duca, A. Hajiyev, & F. Altiparmak (Eds.), *Proceedings of the Fifteenth International Conference on Management Science and Engineering Management. ICMSEM 2021. Lecture Notes on Data Engineering and Communications Technologies* (Vol. 79). Springer. https://doi.org/10.1007/978-3-030-79206-0_64

Jayanthi, G., & Jothilakshmi, P. (2019). Prediction of traffic volume by mining traffic sequences using travel time based PrefixSpan. *IET Intelligent Transport Systems, 13*(7), 1199–1210. https://doi.org/10.1049/iet-its.2018.5165

Jayanthi, G., & Jothilakshmi, P. (2021). Traffic time series forecasting on highways - A contemporary survey of models, methods and techniques. *International Journal of Logistics Systems and Management, 39*(1), 77–110.

Kamarianakis, Y., & Prastacos, P. (2003). Forecasting traffic flow conditions in an urban network: Comparison of multivariate and univariate approaches. *Transportation Research Record Journal of Transportation Research Board, 1857*(3), 74–84.

Karlaftis, M. G., & Vlahogianni, E. I. (2011). Statistical methods versus neural networks in transportation research: Differences, similarities and some insights. *Transportation Research Part-C, 19*(3), 387–399.

Kartikay, G., & Niladri, C. (2019). Forecasting through motifs discovered by genetic algorithms. *IETE Technical Review, 36*(3), 253–264.

Lin, J., Keogh, E. J., & Wei, L. D. (2007). Experiencing SAX: A novel symbolic representation of time series. *Data Mining and Knowledge Discovery, 15*(2), 107–144.

Lv, Y., Duan, Y., & Kang, W. (2015). Traffic flow prediction with big data: A deep learning approach. *IEEE Transaction on Intelligent Transport System, 16*(2), 865–873.

Ma, X., Yu, H., & Wang, Y. (2015). Large-scale transportation network congestion evolution prediction using deep learning theory. *PLoS One, 10*(30), 1–17.

Ma, X., Dai, Z., He, Z., Ma, J., Wang, Y., & Wang, Y. (2017). Learning traffic as images: A deep convolutional neural network for large-scale transportation network speed prediction. *Sensors, 17*(4), 818–828.

Min, W., & Wynter, L. (2011). Real-time road traffic prediction with spatio-temporal correlations. *Transportation Research Part C, 19*(4), 606–616.

Nejad, M. M., Mashayekhy, L., Chinnam, R. B., & Anthony, P. (2017). Hierarchical time-dependent shortest path algorithms for vehicle routing under ITS. *IIE Transactions, 48*(2), 158–169.

Park, D., & Rilett, L. R. (1999). Forecasting freeway link travel times with a multilayer feed forward neural network. *Computing Civil Infrastructure Engineering, 14*(5), 357–367.

Seuring, S. (2013). A review of modeling approaches for sustainable supply chain management. *Decision Support Systems, 54*(4), 1513–1520.

Shi, Y., Deng, M., Gong, J., Lu, C., Xuexi, Y., & Liu, H. (2019). Detection of clusters in traffic networks based on spatio-temporal flow modeling. *Transactions in GIS, 23*(2), 312–333.

Wang, C., & Ye, Z. (2015). Traffic flow forecasting based on a hybrid mode. *Journal of Intelligent Transport System, 20*(5), 428–437.

Wangyang, W., Honghai, W., & Huadong, M. (2019). An AutoEncoder and LSTM-based traffic flow prediction method. *Sensors, 19*(2946), 1–16.

Wu, S., Yang, Z., Zhu, X., & Yu, B. (2014). Improved k-NN for short-term traffic forecasting using temporal and spatial information. *Journal of Transportation Engineering, 140*(7), 1–9.

Xu, Y., Chen, H., Kong, Q., Zhai, X., & Liu, Y. (2015). Urban traffic flow prediction: A spatio-temporal variable selection-based approach. *Journal of Advanced Transportation, 50*(4), 489–506.

Zhang, Y., & Zhang, Y. (2016). A comparative study of three multivariate short-term freeway traffic flow forecasting methods with missing data. *Journal of Transportation System, 20*(3), 205–218.

Zhao, Z., Chen, W., Wu, X., Chen, P. C. Y., & Liu, J. (2017). LSTM network: A deep learning approach for short-term traffic forecast. *IET Intelligent Transport Systems, 11*(2), 68–75.

A Robust Prognostic Indicator for Renewable Energy Fuel Cells: A Hybrid Data-Driven Prediction Approach

Daming Zhou, Zhuang Tian, and Jinping Liang

Abstract As a power generation device, fuel cells are now widely studied as a new type of energy device due to their high energy density and non-pollution advantages. However, the large-scale industrialization of fuel cells is still difficult to realize. One of the important reasons is that its aging failure problem can lead to the degradation of performance and even the decay of useful lifetime.

Prognostic and health management (PHM) is an effective technology to make reasonable predictions of fuel cell lifetime and state of health (SOH) to prevent economic loss and safety hazards due to aging failure. Prognostic is an important component of PHM, and it can predict the subsequent data trends of fuel cells using known measured data such as voltage, power, etc., and thus predict the SOH of fuel cells in the future period. This chapter first develops a hybrid prediction method with a state space model and a data-driven method. Then a prediction method with sliding prediction length is proposed. Finally, the accuracy and reliability of the hybrid method are verified. The main contributions of this chapter are as follows:

1. The proposed hybrid prediction method combines the advantages of the respective prognostic approaches, thus being able to fully make best the advantages of the state space model and data-driven prediction method. In this case, the hybrid prediction method can accurately predict linear degradation trends and the local fluctuations and nonlinear characteristics. Thus, the method compensates for the drawbacks of the single prediction method and has higher prediction accuracy.
2. The sliding prediction length method can update the aging data in the multi-step prediction process in time to ensure the data source of the training set. In addition, the method facilitates the assignment of weight factors for the fusion of different prediction methods to obtain better prediction accuracy.
3. A comprehensive comparison experiment is designed to verify the advancement of the proposed hybrid prediction method aiming at the whole dataset range and

D. Zhou (✉) · Z. Tian · J. Liang
School of Astronautics, Northwestern Polytechnical University, Xi'an, People's Republic of China
e-mail: daming.zhou@nwpu.edu.cn

F. P. García Márquez, B. Lev (eds.), *Sustainability*, International Series in Operations Research & Management Science 333,
https://doi.org/10.1007/978-3-031-16620-4_10

sliding prediction length range. It provides a feasible solution for the aging prediction method of fuel cells, especially for the multi-step prediction under actual operating conditions, thus avoiding the risks caused by the sudden degradation of fuel cells in actual operation.

Keywords Sustainability · Prognostic and health management (PHM) · State space model · Machine learning · Aging prediction · Maintenance management

1 Introduction

Due to the pollution and unsustainability of traditional fossil energy sources, hydrogen energy is widely studied as a clean and sustainable energy source. As a kind of downstream equipment of hydrogen energy industry chain, the advantages of proton exchange membrane fuel cell (PEMFC) compared with a traditional internal combustion engine are mainly in the following aspects (Lü et al., 2018):

1. Fuel flexibility. It can use both pure hydrogen and conversion fuel. The raw materials come from a wide range of sources and are obtained through the processing of water, oil, natural gas, coal, biogas, and methanol, etc.
2. Environmentally benign features. Its discharge is pure water (H_2O) and does not emit harmful gases (SO_2, NO_2).
3. There is no noise in the electrochemical reaction inside the stack.
4. Higher energy conversion efficiency (64%), favorable power-to-weight ratio (0.86 kW/kg).
5. Sustainable power supply. PEMFC is not an energy storage device, but a power generation device, which can continuously generate electricity as long as the raw materials are continuously supplied, and the electrical performance is stable.
6. Stable and reliable operation. Low operating temperature (50–80 °C), quick start-up (<30 s from −20 °C), quick fueling (usually <20 min).
7. Compactness and easy scaling between power and capacity.

Although PEMFC is currently considered as a competitive type of energy device (Thomas et al., 2020; Rath et al., 2019). However, one of the main reasons why PEMFC has not been industrialized on a large scale is that PEMFC reactors are prone to degradation due to the stringent conditions of internal chemical reactions. Specific factors that lead to degradation include: harsh operating environment (Stiller et al., 2006), catalyst failure, reverse pole phenomenon (Fowler et al., 2002), bipolar plate failure, etc. (Wu et al., 2008). These factors that lead to fuel cell degradation are diverse and unpredictable. At present, there is no model that can reflect the degradation mechanism of fuel cells comprehensively and accurately (Jouin et al., 2014; Bressel et al., 2016; Roshandel & Parhizkar, 2016).

To avoid the loss caused by sudden aging of PEMFC and to minimize the maintenance cost, prognostic and health management (PHM) is widely studied as an emerging research direction (Chen et al., 2016; Cheng et al., 2018; Zhou et al., 2016a). The specific flow chart of PHM is shown in Fig. 1. As can be seen from

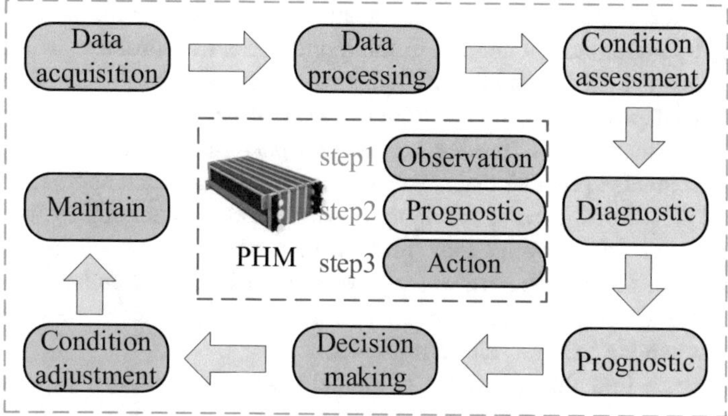

Fig. 1 The workflow of PHM

Fig. 1, prognostic is an important component step of PHM. It can use known monitoring data such as voltage, power, and other data to predict the subsequent data trend of the fuel cell, and then predict the SOH of the fuel cell in the future period, so as to ensure that the fuel cell will not degrade suddenly in a short period of time in the future and ensure the normal operation of the fuel cell (Jouin et al., 2013).

The main prediction approaches for fuel cells are model-based and data-driven methods. The model-based methods usually use a large number of physical equations to construct the aging model of fuel cells and then adopt methods such as state space model to estimate the state quantities. Zhou et al. (2017a) propose a prognostic method based on a multi-physics domain aging model, based on the aging model, both the parameters of the model and the quantity of state are updated in real time by using particle filtering. This prediction method is also applicable in the aging prediction of lithium-ion batteries (Miao et al., 2013; Wang et al., 2013; He et al., 2011). Bressel et al. (2016) combine extended Kalman filter with extrapolation method to predict the aging phenomenon of PEMFC based on the degradation model. Chen et al. (2019a) also use the model-based approach to forecast the decay behavior under different fuel cell parameter configurations. However, all of the above single model-based approaches have the drawbacks of limited model accuracy as well as generality, and all of the above models focus on the electrical domain, ignoring the effects on materials, mechanics, and other aspects. It can be seen from the above methods that it is very difficult to build a model that comprehensively reflects the internal aging mechanism of fuel cells.

Different from the model-based methods, data-driven methods use stochastic processes, mathematical statistics, probability distributions, and other mathematical methods to train large amounts of input and output data to establish input–output relationships. The specific input–output relationship is a "black box" model. It can simplify the study of the internal mechanism of the model. In this case, this approach is useful for predicting nonlinear aging data for which it is difficult to establish specific empirical formulas. Marra et al. (2013) present a neural network prediction method to successfully predict the degradation trajectory of solid oxide fuel cells. Similarly, this approach is applied to the aging prediction problem of lithium-ion batteries (Lin et al., 2012) and PEMFC (Silva et al., 2014). Mao and Jackson (2016a) develop an adaptive neuro-fuzzy inference system (ANFIS) with sensor optimal selection algorithm to describe the degradation phenomenon of fuel cells, and the results show that the prediction accuracy of the method can reach approximately 1%. Ibrahim et al. (2016a) predict the aging phenomenon of fuel cells by using signal processing approach, wavelet transform (WT) method is adopted to process the aging data, and finally, the performance prediction of PEMFC under various operating conditions is achieved. In Sect. 2, a detailed literature review of various prediction methods for fuel cells is presented.

It is easy to find that the focus of model-based and data-driven methods are different, and the characteristics of each as well as the application scenarios are also different. For model-based methods, such as the state space model prediction method (Miao et al., 2013; Wang et al., 2013; He et al., 2011), it is difficult for this method to capture local fluctuations and nonlinear data in the fuel cell aging process. However, for the overall trend, this method is able to make effective predictions by extrapolation and parameter identification. In contrast, for data-driven methods, such as nonlinear autoregressive neural network (NARNN) methods are able to describe the local fluctuations and nonlinear characteristics of fuel cells well. And this method is effective for dealing with short-term nonlinear series predictions (Benmouiza & Cheknane, 2013; Ibrahim et al., 2016b; Arbain & Wibowo, 2012).

In order to combine the advantages of the above two methods, the overall degradation trend of the fuel cell aging data and the accuracy of the prediction results of the local fluctuation characteristics are balanced. It is a novel idea to fuse the two methods to obtain a more superior hybrid method.

In this section, a hybrid prediction method based on model-based and data-driven methods is firstly proposed. Then a prediction method based on sliding prediction length is designed and the weight factors of each method are reasonably assigned. Finally, the superiority of the proposed method is verified from various aspects. The main contributions of this chapter are summarized as follows:

1. A novel hybrid prediction method is developed. The method is capable of accurately predicting the overall fuel cell aging trend as well as the local fluctuation phenomenon. The proposed method can be used for any type of fuel cells, compensating for the poor generality of a single prediction method.
2. A sliding prediction length method is devised. The method can ensure the continuous updating of training data during the prediction process, thus ensuring the accuracy of the state space model and thus improving the prediction accuracy. In addition, the method is conducive to the reasonable allocation of weight factors for different prediction methods and improves the robustness of the hybrid method.
3. A comprehensive experimental validation of the proposed method is carried out. The prediction performance of different methods is compared at multiple levels from the whole aging range and each sliding prediction length range. The

experimental results verify the superiority of the performance of the proposed method.

The chapter is organized as follows: Section 2 gives a detailed literature review of the model-based and data-driven prognostic methods. In Sect. 3, the state space model prediction method and NARNN prediction method are, respectively, introduced at first. Then, the sliding prediction length method and the weight factors assignment approaches are developed. Comprehensive comparative experimental verification is shown in Sect. 4. Section 5 presents the final conclusions of this work.

2 Literature Review

In this section, the aging prediction methods are classified. Then, various prediction methods are reviewed in detail. The specific classification of prediction methods is shown in Fig. 2. The characteristics of all methods in the review are shown in Table 1.

2.1 *Model-Based Method*

The model-based method predicts the aging behavior of fuel cells by building physical and chemical models. The method is generally divided into physical model and system state observer methods.

A. Physical Model Method

The physical model method is based on the physical characteristics of fuel cells. This method establishes a complex physical relationship coupling equation and predicts aging from the perspective of model operation mechanism. Robin et al. (2013) propose a theoretical model of platinum (Pt) surface degradation. The operation and aging dynamic phenomena of fuel cells are analyzed under two numerical

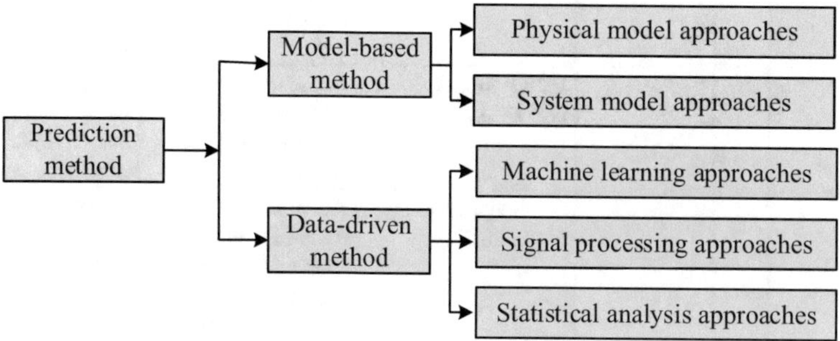

Fig. 2 The classification of prediction methods

Table 1 Literature review of recent approaches for fuel cell aging prediction

Approach category		Subclass	Prediction approaches for fuel cell aging	Upside	Downside
Model-based methods	(1)	Physical model	Pt surface degradation model (Robin et al., 2013); Pt dissolution degradation model (Ao et al., 2021a)	Theoretical description of the actual physical degradation phenomena	Currently not fully understood; significant computational burden
			ECSA degradation model (Polverino & Pianese, 2016)	Computational efficiency due to reduced complexity model	Relatively simple model; currently not fully understood; significant computational burden
			Reconstructed lifetime prediction model (Hu et al., 2018)	Computational efficiency due to reduced complexity model	No theoretical description of physical degradation phenomena; sensitive to aging data quality and quantity
	(2)	State space model method	Direct fitting method for empirical model (Zhang et al., 2017); Direct fitting method for semi-empirical model (Ou et al., 2021)	Simple and easy implementation	Model parameters cannot be updated in real time; difficult to capture nonlinear trends
			EKF model (Bressel et al., 2016; Pan et al., 2020)	Accuracy, dynamic, closed-loop and can be online implementation	Large amount of calculation; difficult to capture nonlinear trends
			UKF model (Chen et al., 2019b)	Small calculation amount, dynamic, easy for online implementation	Relatively poor prediction accuracy and difficulty to capture nonlinear trends
			FDKF model (Ao et al., 2021b)	Accuracy and calculation speed are considered at the same time	Complex program design and difficult to capture nonlinear trends
			PF model (Zhou et al., 2017a)	Accuracy, dynamic, closed-loop and can be online implementation	Large amount of calculation; difficult to capture nonlinear trends

(continued)

Table 1 (continued)

Approach category		Subclass	Prediction approaches for fuel cell aging	Upside	Downside
			UPF model (Chen et al., 2017); APF model (Kimotho et al., 2014); RPF model (Cheng et al., 2018)	Targeted improvement of PF in calculation speed or accuracy	Complex program design and difficult to capture nonlinear trends
Data-driven methods	(1)	Machine learning method	SAE-DNN (Liu et al., 2019a); MC-DNN (Wang et al., 2021)	Good ability of nonlinear; simple and easy implementation	Sensitive to the aging data quality, quantity, and over-fitting
			ELM (Xue et al., 2016); WT-ELM-GA (Chen et al., 2019a)	Fast calculation speed and good generalization ability	The correlation of sequences is weak.
			ESN (Morando et al., 2013); MR-ESN (Mezzi et al., 2018); EESN (Li et al., 2019); MIMO-ESN (Hua et al., 2020); LASSO-ESN (He et al., 2021)	Good ability of nonlinear; simple and easy implementation; strong correlation in time dimension	The prediction accuracy depends on the complexity of the network; large amount of calculation; sensitive to the parameter
			LSTM (Liu et al., 2019b); G-LSTM (Ma et al., 2018); S-LSTM (Wang et al., 2020a); BILSTM-AT (Wang et al., 2020b); A-LSTM (Zuo et al., 2021); NSD-LSTM (Wang et al., 2022); MIMO-LSTM (Zhang et al., 2021); MTW CNN-BLSTM Ensemble (Xia et al., 2020)	Good ability of nonlinear; simple and easy implementation; solve the problem of gradient disappearance and gradient explosion	The prediction accuracy depends on the complexity of the network; large amount of calculation; sensitive to the parameter
			RVM (Wu et al., 2015, 2016a, b); LSSVM-RPF (Cheng et al., 2018); SVR (Chen et al., 2022)	Simple and easy implementation	Sensitive to the aging data quality, quantity, and over-fitting
	(2)				

(continued)

Table 1 (continued)

Approach category		Subclass	Prediction approaches for fuel cell aging	Upside	Downside
		Signal processing method	WT-ELM (Javed et al., 2015a, b, 2016); WT-NARX (Chen et al., 2021); WT-GMDH (Liu et al., 2017); DWT-ARIMA (Ibrahim et al., 2016a); DWT-EESN (Hua et al., 2021a); WT-ESN (Hua et al., 2021b)	Simple and easy implementation; suitable for non-stationary time series	Sensitive to the aging data type and quantity; data pre-process, further prediction method is needed
	(3)	Statistical method	ARMA (Detti et al., 2019); ARMA-TDNN (Zhou et al., 2018)	Good generalization capability; simple and easy implementation	Stationary time series is a necessary requirement; poor nonlinear mapping ability
			GP (Zhu & Chen, 2018); SSA-DGP (Xie et al., 2020); VAE-DGP (Deng et al., 2022)	Explicit probabilistic formulation; a general framework of probabilistic regression	Significant computational burden for large datasets; sensitive to the aging data quality and quantity
			AFNIS (Silva et al., 2014; Mao & Jackson, 2016b); ANFIS-FCM (Liu et al., 2018); ECGM (Zhou et al., 2019); GNNM (Chen et al., 2019c)	Simple and easy implementation	Not good for long-term prediction; sensitive to data quality and quantity

models. Ao et al. (2021a) develop a catalyst degradation model based on the Pt dissolution and Ostwald ripening theory. The model describes catalyst degradation phenomena from the perspective of microscopic particles to predict fuel cell performance. Polverino and Pianese (2016) establish a theoretical model based on the electrochemical surface area (ECSA), an empirical formula for voltage decay is given. Based on the degradation mechanism of ECSA and resistance, Hu et al. (2018) propose a reconstructed fuel cell lifetime prediction model to estimate the performance degradation during demonstration and conduct experiments with dynamic conditions of urban buses. The results show that the prognostic error of the method is within 1%.

B. State Space Model Method

Different from the physical model method, the state space model method is to establish the system state equation according to the empirical model of fuel cell, and then the model parameter identification method is adopted to predict the aging behavior. Zhang et al. (2017) propose an empirical lifetime prediction model. The empirical model establishes the mathematical relationship between the voltage decay rate and the load curve, and a direct fitting method is used to identify the model aging coefficient. Similarly, Ou et al. (2021) also present the direct fitting method to identify the aging parameters of the empirical model and achieve the aging prediction of the fuel cell. However, the prediction accuracy of direct fitting method is low. This is because the model parameters are dynamic, and the dynamic parameters cannot be known by the direct fitting method.

In order to increase the accuracy of the direct fitting methods, the parameters of filtering methods are adaptively updated. Bressel et al. (2016) develop an empirical degradation model and adopt Extended Kalman Filter (EKF) to estimate the unknown parameters, achieving fast and accurate prediction of RUL. Pan et al. (2020) present an adaptive Kalman filtering method to predict the overall degradation trend. However, the calculation process of EKF is complicated. Chen et al. (2019b) develop an unscented Kalman filtering (UKF) method to address the computational complexity of EKF. However, the computational accuracy of UKF is reduced compared to EKF in the same condition. In order to ensure the calculation accuracy and calculation speed at the same time, Ao et al. (2021b) demonstrate a frequency domain Kalman filter (FDKF) algorithm to process data in groups and verify the calculation accuracy and calculation speed of the method with aging data under static and dynamic currents. In addition, different from KF framework, Zhou et al. (2017a) present a particle filter (PF) method based on the Monte Carlo method. This method develops a multi-physical aging model and the fuel cell stack voltage is predicted using the model parameters updated in the prediction stage. Besides, many improved PF methods, such as UPF (Chen et al., 2017), APF (Kimotho et al., 2014), RPF (Cheng et al., 2018), etc., have also been successfully applied in fuel cell aging prediction, and all of them have shown high prediction accuracy.

The model-based method does not need a large amount of prior data. However, the electrochemical reaction and degradation mechanism in fuel cells are complicated. Model-based methods depend heavily on model accuracy. Therefore, the model-based aging prediction method is difficult and challenging.

2.2 *Data-Driven Method*

The data-driven method adopts mathematical statistics or machine learning approaches to directly predict the aging behavior of fuel cells on the basis of prior data. This method does not need to build complex physical models but requires a large amount of training data. This method can be divided into: (1) machine learning

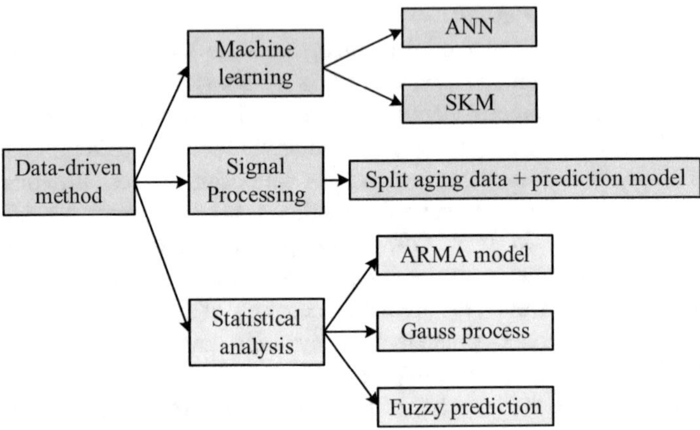

Fig. 3 Classification of data-driven methods

method; (2) signal processing method; and (3) statistical method. The classification of specific data-driven method is shown in Fig. 3.

A. Machine Learning Method

Machine learning method can capture nonlinear characteristic well, and it is often used in prediction problems with missing or complex models. Machine learning method can be divided into two categories: artificial neural network (ANN) and sparse kernel machine (SKM).

1. Artificial neural network

   ANN is composed of a large number of interconnected processing units, it can adaptively process nonlinear time series (Tealab et al., 2017). According to different network models, ANN can be divided into deep neural network (DNN), extreme learning machine (ELM), recurrent neural network (RNN), and others. At present, the ANN method is widely used in fuel cell aging prediction problems (Kui et al., 2018).

   DNN can obtain the features of original data from linear and nonlinear operations by establishing multiple hidden layers, and it is often used in time series forecasting problems (Zraibi et al., 2021). Liu et al. (2019a) propose a method based on the sparse autoencoder and deep neural network (SAE-DNN). The Gaussian-weighted moving average filter is used to remove the noise component in the aging data, and the DNN is applied to predict the aging data after filtering the noise. The results show that DNN has higher prediction accuracy than K-nearest neighbor and support vector regression machine method. Similarly, Wang et al. (2021) develop a hybrid method of Monte Carlo dropout approach and deep neural network (MC-DNN). The MC method is used to obtain the prediction interval, and the DNN method is adopted to predict the aging trend. This method also achieves high prediction accuracy.

   Different from DNN, extreme learning machine (ELM) is a type of single hidden layer feedforward neural network (SLFN) in terms of network structure.

ELM randomly selects nodes and analytically determines the output weights of SLFN. This method has extremely fast learning speed and generalization performance (Huang et al., 2006). On the basis of ELM, Xue et al. (2016) implement the fuel cell RUL estimation using an ELM method. Chen et al. (2019a) develop a combined algorithm of wavelet transform and extreme learning machine (WT-ELM), and then the genetic algorithm (GA) is adopted to optimize ELM network parameters. This method further improves the prediction accuracy of single ELM method. Although ELM has a fast-computing speed, it lacks a feedback loop mechanism and internal memory, so the ELM has weak correlation in time dimension.

Compared with ELM, recurrent neural network (RNN) considers real-time inputs as well as previously received inputs. It has a strong correlation in time dimension, thus suitable for dealing with the prediction of time series (Zhang & Man, 1998). As an RNN structure, the echo state network (ESN) is first applied to the RUL prediction of fuel cells by Morando et al. (2013). Subsequently, Mezzi et al. (2018) propose a multi-reservoir echo state network (MR-ESN) method and verify the accuracy of the proposed method from the perspectives of prediction accuracy and robustness. In order to further enhance the adaptability of aging prediction, Li et al. (2019) develop an ensemble echo state network (EESN), and the results show that the predicted results change with different ensembles. In addition, Hua et al. (2020) present a multiple-input multiple-output echo state network (MIMO-ESN) that jointly adopt voltage, current, temperature and reactant pressure for fuel cell aging prediction. The experiments are carried out with static and dynamic current conditions, and the results show that the MIMO-ESN has higher prediction accuracy than the single-input single-output ESN. In a recent study, in order to optimize the parameters of ESN, He et al. (2021) propose a least absolute shrinkage and selection operator-echo state network (LASSO-ESN). The method achieves long-term aging prediction of fuel cells.

As an improved method of RNN, long short-term memory (LSTM) can solve the problem of RNN gradient disappearance and gradient explosion, and it is widely used in fuel cell aging prediction. Liu et al. (2019b) apply the LSTM algorithm to the RUL prediction problem. The experimental results show that the prediction accuracy of LSTM is much higher than back propagation neural network (BPNN). In order to optimize the LSTM network structure, Ma et al. (2018) develop a grid long short-term memory (G-LSTM) method. The prediction accuracy of this method is higher than the single LSTM method in the four aging datasets from the experiment. Similarly, Wang et al. (2020a) present a stacked long short-term memory (S-LSTM) algorithm and adopt a differential evolution algorithm to optimize the model parameters. Afterward, an attention mechanism is added and a bidirectional long short-term memory algorithm with attention mechanism (BILSTM-AT) model is proposed to further improve the prediction accuracy (Wang et al., 2020b). On this basis, Zuo et al. (2021) develop an attention-based long short-term memory (A-LSTM) method and conduct a 1000-h aging test experiment to verify the accuracy of the proposed method. In addition, there are many improved methods of LSTM applied in fuel cell aging prediction, such as navigation sequence driven long short-term memory (NSD-LSTM) (Wang et al., 2022), multi-step input and multi-step output long short-term memory (MIMO-LSTM) (Zhang et al., 2021), ensemble framework based on convolutional bi-directional long short-term memory with multiple time windows (MTW CNN-BLSTM Ensemble) (Xia et al., 2020).

2. Sparse kernel machine (SKM)

   Sparse kernel machine is a kernel-based algorithm with sparse solution, and it is a branch of machine learning algorithm. This method mainly includes support vector machine (SVM) and relevance vector machine (RVM) (Rao & Man-Wai, 2016). Wu et al. (2015) apply RVM algorithm to fuel cell aging prediction and accurately predict the aging trend. Subsequently, a modified RVM algorithm is developed for the characteristics of aging data, and a new RVM correction formula is deduced for fuel cell aging prediction (Wu et al., 2016a). However, it is difficult for a single RVM algorithm to capture the local nonlinear characteristics in the aging process. In order to better describe nonlinear performance in the aging process, Wu et al. (2016b) develop an adaptive RVM algorithm to modify the design matrix and verify the prediction accuracy improvement of the nonlinear characteristics from single-step and multi-step prediction ways. Cheng et al. (2018) propose a method combining the least square support vector machine (LSSVM) and the regularized particle filter (RPF). This method provides the uncertainty representation of RUL with probability distribution. Compared with the single RPF algorithm, LSSVM algorithm improves the nonlinear characteristics capture ability of the prediction model. In addition, Chen et al. (2022) present a support vector regression (SVR) algorithm for fuel cell aging prediction and adopt the Grey Wolf Optimizer (GWO) to optimize the parameters of SVR. The aging datasets of static and dynamic current are used. The results show that the mean absolute percentage error (MAPE) of this method is less than 0.003 and the RUL of 492 h can be predicted.

B. Signal Processing Method

The signal processing method splits or filters the original aging data to eliminate irrelevant information and noise, thus facilitating the implementation of the prediction method (Hua et al., 2022). Signal processing method is an important tool for fuel cell aging prediction, and it shows a good prediction accuracy.

Javed et al. (2015a, b, 2016) propose a wavelet transform and extreme learning machine (WT-ELM) method. The WT improves the ability of forecasting method by processing uncertain aging data, and the ELM is adopted to predict the processed aging data components. The aging datasets in static and dynamic current verify the ability of the method to predict long-term aging behavior. Chen et al. (2019a) implement GA to optimize the global parameters on the basis of WT-ELM and consider the effects of PEMFC load current, relative humidity, temperature, and hydrogen pressure in aging process. Subsequently, a hybrid method of wavelet transform and nonlinear autoregressive exogenous neural network (WT-NARX) is developed, this method further adds the historical state information into the prediction network (Chen et al., 2021). Liu et al. (2017) develop a group method of data handling approach based on wavelet transform (WT-GMDH). The WT splits original aging data into multiple sub-waveforms, and then predicts each sub-waveform separately by GMDH. This method can effectively reduce the vibration problem at the beginning of each experiment cycle. However, these methods can only make short-term predictions. Ibrahim et al. (2016a) propose a method that combines discrete wavelet transform (DWT) and various prediction algorithms. This method uses Auto-Regressive Integrated Moving Average (ARIMA) and polynomial regression method to predict the sub-waveform separately. It can effectively solve the problem of low long-term prediction accuracy of RUL. Hua et al. (2021a) develop a method combining DWT with EESN. This method is used to deal with the multi-time scale characteristics of relative power-loss rate health indicators and fuse the prediction results from different time scales. The results show that this method has higher prediction accuracy than the traditional ESN. In addition, another work of the author uses GA to optimize the key parameters of ESN. This method also improves prediction accuracy (Hua et al., 2021b).

C. Statistical Method

The statistical method is to establish a random statistical model on the basis of the prior data to describe the uncertainty of the whole sequence. Then the stochastic process or probability distribution is used to describe the subsequent changes of the model. There are many uncertain factors in fuel cell aging data, it is suitable to use statistical method to predict the aging data with the theory of statistical method (Hu et al., 2020). Statistical methods can be divided into autoregressive moving average model (ARMA) method, Gaussian process method and fuzzy prediction method.

1. Autoregressive moving average model (ARMA)

   ARMA model is a statistical method, which combines AR model and MA model. It can effectively deal with the prediction of time series. Detti et al. (2019) apply ARMA model to fuel cell aging prediction, and the result shows that ARMA can accurately predict the linear aging trend. However, this method cannot accurately capture the nonlinear trend. Zhou et al. (2018) develop an ARMA and time delay neural network (ARMA-TDNN) method, which adopts ARMA to predict the linear component of aging data and TDNN to predict the nonlinear component. This method combines the advantages of statistical model and machine learning method thus effectively improving the prediction accuracy.
2. Gaussian Process (GP)

   Gaussian process is the joint distribution of all random variables in time domain, and it is widely used in multi-step prediction of time series (Girard et al., 2003; Brahim-Belhouari & Bermak, 2004). Zhu and Chen (2018) develop a state space method based on GP to infer the aging trend of fuel cells. Xie et al. (2020) propose a method of singular spectrum analysis and deep Gaussian process (SSA-DGP), which effectively eliminates the peak phenomenon caused by high-frequency noise in the aging prediction process. Deng et al. (2022) present two statistical models: sparse pseudo-input Gaussian process (SPGP) and variable auto-encoded deep Gaussian process (VAE-DGP). Long-term aging experiments are carried out with static and dynamic current conditions, respectively. The results show that SPGP has higher prediction accuracy in large samples, and VAE-DGP is more suitable for aging prediction in small samples.

3. Fuzzy prediction

   As a statistical method, fuzzy prediction method can make a long-term fuzzy description of uncertain problems. Jang (1993) propose an adaptive neuro-fuzzy system (ANFIS) method based on fuzzy prediction method, which can effectively deal with the prediction of time series. On the basis of this theory, Silva et al. (2014) adopt ANFIS method and analyze the selection of the model parameters. By extracting the aging data under normal operation and removing external disturbance, the prediction results can follow the aging trend in real time. Mao and Jackson (2016b) compare PF, NN, and ANFIS algorithms with average prediction error and calculation cost. The results show that ANFIS has the best performance in both aspects. Liu et al. (2018) develop an ANFIS with fuzzy c-means (ANFIS-FCM) strategy, and adopts particle swarm optimization (PSO) algorithm to select the relevant parameters. The results show that the root mean square error (RMSE) of the proposed method in single-step prediction is only 0.62%.

   Grey model (GM) is also developed based on fuzzy prediction method. Because of the original error of GM, Zhou et al. (2019) develop an error correction grey model (ECGM) to decrease the original error of GM and improve the accuracy of the prediction model. Chen et al. (2019c) propose a grey neural network (GNNM) algorithm, and apply PSO to optimize the parameters of the network, considering the aging factors from various working conditions. The results show that this method performs high accuracy in the prediction of small sample aging of fuel cells.

   Data-driven method avoids the establishment of complex physical model, and its prediction model is a "black box." It can accurately track the linear and nonlinear trends of aging data, when training time is enough. However, in order to ensure the calculation speed, the prediction accuracy and calculation time have to be compromised. In addition, the parameter selection of neural network model is a major factor for the prediction accuracy. Therefore, a lot of debugging or optimization algorithm is necessary for appropriate model parameters.

3 Hybrid Prediction Method

In this section, the state space model prediction method is first introduced and the concrete implementation steps of this method are described in detail. Then, a data-driven prognostic method is introduced in detail. Finally, a hybrid prediction method is developed based on the sliding prediction length method.

The state space model prediction method is able to track the linear decay trend of the fuel cell more accurately. For local fluctuations in the decay process and sudden changes at the beginning of each experimental cycle, the data-driven approach has more obvious advantages.

3.1 State Space Model Prognostic Approach

In this section, a state space model prediction method is adopted to predict the overall trend in the fuel cell decay process. First, the aging data of fuel cells are fitted, and the linear aging process of fuel cells can be obtained according to the fitting results (Miao et al., 2013; Wang et al., 2013; He et al., 2011). The specific process is shown in Eq. (1):

$$U_c(t) = a(t) \cdot \exp(b(t) \cdot t) + c(t) \cdot \exp(d(t) \cdot t), \tag{1}$$

where a, b, c, and d are the parameters of empirical formula, the parameters can be determined according to the identification method. $U_c(t)$ is the value of the output voltage of fuel cell. In order to further identify the unknown parameters for a reliable model, state space model method is adopted to identify the unknown parameter. The state space model of fuel cell can be expressed as the following form:

$$\begin{cases} x(t) = x(t-1) + Q_{t-1} \\ y(t) = a(t) \cdot \exp(b(t) \cdot t) + c(t) \cdot \exp(d(t) \cdot t) \end{cases}, \tag{2}$$

where $x(t)$ is the state variable at state t; $y(t)$ is the observed variable obtained by the sensor; Q_{t-1} represents process noise.

Since the state space model is not linear, nonlinear parameter identification or estimation methods should be used to deal with the state space model. The particle filter method can express the posterior probability distribution based on the observed and control quantities more accurately and can solve the Monte Carlo problem effectively (Zhou et al., 2016b). Therefore, the particle filter method is used to identify the parameters in the state space model and predict the state variables to obtain the overall degradation trajectory of the fuel cell. The flow chart of the state space model prediction method is shown in Fig. 4, followed by a detailed description of each step of the prediction method.

A. Initialization of State Space Model Prediction Method

The initialization of state model prediction method mainly includes three key processes:

1. Initialization of critical points of training set and prediction set: the total number of time points of all degradation experimental data is initialized to N_{com}. The

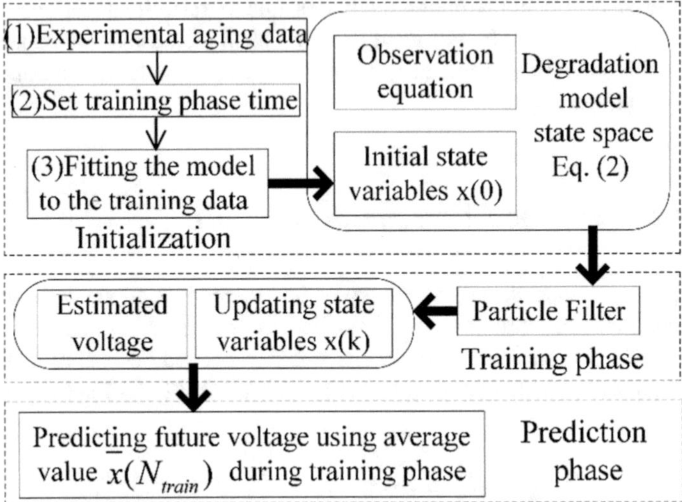

Fig. 4 The flow chart of state space model prediction method (Zhou et al., 2017b)

critical time point between the training set and the prediction set is set to N_{cri}. The length of the training set is N_{cri} and the length of the prediction set is $N_{\text{com}} - N_{\text{cri}}$.

2. Initialization of state variables: the initial values of state variables are determined by the fitting results of experimental aging data.
3. Initialization of particle number: too large particle number will lead to slow computation; too small particle number will lead to a lack of particle diversity and unreliable identification results. Therefore, it is necessary to weigh these two factors and set the appropriate number of particles.

B. Training Phase of State Space Model Method

After the initialization of the model, particle filtering is used to continuously use the data in the training set as observations to predict the state quantities and model parameters at the next moment. In this process, the variance of the process noise Q_t has an important impact on the accuracy of parameter identification and model update. Too large a value of Q_t may lead to slow convergence of the prediction process and increased computational time; too small a value of Q_t may lead to overfitting of the prediction results. Therefore, the variance value of Q_t must be carefully determined.

In addition, in the parameter identification pre-update stage of the model, corrections are to be made based on the observed and estimated quantities, so that the weights of each particle are reassigned and resampled to estimate the state quantities at the next moment. The specific implementation process can be found in Ref. Zhou et al. (2016b, 2017a).

C. Prognostic Phase of State Space Model Method

In the prediction phase, a prediction particle can be obtained for each particle based on the state transfer equation of the trained model plus a control volume. Each predicted particle is then evaluated and a larger weight is set for the particles that are close to the true value. In setting the weights, a resampling method is used to ensure the diversity of particles. It is worth noting that the initial state volume in the prognostic process is the average of the state volume in the training phase, as shown in Eq. (3):

$$\overline{x}_{0-N_{\text{cri}}} = \frac{\sum_{t=0}^{N_{\text{cri}}} \widehat{x}_t}{n_{\text{cri}}}, \tag{3}$$

where n_{cri} is the number of the points of training set. Then, the resampled particles are brought into the model to obtain the predicted values for the next moment. The process is shown in Eq. (4):

$$\widehat{y}_t = \overline{a}_{0-N_{\text{cri}}} \cdot \exp\left(\overline{b}_{0-N_{\text{cri}}} \cdot t\right) + \overline{c}_{0-N_{\text{cri}}} \cdot \exp\left(\overline{d}_{0-N_{\text{cri}}} \cdot t\right). \tag{4}$$

3.2 Data-Driven Prediction Method

Since the characteristics of local fluctuations in the fuel cell aging process are difficult to be described by empirical formulas, the functional relationship of its input and output is difficult to be obtained. Therefore, the state space model prediction method is not suitable for the prediction of local nonlinearities and fluctuations. Compared with the state space model prediction method, the data-driven method is able to train the input–output relationship by a large amount of known data without establishing a specific functional model to map the input to the output, thus effectively predicting the complex local fluctuation situation.

As a data-driven method, the neural network approach is able to build a reliable "black box" model by building hidden layers and setting multiple neurons to continuously train the input and output. For the local fluctuation features of the fuel cell aging data, the neural network approach can be used to continuously train these known features and effectively predict the unknown features by the trained neural network model. As described in Sect. 2, there are various neural network approaches, among which NARNN is used to predict the local fluctuation trend of aging data.

A. Nonlinear Autoregressive Neural Network Model

NARNN has the advantage to describe the local fluctuation characteristics due to the unique delay unit mechanism and the ability to provide real-time feedback on the accuracy of the output with aspect to the next moment of input (Benmouiza & Cheknane, 2013; Ibrahim et al., 2016b; Arbain & Wibowo, 2012). The flowchart of the algorithm for the prediction of fuel cell aging data using the NARNN method is shown in Fig. 5.

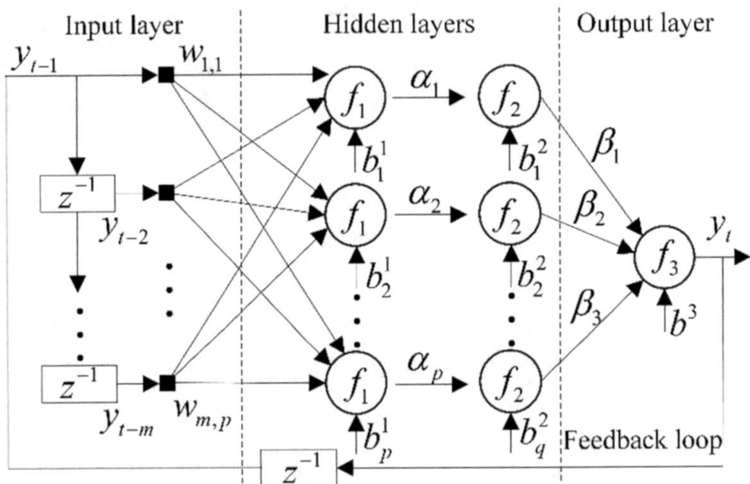

Fig. 5 The topological graph of NARNN

As can be seen from Fig. 5, the NARNN model has a delay layer, and its input is a vector of previous output values (Benmouiza & Cheknane, 2013; Chow & Leung, 1996). The specific expression form is shown in Eq. (5):

$$y(t) = f_3\left(w^3 f_2\left(w^2 f_1\left(w^1 U_t + b^1\right) + b^2\right) + b^3\right), \tag{5}$$

where $y(t)$ is the prognostic output by the network; U_t is the input value and the previous output value; w and b represent the weights and bias for the unit of the hidden layer.

B. The Specific Process of NARNN Prediction Method

The flow chart of the specific process of NARNN prediction method is shown in Fig. 6. NARNN is divided into three main phases: training phase, evaluation phase, and prediction phase. In the training phase, initialization is first performed to set up the hidden and delay layers of the network as well as the weights factor of each layer. Compared with the traditional accelerated training convergence algorithm, Levenberg–Marquardt can make a linear approximation to the parameters to be evaluated in their neighborhoods, ignoring derivative terms above the second order, thus transforming the nonlinear problem into a linear least squares problem and thus reducing the time required for convergence (Çay et al., 2013). Therefore, the Levenberg–Marquardt algorithm is used in the training process.

In addition, to prevent overfitting or divergence during the training process, the data obtained from the training phase should be evaluated. When the mean square error (MSE) between the output obtained during the training phase of NARNN and the output of the actual training set is no longer decreasing, the prediction phase is entered. Although this approach leads to a limitation in the accuracy of the training

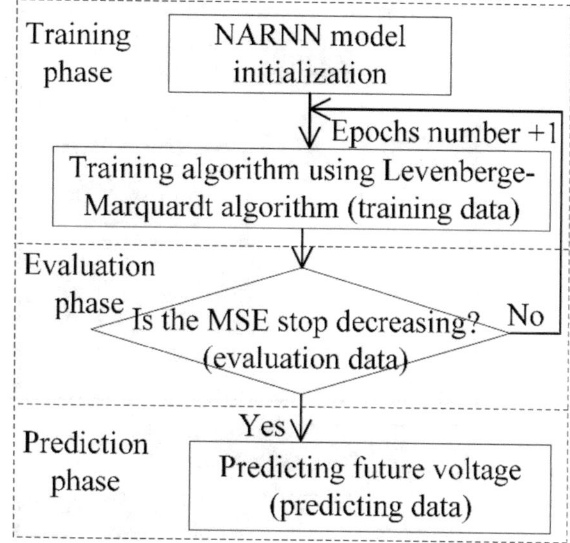

Fig. 6 The flow chart of NARNN method (Zhou et al., 2017b)

phase, this approach will shorten the time of the training process and at the same time enhance the generalization ability of NARNN in the prediction process, which is beneficial for the NARNN model to predict unknown datasets (Marra et al., 2013). Finally, a multi-step prognostic approach is adopted for the prediction phase.

3.3 *Hybrid Prognostic Approaches*

It can be found that suitable fusion methods can substantially improve the prediction accuracy based on the collection of different methods for fuel cell aging prediction and the comparison of their results (Luo et al., 2009; Chen & Vachtsevanos, 2012; Sankavaram et al., 2009). The hybrid prediction method can make the prediction accuracy perform optimally by combining the advantages between different prediction algorithms and designing a reasonable fusion method. Therefore, compared with a single prediction method, the prediction results of the hybrid method obviously have higher accuracy and robustness.

It should be noted that designing a reasonable fusion way is the key to hybrid methods. It is significant to make a detailed analysis of the characteristics of each prediction algorithm to be fused. Based on the analysis in the previous subsection, the state space model prediction method is able to track the linear decay trend of the fuel cell more accurately, but it is difficult to predict the local oscillations and fluctuations of the aging data. In contrast, the NARNN prediction method can predict local fluctuations and sudden changes at the beginning of each experimental cycle more accurately, but it is difficult to predict the overall degradation trajectory accurately. In order to fully combine the advantages of both methods, the prediction

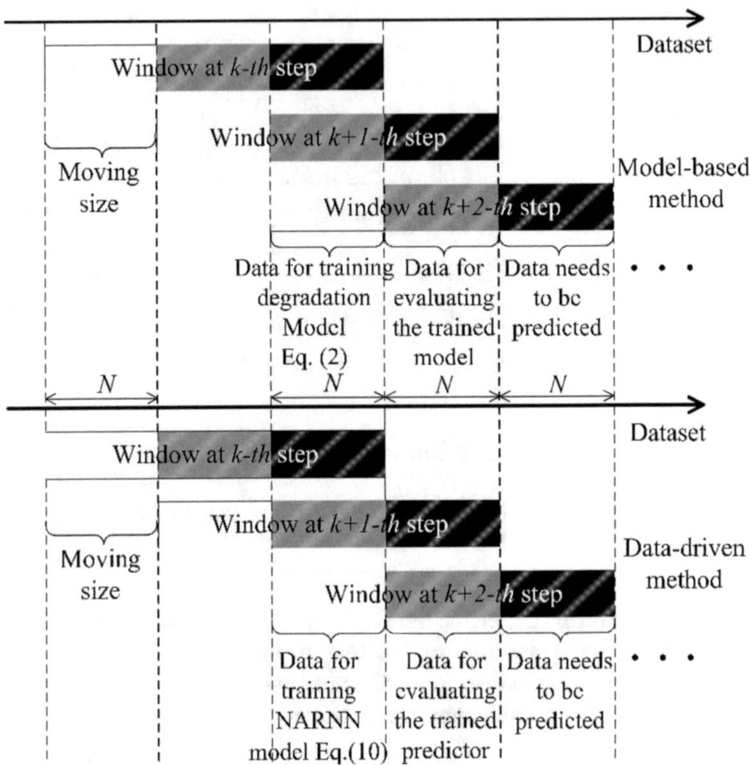

Fig. 7 The specific flow of sliding prediction length method (Zhou et al., 2017b)

length of each step and the weights of each prediction method should be designed thoughtfully. In this paper, a sliding prediction length method is proposed to ensure the prediction accuracy of the hybrid method.

A. Sliding Prediction Length Method

In order to ensure the update of the data source of the training set in the prediction process, the optimal weight factors are assigned to different prediction methods in each prediction process to obtain the highest prediction accuracy. The sliding prediction length method is developed for hybrid prediction of fuel cell aging process. The specific flow of this method is shown in Fig. 7.

It can be seen from Fig. 7 that the whole sliding prediction length method is divided into three stages, where N is the length of each sliding step for prediction process as well as the length of the training set. In this way, it can be ensured that each time the state space model method and NARNN method keep the same prediction step size for fusion. This method can ensure the continuous updating of training data in the prediction process and make sliding step-by-step predictions. In addition, this method is conducive to the rational distribution of the weight factors of different prediction methods and improves the robustness of the hybrid method.

B. Weight Factors Assignment

In order to integrate the two prediction methods more rationally, the prediction results of the two prediction methods are compared with the actual measured voltage results after each step prediction, and the deviation values are recorded. Through the comparison, the prediction method with smaller deviation value is assigned a larger weight for the next prediction step. The weights are assigned in Eq. (6):

$$w_{p,k} = \frac{1}{\sum_{i=1}^{N} \sqrt{\left(y_{\mathrm{eva,p},k}(i) - \widehat{y}_{\mathrm{eva},p,k}(i)\right)^2}}, \tag{6}$$

where k is the actual prediction section; $p \in \{1, 2\}$ represents the two different prediction method; $y_{\mathrm{eva},\ p,\ k}$ is the monitoring voltage, $\widehat{y}_{\mathrm{eva},p,k}$ represents the predicted values by the single prediction method.

After calculating the weights, it is necessary to normalize the weight factors of the two methods so as to further integrate the two methods, the process of weight factors normalization is as follows:

$$w_{\mathrm{norm},p,k} = \frac{w_{p,k}}{\sum_{p=1}^{P} w_{p,k}}, \tag{7}$$

where the $w_{\mathrm{norm},\ p,\ k}$ is the normalized weight factor. According to the result of weight factor normalization, the final fusion mode can be obtained as follows:

$$y_{\mathrm{fus},k} = \sum_{p=1}^{P} y_{\mathrm{pre},p,k} \cdot w_{\mathrm{norm},p,k}, \tag{8}$$

where $y_{\mathrm{pre},\ p,\ k}$ is the prediction output voltage of two approaches. The proposed hybrid prediction method is shown in Fig. 8.

4 Experimental Results

In this section, the fuel cell configuration and operating conditions for the aging experiments are shown first. The specific parameter configuration is shown in Table 2, and the aging dataset of the fuel cell is shown in Fig. 9. Then the proposed hybrid method is experimented and compared with the results of some other advanced algorithms to verify the accuracy of the prediction results of the proposed method.

The prediction results of the proposed hybrid method of state space model and NARNN and sliding prediction length are shown in Fig. 10. It can be seen from Fig. 10 that the prediction results of the proposed hybrid method have higher

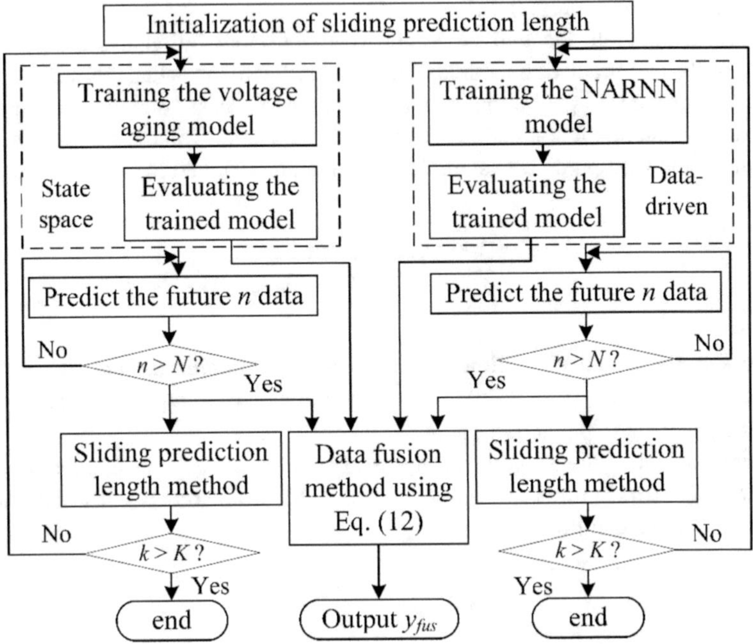

Fig. 8 The flow chart of the hybrid prediction method (Zhou et al., 2017b)

Table 2 Operation conditions: 600 W fuel cell stack

Operating environment	Specific parameters
Stack type	600 W PEMFC assembled at FCLAB
Number of cells	5
Operation mode	Recirculation mode
Air supply	Air boiler
Cooling	Cooling water system
Operating hours	1200 h
Active area	100 cm^2
Stack current density	0.70 A/cm^2

accuracy compared to the single state space model prediction method and the NARNN prediction method. This is mainly due to the fact that the proposed hybrid method can combine the advantages of both above methods and capture both linear aging trajectories as well as local nonlinear characteristics.

The average prediction error results of the proposed hybrid method and other advanced algorithms are shown in Table 3. From Table 3, it can be seen that the proposed hybrid method improves the prediction accuracy by 26% compared to the single state space model prediction method and 22% compared to the single NARNN prediction method. This further demonstrates that the proposed hybrid method is indeed better than the single prediction method. In addition, the proposed

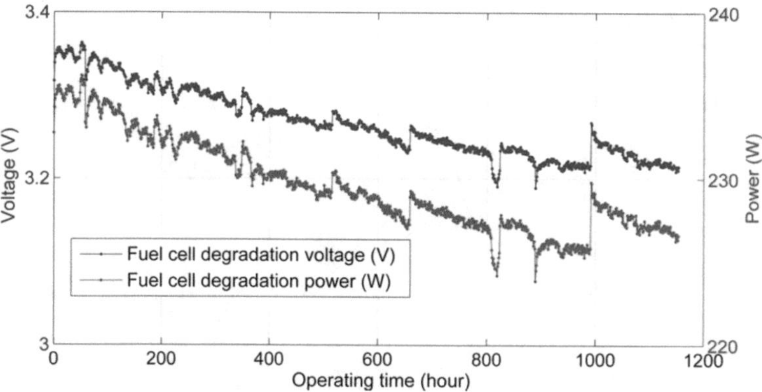

Fig. 9 The experiment dataset with static current density

hybrid method improves the prediction accuracy by 21% compared to the ANFIS algorithm. It should be noted that the ANFIS algorithm proposed by Mao and Jackson (2016a) has a longer training set (0–825 h), while the proposed method only needs (0–330 h) to provide a small amount of data to obtain higher accuracy prediction results. Moreover, the proposed sliding prediction length method can also continuously update the prediction data and dynamically update the model parameters, which can fully guarantee the reliability of the model. The above experimental results and analysis verify the superiority of the proposed hybrid method.

In order to further verify the prediction performance of the proposed method aiming at each sliding prediction length, the proposed method is compared with a single prediction method by sliding prediction step, and the comparison results are shown in Fig. 11. It can be seen from Fig. 11 that the prediction accuracy of the proposed hybrid method is obviously higher than the single prediction method in the same way. In addition, the prediction results of auto-regressive integrated moving average (ARIMA) (Ibrahim et al., 2016a) method are compared with the proposed method. Among them, the sliding prediction step is 168 h, and the specific ARIMA method formula is as follows:

$$y_t = c + \alpha_1 y_{t-1} + \alpha_2 y_{t-2} + \ldots + \alpha_p y_{t-p} + \theta_1 u_{t-1} + \theta_2 u_{t-2} + \ldots + \theta_q u_{t-q} + u_t, \tag{9}$$

where p is the autoregressive polynomial order, q is the moving average order, y_t is a univariate time series, α_1, α_1, ..., α_p and θ_1, θ_2, ..., θ_q are the parameters (or coefficients) of the model, and u_t is a white noise. d in ARIMA(p, d, q) is the differentiation order when the time series is not stationary.

As shown in Table 4, the proposed method has the highest accuracy of prediction results throughout the experimental cycle. Compared with the single state space model, NARNN and ARIMA prediction methods, the proposed hybrid method has improved the prediction accuracy by 20.7%, 19.4%, and 18.5%, respectively. The experimental results further validate the accuracy of the proposed method.

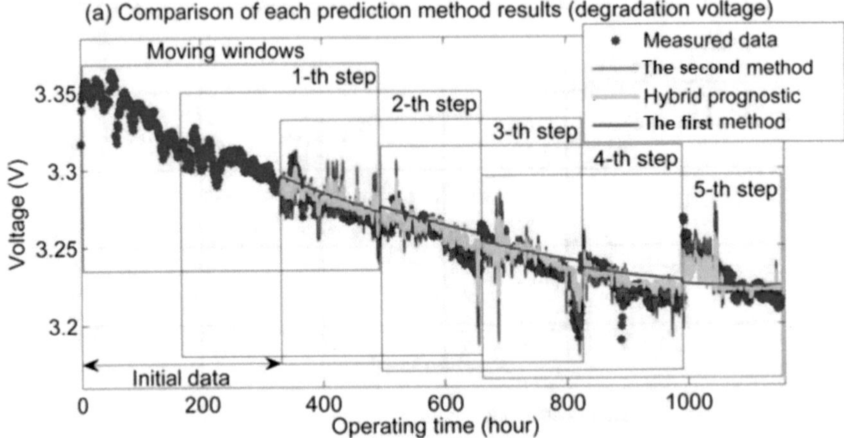

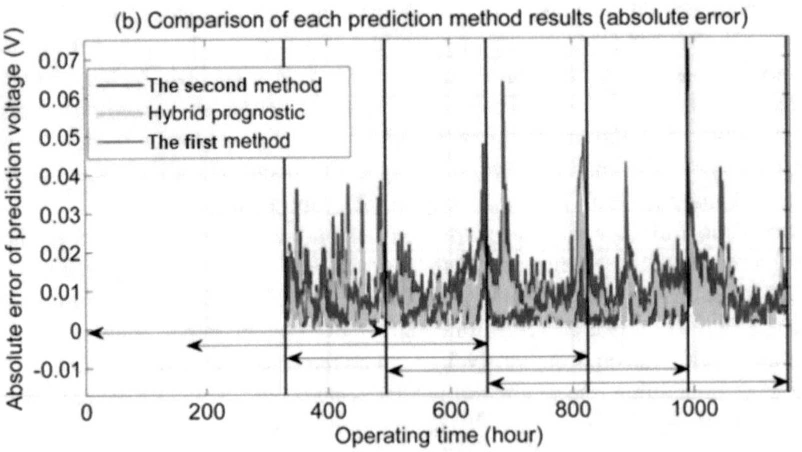

Fig. 10 The prediction results of three methods (whole dataset range)

Table 3 Comparison of mean prediction error between the proposed methods and ANFIS algorithm (Mao & Jackson, 2016a)

	State space model method	Machine learning method	Hybrid prognostic method	ANFIS algorithm (Mao & Jackson, 2016a)
Mean prediction error	0.0093	0.0089	0.0069	0.0087

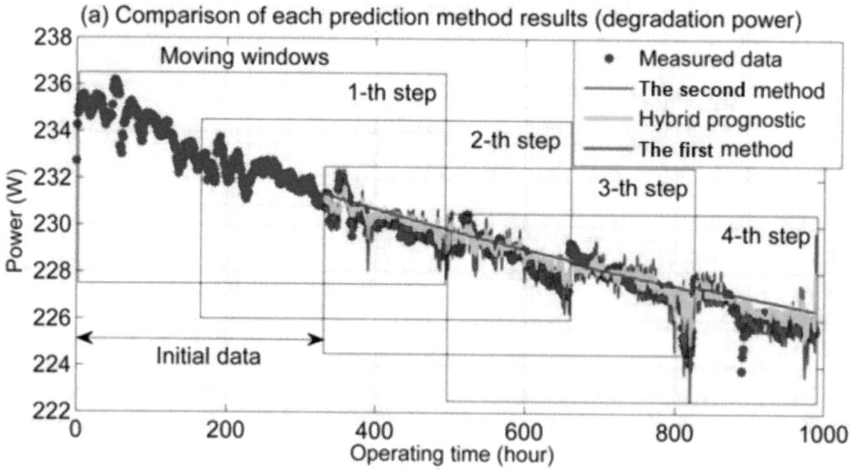

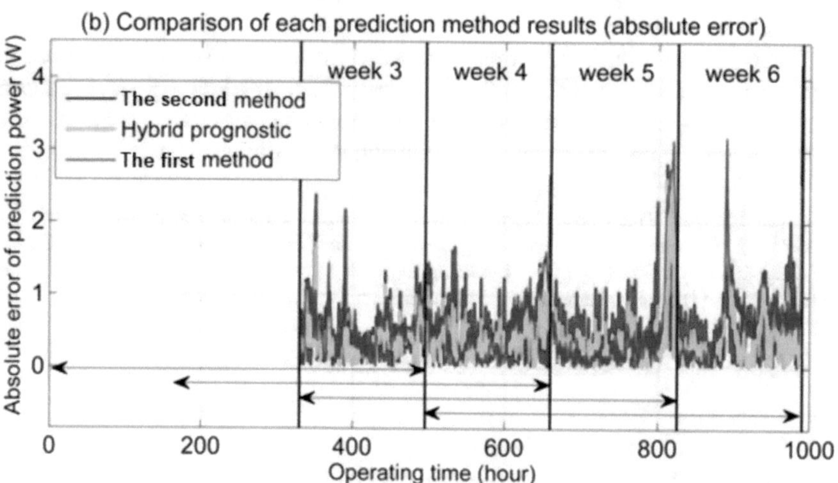

Fig. 11 The prediction results of three methods (sliding prediction length range)

5 Conclusion

In order to avoid the economic losses and potential safety hazards caused by the sudden degradation of fuel cells, prognostic and health management (PHM) has been widely studied. As an important part of PHM, prognostic can judge the aging behavior of fuel cells in advance, so as to avoid the occurrence of risks.

In this chapter, a novel hybrid prognostic method is proposed. This method can accurately predict the overall aging trend and local fluctuation of fuel cells. It is suitable for any type of fuel cell aging data, which makes up for the poor universality of single prediction method.

Table 4 Comparison of RMSEs between the proposed methods and ARIMA algorithm (Ibrahim et al., 2016a)

	State space model method	Data-driven method	Hybrid prognostic method	ARIMA algorithm (Ibrahim et al., 2016a)
Week 3	0.6184	0.5832	0.4778	0.4
Week 4	0.6458	0.6377	0.5145	0.6
Week 5	0.8026	0.8571	0.6633	0.7
Week 6	0.8027	0.7523	0.6265	1.1
Total 4 weeks	0.7198	0.7075	0.5705	0.7

Moreover, a prediction method with sliding prediction length is developed. This method can ensure the continuous updating of training data in the prediction process, thus ensuring the accuracy of the state space model, and further improving the prediction accuracy. In addition, this method is conducive to the rational assignment of the weight factors of different prediction methods and improves the robustness of the hybrid method.

Finally, the prediction performance of different methods is compared from the whole aging dataset range and each sliding prediction length range, and the superiority of the proposed method is verified. The research results can provide ideas for multi-step prediction of fuel cell aging in engineering applications.

Acknowledgments This work is supported by: (1) National Natural Science Foundation of China (51977177); (2) Shaanxi Province Key Research and Development Plan 2022QCY-LL-11, 2021ZDLGY11-04; (3) Basic Research Plan of Natural Science in Shaanxi Province (2020JQ-152); (4) the Fundamental Research Funds for the Central Universities (D5000210763).

References

Ao, Y., Chen, K., Laghrouche, S., & Depernet, D. (2021a). Proton exchange membrane fuel cell degradation model based on catalyst transformation theory. *International Journal of Hydrogen Energy, 21*, 254–268.

Ao, Y., Laghrouche, S., Depernet, D., & Chen, K. (2021b). Proton exchange membrane fuel cell prognosis based on frequency-domain Kalman filter. *IEEE Transactions on Transportation Electrification, 7*, 2332–2343.

Arbain, S. H., & Wibowo, A. (2012). Neural networks based nonlinear time series regression for water level forecasting of Dungun River. *Journal of Computer Science, 8*, 1506.

Benmouiza, K., & Cheknane, A. (2013). Forecasting hourly global solar radiation using hybrid k-means and nonlinear autoregressive neural network models. *Energy Conversion and Management, 75*, 561–569.

Brahim-Belhouari, S., & Bermak, A. (2004). Gaussian process for nonstationary time series prediction. *Computational Statistics & Data Analysis, 47*, 705–712.

Bressel, M., Hilairet, M., Hissel, D., & Bouamama, B. O. (2016). Extended Kalman filter for prognostic of proton exchange membrane fuel cell. *Applied Energy, 164*, 220–227.

Çay, Y., Korkmaz, I., Çiçek, A., & Kara, F. (2013). Prediction of engine performance and exhaust emissions for gasoline and methanol using artificial neural network. *Energy, 50*, 177–186.

Chen, C., & Vachtsevanos, G. (2012). Orchard M. *Machine remaining useful life prediction: An integrated adaptive neuro-fuzzy and high-order particle filtering approach., 28*, 597–607.

Chen, B., Wang, J., Yang, T., Cai, Y., Zhang, C., Chan, S. H., et al. (2016). Carbon corrosion and performance degradation mechanism in a proton exchange membrane fuel cell with dead-ended anode and cathode. *Energy, 106*, 54–62.

Chen, J., Zhou, D., Lyu, C., & Lu, C. (2017). A novel health indicator for PEMFC state of health estimation and remaining useful life prediction. *Energy Science and Engineering, 42*, 20230–20238.

Chen, K., Laghrouche, S., & Djerdir, A. J. (2019a). Degradation model of proton exchange membrane fuel cell based on a novel hybrid method. *Applied Energy, 252*, 113439.

Chen, K., Laghrouche, S., & Djerdir, A. (2019b). Fuel cell health prognosis using Unscented Kalman Filter: Postal fuel cell electric vehicles case study. *Engineering Sciences, 44*, 1930–1939.

Chen, K., Laghrouche, S., & Djerdir, A. (2019c). Degradation prediction of proton exchange membrane fuel cell based on grey neural network model and particle swarm optimization. *Energy Conversion and Management, 195*, 810–818.

Chen, K., Laghrouche, S., & Djerdir, A. (2021). Prognosis of fuel cell degradation under different applications using wavelet analysis and nonlinear autoregressive exogenous neural network. *Renewable Energy, 179*, 802–814.

Chen, K., Laghrouche, S., & Djerdir, A. (2022). Remaining useful life prediction for fuel cell based on support vector regression and Grey wolf optimizer Algorithm. *IEEE Transactions on Energy Conversion*. https://doi.org/10.1109/TEC.2021.3121650

Cheng, Y., Zerhouni, N., & Lu, C. (2018). A hybrid remaining useful life prognostic method for proton exchange membrane fuel cell. *International Journal of Hydrogen Energy, 43*, 12314–12327.

Chow, T. W., & Leung, C.-T. (1996). Neural network based short-term load forecasting using weather compensation. *IEEE Transactions on Power Systems, 11*, 1736–1742.

Deng, H., Hu, W., Cao, D., Chen, W., Huang, Q., Chen, Z., et al. (2022). Degradation trajectories prognosis for PEM fuel cell systems based on Gaussian process regression. *Energy, 244*, 122569.

Detti, A. H., Steiner, N. Y., Bouillaut, L., Same, A., & Jemei, S. (2019). Fuel cell performance prediction using an AutoRegressive moving-average ARMA model. In *2019 IEEE Vehicle Power and Propulsion Conference (VPPC)*. IEEE, pp. 1–5.

Fowler, M. W., Mann, R. F., Amphlett, J. C., Peppley, B. A., & Roberge, P. R. J. (2002). Incorporation of voltage degradation into a generalised steady state electrochemical model for a PEM fuel cell. *Journal of Power Sources, 106*, 274–283.

Girard, A., Rasmussen, C., Candela, J. Q., & Murray-Smith, R. (2003). Gaussian process priors with uncertain inputs application to multiple-step ahead time series forecasting. *Advances in Neural Information Processing Systems, 15*.

He, W., Williard, N., Osterman, M., & Pecht, M. (2011). Prognostics of lithium-ion batteries based on Dempster–Shafer theory and the Bayesian Monte Carlo method. *Journal of Power Sources, 196*, 10314–10321.

He, K., Mao, L., Yu, J., Huang, W., He, Q., Jackson, L., et al. (2021). Long-term performance prediction of PEMFC based on LASSO-ESN. *Engineering Material Science, 70*, 1–11.

Hu, Z., Xu, L., Li, J., Ouyang, M., Song, Z., Huang, H., et al. (2018). A reconstructed fuel cell life-prediction model for a fuel cell hybrid city bus. *Energy Conversion and Management, 156*, 723–732.

Hu, X., Xu, L., Lin, X., & Pecht, M. (2020). Battery lifetime prognostics. *Joule, 4*, 310–346.

Hua, Z., Zheng, Z., Péra, M.-C., & Gao, F. (2020). Remaining useful life prediction of PEMFC systems based on the multi-input echo state network. *Applied Energy, 265*, 114791.

Hua, Z., Zheng, Z., Pahon, E., Péra, M.-C., & Gao, F. (2021a). Multi-timescale lifespan prediction for PEMFC systems under dynamic operating conditions. *IEEE Transactions on Transportation Electrification, 8*, 345–355.

Hua, Z., Zheng, Z., Pahon, E., Péra, M.-C., & Gao, F. (2021b). Lifespan prediction for proton exchange membrane fuel cells based on wavelet transform and Echo state network. *IEEE Transactions on Transportation Electrification, 8*(1).

Hua, Z., Zheng, Z., Pahon, E., Péra, M.-C., & Gao, F. (2022). A review on lifetime prediction of proton exchange membrane fuel cells system. *Journal of Power Sources, 529*, 231256.

Huang, G.-B., Zhu, Q.-Y., & Siew, C.-K. (2006). Extreme learning machine: theory and applications. *Neurocomputing, 70*, 489–501.

Ibrahim, M., Steiner, N. Y., Jemei, S., & Hissel, D. (2016a). Wavelet-based approach for online fuel cell remaining useful lifetime prediction. *IEEE Transactions on Industrial Electronics, Instituteof Electrical and Electronics Engineers, 63*, 5057–5068.

Ibrahim, M., Jemei, S., Wimmer, G., & Hissel, D. J. E. P. S. R. (2016b). Nonlinear autoregressive neural network in an energy management strategy for battery/ultra-capacitor hybrid electrical vehicles. *Electric Power Systems Research, 136*, 262–269.

Jang, J. S. R. (1993). ANFIS: adaptive-network-based fuzzy inference system. *IEEE Transactions on Systems Man & Cybernetics, 23*, 665–685.

Javed, K., Gouriveau, R., Zerhouni, N., & Hissel, D. (2015a). Improving accuracy of long-term prognostics of PEMFC stack to estimate remaining useful life. In *2015 IEEE international conference on industrial technology (ICIT)*. IEEE, pp. 1047–1052.

Javed, K., Gouriveau, R., Zerhouni, N., & Hissel, D. (2015b). Data-driven prognostics of proton exchange membrane fuel cell stack with constraint based summation-wavelet extreme learning machine. In *International Conference on Fundamentals and Development of Fuel Cells*.

Javed, K., Gouriveau, R., Zerhouni, N., & Hissel, D. (2016). Prognostics of proton exchange membrane fuel cells stack using an ensemble of constraints based connectionist networks. *Journal of Power Sources, 324*, 745–757.

Jouin, M., Gouriveau, R., Hissel, D., Péra, M.-C., & Zerhouni, N. (2013). Prognostics and health management of PEMFC–state of the art and remaining challenges. *International Journal of Hydrogen Energy, 38*, 15307–15317.

Jouin, M., Gouriveau, R., Hissel, D., Péra, M.-C., & Zerhouni, N. J. (2014). Prognostics of PEM fuel cell in a particle filtering framework. *International Journal of Hydrogen Energy, 39*, 481–494.

Kimotho, J. K., Meyer, T., & Sextro, W. (2014). PEM fuel cell prognostics using particle filter with model parameter adaptation. In *2014 International Conference on Prognostics and Health Management*. IEEE, pp. 1–6.

Kui, C., Laghrouche, S., & Djerdir, A. (2018). Proton exchange membrane fuel cell degradation and remaining useful life prediction based on artificial neural network. In *2018 7th International Conference on Renewable Energy Research and Applications (ICRERA)*. IEEE, pp. 407–411.

Li, Z., Zheng, Z., & Outbib, R. (2019). Adaptive prognostic of fuel cells by implementing ensemble echo state networks in time-varying model space. *Engineering Sciences, 67*, 379–389.

Lin, H.-T., Liang, T.-J., & Chen, S.-M. (2012). Estimation of battery state of health using probabilistic neural network. *IEEE Transactions on Industrial Informatics, 9*, 679–685.

Liu, H., Chen, J., Hou, M., Shao, Z., & Su, H. (2017). Data-based short-term prognostics for proton exchange membrane fuel cells. *International Journal of Hydrogen Energy, 42*, 20791–20808.

Liu, H., Chen, J., Hissel, D., & Su, H. (2018). Short-term prognostics of PEM fuel cells: A comparative and improvement study. *IEEE Transactions on Industrial Electronics, 66*, 6077–6086.

Liu, J., Li, Q., Han, Y., Zhang, G., Meng, X., Yu, J., et al. (2019a). PEMFC residual life prediction using sparse autoencoder-based deep neural network. *IEEE Transactions on Transportation Electrification, 5*, 1279–1293.

Liu, J., Li, Q., Chen, W., Yan, Y., Qiu, Y., & Cao, T. (2019b). Remaining useful life prediction of PEMFC based on long short-term memory recurrent neural networks. *International Journal of Hydrogen Energy, 44*, 5470–5480.

Lü, X., Qu, Y., Wang, Y., Qin, C., & GJE, L. (2018). A comprehensive review on hybrid power system for PEMFC-HEV: Issues and strategies. *Management, 171*, 1273–1291.

Luo, J., Namburu, M., Pattipati, K. R., Qiao, L., & Chigusa, S. (2009). Integrated model-based and data-driven diagnosis of automotive antilock braking systems. *IEEE Transactions on Systems, Man, and Cybernetics - Part A: Systems and Humans, 40*, 321–336.

Ma, R., Yang, T., Breaz, E., Li, Z., Briois, P., & Gao FJAe. (2018). Data-driven proton exchange membrane fuel cell degradation predication through deep learning method. *Applied Energy, 231*, 102–115.

Mao, L., & Jackson, L. (2016a). Selection of optimal sensors for predicting performance of polymer electrolyte membrane fuel cell. *Journal of Power Sources, 328*, 151–160.

Mao, L., & Jackson, L. (2016b). Comparative study on prediction of fuel cell performance using machine learning approaches. In *Proceedings of the International MultiConference of Engineers and Computer Scientists*, p. 825.

Marra, D., Sorrentino, M., Pianese, C., & Iwanschitz, B. (2013). A neural network estimator of solid oxide fuel cell performance for on-field diagnostics and prognostics applications. *Journal of Power Sources, 241*, 320–329.

Mezzi, R., Morando, S., Steiner, N. Y., Péra, M. C., Hissel, D., & Larger, L. (2018). Multi-reservoir echo state network for proton exchange membrane fuel cell remaining useful life prediction. In *IECON 2018-44th Annual Conference of the IEEE Industrial Electronics Society*. IEEE, pp. 1872–1877.

Miao, Q., Xie, L., Cui, H., Liang, W., & Pecht, M. (2013). Remaining useful life prediction of lithium-ion battery with unscented particle filter technique. *Microelectronics Reliability, 53*, 805–810.

Morando, S., Jemei, S., Gouriveau, R., Zerhouni, N., & Hissel, D. (2013). Fuel cells prognostics using echo state network. In *IECON 2013-39th Annual Conference of the IEEE Industrial Electronics Society*. IEEE, pp. 1632–1637.

Ou, M., Zhang, R., Shao, Z., Li, B., Yang, D., Ming, P., et al. (2021). A novel approach based on semi-empirical model for degradation prediction of fuel cells. *Journal of Power Sources, 488*, 229435.

Pan, R., Yang, D., Wang, Y., & Chen, Z. (2020). Performance degradation prediction of proton exchange membrane fuel cell using a hybrid prognostic approach. *ISA Trans, 45*, 30994–31008.

Polverino, P., & Pianese, C. (2016). Model-based prognostic algorithm for online RUL estimation of PEMFCs. In *2016 3rd Conference on Control and Fault-Tolerant Systems (SysTol)*. IEEE, p. 599–604.

Rao, W., & Man-Wai, M. A. K. (2016). Sparse kernel machines with empirical kernel maps for PLDA speaker verification. *Preprint, 38*, 104–121.

Rath, R., Kumar, P., Mohanty, S., & Nayak, S. K. (2019). Recent advances, unsolved deficiencies, and future perspectives of hydrogen fuel cells in transportation and portable sectors. *International Journal of Energy and Research, 43*, 8931–8955.

Robin, C., Gerard, M., Franco, A. A., & Schott, P. (2013). Multi-scale coupling between two dynamical models for PEMFC aging prediction. *International Journal of Energy and Power Engineering, 38*, 4675–4688.

Roshandel, R., & Parhizkar, T. (2016). Degradation based optimization framework for long term applications of energy systems, case study: solid oxide fuel cell stacks. *Energy, 107*, 172–181.

Sankavaram, C., Pattipati, B., Kodali, A., Pattipati, K., Azam, M., Kumar, S., et al. (2009). Model-based and data-driven prognosis of automotive and electronic systems. In *2009 IEEE International Conference on Automation Science and Engineering*. IEEE, pp. 96–101.

Silva, R., Gouriveau, R., Jemei, S., Hissel, D., Boulon, L., Agbossou, K., et al. (2014). Proton exchange membrane fuel cell degradation prediction based on adaptive neuro-fuzzy inference systems. *International Journal of Hydrogen Energy, 39*, 11128–11144.

Stiller, C., Thorud, B., Bolland, O., Kandepu, R., & Imsland, L. J. (2006). Control strategy for a solid oxide fuel cell and gas turbine hybrid system. *Journal of Power Sources, 158*, 303–315.

Tealab, A., Hefny, H., & Badr, A. (2017). Forecasting of nonlinear time series using ANN. *Future Computing and Informatics Journal, 2*, 39–47.

Thomas, J. M., Edwards, P. P., Dobson, P. J., & Owen, G. P. (2020). Decarbonising energy: The developing international activity in hydrogen technologies and fuel cells. *Journal of Energy Chemistry, 51*, 405–415.

Wang, D., Miao, Q., & Pecht, M. (2013). Prognostics of lithium-ion batteries based on relevance vectors and a conditional three-parameter capacity degradation model. *Journal of Power Sources, 239*, 253–264.

Wang, F.-K., Cheng, X.-B., & Hsiao, K.-C. (2020a). Stacked long short-term memory model for proton exchange membrane fuel cell systems degradation. *Journal of Power Sources, 448*, 227591.

Wang, F.-K., Mamo, T., & Cheng, X.-B. (2020b). Bi-directional long short-term memory recurrent neural network with attention for stack voltage degradation from proton exchange membrane fuel cells. *Journal of Power Sources, 461*, 228170.

Wang, F.-K., Amogne, Z. E., & Chou, J.-H. (2021). A hybrid method for remaining useful life prediction of proton exchange membrane fuel cell stack. *IEEE Access, 9*, 40486–40495.

Wang, C., Li, Z., Outbib, R., Dou, M., & Zhao, D. (2022). A novel long short-term memory networks-based data-driven prognostic strategy for proton exchange membrane fuel cells. *Engineering Sciences, 47*, 10395–10408.

Wu, J., Yuan, X. Z., Martin, J. J., Wang, H., Zhang, J., Shen, J., et al. (2008). *A review of PEM fuel cell durability: Degradation mechanisms and mitigation strategies., 184*, 104–119.

Wu, Y., Breaz, E., Gao, F., & Miraoui, A. (2015). Prediction of PEMFC stack aging based on relevance vector machine. In *2015 IEEE Transportation Electrification Conference and Expo (ITEC)*. IEEE, pp. 1–5.

Wu, Y., Breaz, E., Gao, F., & Miraoui, A. (2016a). A modified relevance vector machine for PEM fuel-cell stack aging prediction. *Processes, 52*, 2573–2581.

Wu, Y., Breaz, E., Gao, F., Paire, D., & Miraoui, A. (2016b). Nonlinear performance degradation prediction of proton exchange membrane fuel cells using relevance vector machine. *IEEE Transactions on Energy Conversion, 31*, 1570–1582.

Xia, T., Song, Y., Zheng, Y., Pan, E., & Xi, L. (2020). An ensemble framework based on convolutional bi-directional LSTM with multiple time windows for remaining useful life estimation. *Computers in Industry, 115*, 103182.

Xie, Y., Zou, J., Peng, C., Zhu, Y., & Gao, F. (2020). A novel PEM fuel cell remaining useful life prediction method based on singular spectrum analysis and deep Gaussian processes. *Sustainability, 45*, 30942–30956.

Xue, X., Hu, Y., & Qi, S. (2016). Remaining useful life estimation for proton exchange membrane fuel cell based on extreme learning machine. In *2016 31st Youth Academic Annual Conference of Chinese Association of Automation (YAC)*. IEEE, pp. 43–47.

Zhang, J., & Man, K.-F. (1998). Time series prediction using RNN in multi-dimension embedding phase space. In *SMC'98 Conference Proceedings 1998 IEEE International Conference on Systems, Man, and Cybernetics (Cat No 98CH36218)*. IEEE, pp. 1868–1873.

Zhang, X., Yang, D., Luo, M., & Dong, Z. (2017). Load profile based empirical model for the lifetime prediction of an automotive PEM fuel cell. *International Journal of Hydrogen Energy, 42*, 11868–11878.

Zhang, Z., Wang, Y.-X., He, H., & Sun, F. (2021). A short-and long-term prognostic associating with remaining useful life estimation for proton exchange membrane fuel cell. *Applied Energy, 304*, 117841.

Zhou, D., Yu, Z., Zhang, H., & Weng, S. (2016a). A novel grey prognostic model based on Markov process and grey incidence analysis for energy conversion equipment degradation. *Energy, 109*, 420–429.

Zhou, D., Zhang, K., Ravey, A., Gao, F., & Miraoui, A. (2016b). Online estimation of lithium polymer batteries state-of-charge using particle filter-based data fusion with multimodels approach. *Engineering Sciences, 52*, 2582–2595.

Zhou, D., Wu, Y., Gao, F., Breaz, E., Ravey, A., & Miraoui, A. (2017a). Degradation prediction of PEM fuel cell stack based on multiphysical aging model with particle filter approach. *Energy, 53*, 4041–4052.

Zhou, D., Gao, F., Breaz, E., Ravey, A., & Miraoui, A. (2017b). Degradation prediction of PEM fuel cell using a moving window based hybrid prognostic approach. *Energy, 138*, 1175–1186.

Zhou, D., Al-Durra, A., Zhang, K., Ravey, A., & Gao, F. (2018). Online remaining useful lifetime prediction of proton exchange membrane fuel cells using a novel robust methodology. *Journal of Power Sources, 399*, 314–328.

Zhou, D., Al-Durra, A., Zhang, K., Ravey, A., & Gao, F. (2019). A robust prognostic indicator for renewable energy technologies: A novel error correction grey prediction model. *IEEE Transactions on Industrial Electronics, 66*, 9312–9325.

Zhu, L., & Chen, J. J. E. (2018). Prognostics of PEM fuel cells based on Gaussian process state space models. *Energy, 149*, 63–73.

Zraibi, B., Okar, C., Chaoui, H., & Mansouri, M. (2021). Remaining useful life assessment for lithium-ion batteries using CNN-LSTM-DNN hybrid method. *The Korean Institute of Power Electronics, 70*, 4252–4261.

Zuo, J., Lv, H., Zhou, D., Xue, Q., Jin, L., Zhou, W., et al. (2021). Deep learning based prognostic framework towards proton exchange membrane fuel cell for automotive application. *Applied Energy, 281*, 115937.

An Assessment of Electricity Markets in Turkey: Price Mechanisms, Regulations, and Methods

Hakan Acaroğlu

Abstract This chapter is focused on the price determination problem of the Turkish day-ahead electricity market. Management science techniques are used to analyze the problem. An optimization objective function is defined and explained. Once the problem formulation is stated, and the price is compared to the market clearing price, there is an acceptance guarantee for the affordable block, flexible, and, specifically, paradoxical bids (offers). The objective function of the economic model has been determined by maximizing the total market surplus as a requirement of a market that operates efficiently. The goal is to find the best solution using the algorithms and procedures that have been developed. The problem of price determination in the day-ahead electricity markets includes many research subjects and development areas. With the ongoing researches in this field, it will be useful to academically examine both the suggestions from the sector and the applications in the developed markets. Thereby, some of the applications and simulations can be put into practice.

Keywords Electricity markets · Day-ahead market · Optimization · Price mechanisms · Regulations

1 Introduction

Power markets have become deregulated around the world over the last two decades, since the structure of electricity as a commodity has been changed from government-controlled (i.e., monopolistic) to competitive market and it began to be traded under these rules in many countries (Mayer & Trück, 2018; Acaroğlu & García Márquez, 2021; Weron, 2006). Turkey is one of those countries whose energy policy futures a competitive, liberal, non-discriminatory, transparent, and stable electricity market while maintaining a reliable, cost-effective, high-quality, and continuous supply

H. Acaroğlu (✉)
Faculty of Economics and Administrative Sciences, Department of Economics, Eskisehir Osmangazi University, Eskisehir, Turkey
e-mail: hacaroglu@ogu.edu.tr

F. P. García Márquez, B. Lev (eds.), *Sustainability*, International Series in Operations Research & Management Science 333,
https://doi.org/10.1007/978-3-031-16620-4_11

(Somay et al., 2021). The Turkish government has been attempting to achieve these goals by encouraging private investment in the electricity market as well as reforming energy legislation.

The development of the Turkish electricity market can be categorized into three phases: (1) The introduction of the Electricity Market Law (corresponds to the 1920s–1960s, and represented by the early stage), (2) the liberalization of the increased market (corresponds to the 1960s–2000s, and represented by structuring stage), and (3) the growth of the market (corresponds to the 2000s to present, and represented by the growth stage) (PwC Turkey and Presidency of the Republic of Turkey, Investment Office, 2021).

This chapter focuses on an economic model that is developed to solve the bid matching (i.e., the term bid matching is used to define the match price and match quantity for each bid; the matched quantity represents the accepted portion of the offered quantity; and the matching price represents the unit price determined to buy or sell this quantity) and the price determination problem in accordance with the above-mentioned stages, and specifically on the growth stage in the Turkish day-ahead electricity market (DAM) (EPİAŞ, 2016).

The motivation for this subject is to explain the price mechanisms, regulations, and methods of Turkey's electricity markets, as well as to shed light on the current working principles of Turkish DAM using management science techniques. During this process, the techniques of pretreatment, heuristic algorithm, optimization, and postprocessing are explained step by step with their subtitles. This chapter aims to combine economics (see Ref. Thomaidis & Biskas, 2021) on linear and non-linear panel models for investigating pricing mechanisms) and management science techniques on a critical issue: the Turkish DAM price determination problem. An optimization approach is followed and an objective function is theoretically explained after its definition. The problem is defined based on the balancing and settlement regulation; the structure of the DAM bids; the evaluation of the bid types (i.e., single, block, and flexible bids). However, the details of the formulation are not provided and they are beyond the scope of this work (see Ref. EPİAŞ, 2016; Derinkuyu et al., 2020 for the details of the formulation).

This chapter also presents a review on the latest developments in the Turkish DAM considering management science techniques and updated references. The advances in the Turkish electricity market and regulations are discussed, and they are further explained by the observations and findings. The below-mentioned items summarize the content of this work:

- After comparing the price with the market clearing price (MCP) there is an acceptation guarantee for the affordable block, flexible, and specifically, paradoxical offers.
- The objective function of the economic model has been determined by maximizing the total market surplus as a requirement of an efficiently operating market.
- The goal is finding an optimal solution with the developed algorithms and procedures.

Nonetheless, the problem of price determination in DAMs includes many research subjects and development areas. With the ongoing researches in this field, it will be useful to academically examine both the suggestions from the sector and the applications in the developed markets. Thereby, some of the applications and simulations can be put into practice.

The chapter is presented as follows: Section 2 presents the Turkish electricity markets through its mechanisms, and current regulations. Section 3 includes a theoretical assessment of the Turkish DAM, as well as tables and figures. Section 4 deals with the definition of the problem, as well as an explanation of the model and the demand and supply forces. Section 5 discusses the price determination methods used in the Turkish DAM using management science techniques. Section 6 contains the conclusions as well as the proposed future studies.

2 The Turkish Electricity Markets

In Turkey, the Ministry of Energy and Natural Resources (MENR), also known as the strategy and policy maker, is in charge of designing and implementing electricity market policies. In co-ordination and co-operation with MENR the regulatory body for the electricity market is the Energy Market Regulatory Authority (EMRA), which is referred as the independent regulator. EMRA is responsible for supervising market participants' activities; granting and renewing licenses; preparing, developing, and executing legislation; preparing, regulating, and executing tariffs; establishing performance standards; auditing licenses; and providing that electricity market activities are in consent with the Electricity Market Law. Thereby, EMRA closely follows and audits the license holders (i.e., generators, transmission operators (TSOs), market and distribution system operators (DSOs) (IEA, 2021).

2.1 *Mechanisms*

Under the auditing of EMRA, the Turkish Electricity Transmission Company (TEİAŞ) which is referred to as the transmission operator (TSO), has a monopoly power over transmission activities. The real-time markets, balancing markets, and ancillary services are operated by the state-owned TEİAŞ. The distribution operations are also under the control of the Turkish Electricity Distribution Company (TEDAŞ); however, the state-owned TEDAŞ is not responsible for the retail activities which are conducted by the authorized suppliers (IEA, 2021).

The Turkish electricity market is primarily based on bilateral contracts on a spot market platform and a balancing mechanism. The day-ahead (i.e., the DAM mechanism was entered into force in 2011), intra-day, and balancing power markets are components based on transparency, competition, and integrity. In addition, a power futures market was set to open in 2020. The DAM and intra-day electricity markets

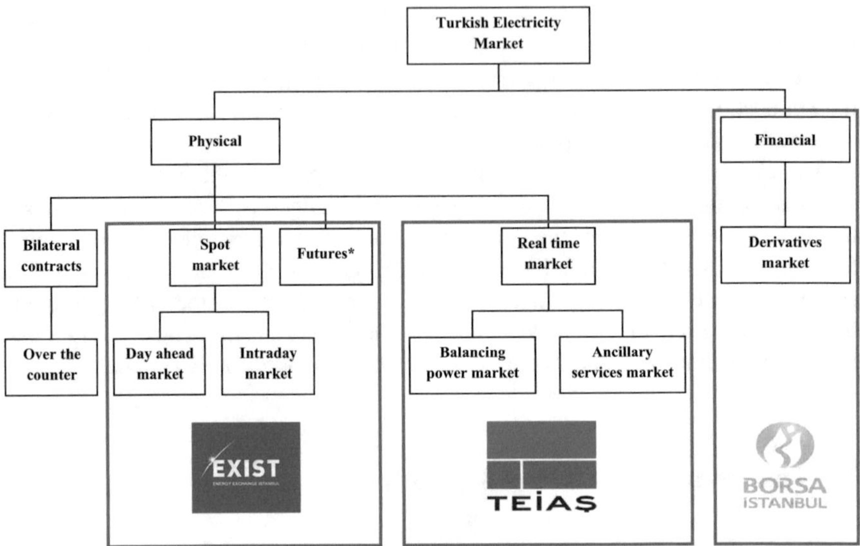

Fig. 1 Electricity trade mechanism in Turkey. Note*: The power futures market was planned for operating in 2020. Source: Adapted from PwC Turkey and Presidency of the Republic of Turkey, Investment Office (2021), IEA (2021)

are operated by the Energy Exchange Istanbul (EXIST) which was established in 2015 with Laws No. 6446 and 6102. EXIST is responsible for planning, establishing, developing, and operating energy markets in a secure way without any discrimination among providers (IEA, 2021). The Turkish Electricity Markets' trade mechanism is shown in Fig. 1.

2.2 *Market Structure Through Its Components*

Turkey allowed the private sector to participate in the electricity markets in 2001. As a result, the components of the electricity supply chain, generation, distribution, and supply, are now carried out by both state-owned (i.e., 21%) and private companies (i.e., 79%) (IEA, 2021).

2.2.1 Generation

The Electricity Production Company (EÜAŞ), also known as the public generation, is the state-owned company in charge of electricity generation operations. The public (approximately 24%), private companies (approximately 73%, and power plants privatized under the transfer operation rights and build-operate transfer power plants (approximately 3%) participate in this activity with an EMRA generation license (Somay et al., 2021).

2.2.2 Transmission

TEİAŞ and Energy Markets Operations Inc. (EPİAŞ), are responsible for the transmission system. While TEİAŞ is responsible for the balancing power market and ancillary services, EPA, also known as the market operator, is in charge of the DAM, intra-day, the newly established YEK-G (Renewable Energy Guarantees of the Origin System & Organized Market), and the power future markets (Somay et al., 2021).

2.2.3 Distribution

After the privatization of all the distribution companies in 2013, distribution licenses were obtained by EMRA. The distribution activities are conducted by the private sector now; however, it is important to note that their assets are currently owned by the state-owned Turkish Electricity Distribution Company (TEDAŞ) (Somay et al., 2021).

2.2.4 Supply

The retail electricity market is operated by the private sector, which is licensed by EMRA. Incumbent companies that are established by each distribution company sell electricity to non-eligible consumers (i.e., the threshold for being an eligible consumer is 1200 kWh for 2021) (Somay et al., 2021). The terms and conditions of the sales are regulated by EMRA.

3 An Assessment for the Turkish Day Ahead Electricity Market

The price of electricity in the regulated electricity markets (i.e., spot and future markets) has a specific importance for both the market participants and managers (Akbary et al., 2019; Lu et al., 2021). For instance, European DAMs organize spot markets for trading electricity between market participants a day prior to the actual consumption and generation with a two-sided, double-blind auction (Derinkuyu et al., 2020). Sellers are able to submit several types of bids in the combination of various quantities and prices for specific periods of delivery time. The market operator solves the MCP problem by maximizing social welfare through surplus (Ceyhan et al., 2021; Nasiri et al., 2021).

There are various models and MCP algorithms that are used in the design of DAMs (Shariat Torbaghan et al., 2021; Li & Becker, 2021; Sarıca et al., 2012; Chatzigiannis et al., 2016). A software has been developed since 2016 in the DAM

Table 1 The Turkish DAM's daily operations

Time interval (h)	Operation
00:00–17:00	The bilateral agreements are entered into the system for the next day by the market participants.
00:00–12:30	The bids for the upcoming day are submitted by DAM participants.
12:30–13:00	The bids are validated and the control payments are checked. The market operator has the right for calling the participant for confirmation if there is an uncommon bid submission.
13:00–13:10	The optimization toll determines the MCPs.
13:10–13:30	Objections to the bid matching are received and results are published.
13:30–14:00	Objections are resolved after an evaluation.
14:00	Average hourly prices and MCPs are finalized, and trade volumes are announced.

Source: Adapted from Derinkuyu et al. (2020)

of EPİAŞ, which is the market operator in the Turkish electricity sector. The software receives bids from market participants and determines the MCP per hour and the match quantity and price for each bid. By this process, it uses multiple algorithms and mathematical models are used to obtain the optimal solution. As a result, if there is limited time for problem-solving, the software will use heuristic algorithms to find a solution that meets the constraints first. It then tries to find the optimal solution which minimizes or maximizes a function subject to various constraints (Marugán et al., 2018) using a mathematical program solver.

A valid offer that can be submitted in the market should comply with the features and constraints determined by the relevant legislation (see (Gökgöz & Atmaca, 2017) for the historical background, the current status, and the structure of the Turkish electricity market). For the types and structures of offers in the Turkish DAM, there are three different bid types: single (hourly); block; and flexible (Derinkuyu, 2015). Each offer consists of at least one price-quantity pair. Price-quantity pairs can be offered in two different directions, either buying or selling. If the provided quantity is negative, this pair is in the sell direction, if it is positive, it is in the buy direction. The participant who provides a binary in which p represents the price and q represents the quantity indicates that he or she has provided the maximum/minimum p unit price (/MWh) to buy/sell q quantity. The quantities can be represented as volume (Sahin, 2018) and they are expressed in Lot units and one Lot equals 0.1 MWh. Table 1 illustrates the daily operations in the Turkish DAM.

3.1 *Single Offers (Hourly Bids)*

Participants make offers as bid price-quantity pairs (breakpoint) for the hours (for one time period only) of the next day to buy or sell electricity. The amounts of price breakdowns on the sell side are considered as negative. Due to the nature of the

Table 2 A single bid example of participant 1

Price (₺/MWh)	0	1000
Quantity (Lot)	200	200

Table 3 A single bid example of participant 2

Price (₺/MWh)	0	50	100	150	200	10,000
Quantity (Lot)	200	200	100	0	−100	−200

Table 4 A single bid example of participant 3

Price (₺/MWh)	0	150	75	300	2000
Quantity (Lot)	0	−100	−50	−200	−200

economy, market players want to buy the goods at low prices and sell them at high prices. Therefore, as price breakdowns increase in single offers, the corresponding quantities should decrease (i.e., the buying quantity should decrease and the absolute value of the selling quantity should increase). Examples of single offers are presented in Tables 2, 3, and 4.

Table 2 shows a single bid example of a participant declaring that she/he will buy 200 lots regardless of the MCP for the period in which the bid is valid. Such bids are called price-independent hourly bids. A participant who makes a single offer may sell less than the sales amount (if any) offered for this price breakdown if the MCP is at the minimum limit in the period in which the offer is valid. She/He says he can buy in quantity. For example, first participant in Table 2 states that she/he can buy less than 200 lots if the MCP in the relevant period is 1000 TL(₺)/MWh.

In Table 3, there is an offer with six price levels. In this table, the second participant states that she/he will buy when the unit price announced in the period in which the offer is valid is 150 TL or less, and she/he will sell when it is higher than that.

Table 4 shows hourly selling offers with five price levels.

Although single bidding is a set of price-quantity pairs, it allows for linear interpolation to fill in between breakpoints. The result is a piecewise linear function. The matching quantity for a single bid is also determined as the amount corresponding to the MCP. Figure 2 depicts the supply curve for the third participant. According to Fig. 2, when the MCP is 75 ₺/MWh, the Matching Quantity is 50 lots for this participant. Single bids are frequently placed by solar, wind, hydroelectricity plants, as well as distributors and retailers (Derinkuyu et al., 2020).

3.2 *Block Offers*

Block offers are the second most prevalent type of offer in the market after single offers. Block bids can be considered as consecutive single bids that cannot be broken down on the timeline. However, in this offer type, there is only one price-quantity pair for the period of validity. In this case, the price offered is provided on the condition of buying/selling the whole amount. In addition to the price and quantity

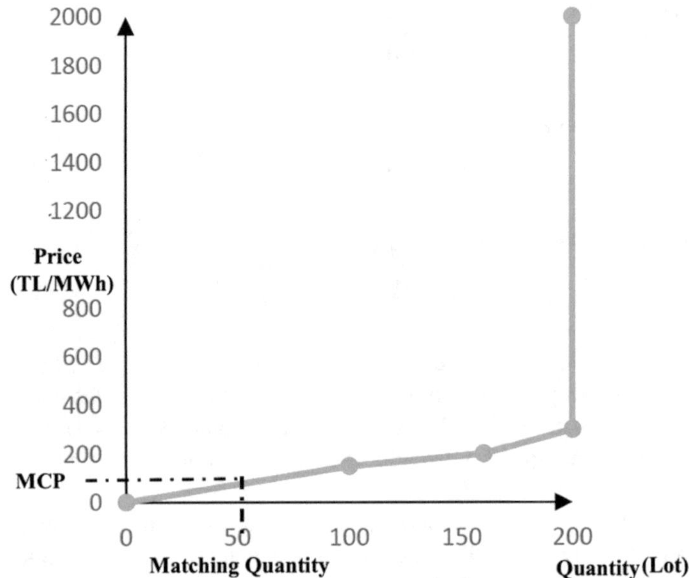

Fig. 2 The piecewise linear supply curve of third participant. Source: Adapted from EPİAŞ (2016)

Table 5 An example for block offers

Periods (h)	Price (₺/MWh)	Quantity (Lot)
0–7	170	−500
10–24	130	350

information, the number of consecutive hours to buy/sell electricity for the next day is also specified. A block bid is either fully accepted or rejected for the time it is active which means that it is not interpolated. In addition to this, block offers can cause non-convexity, therefore, the auction problem can be nondeterministic polynomial time (NP)-Hard (i.e., NP is the set of decision problems that can be solvable in polynomial time by nondeterministic Turing machine) to solve (Derinkuyu, 2015). However, there are developed approaches for solving this problem. For instance, while the European markets solve true linear pricing, the US markets prefer deviating from linear pricing (van Vyve, 2011).

For example, in the case of block offers, Table 5 shows that the bidder is intended to sell 500 Lots for each period ranging from 0 to 7 (i.e., 3500 Lots in total) if the average clearing price (ACP) regarding that period is greater than 170 ₺//MWh, while seek to buy 350 Lots for each period ranging from 10 to 24 if the ACP for these periods is less than 130 ₺//MWh.

Linked block bids, which are a type of block bids, are also a type of bid used in the DAM. If a block bid is linked to another block bid, the latter linked block bid is referred to as the child bid, and the former linked block bid is referred to as the mother bid. As a result, if the mother's offer is rejected, the child block offer is also rejected. In this block bid type, up to 3 block bids can be linked together. Also,

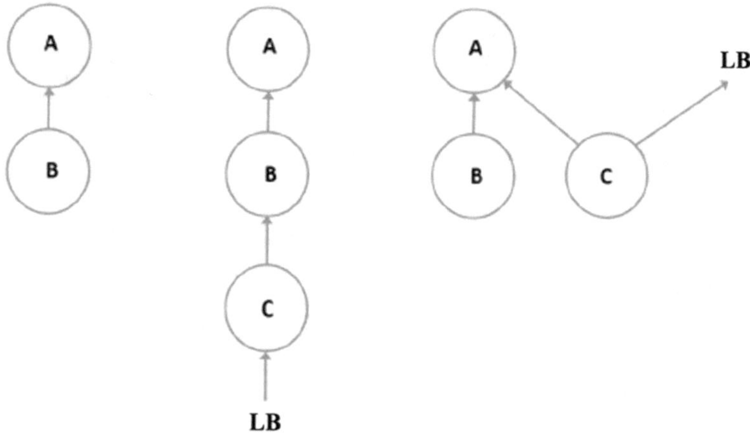

Fig. 3 Possible set of linked blocks (LB) offer sets. Source: Adapted from EPİAŞ (2016)

Table 6 Linked block offer example

Block offer	Time	Price (₺/MWh)	Quantity (Lot)	Linked block
A	1–6	100	−100	–
B	3–10	50	−200	A

linked block proposals cannot be linked in a loop. For example, if block bid A is tied to block bid B, then block bid B cannot be tied to block bid A. However, interlinked block bids must be in the same direction (sell or buy). Figure 3 shows three sets of alternative linked block bids that can be placed on the market.

Two block bids are presented in Table 6, A represents the mother block bid, and B represents the child block bid. In this case, the A block offer should also be accepted in order for the B block offer to be accepted. Block bids are frequently placed by natural gas and coal-powered plants likewise by electricity retailers and large industrial consumers (Derinkuyu et al., 2020).

3.3 *Flexible Offers*

Flexible offers only have price and quantity information. A flexible offer is given hourly and can be accepted at any time of the day. It can only be given in the sales direction in the current market. Flexible offers are either completely accepted or completely rejected. The time the offer is accepted does not have to be the hour with the highest MCP, but if the bid price is below the highest MCP, the offer is accepted for a suitable hour.

According to two flexible offers that are shown in Table 7, if the maximum MCP exceeds 300 ₺/MWh then the market participant is willing to sell 400 Lot for one

Table 7 An example for flexible offers

Price (₺/MWh)	Quantity (Lot)
300	−400
100	300

Table 8 A possible paradoxical block case with an example

Bids (hourly)	**Pairs (price-quantity)**					
Price (₺/MWh)	0	100	200	300	400	800
Quantity (hour 1)	200	150	0	−75	−150	−400
Quantity (hour 2)	200	150	0	−75	−150	−400
Block bid	**Pairs (price-quantity)**					
Price (₺/MWh)					250	
Quantity (hours 1 and 2)					75	

time period. Likewise, if the minimum MCP is below 100 ₺/MWh, then the market participant is willing to buy 300 Lot for one period. Flexible bids are typically placed by consumption units and flexible generation, such as cement producers, fuel-oil, and diesel-powered plants (Derinkuyu et al., 2020).

In practice, paradoxical cases may arise when fully accepting or rejecting block bids. For instance, Table 8 shows two hourly bids and one demand block bid through a simple example. The demand block bid for buying 75 MWh at the price of 250 ₺/MWh, covering hours 1 and 2. If the block bid is rejected, then the MCP for both hours (1 and 2) is 200 ₺/MWh. At the first stage, the average MCP is lower than the block bid price (i.e., 200 ₺/MWh vs. 250 ₺/MWh), the block bid is allowed to be accepted. However, in the case of the block is accepted, the average MCP is now 300 ₺/MWh, and with this price the block bid is not allowed to be accepted or it should be rejected. This event is called as a paradoxical block, the feasibility of the block is neither accepted nor rejected (Derinkuyu et al., 2020).

4 Definition of the Problem

4.1 Model

In the day-ahead electricity markets, and participants submit different types of bids for different hours of the next day. The market operator is responsible for finding a valid purchase or sale amount for each offer. This amount is also called the matching quantity. The market operator finds the matching quantity to increase the daily market surplus. During this process, they should be careful that the total supply and demand quantities matched for each hour are equal to each other (EPİAŞ, 2016).

The acceptance status of each type of offer is subject to some constraints. For hourly bid or sell bids, this rule is the quantity in the piecewise linear function corresponding to the calculated MCP value of the relevant hour. Since block bids

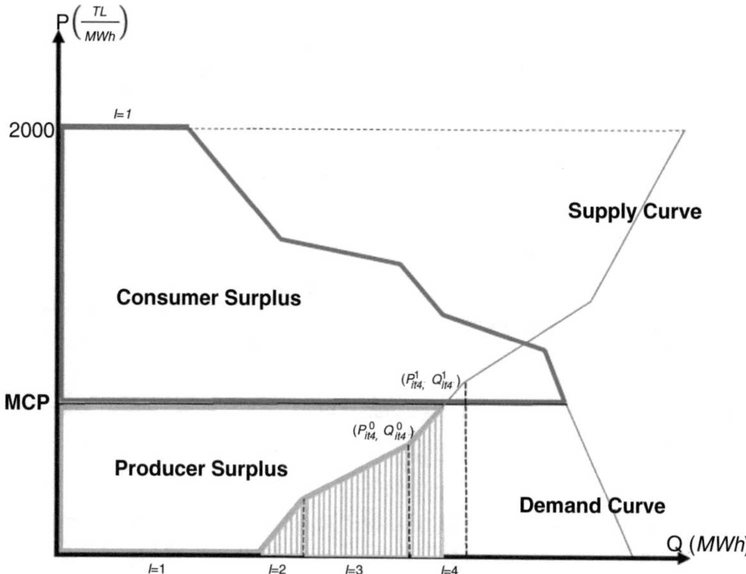

Fig. 4 Supply and demand curves obtained from hourly offers. Source: Adapted from EPİAŞ (2016), Derinkuyu et al. (2020)

cover more than one period, the bid price for some periods is appropriate according to the MCP value, while it may not be suitable for some periods. Thus, a block bid in the sell direction is accepted when the bid price is less than or equal to the average MCP of the periods it covers, whereas a block bid in the buy direction is accepted when the bid price is higher or equal to the average MCP of the periods it covers. In flexible sales offers, if the price of the offer is lower than or equal to the highest MCP announced, this offer is accepted. However, due to the nature of the problem, in some cases, even if the relation between the relevant bid price and MCP for block and flexible bids is different from the one described above, these bids may be accepted. Such offers are also called accepted offers due to the paradox. Acceptance of such offers will result in the formation of a difference quantity (EPİAŞ, 2016).

The objective function of the problem is maximizing the daily market surplus. The daily market surplus is equal to the sum of the daily consumer and producer surpluses. A participants' consumer surplus is the difference between the amount offered to the market for the purchase quantity resulting from the matching and the quantity it will pay to receive this amount. A participants' producer surplus is the difference between the amount she/he will receive for the quantity of sales resulting from the matching and the quantity she offers to the market to sell this quantity (EPİAŞ, 2016).

The market surplus is calculated using hourly, block, and flexible bids. Figure 4 shows the consumer and producer surplus resulting from matching hourly bids for an hour. In order to find these values, an hourly bid must be decomposed into demand and supply curves. Consumer surplus is the area between the demand curve, MCP

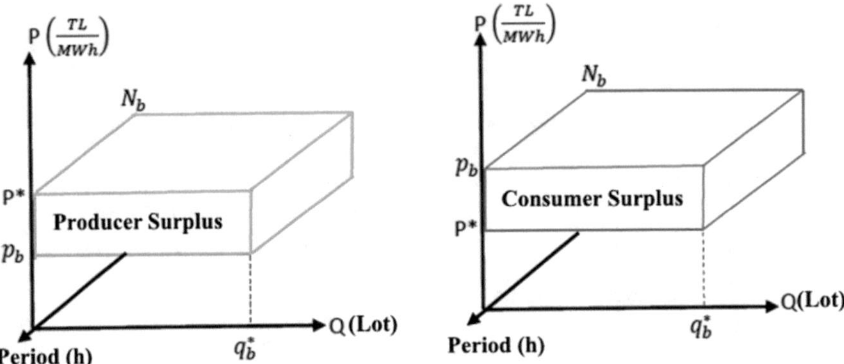

Fig. 5 The producer and consumer surplus obtained from accepted block offers. Source: Adapted from EPİAŞ (2016)

and P-Q axes (blue polygon), while producer surplus is the area between the supply curve, MCP, and P-Q axes (green polygon). The shaded area in Fig. 4 corresponds to the sales amount declared by the participant to sell this matching amount. The amount of producers' surplus in Fig. 4 can be geometrically calculated from the difference between the rectangular area (i.e., MCP * the matching quantity − the shaded area). It should be noted that due to the accepted block and flexible offers, MCP may not appear at the intersection of the demand and supply curves of the combined hourly offers (EPİAŞ, 2016).

Hourly offers are segmented as they are included in the mathematical model. The segment denotes the region between two consecutive breakpoints, which is formed after the decomposition of the one-hour offer into the supply and demand curves. If the minimum sales quantity or the minimum purchase quantity of an hourly offer is different from zero, the area between the minimum price breakpoints of the supply curve or the maximum price breakdown points of the demand curve and the price axis also forms a segment (EPİAŞ, 2016).

Figure 5 shows the producer and consumer surplus obtained from block offers. P* is equal to the average MCP of the hours it is valid for block offers. The excess from an accepted block bid is the volume of the created rectangular prism. This volume is equal to the difference between the average MCP and the bid price multiplied by the matching quantity (q_b^*) and the total number of periods in which the block bid is active (N_b) (EPİAŞ, 2016).

The results released to the market should meet the following characteristics (EPİAŞ, 2016):

1. Supply must equal demand in each period.
2. The matching quantity of an hourly bid is the amount corresponding to the MCP at the hour specified in that offer. If there is no price breakdown corresponding to the MCP in the hourly bid, the matching quantity is found according to the linear interpolation rule.

3. Block bids are either completely accepted or completely rejected; partial acceptance is not allowed. If a block bid is accepted, it is accepted for all periods in which it is valid.
4. Flexible offers are either completely accepted or completely rejected.
5. Block bids on the sell side are accepted if they bid less than or equal to the average MCP of the hours in which they are valid. Block bids on the buyer side are accepted if they bid higher than or equal to the average MCP of the hours in which they are valid. This rule does not apply to child blocks.
6. The child block proposal cannot be accepted until the mother's block proposal is accepted.
7. A flexible bid should be accepted if it is lower than or equal to the highest MCP.
8. A flexible bid is accepted for a maximum of one period.

While all of these constraints should be satisfied in the results described, in some cases no result that satisfies these constraints may be found. In excess of demand or supply, the results disclosed do not have to meet the second, fifth, and seventh constraints. This model finds matching quantities and prices with high precision.

4.2 Demand or Excess Supply

In the case of excess demand (lack of energy) or excess supply (excess energy), the supply and demand in the system cannot be equalized within the price limits currently valid in the market and in a way that satisfies all the constraints specified in the previous section. In order to achieve the supply-demand balance, hourly offers are interrupted and the acceptance condition in block and flexible offers may not be met (EPİAŞ, 2016).

In the case of excess energy, the MCP of that hour comes out at the minimum price. In this case, the aggregate demand offered to the market at the minimum price, Demand (*Pmin*), is less than the aggregate supply at this price, Supply (*Pmin*). The energy surplus in an hour is the difference between supply and demand corresponding to the minimum price at that hour, Supply (*Pmin*) – Demand (*Pmin*) (EPİAŞ, 2016).

In the case of a lack of energy, the MCP of that hour is *Pmax* at the maximum price. In this case, the total supply offered to the market at the maximum price, Supply (*Pmax*), is less than the aggregate demand at this price, Demand (*Pmax*). The lack of energy in an hour is the difference between demand and supply corresponding to the maximum price at that hour, Demand (*Pmax*) – Supply (*Pmax*) (EPİAŞ, 2016).

5 Solution Methods with Management Science Techniques

In this section, the method that is followed to solve the problem described in the previous section is provided. This method consists of four main modules as pretreatment (pre-processing), heuristic, optimization, and postprocessing algorithms as it is shown in Fig. 6. In each module, different algorithms are designed specifically for this problem. In addition to the algorithms in these modules, validation and repair algorithms that are not specific to a single module and used in multiple modules have also been created. While the first three modules run sequentially, the algorithms of the non-reporting postprocessing module can run at any point in the program. Reporting is the last step in the solver process. In this step, the solution found is converted into a format to be released to the market (EPİAŞ, 2016).

5.1 Pretreatment

The aggregation specified in this module is the conversion of single offers from each hour into a single hourly offer. The purpose of variable elimination is to reduce the problem size so that it does not lose the optimal value (EPİAŞ, 2016).

5.2 Heuristic Algorithm

Two heuristic algorithms are run in parallel to provide the optimization module with an initial solution based on a high total surplus. Heuristic algorithms try to find out which block bid is accepted and which block bid is rejected, resulting in the best overall surplus. In order to achieve this, it pretends that flexible offers do not exist (or are rejected). After the most suitable block combination is found, flexible bids are added to this solution. Among the solutions including the acceptance/rejection

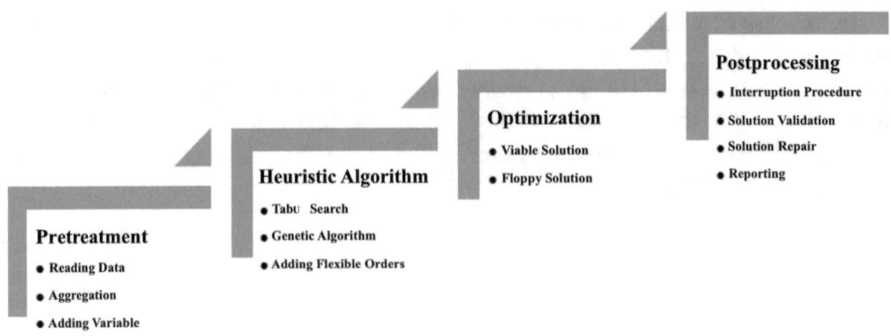

Fig. 6 Solution management flowchart. Source: Adapted from EPİAŞ (2016)

decisions of flexible offers, the solution with the highest total surplus is selected and provided to the optimization tool as the initial solution. If the optimization module cannot find a better solution than this solution in the given time, the best solution found by the heuristics is treated as the final solution (EPİAŞ, 2016). Specifically, the tabu search algorithm is preferred for the calculations as can be seen in Fig. 6. Tabu search algorithm is a metaheuristic search method that conducts a local search in the search space. However, in contrast to local search algorithms, tabu search methods prohibit the usage of previously visited solutions that are included and this prevents the search to be stopped at the local optimums (Derinkuyu et al., 2020). Additionally, genetic algorithms which are population-based methods are applied to the search. These algorithms aim to improve the quality of the solution set. The algorithm works with a convergence criterion stops with the best solution (Derinkuyu et al., 2020). After the defined algorithms are applied a solution occurs and with a new algorithm the flexible orders are added into the solution set (Derinkuyu et al., 2020).

5.3 *Optimization*

The solution deriving from the heuristic algorithms is given to the mathematical program solver and it is aimed to find the best solution in a determined time. Constrained and unconstrained (the model without constraints with items 5 and 7 in Sect. 4.1) are both solved by this solver. If the unconstrained model satisfies constraints 5 and 7, this solution is considered the final solution. Otherwise, the constrained model is solved, the solution of the unconstrained model is repaired to meet the constraints of items 5 and 7, and these two solutions are compared with each other and the solution with the highest total surplus is explained (EPİAŞ, 2016).

5.4 *Postprocessing*

This module includes the outage procedure, solution verification and repair algorithms, and reporting in case of excess supply or demand. Verification algorithms check that the solutions produced by the heuristics and optimization modules comply with the constraints specified in Sect. 4.1. If a generated solution does not meet these constraints, the solution is made feasible using the repair algorithm. When there is an excess of supply or demand, some of the constraints (fifth and seventh constraints) specified in Sect. 4.1 are relaxed and an attempt is made to achieve supply and demand equality (EPİAŞ, 2016).

6 Conclusions

This work explains the price determination solution of the Turkish electricity trade in the day-ahead market. The considered solution is the subject of management science techniques and is shown in a management flowchart in the steps of: pretreatment; heuristic algorithm; optimization; and postprocessing. An objective function of the economic model is defined and explained in the concept of an optimization and determined by maximizing the total market surplus as a requirement of an efficiently operating market. The formulation of the solution is expressed based on: a balancing mechanism with the related settlement regulation, structure, and procedure of the day-ahead market bids; the bid types (i.e., single, block, and flexible bids) with paradoxical cases. When the price is compared to the market clearing, the affordable block and flexible offers are guaranteed to be accepted.

The aim is finding an optimal solution with the related procedures and developed algorithms. The findings of the research reveal that:

- Management science techniques are the prominent instruments in the development of the electricity markets. Specifically, an optimization algorithm based on an objective function of an economic model would improve the total outputs.
- There is an acceptance guarantee for affordable single, block, and flexible offers after the market operator compares the price to the market clearing price. This is the basis on which market participants take their positions. Each energy type, however, has its own set of characteristics.
- Electricity markets should be designed to accommodate various types of energy. Renewable energy sources, in particular, may have unpredictable outputs due to their nature. Recent studies, however, have integrated stochastic programming approaches and concepts into DAMs for the purpose of dealing with the uncertainties caused by renewable energy.
- Those characteristics should be used to organize the different types of bids. The convenience of the producers can be increased by developing alternative options. Unlike European DAMs, Turkish DAMs evaluate paradoxical bids resulting from the non-convex nature of the optimization problem. Recent research, however, proposes market clearing models on this issue for European DAMs.
- The software and developed algorithms are used for more accurate, precise, and quick numerical calculations based on the supply and demand laws. Nonetheless, those should be improved due to changing conditions and regulations. Therefore, the improvement of this software is essential and the domestic possibilities of the countries can provide this improvement. This would decrease developing countries' foreign dependency on energy.

The subject of price determination in day-ahead electricity markets consists of many development research areas. For instance, for conventional energy (i.e., natural gas, coal, liquid fuels), and for renewable energy (i.e., solar, wind, geothermal, biomass), there are various possible market structure types. Therefore, the applications and simulations that include case studies will be useful in the development of markets. Specifically, we can listen to the solar and wind energy producers and the feedback from them may shed light for the regulations and new research considering these feedbacks may provide a guideline for the market participants.

References

Acaroğlu, H., & García Márquez, F. P. (2021). Comprehensive review on electricity market price and load forecasting based on wind energy. *Energies, 14*(22), 7473.

Akbary, P., Ghiasi, M., Pourkheranjani, M. R. R., Alipour, H., & Ghadimi, N. (2019). Extracting appropriate nodal marginal prices for all types of committed reserve. *Computational Economics, 53*(1, 1), –26. https://doi.org/10.1007/s10614-017-9716-2

Ceyhan, G., Köksalan, M., & Lokman, B. (2021). Extensions for benders cuts and new valid inequalities for solving the European day-ahead electricity market clearing problem efficiently. *European Journal of Operational Research, 300*, 713. https://doi.org/10.1016/j.ejor.2021.10.007

Chatzigiannis, D. I., Dourbois, G. A., Biskas, P. N., & Bakirtzis, A. G. (2016). European day-ahead electricity market clearing model. *Electric Power Systems Research, 140*, 225–239. https://doi.org/10.1016/j.epsr.2016.06.019

Derinkuyu, K. (2015). On the determination of European day ahead electricity prices: The Turkish case. *European Journal of Operational Research, 244*(3), 980–989. https://doi.org/10.1016/j.ejor.2015.02.031

Derinkuyu, K., Tanrisever, F., Kurt, N., & Ceyhan, G. (2020). Optimizing day-ahead electricity market prices: Increasing the total surplus for energy exchange Istanbul. *Manufacturing & Service Operations Management, 22*(4), 700–716. https://doi.org/10.1287/msom.2018.0767

EPİAŞ. (2016). *Gün Öncesi Elektrik Piyasası, Piyasa Takas Fiyatı Belirleme Yöntemi*. Enerji Piyasaları Anonim Şirketi.

Gökgöz, F., & Atmaca, M. E. (2017). Portfolio optimization under lower partial moments in emerging electricity markets: Evidence from Turkey. *Renewable and Sustainable Energy Reviews, 67*, 437–449. https://doi.org/10.1016/j.rser.2016.09.029

IEA. (2021). Turkey 2021. In *Energy Policy Review*. IEA [Online]. Accessed from https://www.iea.org/reports/turkey-2021

Li, W., & Becker, D. M. (2021). Day-ahead electricity price prediction applying hybrid models of LSTM-based deep learning methods and feature selection algorithms under consideration of market coupling. *Energy, 237*, 121543. https://doi.org/10.1016/j.energy.2021.121543

Lu, X., Yang, Y., Wang, P., Fan, Y., Yu, F., & Zafetti, N. (2021). A new converged emperor penguin optimizer for biding strategy in a day-ahead deregulated market clearing price: A case study in China. *Energy, 227*, 120386. https://doi.org/10.1016/j.energy.2021.120386

Marugán, A. P., Márquez, F. P. G., Perez, J. M. P., & Ruiz-Hernández, D. (2018). A survey of artificial neural network in wind energy systems. *Applied Energy, 228*, 1822–1836. https://doi.org/10.1016/j.apenergy.2018.07.084

Mayer, K., & Trück, S. (2018). Electricity markets around the world. *Journal of Commodity Markets, 9*, 77–100. https://doi.org/10.1016/j.jcomm.2018.02.001

Nasiri, N., Zeynali, S., Ravadanegh, S. N., & Marzband, M. (2021). A hybrid robust-stochastic approach for strategic scheduling of a multi-energy system as a price-maker player in day-ahead wholesale market. *Energy, 235*, 121398. https://doi.org/10.1016/j.energy.2021.121398

PwC Turkey and Presidency of the Republic of Turkey, Investment Office. (2021). *Overview of the Turkish Electiricty Market*. [Online]. Accessed from https://www.pwc.com.tr/overview-of-the-turkish-electricity-market

Sahin, C. (2018). Consideration of network constraints in the Turkish day ahead electricity market. *International Journal of Electrical Power & Energy Systems, 102*, 245–253. https://doi.org/10.1016/j.ijepes.2018.04.027

Sarıca, K., Kumbaroğlu, G., & Or, I. (2012). Modeling and analysis of a decentralized electricity market: An integrated simulation/optimization approach. *Energy, 44*(1), 830–852. https://doi.org/10.1016/j.energy.2012.05.009

Shariat Torbaghan, S., et al. (2021). Designing day-ahead multi-carrier markets for flexibility: Models and clearing algorithms. *Applied Energy, 285*, 116390. https://doi.org/10.1016/j.apenergy.2020.116390

Somay, S., Samlı, Z., Dağlı, S., & Sabri Kaya, P. (2021). *Electricity regulation in Turkey: Overview*. Thomson Reuters.

Thomaidis, N. S., & Biskas, P. N. (2021). Fundamental pricing laws and long memory effects in the day-ahead power market. *Energy Economics, 100*, 105211. https://doi.org/10.1016/j.eneco.2021.105211

van Vyve, M. (2011). *Linear prices for non-convex electricity markets: Models and algorithms*. Université catholique de Louvain, Center for Operations Research and Econometrics (CORE) [Online]. Accessed from https://EconPapers.repec.org/RePEc:cor:louvco:2011050

Weron, R. (2006). *Modeling and forecasting electricity loads and prices: A statistical approach (Wiley finance series)*. Wiley.

The Role of DevOps in Sustainable Enterprise Development

Zorica Bogdanović, Marijana Despotović-Zrakić, Dušan Barać, Aleksandra Labus, and Miloš Radenković

Abstract This chapter considers the problem of sustainable enterprise development from the perspective of the organization of software development. The goal is to provide a comprehensive analysis of the characteristics of the DevOps approach and present how it enables organization of continuous software development. By utilizing these DevOps principles, an enterprise can quickly adapt to ever-changing business requirements while fostering an environment capable of sustainable growth. This growth is primarily enabled by a shift in culture, where teams are encouraged to be functionally diverse which allows them to consider the whole lifecycle of an activity, rather than one single aspect. A case study of a university research and development center is presented to demonstrate the peculiarities arising from the DevOps approach. Conclusions and recommendations for future works are drawn from the literature analysis and own experiences.

Keywords DevOps · IT management · Information system development · Business-IT alignment · Sustainable software development

1 Introduction

Nowadays, numerous enterprises are intensively adopting Industry 4.0 technologies, with the goal to increase speed, efficiency, and quality of products and services, as well as customer satisfaction (Park & Huh, 2018) In order to be effective, the transition to the Industry 4.0 must be based on the sustainable enterprise development and complete alignment between business and information technologies (business-IT alignment). The result of this alignment is that the information system

Z. Bogdanović (✉) · M. Despotović-Zrakić · D. Barać · A. Labus
Faculty of Organizational Sciences, University of Belgrade, Belgrade, Serbia
e-mail: zorica@elab.rs; maja@elab.rs; dusan@elab.rs; aleksandra@elab.rs

M. Radenković
School of Computing, Union University, Belgrade, Serbia
e-mail: mradenkovic@raf.edu.rs

F. P. García Márquez, B. Lev (eds.), *Sustainability*, International Series in Operations Research & Management Science 333,
https://doi.org/10.1007/978-3-031-16620-4_12

and software development are both in accordance with the lifecycle of the enterprise (Ali et al., 2020; Erich et al., 2017). As adoption of Industry 4.0 increases, traditional approaches to IT management and software development, such as waterfall and even common agile methodologies are falling short and cannot ensure a quick and continuous delivery of reliable and scalable software services (Hemon et al., 2020). Therefore, the practitioners are encouraged to rely on new approaches, based on lean production principles and changes in culture where all the employees work in unison to enhance business processes and increase value to consumers (Ebert et al., 2016). A new approach called DevOps has emerged by tackling the specific challenges of aligning the software development process (Dev) and IT operations (Ops) with business development (Ebert et al., 2016; Leite et al., 2019). Most of researchers and practitioners consider DevOps only as an approach to the organization of software development process, using the concepts of continuous development, integration, testing, delivery, and monitoring (Fitzgerald & Stol, 2017). However, the comprehensive analysis of the DevOps principles leads to the conclusion that it originates from the concepts of sustainable development. Through sustainable development, traditional borders between management and IT are removed and the software development is put fully in line with the lifecycle of the enterprise (Gruhn & Schäfer, 2015).

The goal of this article is to provide a critical analysis of the results of previous approaches to information system development and present the characteristics of the DevOps practice that enables continuous software development. We present implementation aspects and provide a set of recommendations in the process of transition to DevOps.

The rest of the chapter is organized as follows. Section 2 discusses the aspects of business-IT alignment. In Sect. 3, the evolution of IT project management and software development paradigms are presented to explain both organizational and technical background of DevOps. Section 4 gives different details on the DevOps concepts, process, tools, and implications. In Sect. 5, a case study on introduction DevOps approach to one small IT organization is presented. Finally, discussions, implications, and conclusions are introduced.

2 Business: IT Alignment

In current digital economy, software applications are not just a part of the business. The Internet has irreversibly transformed the world and its industries, and software is no longer just the means to support a business, it has become an integral part of it. Companies provide value to their customers through software that is delivered as an online service through any type of device. Therefore, endeavors must be undertaken to ensure that the application meets the needs and expectations of the end users. In order to stay competitive in these circumstances, companies have to align their business and IT strategies and operations (Mekonnen et al., 2021). The concept of business-IT alignment originated from the 1990s. Many of the existing

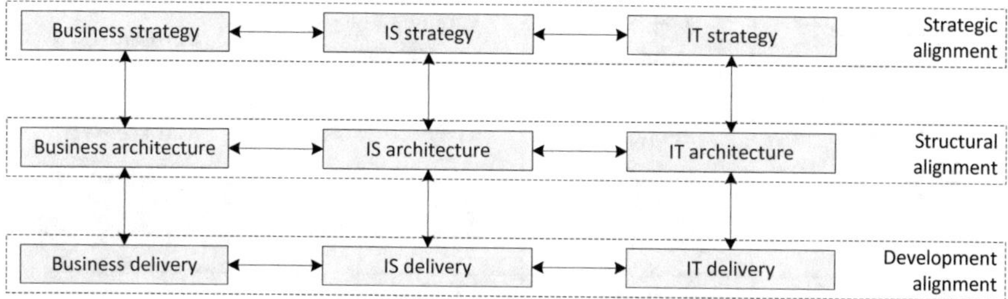

Fig. 1 Business-IT alignment on strategic and operational levels (Zhang et al., 2018)

approaches to business-IT alignment rely on traditional enterprise information systems architectures, such as service-oriented architecture (Zhang et al., 2018). In most of them, the general idea of business-IT alignment is to find ways to employ IT effectively to support the business needs across the organization. However, the fast development of IT has posed new challenges to businesses, requiring more agility, flexibility, and efficacy in addressing the needs and expectations of various stakeholders. The companies have started to explore sustainable ways to align their business and IT. Figure 1 presents various views on the business-IT alignment elements. The companies need to implement the changes on strategic, structural, and development levels, and adjust their structure and organizational culture in order to implement a successful digital transformation (Jonathan et al., 2021).

The term business-IT alignment can be defined as a way of applying information technology in an appropriate and timely way in harmony with business strategies, goals and needs, and literature has confirmed its significance for organizational performance (Zhang et al., 2018; Jonathan et al., 2021). However, there have not been many studies regarding the aspects of business-IT alignment in the context of modern-day digital transformation, where digital services evolved into core services for numerous companies. This evolution is enhanced by DevOps, defined as a set of tools and techniques specifically suited for continuous development of services. The understanding of the evolution of software development and IT project management is required to comprehend the specifics of business-IT alignment in the DevOps context.

3 Evolution of IT Project Management and Software Development Paradigms

The approaches to IT project management evolved, as the organization theory and practices developed. Along with the approaches to IT project management, the software architectures were developed trying to fulfill the requirements of scalability, reliability, and globally distributed systems, using the available technologies. Figure 2 shows the overview of the dominant approaches to organize the software

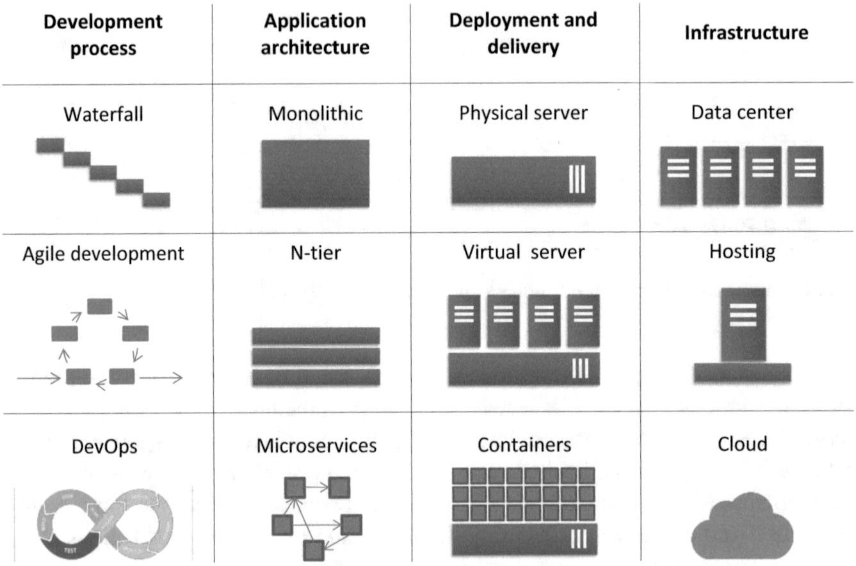

Development process	Application architecture	Deployment and delivery	Infrastructure
Waterfall	Monolithic	Physical server	Data center
Agile development	N-tier	Virtual server	Hosting
DevOps	Microservices	Containers	Cloud

Fig. 2 The evolution of IT project management, software architecture, deployment and infrastructure approaches

development process with corresponding software architectures, deployment and infrastructure approaches.

In the period 1960s–2000s, software was developed using the waterfall and, later, agile methods (Raj & Sinha, 2020; Banica et al., 2017). Waterfall method is a sequential method of software development. Within this model, the software development is linear, organized in several phases, where the outputs of each phase represent the input into the next, and the return to the previous phases is not allowed. The software is completed and delivered when all the features have been implemented. Typically, waterfall produces stable software solutions, but in a long term, this method cannot adapt to the ever-changing user requirements. Therefore, agile approaches appear in the 1990s.

Agile software development is an iterative method, where the development of individual features is done in iterations. The prototypes are generated in the early stages of development, and the delivery of usable software applications can be done after the key feature have been developed and tested. Lower priority features are implemented later and delivered through software updates. The result is a quickly developed software that in terms of quality and security, is not comparable to waterfall-based development (Gall & Pigni, 2021).

Previous approaches in software development tried to contribute to solving these requirements using the principles of object-oriented development and relational databases, although with limited success. Currently, companies of all sizes are adopting business strategies and models that require software to be globally available and highly scalable. The databases are turning to big data, where high volumes of unstructured and semi-structured data have to be efficiently stored, searched, and analyzed. For developing these modern information systems, linearity of the waterfall and cross-functionality of the agile development cannot provide quick and continuous delivery of high-quality software solutions (Ward & Legorreta, 2009). This is where DevOps rises to the challenge of overcoming the drawbacks of waterfall and agile approaches, while taking the best of both worlds. DevOps could be considered as a natural evolution of agile methods where fast project solutions are maturing into core services of companies. This maturing is due to the rising requirements of stakeholders where value must be provided continuously while still maintaining its speed and quality.

Before the 2000s, the term of "lean production" has already been known. Lean production refers to the organizational culture where all the employees actively seek ways of improving processes and delivering value to clients (Hart, 2012). Speed is the most important requirement, even if the consequence is lower quality of the final product. Therefore, the main goal is to shorten time between the request and delivery, leading to increased productivity, quality and customer satisfaction (Bierwolf et al., 2017; Díaz et al., 2019). Considering this, the practitioners are constantly investigating ways to:

- Keep inventory at minimum, both for raw materials and final products.
- Shorten order queues. Ideally, each order should be immediately processed and delivered in the shortest time.
- Increase efficiency in production. For this purpose, business process reengineering and increased automation are needed.

Regarding IT operations, cloud computing models, including Infrastructure as a Service (IaaS), Platform as a Service (PaaS) and Software as a Service (SaaS) have reached maturity and have enabled fast set-up of the infrastructure for any software service. The tools for continuous integration have emerged, leading to the regulation of relations between software development and IT operations (Lwakatare et al., 2016a; Virmani, 2015). These regulations primarily focused on communication and cooperation within multifunctional teams. Some of the other notable concepts that increase the overall software production of DevOps include: stable work environment, super-fast delivery, collaboration, time optimization, permanent optimization, among others.

Software architectures have also changed along with the development of IT project management approaches. The three most used approaches for developing software in internet environments can be identified as follows:

- Event-driven development
- Service-oriented development
- Process-oriented development

Event-Driven Software Development

The event-driven approach is based on the principle of diffuse event broadcasting. The components of the software system generate and send messages to the event bus, while other components consume those events. Message generation and joining process are asynchronous. After the component generates a message, it moves on to another job, without the need to wait for the other components to receive the message. The basic building block of any event-based system is the event. The event represents any change in the computer program or its environment, regardless of whether it occurred in the system or as a reaction to the actions of users. Events can be a click on an ad, a specific moment of time, a change in the value of a variable, etc.

Event-driven software development is proving to be popular despite its age, and it can be attributed to the ease modeling of complex processes through Publish/ Subscribe methods. This ease of use and speed of software development, are resulting in Event Driven software development principles being used in new and innovative ways, such as microservice environments, etc.

Service-Oriented Development

The basic building block of any service-oriented software is service. A service is a collection of events that changes the state of an entity. Services can be made up of a series of smaller services distributed in different places while functioning as a whole, such a way of organization is microservice software architecture. Each microservice fulfills a certain required process, manages its own separate database, and is present at all times. The advantage of the microservices architecture is that it enables the system to operate at full capacity, even when problems arise in individual services. Likewise, decoupling of services allows for much higher scalability, allowing each service to be replicated without much concern of performance. The microservice architecture is the dominant one in the DevOps community, since services as a building block allow the most flexibility and ease of development/testing/deployment required for continuous development (Bass et al., 2015; Balalaie et al., 2016; Hall, n.d.).

Process-Oriented Development

Process-oriented development takes place through processes, combining services into processes in order to follow a certain phenomenon from creation to completion. Processes can be distributed, leading to a data management problems. Some of the most important patterns for describing these processes are distributed (choreography) and centralized decision-making (orchestration).

The popularity of process-oriented development is mostly attributed to its close connection to business processes, where processes can be accurately modeled to represent any business context and even used to generate perfectly functional applications from the model itself. Process-oriented development in the DevOps context is needed when the individual microservices have to be orchestrated or choreographed to fully support a certain business process.

4 DevOps Concepts

The term DevOps stands for a combination of development (Dev) and IT operations (Ops), with the addition of quality assurance (Fig. 3). It incorporates a set of principles, methods and tools for continuous software development, in all its phases, starting from analysis, design, development, integration, testing, delivery, and monitoring (Leite et al., 2019). The term DevOps has emerged as an approach to the organization of software development and IT operations, but it has evolved to stand for the fully aligned business and IT, where both software development and IT operations are able to be in line with the business needs. This is achieved by enforcing a permanent communication and collaboration of all the stakeholders, continuous development and integration of software components, and a high level of automation (Zhu et al., 2016).

In order to introduce DevOps into a company, changes in organizational structure and culture are needed (Ward & Legorreta, 2009). The traditional approach, where development and operations teams work separately, can no longer be applied, and their different goals, e.g., providing new value to customers for development, security, stability, and scalability for operations, need to be unified toward a joint cause (Shahin & Babar, 2020; Hemon-Hildgen et al., 2020; Wiedemann et al., 2019). Figure 4 illustrates the traditional differences between development and operations teams, in comparison to the DevOps approach.

4.1 *Software Development Process in DevOps*

DevOps is mainly harnessed in the development of web, mobile, and cloud-based applications and large-scale distributed applications (Agrawal & Rawat, 2019). It consists of multiple phases that overlap and are performed in an infinite loop (Fig. 5). In the further text, we briefly describe the main phases of the software development process in the DevOps approach: (1) planning, (2) Coding, (3) Building, (4) Testing, (5) Release, (6) Deployment, (7) Operating, and (8) Monitoring (Fitzgerald & Stol, 2017; Pennington, 2019).

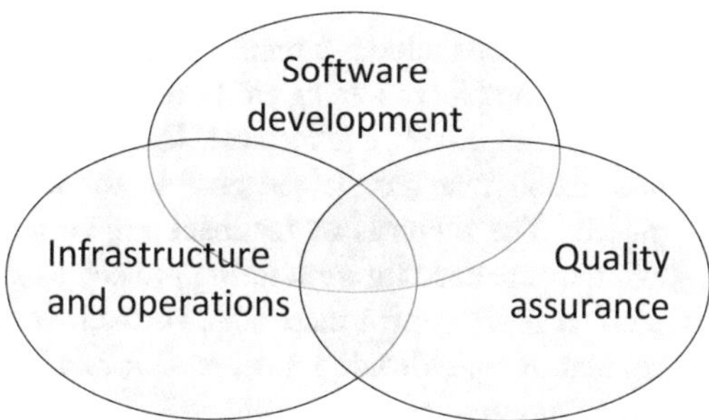

Fig. 3 Overview of DevOps disciplines

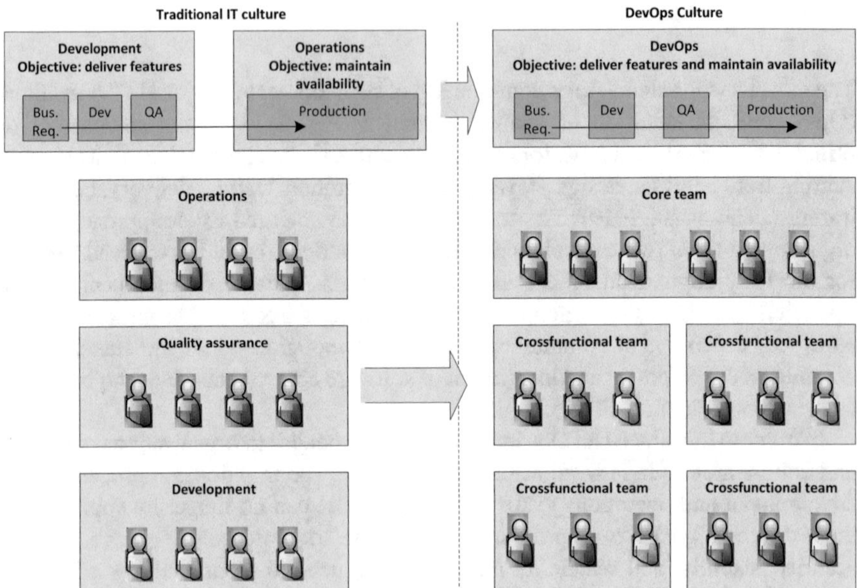

Fig. 4 Difference between traditional and DevOps culture

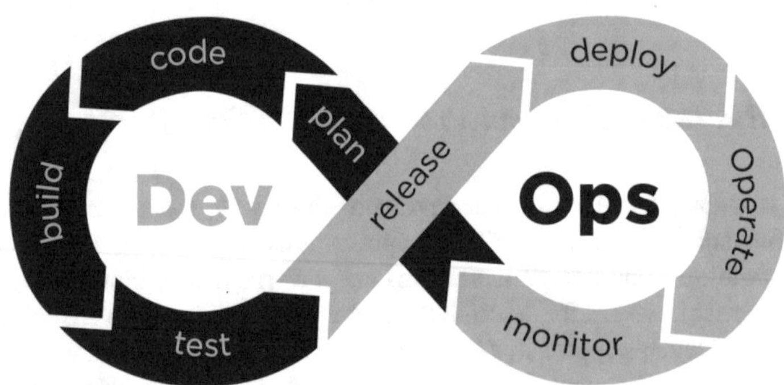

Fig. 5 DevOps infinite loop

1. *Planning*

   This phase includes a preparation for the development. The implementation of the plan (project roadmap, etc.) can be monitored using project management tools such as Jira, Slack, Mattermost, OpenProject, or Asana. The project manager or production manager is responsible for this phase, depending on the type of project. The requests or feedback are collected from the stakeholders or users and they are used for monitoring project progress, problems, and milestones. The product development map can be broken down into epics and user stories. In the context of agile development, an epic is a large body of work that can be broken down into smaller and specific tasks—user stories. Epics and user stories form the backlog that portrays customer requirements. Backlog is used for planning sprints and assigning tasks to the team members.

2. *Coding (development)*

   In addition to the common set of tools used by the developer, the team uses a set of add-ons installed in their development environments to help in the implementation of consistent and secure code. This helps developers learn to write code in an effective way, while supporting collaboration. These tools also help to solve problems that may later occur in the flow.

3. *Building*

   Once the programmer completes the task, the code is stored in a shared source code repository. The developer submits a request to merge the new code with a common code base. Another developer reviews the changes and the request is approved if there are no issues. This step is usually fast but it is critical to detect the problem early. At the same time, the withdrawal request runs a process that builds the program and runs a series of tests from the beginning to the end of the integration to identify problems. If the connection or any test fails, the withdrawal request fails, and the developer is informed that the problem needs to be solved. Continuous code building can significantly reduce integration problems because the new code is tested frequently, and potential integration problems can be quickly identified and solved.

4. *Testing*

   Once the build is successful, detailed testing is automatically started. The setup environment can easily be provided as part of the implementation process, using the principles of Infrastructure-as-a-Code and containerized architectures. When the application is ready for testing, it is performed through automated tests. Manual testing is not completely avoided in the DevOps practice, but the focus is on automatic testing. Non-functional testing enables UX/UI testing, and the user can ask any questions or requirements that should be resolved before being introduced to the production environment. At the same time, automated tests can run an application security scan, check for infrastructure changes and compliance, test application performance, or run load testing. This phase can be considered as a test basis that allows new testing to be included without interrupting the work of the programmer or affecting the production environment.

5. *Release*

   The release phase is a turning point in the DevOps flow and it is the point when development is ready for deployment to a production environment. Until this stage, every change of code has passed through manual and automated tests and the operational team can be sure that there will be no problems and regressions.

   It is possible to automatically deploy any development leading to this phase of delivery flow depending on the degree of use of DevOps in the organization. Developers can use the Feature flags to turn off new features so users cannot see them until they are ready for distribution. Alternatively, the organization may have control over the release of product versions and they can schedule a regular release or a new feature. A manual approval process may be added at the publishing stage to allow only certain people in the organization to approve the release to production.

6. *Deployment*

   At this stage, the version is ready and deployed into the production environment. The same Infrastructure-as-a-Code on which the test environment is built can be configured to create a production environment. Once the test environment has been successfully built, it is possible to confirm that the new release will run smoothly in the production environment. The new environment is adjacent to the production environment because features are packed into containers, as microservices. If a problem with the new version is encountered at any time, requests may be redirected back to the old environment until the update is completed.

7. *Operating*

   The new release is operable and user-friendly at this stage. The operational team works intensively, making sure everything runs without errors. The environment can be automatically scaled up or down based on the configuration. The organization has also built a way to collect feedback from users continuously, as well as tools to help collect and link this feedback to future product development. Users know better about their preferences, and they can be considered the best testers as they spend much more time testing the application than DevOps flow.

8. *Monitoring*

   The last phase of the DevOps approach is monitoring and tracking. This phase relies on user feedback collected and data related to user behavior, performance, errors, etc. The DevOps flow can also be monitored, detecting potential bottlenecks that cause issues or affect the productivity of development and operational teams. This information is then passed to the product manager and development team to close the DevOps loop. It is important to note that this process is continuous and the product is constantly evolving during its lifetime and ends only when it ceases to be in use, i.e., when it is no longer useful or necessary.

   DevOps is by definition a natural extension of agile frameworks. The frameworks refer not only to development and operations but also to everything needed to produce value streams, e.g., security, compliance, audit, marketing, among others. Many agile frameworks concepts and principles, e.g., system thinking, incremental-iterative process, quick feedback, directly support DevOps. In addition, the flow of continuous delivery directly supports these business needs. The goal of DevOps in agile frameworks is manyfold: improvement of collaboration through value flow, automating continuous delivery flow, increment of the frequency and quality of application, reduction of risk through safer experimentation, and reduction of product time to market.

There are many challenges in DevOps, as human and technical aspects are interconnected, and operation and development teams often have differing points of view. A culture of collaboration provides an opportunity for people with different knowledge, skills, and abilities to collaborate and create value together (Dornenburg, 2018). Four elements are required to allow effective collaboration: open communication, reconciling responsibilities and incentives, respect, and trust (Rutz, 2019). Open communication helps to avoid conflicts and duplication of work while enabling mutual understanding. Another important aspect is the clear alignment of responsibilities and incentives within and across teams. If a prompt response to the situation is needed, teams with coordinated responsibilities are able to act in the short term.

4.2 DevOps Tools

In addition to the organizational aspects, DevOps must also tackle certain technical requirements that arise from the need of the system to deal with design and transfer of data through its many components. However, an isolated automated system is not sufficient for success even if it is well designed, and must always rely on the human factor that adapts, interacts, and processes data (Cois et al., 2015). In order to achieve all of the above, it is necessary to employ appropriate tools but there is no single tool in DevOps that could cover all aspects such as container management, continuous integration, monitoring, deployment, and testing. This is why large organizations strive to develop their own custom tools, while the ecosystem of DevOps tools is constantly growing and new tools are emerging. The goal of almost every tool is to reduce the time needed to perform manual tasks and/or establish automation. Some of them are specialized for a single phase, while some can be applied across several phases. Some of the tools that have been considered as a key element of the DevOps implementation include (Ebert et al., 2016):

- Project management, communication and collaboration tools, such as Jira, Slack, Trello, Confluence, as well as open-source tools: OpenProject, Mattermost, BigBlueButton
- Version control systems: Git, Subversion.
- Continuous integration (build, integration, orchestration) tools: Maven, Ant, Jenkins.
- Microservice management: Docker, Podman, Kubernetes (Balalaie et al., 2016).
- Server orchestration and configuration: Ansible, Puppet, Chef, Vagrant
- Monitoring: Nagios, Splunk, Sensu, NewRelic.

All the proposed tools are expected to contribute to a better quality of the developed software (Mishra & Otaiwi, 2020).

4.3 Recommendations from Theory and Practice

Numerous authors have conducted a systematic literature review of DevOps scientific and professional texts in recent years, although several recommendations and best practices come from professionals. The main implications of DevOps to sustainable enterprise development include (Díaz et al., 2019; Hall, n.d.; Bajpai, n.d.; Capgemini, 2015):

- A top-down approach—small and self-organizing teams are frequently adopting DevOps approach on their own, that gives good results in start-ups. However, scaling up DevOps to the organizational level is not an easy task, and requires changes in the organizational structure and culture.
- Roles—in the DevOps approach, responsibilities are spread across the team. The leadership is transferred to the younger and middle management, while the traditional roles are being redefined to new ones with significantly extended competencies. New roles include product managers, value stream architects, DevOps engineers, etc.
- Skills—the skillset within the DevOps context is significantly shifting. DevOps tools are booming and new tools for different DevOps phases are appearing every day, so fast learning and adaptation are crucial. In addition, the soft skills that have been identified as increasingly important are creativity, agility, and adaptation to change.
- Incentives—appropriate incentives may be needed in order to ensure a high level of professionalism, accountability, and responsibility within the DevOps practice. Frequently, "intrapreneur" approach is advocated as a means to promote free generation of ideas.
- Observability –The internal functions of a complex system can be better understood by studying its external behaviors. These behaviors can be monitored primarily through logs, traces, and metrics, although maintaining observability in DevOps environments with microservice-based applications is becoming increasingly difficult and teams must adapt new ways of achieving observability, e.g., chaos engineering.
- The right tools—With increasing reliance on tools in DevOps, their selection is becoming increasingly important. Familiarity with tools already in use should always be prioritized over new tools because these not only bring difficulties related to their own adoption but also with the learning of new ways of interacting with other tools. The integration of various tools becomes a necessity for the day-to-day DevOps functions.

5 Case Study: Introducing DevOps Approach into a University R&D Center

5.1 Research Context

The process of introducing the DevOps culture into a research center of an educational institution is presented in this section. The research centers of educational institutions that deal with software development are characterized by:

- A broad portfolio of software projects that need to be realized within a given deadline.
- Requirements to produce permanent innovation, in order to get funding and/or compete in the market.
- Frequent fluctuations of team members, both employees and students, due to the dynamics of IT labor market.
- Heterogeneous technologies and IT components in business and IT ecosystems
- Repetitive tasks and need for automation
- Limited costs of production

The research context is the Laboratory for e-business of the Faculty of Organizational Sciences, University of Belgrade, where numerous projects are being realized within the teaching, scientific and innovation work. The laboratory is not registered as an individual company but its organization and a way of work can be comparable to typical innovation centers, both academic as well as industrial.

5.2 The Process of Introducing DevOps Approach

The process of transitioning to DevOps has been done in two parallel areas, both going through phases shown in Table 1. The phases have been organized in accordance with the DevOps dimensions identified in Lwakatare et al. (2016b).

All the phases included permanent training of all the team members. At the moment of writing this article, phase 1 has been finished, phase 2 is in the middle of implementation and phase 3 is in the preparation stage.

Figures 6 and 7 show examples of Mattermost and OpenProject tools that support the communication and collaboration of DevOps teams. Figure 6 shows an example of a NodeJS project being developed, where the new events on the GitHub repository created by activities of the developers are published on a shared channel. Figure 7 presents a list of projects that have been realized and monitored within the organization.

5.3 Analysis of Results

There are 5 projects conducted within the laboratory, with 58 participants (Table 2). Total of 10 participants are full-time employed in the Laboratory, while 48 are engaged in part-time mode.

Figure 7 presents the team statistics from the Mattermost server, showing an overview of the communication and collaboration activities for the period of one month.

The results presented in Fig. 8, as well as the experiences of participants in the project show that:

Table 1 Phases of introducing DevOps approach

	Infrastructure development	Organizational structure and culture development
Phase 1: Collaboration dimension	1. Set up of communication and collaboration environment, including: – Mattermost – BigBlueButton – OpenProject – Git	Introduction of agile teams with the following roles: – Product owners – Scrum masters – Developers
Phase 2: Automation and culture dimensions	Set up of DevOps tools: – Jenkins – Ansible Transition to microservice architecture of the produced software (containerization) Improving the organization of the version control system, so it enables better tracking of all the projects conducted within the organization.	Extension of team roles with the following: – Portfolio manager – Product managers – QAs – DevOps engineers Introduction of DevOps suitable product development techniques: – Value stream mapping – Customer journey mapping – Design thinking
Phase 3: Measurement and monitoring dimensions	Set up of the monitoring system – Nagios – Grafana	Development of performance indicators Development of incentive-based culture

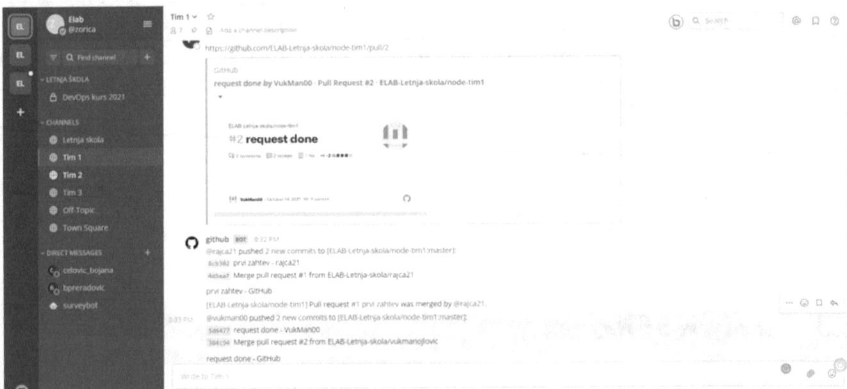

Fig. 6 Mattermost—integration with version control

- All five teams are actively communicating in the development of their projects.
- The number of daily active users is relatively low, but this is in accordance with the scheduled dynamics of the projects; in most cases, part-time team members are working on the projects 1–2 days a week.

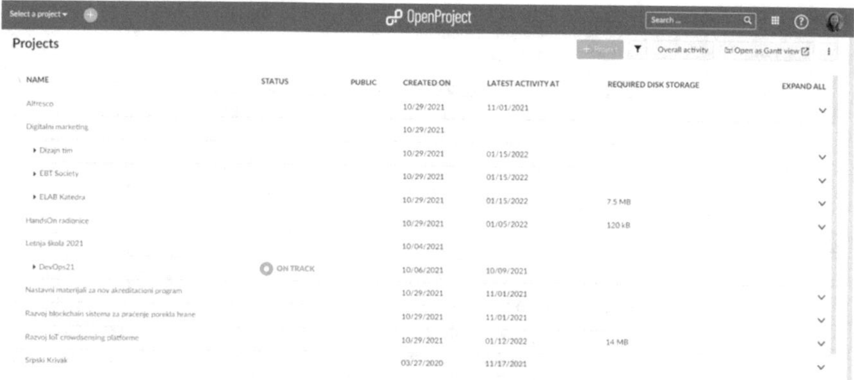

Fig. 7 List of the projects managed within Open project

- From interviews with the team members, it is clear that they are frequently using the means of informal communication (messaging apps, such as Viber, Whatsapp and similar). In future work, it would be necessary to motivate some of the team members to use the communication and collaboration infrastructure of the research center.
- The structure of communication channels should be improved, so as it becomes more natural for the team members. Team members are sometimes more comfortable with individual direct messages.

6 Discussion and Implications

The practice of DevOps encourages efficient and secure delivery of business value to the clients, while including the perspectives of different stakeholders. This given value takes various forms, e.g., new software features, frequent releases and updates, high-quality UX, etc. Fast delivery to end-consumers provides a rapid market entry and competitive advantage (Díaz et al., 2019), and also quick problem solving and resolution of disputes.

It is clear that many scientific articles on DevOps have tried to perform a systematic analysis of review, providing a systematization of DevOps definitions and practices (Zhu et al., 2016; Rutz, 2019; Lwakatare et al., 2016b; Bolscher & Daneva, 2019). However, there are only a few references providing experiences from real-world companies, although market researches showed that many companies are implementing DevOps in various phases (Kersten, 2017). Researchers and market analysts reported that high-performing DevOps organization deploy software more frequently, implement changes in shorter time and recover from outages more quickly, leading to the conclusion that DevOps provides for higher organizational performance. In addition, most of the DevOps articles originate from conferences, while journals and magazines publish lower number of texts (Leite et al., 2019).

Table 2 The list of projects

Project name	Short description	DevOps specifics and experiences
1. Smart city crowdsensing system	The system consisting of various crowdsensing data producers (measuring noise, vibrations, air quality, etc) and consumers (mobile apps, web apps)	The application is organized in a microservice architecture, for all the components to be developed and deployed independently (Labus et al., 2021).
2. Blockchain platform for monitoring organic honey production	The system for monitoring the supply chain of organic honey production, where storing data in the blockchain distributed ledger provides for a possibility to track data from its origin to end consumers. The goal is to reduce the possibility of fraudulent honey production.	The application is organized in a microservice architecture. The application is being developed with a goal to start up a company that will gather all the stakeholders of the organic honey supply chain.
3. Document management system	The system based on Alfresco document management system, but specially customized to enable students and teachers to search students' papers (seminal papers, bachelor, master and PhD theses)	The system needs continuous improvements and adjustments to the process of research and writing a thesis. It has been implemented in microservice architecture.
4. Digital marketing (with 3 subprojects)	The projects include typical tasks: – Development of websites, including web design tasks – Social media marketing tasks – Configuration of various services for digital marketing (newsletters, CRM) – Continuous monitoring – Organization of special events	This is generally not considered as not a typical DevOps project, but it requires continuous development of various components, both software and non-software. Some elements of microservice architecture may be applied. The project requires implementation of DevOps organizational culture more than DevOps software development approach
5. Teaching and learning innovations	The project includes: – Continuous development and improvement of teaching and learning materials – Continuous development and maintenance of software for supporting innovative pedagogical approaches. – Continuous development and monitoring of e-learning infrastructure.	This can be considered as a typical DevOps project in any educational institution, since it has to follow the development of the study programs of the institutions. In general, it does not need to include software development, but in many cases nowadays, various software components are being developed as teaching materials.

Many attempts have been made to systematically summarize the literature on DevOps in recent years. However, methodological approaches to introduce DevOps into an organization have not been consolidated, and not much information has been available on the experiences of companies in transitioning to DevOps. Table 3

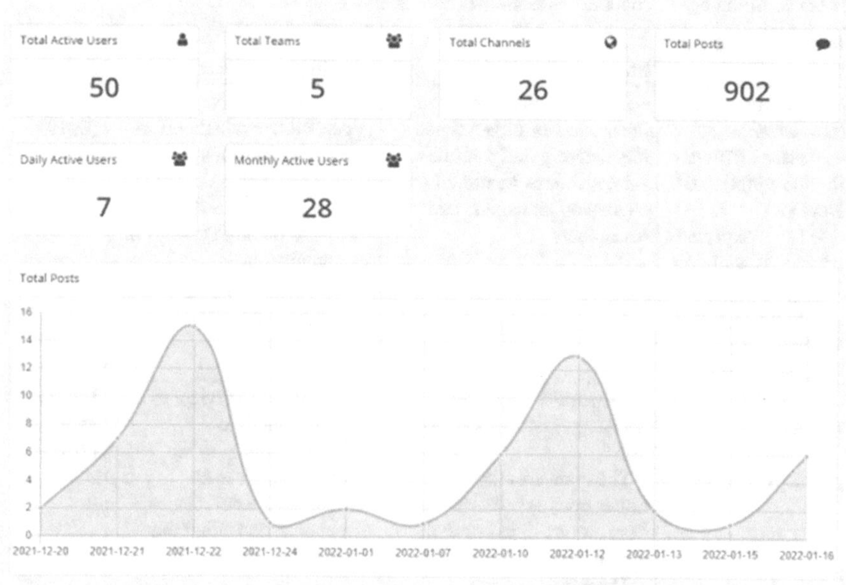

Fig. 8 Team statistics

provides a view on the main results obtained from the literature, with the presentation of the main conclusions and identified open questions.

Finally, in order to support the development of DevOps, the education process also needs to change (Leite et al., 2019). The concepts of DevOps need to be included in curricula of both IT and management fields to prepare students for the labor market. This also requires that methods such as active learning, project-based learning and similar, become standard methods that would support the education completely based on the DevOps principles and methods.

7 Conclusion

The main characteristics of the DevOps approach are presented in this article. The main goal was to point out that the software development process has to be organized to support the development of business. It has presented how the information technology project management approaches have evolved and how the supporting software infrastructures have changed to become suitable for the ongoing development of organization (Raj & Sinha, 2020).

Some experiences from introducing the DevOps approach have also been presented in one university research center. Although this is not a typical company, the obtained results and experiences can be valuable for companies that are transitioning to DevOps. Universities make the transition from teaching traditional

Table 3 Summary of conclusions from literature analysis

Research questions addressed	Main conclusions	Identified open questions	References
– Meaning of the term DevOps – Issues motivating the adoption of DevOps – Main expected benefits of adopting DevOps – Main expected challenges of adopting DevOps	– DevOps considered from the aspects of process, people, delivery and runtime – Implications identified for engineers, managers, and researchers	– More quantitative research is needed to evaluate effectiveness of DevOps – How to redesign systems toward continuous delivery – How to deploy DevOps in an organization – How to qualify engineers for DevOps practice	Leite et al. (2019)
– DevOps definitions – DevOps practices	– Difficulties and inconsistencies in DevOps definitions – The main DevOps dimensions are: collaboration, automation, culture, monitoring, measurement – There are no explicit one-size-fits-all set of practices for adoption or implementation	– Lack of an overview to a DevOps practices – Useful patterns to the DevOps practices are needed, so the companies can choose and adapt those that fit their context	Lwakatare et al. (2016b)
– Software architectural problems – Characteristics of a software architecture for enabling DevOps – Software architecture styles suitable for DevOps	– The focus is on continuous development and deployment – Micro-services have been identified as the dominant architecture foe DevOps	– Need for methods to migrate monolithic code to microservice architectures – Need for framework or guidelines for DevOps software architecture – Need for alignment of short-term software architecture and long-term enterprise architecture	Bolscher and Daneva (2019)

development to DevOps and it is necessary to incorporate DevOps concepts into the university culture and adapt and improve upon it. Therefore, the future of DevOps is not entirely dictated by companies but also by innovations spearheaded by the universities themselves.

Further research will be directed in three directions:

- First, more experiences are needed from organizations that have transitioned to DevOps, in order to obtain a list of recommendations for those that are considering this transition.
- Second, the transition of the presented university research center to DevOps has to be completed, results analyzed and conclusions drawn.
- Finally, a very important aspect is the education for DevOps. It is important not only to teach IT students to use various DevOps tools, but also to educate

managers that DevOps is a way to achieving a sustainable enterprise development, where software can fully comply the business needs.

References

Agrawal, P., & Rawat, N. (2019). Devops a new approach to cloud development & testing. In *2019 International Conference on Issues and Challenges in Intelligent Computing Techniques*, pp. 1–4.

Ali, M., Wood-Harper, T., & Ramlogan, R. (2020, April 29). Value creation through cloud SAAS applications: Business IT alignment in service industries - University of Bolton Institutional Repository (UBIR). In *Proceedings of 25th UK Academy for Information Systems (UKAIS) International Virtual Conference*. Accessed April 22, 2021, from http://ubir.bolton.ac.uk/2779/

Bajpai, G. (n.d.). *DevOps, a guide towards sustainability in software factories*. Capital Carbon Consulting. Accessed January 23, 2022, from https://www.capitalcarbonconsulting.com/devops-a-guide-towards-sustainability-in-software-factories/

Balalaie, A., Heydarnoori, A., & Jamshidi, P. (2016). Microservices architecture enables DevOps: Migration to a cloud-native architecture. *IEEE Software, 33*(3), 42–52. https://doi.org/10.1109/MS.2016.64

Banica, L., Radulescu, M., Rosca, D., & Hagiu, A. (2017). Is DevOps another Project management methodology? *Information Economics, 21*. https://doi.org/10.12948/issn14531305/21.3.2017.04

Bass, L., Weber, I., & Zhu, L. (2015). *DevOps: A software architect's perspective*. Pearson Education.

Bierwolf, R., Frijns, P., & Van Kemenade, P. (2017). Project management in a dynamic environment: Balancing stakeholders. *IEEE European Technology and Engineering Management Summit (E-TEMS), 2017*, 1–6. https://doi.org/10.1109/E-TEMS.2017.8244226

Bolscher, R., & Daneva, M. (2019). Designing software architecture to support continuous delivery and DevOps: A systematic literature review. In *ICSOFT 2019: Proceedings of the 14th International Conference on Software Technologies*, pp. 27–39. https://doi.org/10.5220/0007837000270039

Capgemini. (2015). *A Capgemini architecture whitepaper* (2nd edn). Accessed January 24, 2022, from www.capgemini.com

Cois, C. A., Yankel, J., & Connell, A. (2015, January). Modern DevOps: Optimizing software development through effective system interactions. In *IEEE International Professional Communication Conference*. https://doi.org/10.1109/IPCC.2014.7020388.

Díaz, J., Perez, J. E., Yague, A., Villegas, A., & de Antona, A. (2019). DevOps in practice – A preliminary analysis of two multinational companies. *Lecture Notes in Computer Science, 11915*, 323–330. https://doi.org/10.1007/978-3-030-35333-9_23

Dornenburg, E. (2018). The path to DevOps. *IEEE Software, 35*, 71–75. https://doi.org/10.1109/MS.2018.290110337

Ebert, C., Gallardo, G., Hernantes, J., & Serrano, N. (2016). DevOps. *IEEE Software, 33*, 94–100.

Erich, F. M. A., Amrit, C., & Daneva, M. (2017). A qualitative study of DevOps usage in practice. *Journal of Software: Evolution and Process, 29*, e1885. https://doi.org/10.1002/smr.1885

Fitzgerald, B., & Stol, K. J. (2017). Continuous software engineering: A roadmap and agenda. *Journal of Systems and Software, 123*, 176–189. https://doi.org/10.1016/j.jss.2015.06.063

Gall, M., & Pigni, F. (2021). Taking DevOps mainstream: A critical review and conceptual framework. *European Journal of Information Systems*. https://doi.org/10.1080/0960085X.2021.1997100

Gruhn, V., & Schäfer, C. (2015). *BizDevOps: Because DevOps is not the end of the story* (pp. 388–398). Springer.

Hall, T. (n.d.). *DevOps best practices. Atlassian*. Accessed January 23, 2022, from https://www.atlassian.com/devops/what-is-devops/devops-best-practices

Hart, M. A. (2012). The lean startup: How today's entrepreneurs use continuous innovation to create radically successful businesses. *Journal of Product Innovation Management, 29*, 508–509.

Hemon, A., Lyonnet, B., Rowe, F., & Fitzgerald, B. (2020). From agile to DevOps: Smart skills and collaborations. *Information Systems Frontiers, 22*, 927–945. https://doi.org/10.1007/s10796-019-09905-1

Hemon-Hildgen, A., Rowe, F., & Monnier-Senicourt, L. (2020). Orchestrating automation and sharing in DevOps teams: A revelatory case of job satisfaction factors, risk and work conditions. *European Journal of Information Systems*, 1–26. https://doi.org/10.1080/0960085X.2020.1782276

Jonathan, G. M., Rusu, L., & Van Grembergen, W. (2021). *Business-IT alignment and digital transformation: Setting a research agenda* (1st ed.). Springer.

Kersten, N. (2017). *The 2017 state of DevOps report*. Portland.

Labus, A., Radenković, M., Nešković, S., Popović, S., & Mitrović, S. (2021). IoT crowdsensing system based on data streaming architecture. In *ICMarkTech 21, 2nd Eork. Innovative Business Modelling Applications, Smart Cities*.

Leite, L., Rocha, C., Kon, F., Milojicic, D., & Meirelles, P. (2019). A survey of DevOps concepts and challenges. *ACM Computing Surveys, 52*. https://doi.org/10.1145/3359981

Lwakatare, L. E., Kuvaja, P., & Oivo, M. (2016a). Relationship of DevOps to agile, lean and continuous deployment. *Lecture Notes in Computer Science, 10027*, 399–415. https://doi.org/10.1007/978-3-319-49094-6_27

Lwakatare, L. E., Kuvaja, P., Oivo, M. (2016b). An exploratory study of devops extending the dimensions of DevOps with practices. In *ICSEA 2016 Eleventh International Conference on Software Engineering Advances*, pp. 91–99.

Mekonnen, J. G., Rusu, L., & Van Grembergen, W. (2021.). Business-IT alignment and digital transformation: Setting a research agenda. In *29th International Conference on Information Systems Development*.

Mishra, A., & Otaiwi, Z. (2020). DevOps and software quality: A systematic mapping. *Computer Science Review, 38*, 100308. https://doi.org/10.1016/J.COSREV.2020.100308

Park, S., & Huh, J.-H. (2018). Effect of cooperation on manufacturing IT project development and test bed for successful Industry 4.0 project: Safety management for security. *Processes, 6*, 88. https://doi.org/10.3390/pr6070088

Pennington, J. (2019). *The eight phases of a DevOps pipeline*. Accessed January 23, 2022, from https://medium.com/taptuit/the-eight-phases-of-a-devops-pipeline-fda53ec9bba

Raj, P., & Sinha, P. (2020). Project management in era of agile and Devops methodologies. *International Journal of Scientific and Technology Research, 9*, 1.

Rutz, M. (2019). *DevOps: A systematic literature review*. Fachhochschule Wedel.

Shahin, M., Babar, M. A.. (2020). On the role of software architecture in DevOps transformation: An industrial case study. In *Proceedings of 2020 IEEE/ACM International Conference on Software Systems and Process, ICSSP 2020*. Association for Computing Machinery, pp. 175–184. https://doi.org/10.1145/3379177.3388891.

Virmani, M. (2015). Understanding DevOps & bridging the gap from continuous integration to continuous delivery. *5th International Conference on Innovative Computing Technology, INTECH, 2015*, 78–82. https://doi.org/10.1109/INTECH.2015.7173368

Ward, C., & Legorreta, L. (2009). Beyond waterfall and agile methods: Towards a new contingency model for IT project management. *SSRN Electronic Journal*. https://doi.org/10.2139/SSRN.1400254

Wiedemann, A., Wiesche, M., Gewald, H., & Krcmar, H. (2019). Implementing the planning process within DevOps teams to achieve continuous innovation. *In Proceedings of 52nd Hawaii International Conference on System Sciences*. Accessed from https://hdl.handle.net/10125/60138

Zhang, M., Chen, H., & Luo, A. (2018). A systematic review of business-IT alignment research with Enterprise architecture. *IEEE Access, 6*, 18933–18944. https://doi.org/10.1109/ACCESS.2018.2819185

Zhu, L., Bass, L., & Champlin-Scharff, G. (2016). DevOps and its practices. *IEEE Software, 33*, 32–34. https://doi.org/10.1109/MS.2016.81

Practices and Indicators of Waste and Resource Management in Commercial Buildings

Abimbola Windapo, Dylan Murray, Tristan Hamilton, and Hayden Baum

Abstract This chapter examines the indicators and practices used by facilities managers regarding waste and resource management in commercial buildings and whether these current practices and indicators align with standards impacting facility sustainability and performance. A review of existing literature was undertaken to outline relevant information, which addresses the primary objectives of the report. The research adopted a qualitative approach that employs an embedded case study research design to obtain primary research data from interviews with facilities managers of commercial buildings. Furthermore, secondary data relating to energy, water, and waste was collected from company records. The scope of the study was limited to three Green Building Council of South Africa (GBCSA) Green star-rated (rating of green buildings in South Africa) commercial buildings in Cape Town. The key findings of the research indicate that the level of energy and water consumption is within the prescribed standards. The research concludes that indicators (resource consumption per annum/month/m^2) and practices (continuous service planning, facility plant, and equipment maintenance, accessing tenant requirements) used in waste and resource management by the facilities managers of the sampled commercial buildings are aligned with international and local standards and are adequate in addressing the needs of commercial buildings regarding the management of energy and water resources. Based on these findings, the study recommends that facilities managers use innovative technologies such as metering and the Building Management System Software, and cradle-to-cradle technology to manage the waste generated in commercial buildings. It also recommends that further research be undertaken using a larger sample size to allow for the generalization of the results.

Keywords Energy · Facilities management · GBCSA · Sustainability · Waste management

---

A. Windapo (✉) · D. Murray · T. Hamilton · H. Baum
Department of Construction Economics and Management, Faculty of Engineering and the Built Environment, University of Cape Town, Rondebosch, Cape Town, South Africa
e-mail: abimbola.windapo@uct.ac.za

F. P. García Márquez, B. Lev (eds.), *Sustainability*, International Series in Operations Research & Management Science 333,
https://doi.org/10.1007/978-3-031-16620-4_13

Nomenclature

DEA	Department of Environmental Affairs
FM	Facilities management
GBCSA	Green Building Council of South Africa
ISO	International Standards Organization
NEMA	National Environmental Management Act
NWMS	National Waste Management Strategy
SA	South Africa
SABS	South African Bureau of Standards
SANS	South African National Standards

1 Introduction

Sustainable development is the development that meets the needs of the present without compromising the ability of future generations to meet their own needs (Brundtland, 1987). The issue of sustainability in South Africa is prevalent now more than ever, given the scarcity of electricity (power outages) and water resources and the construction industry's need for sustainable development. The World Green Building Council (WGBC) Report of 2016 indicates that the building and construction sector of the built environment accounts for 36% of global energy use and 39% of energy-related carbon dioxide (Dean et al., 2016). Furthermore, a report released by the World Bank in 2018 indicates that the world produced 2.01 billion tons of solid waste in 2016, estimated to be 0.74 kg of solid waste per person per day (Kaza et al., 2018). Furthermore, the report forecasts that a 70% increase is expected by 2050, which will result in 3.40 billion tons of solid waste. It also concludes that managing waste efficiently is essential for building sustainable and civilized cities; however, it remains a challenge for facility managers in many developing countries and cities.

Facility Management (FM) is a multi-faceted profession that supports essential business functions by facilitating a conducive and comfortable environment for all building users (Abdeen & Sandanayake, 2018). FM practices are viewed as those decisions and subsequent actions undertaken by FM specialists to execute tasks that align with creating a conducive and comfortable environment in a facility (Rondeau et al., 2012). FM indicators such as total energy, water, and waste costs per annum are used to ascertain the performance level of a facility against external standards or internal benchmarks, either at a given point or for a given period (Arendse & Godfrey, 2010). Relevant indicators give valuable insight into any positive or negative performance outcomes and inform management-level decision-making to a large degree (Ogola et al., 2011).

Limited research (see, for example, Abdeen & Sandanayake, 2018; Elmualim et al., 2010) has been undertaken in FM practices and indicators used in

South African buildings, and there is no knowledge of whether these adequately address the issues of excessive waste and resource consumption in commercial buildings. Therefore, this chapter investigates the standard of South Africa's FM practices and indicators used by facilities managers in waste and resource management on commercial buildings. Therefore, the research question asks, "Are the waste management practices and indicators used by facilities managers effective in managing water and energy consumption and waste generation in commercial buildings?"

In order to achieve the aim, the following objectives must be achieved: (1) identifying the *practices* implemented by FM specialists for the management of waste and resources, (2) identifying the *indicators* used by FM specialists in respect of waste and resource management, (3) establishing the level of waste produced and resources consumed in commercial buildings, (4) establishing whether the FM *practices and indicators*, in respect of waste and resource management, align with standards, and (5) finding out whether the levels of waste generation and resource consumption can be attributed to the standard of the practices and indicators used by facilities managers in South Africa. Therefore, to guide the direction of the study, the research proposes that the waste management practices and indicators used by facilities managers effectively manage resource consumption and waste generation in commercial buildings. The scope of the investigation is limited to green star-rated commercial buildings in Cape Town. It focuses on identifying the practices and indicators used only for the FM application to waste and resource management. The study made use of interviews of facilities managers in collecting primary data and secondary data on resource consumption on three commercial buildings.

2 Overview of Practices, Indicators, and Standards Used in Waste and Resource Management

The generation of solid waste is a natural consequence of human life; however, reducing or removing waste is a sturdy way of improving the quality of life. On average, commercial buildings worldwide use approximately 70–300 kWh/m^2 per annum, roughly 10–20 times more than residential buildings (Saidur, 2009). Due to the rapid growth of energy use worldwide, there is a concern about the problem of supplying enough energy to sustain the population growth. Globalization and improvement of the standard of living have led to a pattern of energy consumption that will, without a doubt, exhaust all fossil fuel resources soon (Saidur, 2009). Kyrö et al. (2010) found that facilities management systems with appropriate indicators in place directly influenced LEED points earned by the building they studied in Finland. Therefore, it is essential to understand the practices used by facilities managers in managing this important resource in buildings.

2.1 Practices Used in Waste and Resource Management

There is a lack of explicit definitions as to facilities' management practices. Research conducted by Alexander (2013) indicates that practices implemented by facilities management are defined as the commonly performed activities conducted towards achieving the integrated goals of businesses and facilities. What can be inferred from this definition is that facilities managers are well-trained individuals who regularly perform routine actions to further the interests of stakeholders that use a facility. These routine actions performed can be considered as facility management practices.

2.1.1 Functional Areas

Functional areas are those primary focal areas that facilities managers must attend to at various intervals to ensure the effectiveness of the facilities management function within a facility (Rondeau et al., 2012). Table 1 highlights some of the most important functional areas of facilities managers. These were drawn from literature to create a framework within which facilities managers' practices in managing waste and resources can be expanded upon. Strategic and tactical planning describes the long or short-term nature of a functional area where strategic activities are long-term and tactical actions are short-term (Williams, 2000). Both long and short-term actions can be categorized as proactive or reactive (Shen et al., 2012).

Table 1 shows that the functional areas of facility management are divided into three broad areas of Plan, Operate, and Evaluate. This follows earlier studies by Clements-Croome and Croome (2004), which divides the functional areas of intelligent buildings (that provide stimulating environments for people to work and live in and operate with systems that provide communications and conveniences for various functions to take place) into the design, operation, and management. These phases are like the planning, operating, and evaluation phases, which form part of facilities management, hence their implementation.

Table 1 Functional areas of facility management

Functional areas	Framework	Functional category	Source
Long-term facility planning	Plan	Strategic planning	Rondeau et al. (2012)
Continuous service planning		Tactical planning	Rondeau et al. (2012), Alexander (1994)
Establishing and enforcing policies		Strategic/Tactical planning	Alexander (1994)
Acquiring service agreements	Operate	Support	Alexander (1994)
Installation of new equipment and technology		Support	Rondeau et al. (2012)
Facility plant and equipment maintenance		Maintenance	Rondeau et al. (2012)
Assessing tenant requirements	Evaluate	Evaluation	Alexander (1994)
Continuous service appraisal		Evaluation	Alexander (1994)

Long-Term Facility Planning

benchmarking is a commonly used long-term facility planning technique that allows facilities managers to plan tactical and strategic goals and targets. Benchmarking is a systematic process of evaluation that relies on establishing performance standards to improve services and product offerings over and above best practices (Wauters, 2005; Williams, 2000). Once the standards have been established, facilities managers must also monitor the progress or regression made by comparing performance indicators against benchmarks (Cox et al., 2003).

Continuous Service Planning

Continuous service planning stems directly from continuous service evaluation and is done periodically to adjust for variations in actual building performance (Bucking et al., 2014). Continuous service planning most commonly involves establishing tactical targets aligned with strategic objectives, which aid facilities managers to control cost and quality of service delivery (Amaratunga et al., 2000).

Establishing and Enforcing Policies

Policymakers in facilities management combine organizations' top-tier management and facility managers (Price et al., 2011). The policy creation process is initiated by evaluating building performance at middle and top decision-making levels to establish whether there is a need for corrective action requiring policies (Nutt, 2004). According to the policies, these policies have to be communicated and enforced effectively by the facilities manager (Elmualim et al., 2010).

Acquiring Service Agreements

Facilities management activities require service agreements (Ancarani & Capaldo, 2005). Common practices relating to the negotiation of service agreements involve establishing business and facility support requirements and the corresponding procurement strategies to complement these requirements (Kaya et al., 2005). This process involves conducting detailed facility and business analyzes and procedures such as the establishment and management of support service supply chains (Goyal & Pitt, 2007; Kakabadse & Kakabadse, 2005); issuing of tenders for out-sourced support services (Steane & Walker, 2000); and negotiating contractual terms of service when a negotiated procurement strategy is used (Watermeyer, 2012). Usher (2003) notes that there is no standard FM contractual model as each support service requirement is different from one facility to the other because of its uniqueness.

Installation of New Equipment and Technology

Common practices relating to installing a new facility support functions link closely with acquiring service agreements because industry specialists handle major plant and equipment installations and maintenance. According to Alexander (2013), researching, testing, and piloting new technologies are often-overlooked practices, which must be carried out by facilities managers to continually improve on the efficiency, effectiveness, and sustainability of the services provided (Finch, 2012).

Facility Plant and Equipment Maintenance

The maintenance of existing facility supports functions aim to ensure the efficiency, effectiveness, and sustainability of the services provided by facilities management (Finch, 2012). Continuous checking and monitoring, record keeping of service intervals, and appropriate maintenance scheduling are the primary practices that facilities managers must carry out in respect of maintenance (Lai et al., 2004).

Assessing Tenant Requirements To understand the needs of a facility's tenants, facilities managers conduct tenant surveys to uncover and assess building performance issues highlighted by these surveys (Bordass et al., 2001). This feedback is categorized as external feedback to the facilities management (Abdeen & Sandanayake, 2018).

Continuous Service Appraisal

Facilities managers conduct detailed analyses of service provisions to uncover and assess building performance issues to understand the needs of a facility and its tenants (Bordass et al., 2001). Subsequently, facilities managers must establish whether there is a need for the service issues to be rectified and act proactively or reactively depending on the operation window available (Shen et al., 2012). Because of the detailed series of analyses, the feedback is categorized as internal feedback from within the facilities management function (Abdeen & Sandanayake, 2018).

2.2 *Indicators Used in Waste and Resource Management*

Key Performance Indicators (KPIs) are described as systematic and consistent measures used by managers to understand where current levels of facility performance are about management's targets and objectives (WAREG, 2017); that focus on critical aspects and outcomes (Chan & Chan, 2004). Fitz-Gibbon (1994) describes indicators as a measure of performance.

2.2.1 Benchmark and Performance Indicators

Benchmark Indicators (BIs) and Performance Indicators (PIs) are concepts derived directly from the concepts of Energy Benchmark and Performance Indicators (EnBIs and EnPIs, respectively) as outlined by the International Organization for Standardization (ISO), which was later adopted by the South African Bureau of Standards (SABS). The desired outcome of using EnBIs and EnPIs is to identify, by comparing benchmarked and actual building performance, the areas requiring urgent improvement in a facility. Facilities managers use historical data to establish benchmarks aimed at problematic areas in a facility, following which goal-based targets are established. This process relies heavily on indicators as a means of planning, measuring, and evaluating building performance.

2.2.2 Categorizing Indicators

Alwaer and Clements-Croome (2010) elaborate on four main categories of indicators—mandatory, desired, inspired, and non-active/non-applicable, summarized in Table 2.

2.2.3 Consolidation of Indicators

Various KPIs are used universally by facilities managers, all of which are useful in planning, measuring, and evaluating building performance. These indicators are shown in Table 3 to consolidate several indicators and their categories as given by various sources, which are also provided. The indicators have been distributed according to facilities management planning, monitoring, and evaluation phases.

Table 3 shows that determining the total energy (kWh/kVA) and water (kL) consumed and the tons of waste generated per annum are mandatory indicators at the planning phase; the monthly measurement of the indicators are mandatory at the operation phase. In contrast, measurements of the indicators of energy and water consumed and waste generated annually, monthly or per m^2 are mandatory at the evaluation phase.

Table 2 Categories of indicators and where it is used

Indicator category	Uses
1. Mandatory	It is used to eliminate shortfalls and is compliant with standards and regulations
2. Desired	Used to set building performance targets beyond the standards to account for managers' visions
3. Inspired	Used to inspire the goals and visions of managers
4. Non-Active/Non-Applicable	The scope of tasks to be managed does not require these, or they are unobtainable/measurable

Source: Alwaer and Clements-Croome (2010)

Table 3 Categorized table of indicators and where it is used

Indicators			
Plan	**Energy**	**Category**	**Sources**
	• Total kWh per annum	Mandatory	SABS (2011)
	• Total kVA per annum	Mandatory	SABS (2011)
	• Total energy costs per annum	Desired	Kontokosta (2016)
	• Total energy hours per annum	Desired	Field et al. (1997)
	Water	**Category**	**Sources**
	• kL per annum	Mandatory	SABS (2012)
	• Total water costs per annum	Desired	WAREG (2017)
	• L or kL per sanitary fitting per annum	Mandatory/ Desired	SABS (2012)
	• L or kL per tenant per annum	Desired/ Inspired	WAREG (2017)
	Waste	**Category**	**Sources**
	• Tons of waste generated per annum	Mandatory	DEA (2011)
	• Total waste costs per annum	Desired	DEA (2011)
Operate	**Energy**	**Category**	**Sources**
	• Total kWh per month	Mandatory	SABS (2011)
	• Total kVA per month	Mandatory	SABS (2011)
	• Total energy costs per month	Desired	Kontokosta (2016)
	• Total energy hours per month	Desired	Field et al. (1997)
	Water	**Category**	**Sources**
	• L or kL per month	Mandatory	SABS (2012)
	• Total water costs per month	Desired	WAREG (2017)
	• L or kL per sanitary fitting per month	Mandatory/ Desired	SABS (2012)
	• L or kL per tenant per month	Desired/ Inspired	WAREG (2017)
	Waste	**Category**	**Sources**
	• Tons of waste generated per month	Mandatory/ Desired	DEA (2011)
	• Total waste costs per month	Desired	DEA (2011)
Evaluate	**Energy**	**Category**	**Sources**
	• Total kWh per month/annum per m^2	Mandatory/ Desired	SABS (2015), Bennet and O'Brien (2017)
	• Total kVA per month/annum per m^2	Mandatory/ Desired	SABS (2015), Bennet and O'Brien (2017)
	• Total kWh cost per month/annum per m^2	Desired	Kontokosta (2016)
	• Total kVA cost per month/annum per m^2	Desired	Kontokosta (2016)
	• % Annual costs recovered from tenants	Desired/ Inspired	Kontokosta (2016)
	Water	**Category**	**Sources**
		Mandatory	SABS (2012)

(continued)

Table 3 (continued)

Indicators			
	• Total L or kL per month/annum per m^2		
	• % annual costs recovered from tenants	Desired/ Inspired	WAREG (2017)
	• L or kL per sanitary fitting per month/ annum per m^2	Mandatory/ Desired	SABS (2012)
	Waste	**Category**	**Sources**
	• Tons of waste generated per month/ annum per m^2	Mandatory	DEA (2011)
	• % Annual costs recovered from tenants	Desired	DEA (2011)

Key: *L* litres of water; *SABS* South African Building Standards; *DEA* Department of Environmental Affairs, South Africa; *WAREG* European Water Regulators

2.3 Key Standards and Regulations Used in Waste and Resource Management

Standards are actions or omissions that must be adhered to and are often the means of implementing or enforcing policies at a user level. The key standards and regulations applicable to facilities managers in South Africa include (1) National Environmental Management: Waste Act No. 89 of 2008, (2) National Waste Management Strategy of 2011, (3) SANS 204, Edition 1, 2011—Energy Efficiency in Buildings, (4) SANS 10252, Edition 3, 2012—Water Supply and Drainage for Buildings, (5) SANS and ISO 50001, Edition 1, 2011—Energy management systems—Requirements with guidance for use, (6) SANS and ISO 50004, Edition 1, 2015—Energy management systems—Guidance for implementation, (7) SANS and ISO 50006, Edition 1, 2015—Energy management systems—Measuring energy performance, (8) ISO 46001—Water Efficiency Management Systems and (9) ISO 50047—Determination of Energy Savings.

2.3.1 The National Environmental Waste Management Act and National Waste Management Strategy

The main objective of the National Environmental Management: Waste Act No. 89 of 2008 is to promote the health and well-being of the environment by ensuring that the consumption of natural resources and the generation of waste is minimized or avoided where eradicating waste is not possible, that is reduced, reused, and recycled. When the Act was first released, it proposed the establishment of a National Waste Management Strategy (NWMS) as a means of operationalizing solutions to drive change at operational levels across South Africa.

2.3.2 South African National Standards (SANS)

The SANS body of standards is provided by the South African Bureau of Standards (SABS). It is an extension of government policies to effect changes at the operational level. This section presents all the standards relevant to the topic of the investigation to create a platform from which comparisons can be drawn between the waste and resource management practices and indicators used in South Africa and those used internationally.

SANS 204, Edition 1, 2011—Energy Efficiency in Buildings The SANS 204 defines energy efficiency in building as the minimization of energy consumption while still achieving the desired outcomes (SABS, 2011). Furthermore, SANS 204 specifies several maximum energy uses and demand figures, which must be adhered to. It distinguishes between different buildings and further provides standard figures to account for geographical differences and the energy demand because of external environmental differences. It is important to differentiate between energy demand and energy consumption at this stage. Energy consumption is the total amount of energy used to perform an activity over a given period, typically in hours, while Energy demand is the maximum amount of power recorded during a certain period measured in kW or kVA (Windapo, 2017). A consistent measurement duration must be used when calculating energy demand for billing purposes in commercial buildings.

On average, for commercial buildings such as offices situated in Cape Town (climate zone four as indicated in Annexure A of SANS 204), the total energy demand over 12 months should not exceed 75 VA/m^2 (SABS, 2011). The maximum permissible level of energy consumption for office buildings situated in Cape Town is 185 kWh/m^2 (SABS, 2011). It is also important to note that the unit of measure for energy *consumption* in a facility or building is the same unit used by municipal councils to bill buildings and property managers periodically for their total energy use. As such, it is a best practice that this is the same indicator (kWh) that facilities managers should use to plan and control energy consumption levels to reduce the costs of municipal electricity supply.

SANS 10252, Edition 3, 2012—Water Supply and Drainage for Buildings The SANS 10252 set of standards relates directly to water supply and drainage requirements for all building types, including commercial buildings such as offices and hotels. The standards also suggest water consumption levels per sanitary unit. The SANS 10252 document prescribes that commercial premises such as offices with canteens or kitchen areas are likely to consume significantly more water per square meter per day than offices that do not have a canteen (SABS, 2012). A range of 10–15 l of water per m^2 per day is suggested for offices with canteens, while 7–10 l/m^2/day is the suggested range for offices without canteens.

SANS 50001, Edition 1, 2011—Energy Management Systems—Requirements with Guidance for Use SANS 50001 is a standard adopted from the ISO 50001 document that proposes several best practice guidelines. It provides users with a tool to establish, implement, maintain, and improve an energy management system to increase overall energy performance within an organization, division, or facility. The document highlights that improved energy performance can be obtained by continually improving the five energy-related aspects, including Energy use and Energy consumption.

SANS 50006, Edition 1, 2015—Energy Management Systems—Measuring Energy Performance The establishment of energy baselines and energy performance indicators is one of the practices underpinning energy management systems based on the SANS standards (SABS, 2015). SANS 50006 further highlights indicators that facilities managers use to establish and compare energy use. The indicators that are useful in the planning, operational, and review phase identified in SANS50006 include Energy consumption (measuring reduction in the consumption of Energy—kWh) and Energy demand (monitoring and control of energy stocks and costs—kW).

2.3.3 International Organization for Standardization (ISO) Standards

ISO standards are related to water efficiency management systems, energy management systems which the South African Bureau of Standards has adopted because SANS is widely established.

2.4 *Review of Level of Waste Production and Resource Consumption in Commercial Buildings*

2.4.1 Energy Consumption

Table 4 shows the Energy consumption in commercial buildings in comparison by country.

Table 4 reveals that high-level income countries such as Singapore have low energy consumption per annum. In contrast, upper-middle-income countries such as Malaysia, China, Botswana, and South Africa all have medium to high energy consumption levels per annum. The energy consumption in a commercial building in Singapore was evaluated in a case study by Kua and Wong (2012). The study proved that improved technology and design modifications to old buildings had

Table 4 Energy comparison in commercial buildings by country

Country	Income level	Consumption (kWh/m^2)	Consumption level
Singapore	High	55.2	Low
Malaysia	Upper Middle	130	Medium
China	Upper Middle	153	Medium
South Africa and Botswana	Upper Middle	172–255	High

significant advantages and resulted in 55.2 kWh/m^2 per annum (Kua & Wong, 2012). Also, Saidur (2009) found that the energy intensity for commercial buildings in Malaysia was on average 130 kWh/m^2 per annum, while energy consumption in some buildings was as low as 114 kWh/m^2 per annum.

Jiang and Tovey (2009) found from their study of nine large commercial buildings in Shanghai and Beijing that the average electricity consumption averaged around 153 kWh/m^2, which they found to be approximately five times higher than in residential buildings. While Masoso and Grobler (2010) found from their study of five buildings in Botswana and South Africa situated in dry, hot climates and rely on air conditioners averaged at 172 kWh/m^2 per annum, which was a massive saving from the normal South African standard of 255 kWh/m^2 per annum just by turning off air conditioners and lighting during the night.

2.4.2 Water Consumption

Even though South Africa is ranked as the 30th driest country globally, water consumption is approximately 233 l per capita per day. In contrast, the international usage is around 180 l per person per day (GreenCape, 2018). In a study done by Tabassum et al. (2016), in which the water consumption within the commercial sector in Karachi City, Pakistan (the third driest country) was calculated, it was found that water consumption in Karachi city is at a low of approximately 106 l per person per day which is half the required amount,

3 Research Methodology

Several research assumptions provide the framework in which the research has been placed. The philosophy that underpins the quantitative portions of the research is based on the premise that a single perspective of reality exists and is independent, ordered, and objective (Runeson & Skitmore, 2008). This contrasts with the qualitative portions of the research, which are constructivist. This is because both the researcher and the subject of the research already have some knowledge of what is being researched (Fosnot, 2013). Epistemologically, positivism is preferred due to the need for the quantitative parts of the research to be kept independent from the researchers and their perspectives. Positivism proposes the understanding that the social world is based on facts and objective views (Noor, 2008). This is contrasted by constructivism, which is based on the need for greater understanding and expansion of knowledge based on social interactions between the researcher and the phenomenon that is the subject matter of the research (Hays & Singh, 2011).

The research is situated in the property field and the built environment, with sustainability as the overarching concept guiding the research. The study employs an exploratory case study design employing interviews and secondary data in answering the research question. The study population consisted of Green star-rated commercial buildings located in Cape Town, South Africa. The study used a convenience sampling technique in selecting the sampled building facilities. Using pre-determined case studies is known either as convenience sampling or purposive sampling (Etikan et al., 2016). The three criteria considered in selecting appropriate buildings were (1) that only commercial buildings that were either (2) designed before construction or refurbished post-construction to be rated from 4 to 6-stars by the GBCSA, and (3) were accredited at the time of the research. The three buildings selected for the study based on the criteria are as follows:

- Building 1 is a commercial building situated in the CBD of Cape Town. The space occupied for commercial purposes totals 12,391 m^2 and is currently a GBCSA Existing Building Performance (EBP) four-star rated building. The building was commissioned in 1999. In 2016 a PV solar array was installed in the building to help reduce reliance on council energy supply. In addition, water-cooling systems were installed in 2018 to alleviate high energy demand and consumption levels.
- Building 2 is a commercial building situated in the Century City business precinct near Cape Town. The building was commissioned in 2015 with only one tenant, occupying all the lettable areas for their Western Cape Head Office purposes. The space occupied for commercial purposes is approximately 14,186 m^2 and is currently a GBSCA EBP and As-Built (AB) dual five-star rated building. The building employs an extensive greywater and stormwater system to service its water needs.
- Building 3 is a high-end commercial building situated in the Central Business District (CBD) of Cape Town. The building was commissioned in early 2016, with two tenants occupying the largest portions of the lettable space. The space occupied for commercial purposes is 12,316 m^2 and is currently a GBCSA Existing Building Performance (EBP) four-star rated building.

The primary data collection method employs semi-structured interviews with the facilities managers of the selected buildings. Transcription software was then used to obtain valid primary data from the transcripts on the practices, indicators of waste and resource management used by the facilities managers. Secondary data on the level of water and energy consumption and the waste generated in the buildings was collected from the respondents via email. Each data set is building specific but relates to a singular case study. All interviews followed a strict protocol requiring interviewees to sign consent forms. Furthermore, any potential benefits relating to the study were also highlighted. All interviewees and organizations were kept anonymous. Regarding writing the manuscript, all guidelines for avoiding plagiarism were implemented.

The main limitation of the research conducted was the lack of primary and secondary data on the waste generated in the building. This is mainly due to how waste is collected and handled, using manual practices, which are inherently flawed regarding effective data collection and recording.

4 Results and Discussion

A cross-case analysis was performed to identify which practices and indicators were similar and different amongst the three buildings studied.

4.1 Practices

The qualitative findings relating to practices were acquired by conducting a thematic analysis of transcriptions of the interviews held with facility managers in the case study buildings. The main inferences drawn from the interviews are that the predominant practices lie in managing water while the least practices implemented are in respect of waste. This is due to the drought experienced by Cape Town and the greater Western Cape region in 2017–2018. It is seen that emphasis was placed on the water during these periods. Table 5 presents the findings of the cross-case analysis conducted to identify the similarities in water, energy, and waste management practices across all the three buildings sampled.

4.1.1 Similarities: Practices

The similarities between the buildings' practices are highlighted in Table 5. It is inferred from Table 5 that most of the practices identified stem from the need to plan and operate following standards. Furthermore, planning is a significant focus area in ensuring that a best practice framework is established for practices such as implementing new plant and equipment and other lean facilities management installations. In addition, it is noted that in all buildings, a clear hierarchy exists between the various levels of management structures concerning establishing, communicating and enforcing policies. This indicates a high level of control within the case study. In addition, negotiations with tenants at early stages and working alongside tenants to continually improve service delivery are also the primary focal areas of the facilities managers. In addition, new technology and the maintenance of existing implementations seem to focus mainly on areas of significant concern, such as those areas which show a significant consumption of resources or generation of waste. Regarding waste, it is further noted that not a significant amount of focus is given to waste, but that some priority is given to the separation of recyclable materials from traditional wet and solid waste.

4.1.2 Differences: Practices

The differences between the buildings' practices stem from the need to plan and operate within the facility's specific requirements. Varying degrees of control are

Table 5 Water, energy, and waste management practices in the buildings studied

Functional area	Energy	Water	Waste
Long-term facility planning	Feasibility of future upgrades and installation	Feasibility of future upgrades and installation	Feasibility of future upgrades and installation
			The new sewage treatment facility
Continuous service planning	Desired energy consumption per week, month, annum, or longer	Desired water consumption per week, month, annum, or longer	Desired waste generation per week, month, annum, or longer
	Desired energy demand per week, month, annum, or longer	Scope of service offerings	Scope of service offerings
	Scope of service offerings		
Establishing, communicating, and enforcing policies	**Head office** policies established and communicated to:	**Head office** policies established and communicated to:	**Head office** policies established and communicated to:
	– **Regional** policies established and communicated to:	– **Regional** policies established and communicated to:	– **Regional** policies established and communicated to:
	– **Portfolio** policies communicated to:	– **Portfolio** policies communicated to:	– **Portfolio** policies communicated to:
	– **Facility**-specific policies enforced by:	– **Facility**-specific policies enforced by:	– **Facility**-specific policies enforced by:
	– **Facilities managers**	– **Facilities managers**	– **Facilities managers**
Negotiating service agreements	Acquisition of an out-sourced company to gather data	Acquisition of an out-sourced company to gather data	Acquisition of out-sourced recycling, disposal, and collection company
	Negotiations with tenants	Negotiations with tenants	Negotiations with tenants
New service and technology implementations	Install light dimmers and monitors	Spray taps installed	
Facility plant and equipment maintenance	Generator maintenance scheduling		
	HVAC maintenance scheduling		
	Elevator maintenance scheduling		
Assessing tenant requirements	Surveys	Surveys	Surveys
	Contractual agreements/negotiations	Contractual agreements/negotiations	Contractual agreements/negotiations
Continuous service appraisal			

(continued)

Table 5 (continued)

Functional area	Energy	Water	Waste
	Head office and facility evaluations and analyses (internal)	Head office and facility evaluations and analyses (internal)	Head office and facility evaluations and analyses (internal)
	Tenant survey feedback (external)	Tenant survey feedback (external)	Tenant survey feedback (external)
	Open operations issues portal	Open operations issues portal	Open operations issues portal

required; thus, the monitoring and checking periods are either shorter or longer, where shorter periods indicate a higher need for control. A higher need for control is linked to the need to reduce the consumption of resources. This is seen in Building 3, where electricity usage is high relative to the other buildings. Furthermore, it is noted that there are significantly more similarities than differences in respect of practices. The GBCSA ratings of the three buildings give an insight into the differences in building practices. It was found that the buildings with four Green star ratings utilize more traditional methods (such as innovative software) and have not improved the practices to match those used in the five-star rated building.

4.2 *Indicators of Waste and Resources*

Table 6 indicates the similarities between the indicators used in the three sampled buildings to manage energy, water, and waste. These similarities were the findings of the cross-case analysis conducted.

4.2.1 Similarities: Indicators

It can be inferred from Table 6 that there is an even distribution of indicators across the three main phases—Plan, Measure, and Evaluate. In addition, it is noted that during the planning phases, facilities' managers prefer to plan on a longer-term basis than the basis used to measure facility performance. This is done to set targets at the beginning of a period to control incremental shift towards the bigger goal. In addition, there is a significant focus on costs within all the buildings; this is believed to be a by-product of the depressed economy and the goal of a driving profit as a main focal area. Furthermore, it can be inferred that facilities managers use standardization during the evaluation phases by calculating and comparing resource consumption and waste generation on a per meter square basis. These standardization methods help facilities managers compare like figures with one another and draw accurate comparisons between facilities. This standardization technique allows easy comparison with the indicators noted in the literature and ensures compliance with regulations and standards.

Table 6 Similarities between the water and resource management indicators used in the buildings studied

Plan	Measure	Evaluate
Energy		
kWh per annum	kWh per tenant per month	Total kVA per m^2 per month
kVA per annum	Facility kWh per month	Total energy costs per m^2 per month
Total energy costs per annum	Total energy costs per month	
Water		
kL per annum Water costs per annum	Water costs per month kL per m^2 per month or annum	% Annual costs recovered from tenants
Waste		
Total waste produced per annum	Total waste disposal costs per month	Tons of waste per m^2 per month or annum

4.2.2 Differences: Indicators

The primary reason for differences in the indicators used on the various facilities is the influence of facility-specific goals and targets set by facilities managers in satisfying tenant and business requirements. This also provides context to the varying lengths of time specified for the indicators as these buildings all require varying levels of control by management. It is noted that the more indicators used and the shorter the duration of the indicator, the higher the control requirements are.

The predominant difference in the indicators lies within the Measuring phase, where different indicators were identified for Energy (Facility kWh per week, Tenant kWh per week, HVAC system temperature level, kVA per tenant per month, Facility kVA per month); Water (Facility meter: L or kL per day or week, Council meter: kL per month, L or kL per month/per week); and Waste (Number of Council wheelie bins, total waste generated in 2 weeks, Kg or Tons per week of recycled waste, Tons of waste per month). In addition, the area that facilities managers are noted to focus on is energy, as indicated by the highest number of indicator variations across the three phases.

4.3 Level of Waste Produced and Resources Consumed in the Commercial Buildings

Secondary data is obtained from the commercial buildings for Energy and Water. The annual energy consumption and demand levels for the three buildings studied are shown in Table 7. No levels of waste generation were established. An important additional finding includes the inference that the remote metering systems used provides more accurate, consistent, and thus reliable data when compared to the

Table 7 Levels of energy consumption

Annual energy consumption and demand—internal					
Building	Year	Consumption	Unit	Demand	Unit
1	2016–2017	2,694,600	kWh	6018	kVA
1	2017–2018	2,139,240	kWh	5278	kVA
1	2018–2019	2,002,776	kWh	4975	kVA
2	2017–2018	2,240,823	kWh	5572	kVA
2	2018–2019	2,924,621	kWh	7816	kVA
2	Begin 2019–mid-2019	959,160	kWh	2599	kVA
3	2017–2018	3,032,283	kWh	6719	kVA
3	2018–2019	3,203,376	kWh	6371	kVA

Table 8 Levels of water consumption

Annual water consumption—internal			
Building	Year	Consumption	Unit
1	2016–2017	9120	kL
1	2017–2018	5819	kL
1	2018–2019	4829	kL
2	2017–2018	16,251	kL
2	2018–2019	17,011	kL
2	Begin 2019–mid-2019	8646	kL
3	2017–2018	4943	kL
3	2018–2019	5318	kL

Table 9 Performance of the building in terms of resource consumption

Building	Energy	Water	Ranking
1 (12,391 m^2)	183.91 kWh/m^2—Effective	0.53 kL/m^2—Effective	1
2 (14,186 m^2)	172.94 kWh/m^2—Effective	1.18 kL/m^2—Effective	2
3 (12,136 m^2)	253.15 kWh/m^2—Ineffective	0.42 kL/m^2—Effective	3

council metering, and it is for this reason that the figures presented in Tables 7 and 8 reflect the values obtained from the remote metering and not the council metering.

The annual water consumption levels for all three buildings are presented in Table 8.

Each building was assessed against the standards to assess the cumulative effect of the indicators and practices. If the building showed consumption or demand higher than the standard for more than one period, it was inferred that the practices and indicators were ineffective, while consumption or demand lower than the standards were taken to indicate appropriate facilities management practices and indicators. Table 9 shows the performance of the buildings in terms of resource consumption, and the inferences are drawn.

Building 3 was the only building that did not perform within the standards found in the literature for energy, with an average consumption score of 253.15 kWh/m^2. Based on these findings, the study supports the proposition that the commercial buildings sampled are well managed, the facilities management practices and indicators used are effective in resource management, and by extension, waste generated in the buildings.

4.4 *Discussion of Findings*

A summary of the findings is provided in line with the primary research objectives to frame the findings and provide the basis for formulating conclusions and recommendations.

4.4.1 Facilities Management Practices Used in the Management of Waste and Resources

The research findings indicate that Buildings 1 and 2 have implemented resource and waste management practices but are not as advanced as the practices in the five-star rated Building 3. In addition, current practices focus predominantly on energy and water consumption and demand and less on waste generation. Overall, many practices have been implemented routinely and are continuously being improved and managed to elevate the effectiveness of these practices within the primary functional areas to affect service delivery and sustainability outcomes positively. The practices used in the planning phase of facility management include long-term facility planning, continuous service planning, and establishing, communicating, and enforcing policies. The practices include entering into service agreements, acquisition of new equipment and technology, and facility plant and equipment maintenance at the operating phase. While at the evaluation phase, the practices include accessing tenant requirements and continuous service appraisal. Furthermore, it was revealed that external factors such as rolling blackouts and droughts had affected the decisions of the facilities managers in managing waste and resources. This was highlighted by the significant focus on energy and water management while the focus on waste management lagged.

4.4.2 The Indicators Used by FM Specialists in Waste and Resource Management

Buildings 1 and 2 have implemented resource and waste management indicators, but these indicators are not as advanced as the indicators used in Building 3, the five-star rated building. At the planning stage, the indicators focus primarily on the water consumption being in litres consumed per annum, kWh per annum for energy consumed, and waste in tons per unit per annum. At the operating phase, the indicators focus primarily on the water consumed per month, kWh per month for energy consumed, and waste in the form of tons per unit per month. This is then directly proportional to the cost of each focal area per month. The main difference in the indicators used is the duration of measure, where the operating phase is significantly shorter than the planning phase. This is due to a need to control monthly resource consumption and intensify waste generation costs.

Further key findings showed that the indicators used by facilities managers during the evaluation phase most closely matched the indicators prescribed by the standards. The indicators focus primarily on the water consumption being (1) litres consumed per floor area size per month, (2) kWh per floor area size per month for energy consumed, as well as waste in the form of (3) tons per floor area size per unit per month. In addition, it was noted that the indicators relating to water and energy are more advanced than those used for managing waste. This indicates that the complexity of managing energy and water is higher than that of managing waste.

4.4.3 Level of Waste Produced and Resources Consumed in the Sampled Buildings

The findings, which stem from the secondary data collected from the buildings and presented in Tables 7 and 8, show that the energy consumption levels for Buildings 1 and 2 are comparable to standards found in literature, while that of Building 3 is well above these standards. The findings also show that the water consumption levels for the three buildings are low and are well within the regulation standards. However, no levels of waste generation could be established in the buildings.

Establish Whether the FM Practices and Indicators, Regarding Waste and Resource Management, Align with Standards

The study also sought to know whether the FM practices and indicators used regarding the buildings sampled align with standards. The key suggests that the practices and indicators used by facilities managers align with standards and practice prescribed by the relevant standards in South Africa (National Environmental Management: Waste Act No. 89 of 2008; National Waste Management Strategy of 2011; SANS 204, Edition 1, 2011—Energy Efficiency in Buildings; SANS 10252, Edition 3, 2012—Water Supply and Drainage for Buildings and SANS and ISO 50001). In turn, comparisons between literature (Field et al., 1997; Kontokosta, 2016) and findings revealed many similarities between what was implemented in practice are prescribed in theory by the standards (DEA, 2011; SABS, 2011, 2012; WAREG, 2017).

Find Out Whether the Levels of Waste Generation and Resource Consumption Are Attributable to the Standard of the Practices and Indicators Used by Facilities Managers in South Africa

Findings regarding attributing facilities managers' effectiveness to the standard of their practices could not be ascertained for waste management. However, the primary and secondary research findings (see Table 9) show that facilities managers' indicators and practices show many similarities with both literature (Kua & Wong, 2012; Jiang & Tovey, 2009; Masoso & Grobler, 2010) as well as local and international standards (SABS, 2012). The building practice and indicators were adjudged to effectively manage energy and water consumption in the commercial buildings observed. In addition, FM practices (entering into service agreements, acquisition of new equipment and technology, facility plant and equipment maintenance, accessing tenant requirements and continuous service appraisal) were deemed effective in managing energy and water consumption and demand.

5 Conclusion and Recommendations

The study sought to examine the indicators and practices used by facilities managers regarding waste and resource management in commercial buildings and whether these current practices and indicators are in line with the standards impacting facility, sustainability, and performance. It was found that resource and waste management practices have been implemented across three focus areas: planning, operating, and evaluating within the buildings studied. It also emerged that the facility managers used comprehensive indicators in managing resources and waste generated in the buildings and that resource consumption levels are within the prescribed standards except for Building 3. The level of waste generated in the commercial buildings could not be determined when undertaking this research. The findings also indicate that resource consumption levels in the buildings are within the regulation standards and those outlined in literature except for the energy consumption in Building 3, which was a bit high. It was inferred from the findings that the levels of resource consumption are attributable to the practices and indicators used by Facilities Managers in managing resources—water and energy in the sampled buildings. The alignment of waste-related indicators and practices with standards could not be determined due to a lack of data.

Based on these findings, the study concludes that the comparable levels of resource consumption—water and energy, and waste generation are attributable to the practices and indicators used by Facilities Managers in managing commercial buildings. The study posits that improved metering systems and analysis software implementations are needed along with a more concerted effort to establish new ways of handling, sorting, and disposing of waste. While in respect of energy, techniques and well-trained staff are essential for analyzing metering software and systems raw data when electronic building management software such as Archibus asset management software is installed. Using integrated management systems, the measurement and analysis of waste have not been implemented to their full potential in the buildings studied. The limitation of this study is the small sample size, and the data is non-transferable, and not representative of commercial buildings because of the set criteria used in identifying the sampled buildings. Therefore, more research needs to be done to establish improved ways of measuring and analyzing the waste produced in commercial buildings using a larger sample size.

Acknowledgments The authors are grateful for the contributions made by Dr. Alireza Moghayedi to the research. This work is supported by the N.R.F. (Grant Number-120843). Opinions and conclusions are those of the authors and are not necessarily attributable to the N.R.F.

References

Abdeen, F., & Sandanayake, Y. (2018). Facilities management supply chain: Functions, flows and relationships. *Procedia Manufacturing, 17*, 1104–1111.

Alexander, K. (1994). A strategy for facilities management. *Facilities, 12*(11), 6–10.

Alexander, K. (2013). *Facilities management: Theory and practice*. Routledge.

Alwaer, H., & Clements-Croome, D. (2010). Key performance indicators (KPIs) and priority setting in using the multi-attribute approach for assessing sustainable, intelligent buildings. *Building and Environment, 45*(4), 799–807.

Amaratunga, D., Baldry, D., & Sarshar, M. (2000). Assessment of facilities management performance—What next? *Facilities, 18*(1/2), 66–75.

Ancarani, A., & Capaldo, G. (2005). Supporting decision-making process in facilities management services procurement: A methodological approach. *Journal of Purchasing and Supply Management, 11*(5–6), 232–241.

Arendse, L., & Godfrey, L. (2010). Waste management indicators for a national state of environment reporting. *United Nations Environment Programme (UNEP)*.

Bennet, I. E., & O'Brien, W. (2017). Office building plug and light loads: Comparison of a multi-tenant office tower to conventional assumptions. *Energy and Buildings, 153*, 461–475.

Bordass, B., Leaman, A., & Ruyssevelt, P. (2001). Assessing building performance in use 5: Conclusions and implications. *Building Research & Information, 29*(2), 144–157.

Brundtland, G. H. (1987). Our common future—Call for action. *Environmental Conservation, 14*(4), 291–294.

Bucking, S., Zmeureanu, R., & Athienitis, A. (2014). A methodology for identifying the influence of design variations on building energy performance. *Journal of Building Performance Simulation, 7*(6), 411–426.

Chan, A. P., & Chan, A. P. (2004). Key performance indicators for measuring construction success. *Benchmarking: An International Journal, 11*(2), 203–221.

Clements-Croome, D., & Croome, D. J. (2004). *Intelligent buildings: Design, management and operation*. Thomas Telford.

Cox, R. F., Issa, R. R., & Ahrens, D. (2003). Management's perception of key performance indicators for construction. *Journal of Construction Engineering and Management, 129*(2), 142–151.

DEA. (2011). *National waste management strategy*. Department of Environmental Affairs.

Dean, B., Dulac, J., Petrichenko, K., & Graham, P. (2016). *Towards zero-emission efficient and resilient buildings*. Global Status Report.

Elmualim, A., Shockley, D., Valle, R., Ludlow, G., & Shah, S. (2010). Barriers and commitment of facilities management profession to the sustainability agenda. *Building and Environment, 45*(1), 58–64.

Etikan, I., Musa, S. A., & Alkassim, R. S. (2016). Comparison of convenience sampling and purposive sampling. *American Journal of Theoretical and Applied Statistics, 5*(1), 1–4.

Field, J., Soper, J., Jones, P., Bordass, W., & Grigg, P. (1997). Energy performance of occupied non-domestic buildings: Assessment by analysing end-use energy consumptions. *Building Services Engineering Research and Technology, 18*(1), 39–46.

Finch, E. (2012). *Facilities change management*. Wiley Online Library.
Fitz-Gibbon, C. (1994). Performance indicators, value added and quality assurance. In *Improving education: Promoting quality in schools*. Cassell.
Fosnot, C. T. (2013). *Constructivism: Theory, perspectives, and practice*. Teachers College Press.
Goyal, S., & Pitt, M. (2007). Determining the role of innovation management in facilities management. *Facilities, 25*(1/2), 48–60.
GreenCape. (2018). *Water: Market intelligence report 2018*. GreenCape.
Hays, D. G., & Singh, A. A. (2011). *Qualitative inquiry in clinical and educational settings*. Guilford Press.
Jiang, P., & Tovey, N. K. (2009). Opportunities for low carbon sustainability in large commercial buildings in China. *Energy Policy, 37*(11), 4949–4958.
Kakabadse, A., & Kakabadse, N. (2005). Outsourcing: Current and future trends. *Thunderbird International Business Review, 47*(2), 183–204.
Kaya, S., Heywood, C. A., Arge, K., Brawn, G., & Alexander, K. (2005). Raising facilities management's profile in organisations: Developing a world-class framework. *Journal of Facilities Management, 3*(1), 65–82.
Kaza, S., Yao, L., Bhada-Tata, P., & Van Woerden, F. (2018). *What a waste 2.0: A global snapshot of solid waste management to 2050*. World Bank.
Kontokosta, C. E. (2016). Modeling the energy retrofit decision in commercial office buildings. *Energy and Buildings, 131*, 1–20.
Kua, H. W., & Wong, C. L. (2012). Analysing the life cycle greenhouse gas emission and energy consumption of a multi-storied commercial building in Singapore from an extended system boundary perspective. *Energy and Buildings, 51*, 6–14.
Kyrö, R., Määttänen, E., Aaltonen, A., Lindholm, A.-L., & Junnila, S. (2010). Green buildings and FM—A case study on how FM influences the environmental performance of office buildings. In *CIB W070 conference* (Vol. 13).
Lai, J. H., Yik, F. W., & Jones, P. (2004). Practices and performance of outsourced operation and maintenance in commercial buildings.
Masoso, O., & Grobler, L. J. (2010). The dark side of occupants' behaviour on building energy use. *Energy and Buildings, 42*(2), 173–177.
Noor, K. B. M. (2008). Case study: A strategic research methodology. *American Journal of Applied Sciences, 5*(11), 1602–1604.
Nutt, B. (2004). Infrastructure resources: Forging alignments between supply and demand. *Facilities, 22*(13/14), 335–343.
Ogola, J., Chimuka, L., & Tshivhase, S. (2011). Management of municipal solid wastes: A case study in Limpopo Province, South Africa. In *Integrated waste management—Volume I*. IntechOpen.
Price, S., Pitt, M., & Tucker, M. (2011). Implications of a sustainability policy for facilities management organisations. *Facilities, 29*(9/10), 391–410.
Rondeau, E., Brown, R., & Lapides, P. (2012). *Facility management*. John Wiley & Sons.
Runeson, G., & Skitmore, M. (2008). Scientific theories. In *Advanced research methods in the built environment* (pp. 76–85). Blackwell.
SABS. (2011). *SANS 204—Energy efficiency in buildings* (1st ed.). South African Bureau of Standards.
SABS. (2012). *SANS 10252 Part 1—Water supply installation* (3rd ed.). South African Bureau of Standards.
SABS. (2015). *SANS 50006—Energy management systems—Measuring energy performance using EnB & EnPI—General principles and guidance* (1st ed.). South African Bureau of Standards.
Saidur, R. (2009). Energy consumption, energy savings, and emission analysis in Malaysian office buildings. *Energy Policy, 37*(10), 4104–4113.
Shen, W., Hao, Q., & Xue, Y. (2012). A loosely coupled system integration approach for decision support in facility management and maintenance. *Automation in Construction, 25*, 41–48.

Steane, P., & Walker, D. (2000). Competitive tendering and contracting public sector services in Australia—A facilities management issue. *Facilities, 18*(5/6), 245–255.

Tabassum, R., Arsalan, M. H., & Imam, N. (2016). Estimation of water demand for commercial units in Karachi City.

Usher, N. (2003). Outsource or in-house facilities management: The pros and cons. *Journal of Facilities Management, 2*(4), 351–359.

WAREG. (2017). *An analysis of water efficiency KPIs in WAREG member countries*.

Watermeyer, R. B. (2012). A framework for developing construction procurement strategy. *Proceedings of the Institution of Civil Engineers-Manag Procurement and Law, 165*(4), 223.

Wauters, B. (2005). The added value of facilities management: Benchmarking work processes. *Facilities, 23*(3/4), 142–151.

Williams, B. (2000). *An introduction to benchmarking facilities and justifying the investment in facilities*. Building Economic Bureau Limited.

Windapo, A. O. (2017). Managing energy demand in buildings through appropriate equipment specification and use. *Energy Efficient Buildings,* 163–180.

A Descriptive Comprehension Study on Solar Energy, Solar Products, and Solar Products Marketing in Indian Context

Ansari Sarwar Alam and Arshiya Fathima M.S

Abstract The present research work focuses in detail on the three main components i.e., solar energy, solar product and solar product marketing and tries to connect them to know their applicable part. The research gap of solar product marketing is also well focused. Geographically, India is an ideal country for solar energy. Despite India is having the potential power to tap solar energy at mass level; however, the solar product users are very less, which therefore impacts solar product marketing. The literature shows that this gap is due to the quality of solar products, technology of solar products, unready mindset of consumers and promotions.

The present study is the sum of the aspects of the solar energy sector, marketing of solar products and perspective of Indian consumers, thus this study will benefit the researchers, academicians and business practitioners, this study stands out from the past literature as it covers data from various sources (research papers, government agencies and business sector) from the year 2000 to 2020. Subsequently going through this research work, the researchers, academicians and business practitioners will get to know the status of solar energy and solar product marketing.

Keywords Solar energy · Solar product · Solar product marketing · Energy sector · India

1 Introductory Background of the Study

This era is changing in such a manner that most of the things are dependent on energy. Every country whether it is developed or developing is sustaining on the basis of energy. In fact, a developed country requires energy for sustainability,

A. S. Alam (✉)
Department of Marketing, Universal Business School, Karjat, Raigad, Maharashtra, India

A. Fathima M.S
Department of Management Studies, B.S Abdur Rahman Crescent Institute of Science and Technology, Chennai, Tamilnadu, India

F. P. García Márquez, B. Lev (eds.), *Sustainability*, International Series in Operations Research & Management Science 333,
https://doi.org/10.1007/978-3-031-16620-4_14

whereas a developing country needs energy for development. Energy can be categorized into two forms one which can be renewed "Renewable Energy" and another one which cannot be renewed "Non-Renewable Energy". Renewable energy is generated from natural processes and it can be in different forms, for example sun's light and heat, geothermal energy, wind, tides, water and biomass; these sources of energy are also called alternative sources of energy.

On the other hand, Non-Renewable energy comes from sources that will run out or will not be replenished for thousands or millions of years. The sources of non-renewable energy are fossil fuels (coal, crude oil, natural gas and nuclear energy). The burning of fossil fuels for releasing energy causes harm to the environment. When coal and oil are burned, they released particles that can pollute the air, water and land. According to Burt et al. (2013), "coal combustion pollutants, particulate matter (pm), sulphur dioxide (SO_2) and nitrogen (NO_2) affects the respiratory system, cardiovascular system, reproductive system, neurological problem and life expectancy. In India, the year of life lost is estimated to be 2.5 years".

Renewable energy has many advantages over non-renewable energy and it is considered a green and clean form of energy. All renewable sources of energy, sunlight, wind, water and biomass, generate zero harmful emissions or pollutants in the environment. Thus, it is also called a sustainable source of energy in terms of reducing greenhouse gases and global warming. India has been recorded as one of the fastest-growing among developing countries; hence the consumption of energy is also high. The energy demand is highest in the industry sector, followed by the building sector, transport sector and agriculture sector (International Energy Agency, 2015). Fuelwood and biomass burning is the primary reason for permanent haze and smoke observed in rural and urban areas of India. Fuelwood and biomass cakes are used for cooking and general heating needs in most of the parts of rural India. They are burnt in cooking stoves known as "Chullah" or "Chulha". These cooking stoves are present in over 100 million Indian households and are used two or three times a day daily. The majority of Indians still use traditional fuels such as dried cow dung, agricultural waste and firewood as cooking fuel. These forms of fuels are inefficient sources of energy, as their burning releases high levels of smoke, particulate matter, NO, SO, polyaromatics, formaldehyde, carbon monoxide and other air pollutants. Air pollution is also the main cause of the Asian brown cloud, which is delaying the start of the monsoon. The burning of biomass and firewood will stop unless electricity or clean-burning fuel and combustion technologies become reliably available and widely adopted in rural and urban India. On the other hand, India stands among those countries which can generate energy through renewable energy in mass volume. It has 300 sunny days, which make India among those countries which have a huge source for generating solar energy (Tripathi, 2016). India has potential for the solar market as India is a populous country and has a demand for energy (Bridge to India, 2014).

In India, Around 300 million villages are still not connected with the central grid. The Mini-Grid Renewable Energy Generation and Supply Regulations (MREG & S regulations) on 4 March 2016, issued a draft for mini/micro-grid policy 2016. The electrification in rural areas should focus on sustainability and economic development (Garud & Sharma, 2016). Out of the 1.3 billion people of the world who live without access to electricity, 400 million people live in India (Millinger et al., 2012).

Therefore renewable energy is the need of the hour in the Indian context; the most viable source of renewable energy for India is solar energy. About 50% of the world's population resides in rural areas, and in developing countries, the rural population is more as compared to the developed countries. Targeting the rural market gives a great potential for solar PV, as 200 million people live in the rural areas of India (Solanki & Mudaliar, 2010). Based on the literature review, it has been found that among both groups of consumers (who used solar products early and who used the solar products in the majority) most of them had a positive perception about solar products. Some consumers stated that the financial, economic and aesthetic characteristics limit its adoption; others are prepared to pay the price of the solar water heater if the technology is of a high standard. The critical points for promoting solar products are cost and awareness. If an individual is persuaded properly, then there are chances of solar PV adoption (Rai & Sharma, 2014). India is unable to meet the rising power demands; power cuts and unavailability of power supply are common problems in all the areas of our country.

The best solution for this is to promote the usage of solar energy and other renewable sources of energy. The solar rooftop segment in India is still in its nascent stage but is growing fast due to the great demand for energy and its potential. Because of the high electricity bill, power shortage and the "go green" trend, the solar inverter is gaining great acceptance among the residents of India (Gupta, 2015). The need for dissemination programmes and the issues that arise in the dissemination of solar PV is that Product should be produced to meet the requirement of the end-user, promotion should focus on benefits than on subsidies, promotion should be extended beyond the government agencies and word of mouth communication should be encouraged; need and affordability should be the criteria for target group identification and promotion should be made to target groups (Velayudhan, 2003).

2 Energy Scenario: World and India

According to the statistics given in the World Energy Council Report, World Energy Scenario—2016, new technology will drive industrialized economies to transition more quickly into service and sustainable growth; hence, electricity demand is expected to double by 2060 due to urban lifestyle. To meet this demand renewable energy sources are required. Fossil fuels demand is expected to decrease by 50–70% (World Energy Council, 2016). The growth of non-fossil energy sources will dominate electricity generation by the year 2060 and it will be driven by solar and wind energy and there will be a reduction in carbon emission from the year 2020.

Figure 1 factors shaped world energy shows the different factors that are interconnected to shape the world energy

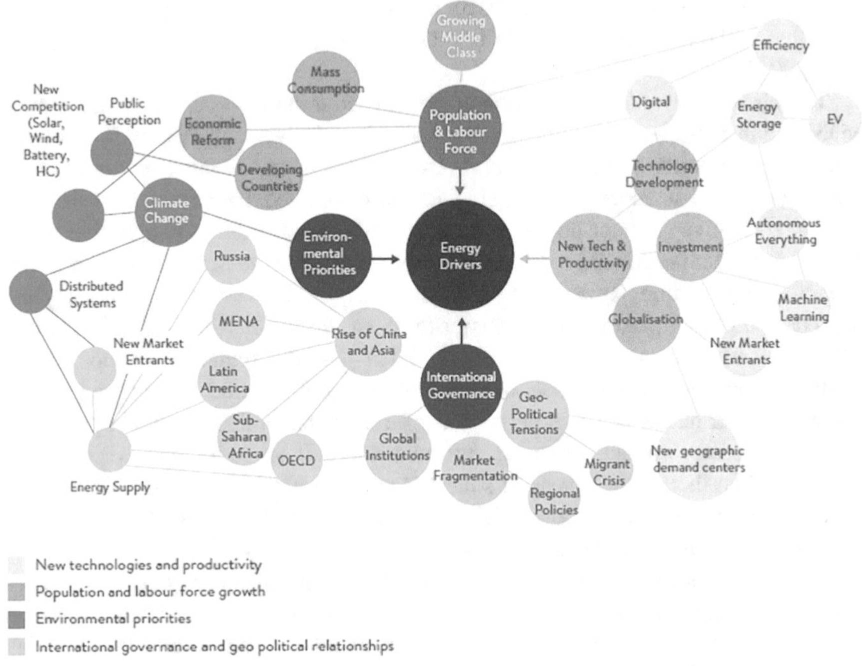

Fig. 1 Factors shaped world energy. Source: World Energy Council (2016)

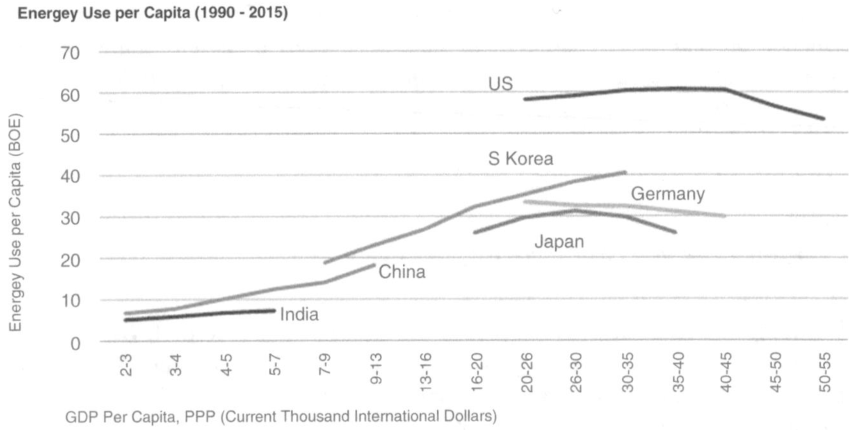

Fig. 2 Energy use and income per capita (1990–2015). Source: IMF, World Bank, BP conversion, Accenture Analysis

Figure 2 shows energy use per capita on *Y*-axis and GDP per capita, PPP (current thousands in international dollars). It demonstrates that the USA is at the first position in energy use followed by South Korea, Germany, Japan, China and India. India stands at the last position but its energy use percentage is increasing.

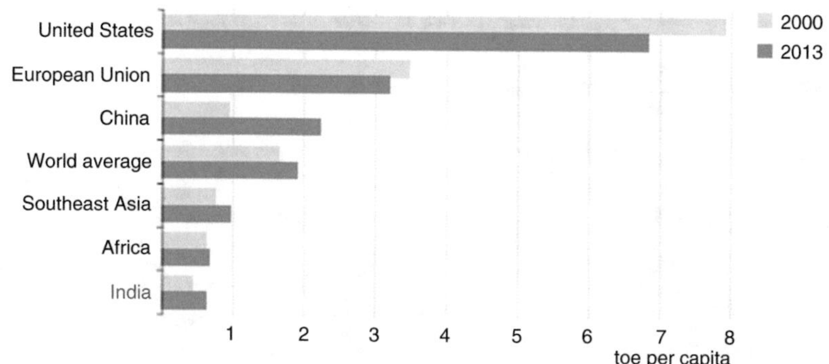

Fig. 3 Per capita energy consumption in India and selected regions. Note: toe—tonnes of oil equivalent. Source: International Energy Agency (2015)

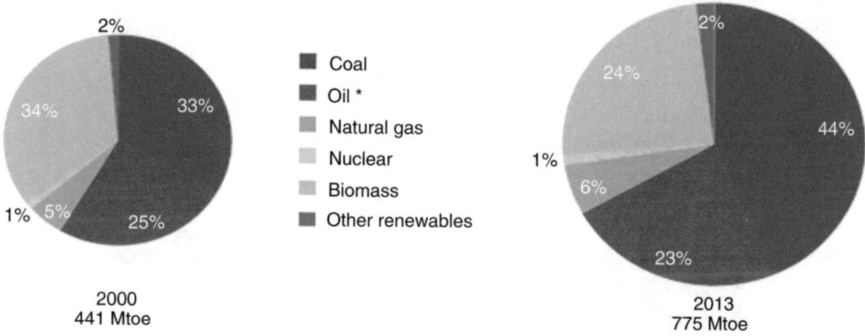

Fig. 4 Primary Energy demand in India by fuels. Source: International Energy Agency (2015)

Figure 3 shows the comparative study of per capita energy consumption of different countries in the years 2000 and 2013. It represents that USA and European Union energy consumption demand has decreased in the year 2013 as compared to the year 2000; Whereas China, World average, Southeast Asia, Africa and India energy consumption demand has increased.

From Fig. 4 one can easily understand that the demand of energy from different sources in the year 2000 are: Coal = 33%, Oil = 25%, Biomass = 34%, Natural Gas = 5%, other renewable = 2% and nuclear = 1%, Whereas as all of the represented sources are possessing different percentage in the year 2013, Coal = 44%, Oil = 23%, Biomass = 24%, other renewable = 2% and Nuclear = 1%. There is a decrease in the percentage of Biogas and Oil energy sources in comparison to the energy demand in the year 2000. While the percentage of other renewable and nuclear fuels has not changed, coal and natural gas fuels percentage have increased in the year 2013 (International Energy Agency, 2015).

In Fig. 5, a comparison of the energy demand of fuels by selected end-user sectors, i.e. Industry, Buildings, Transport and Agriculture sectors in India has been shown. It shows that coal and electricity demand in the industrial sector and

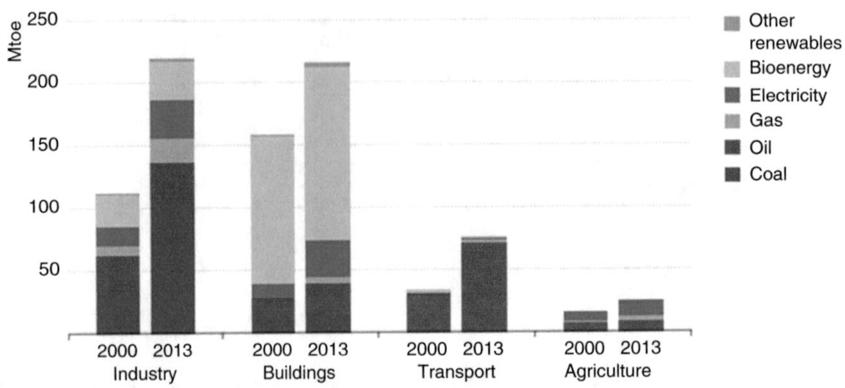

Fig. 5 Energy demand by fuel in selected end use sectors in India. Source: International Energy Agency (2015)

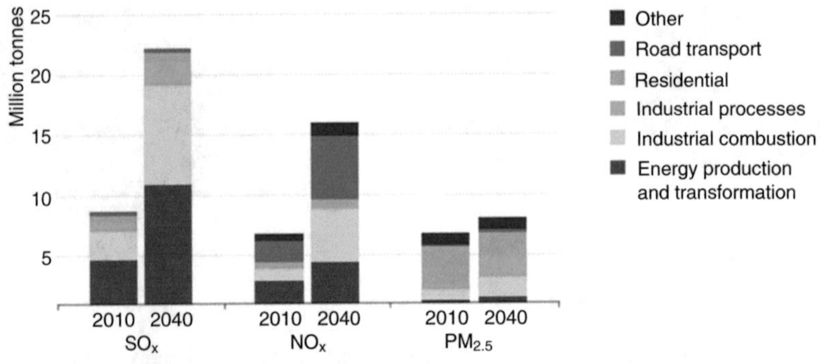

Fig. 6 Emission of NO_2, SO_2 and pm 2.5 by sectors 2010–2040. Source: International Energy Agency (2015)

building sectors have increased, whereas oil and electricity demand is increasing in the transport and agricultural sectors from the year 2000 to 2013.

Based on Fig. 6 it can be seen that Air pollution is expected to increase day by day in India from the year 2010 to 2040, which is hazardous to the eco-system. Sulphur oxide (SO), Nitrogen Oxide (NO) and particulate matter (pm 2.5) emerging from different sectors (Road Transport, Residential, Industrial Processes, Industrial Combustion, Energy production and transformation and others) are showing significant growth. This will pose more harm to the environment. (International Energy Agency, 2015) Power Sector at a Glance in India Statistics from the Ministry of Power in India claims the following percentage of various power sectors.

Table 1 shows that the private sector requires more energy in comparison to the central and state sector and the state sector is also surpassing the central sector.

Table 2 shows the fuel-wise installation capacity in different power sectors (Total thermal, Coal, Gas, Oil, Hydro (renewable), and Nuclear and Renewable energy

Table 1 Total Installed capacity—sectorwise (As on 31.01.2017)

Sector	MW	% of total
State Sector	103,212	32.80
Central Sector	76,852	24.43
Private Sector	134,578	42.77
Total	314,642	

Table 2 Total installed capacity—fuel-wise (As on 31.01.2017)

Fuel	MW	% of Total
Total Thermal	214,655	68.2
Coal	188,488	59.9
Gas	25,329	8.1
Oil	838	0.3
Hydro (Renewable)	44,189	14.0
Nuclear	5780	1.8
RES (MNRE)	50,018	15.9
Total	314,642	

Renewable Energy Sources (RES) include SHP, BG, BP, U & I and Wind Energy, *SHP* small hydro project, *BG* biomass gasifier, *BP* biomass power, *U & I* urban & industrial waste power, *RES* renewable energy sources
Source: Ministry of Power, Government of India

sources installed by MNRE). It shows that Total thermal, Coal and RES fuels installed capacity are higher in comparison to Gas, Oil, Hydro and Nuclear.

3 Renewable Energy Revolution

The first Renewable Energy Global Investor Meet and Expo (RI-Invest, 2015) were organized by the MNRE from 15–17, Feb. 2015. The aim of this conference was "moving from Megawatt to Gigawatt". MNRE includes harnessing renewable energy and usage of RE in Urban, Industrial and Commercial applications. RE contributes 12.96% to the national electricity installed capacity. RE-based decentralized and distributed applications are very environment friendly. It has re-educated the social and economic problems in rural India and also created an opportunity for economic activities at the village level. The JNNSM (Jawaharlal Nehru National Solar Mission) laid down the vision and ambition for the future. REC (Rural Electrification Certificates) mechanism helps in the creation of pan-India RE (Renewable Energy) market.

4 Energy Security Affair

The main source of energy generation in India is coal for the last many decades, which is being utilized in most of the sectors very frequently. At present around 60% of India's power generation capacity is dependent on coal. India's coal import and oil dependencies have increased by 28% in all the sectors of energy utilization.

Table 3 Ernst and Young LLP's attractive Index-2013

Rank	Country	All renewable	Wind Index	Solar Index
1	USA	75.4	68.8	78.0
2	China	71.9	76.7	79.6
3	Germany	69.6	58.4	59.6
4	UK	62.1	58.8	38.9
5	Japan	61.8	43.7	56.8
6	Australia	61.3	46.2	57.2
7	Canada	59.3	52.5	46.1
8	France	56.9	47.3	48.3
9	India	56.2	50.5	60.6
10	Italy	54.4	37.3	50.3
11	Belgium	53.0	42.5	35.3
12	South Korea	52.2	39.9	41.7
13	Spain	51.7	36.0	45.5
14	Denmark	51.3	46.0	24.9
15	Brazil	50.9	47.4	46.9

Source: Ernst and Young LLP

5 Energy Shortage

Despite the increase in installed capacity by 110 times in 62 years, India is still not in a position to meet the peak electricity demand. (Renewable Energy Global Investor Meet and Expo, 2015) According to Ernst & Young Report: Mapping India's RE growth potential: Status and outlooks—2013, the target for the 12th 5-year plan was to take the total RE capacity to almost 55 GW by the end of the fiscal year 2017. India ranks 9thin the overall Ernst and Young LLP's (Limited Liability Partnership) recent renewable attractiveness index. Whereas the USA, China, Germany, UK, Japan, Australia, Canada and France are still ahead of India (Ernst & Young Report, 2013) (Table 3).

6 Solar Energy in India

Krishnan (2013) mentioned in an article, "Solar Energy: A viable alternative for India", India is endowed with very good natural energy resources. The average intensity of solar radiation received in India is approximately 200 MW/km. Even if 10% of the available area can be used, the expected energy generation would approximately be 8 million MW (which is equivalent to 5909 million tons of oil) per year. Solar energy is renewable energy's fastest-growing sector with its demand and utilization evidently increasing. In January (winter) the southern peninsula receives above 4.5 kWh/m^2/day reaching a maximum of 5.5 kWh/m^2/day in the Western plains and Ghats region. A major part of India receives above 5 kWh/m^2/

day Western Himalayas (Himachal Pradesh, Jammu Kashmir and Uttarakhand) and Eastern Himalayas (Assam, Arunachal Pradesh, Nagaland) receive insolation in of 3–4 kWh/m^2/day. During summer in the month of April and May, more than 90% of the country receives insolation above 5 kWh/m^2/day and a maximum recorded 7.5 kWh/m^2/day in the western states. Monsoon clouds that originate from central Asia in October bring the global insolation below 4 kWh/m^2/day in the lower gangetic and east coast plains. The Himalaya foothills, plains, central and western dry zones receive above 4.7 kWh/m^2/day as the Himalayas act as a barrier to this winter monsoon and allow only dry winds to the Indian mainland. These observed seasonal variations of global insolation throughout the country conform from the earlier investigations based on 18 surface solar radiations.

Figures depicts the monthly average insolation map of India with isohels from (Jan to Apr) Fig. 7, *(May to Aug)* Fig. 8, (Sep to Dec) Fig. 9, and monthly average insolation map of India with Isohels with solar hotspots (Fig. 10).

The author also mentioned the state government programme launched in Punjab for lighting the educational institute by generating solar power through the rooftop of

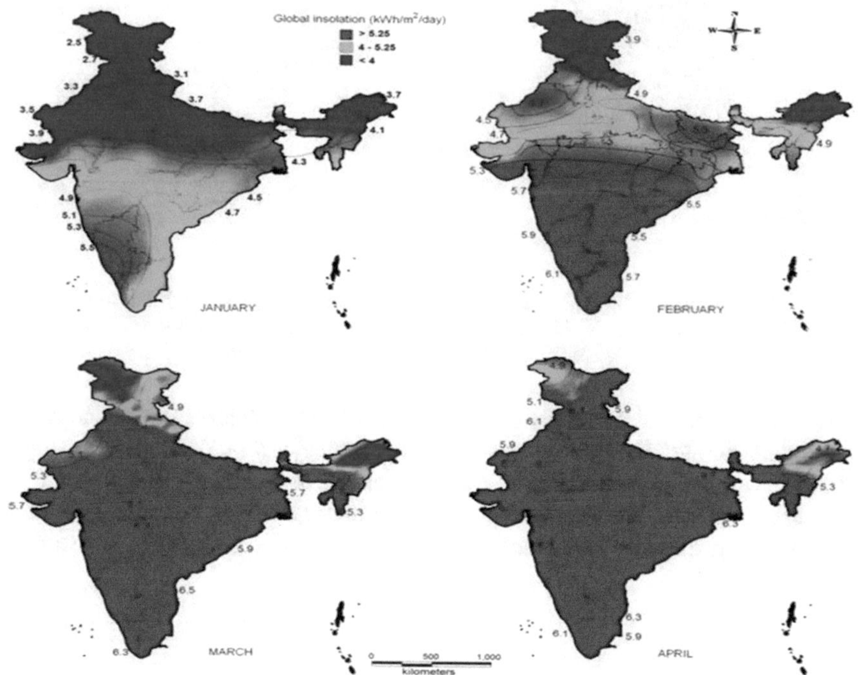

Source: Krishnan S. (2013) Solar Energy: A viable alternative for India. BIOINFO Renewable & Sustainable Energy, ISSN: 2249-1694 & EISSN: 2249-1708, Volume 3, Issue 1, pp.-147-152.

Fig. 7 Monthly average Insolation map of India with isohels (Jan to Apr). Source: Krishnan S. (2013) Solar Energy: A viable alternative for India. BIOINFO Renewable & Sustainable Energy, ISSN: 2249-1694 & EISSN: 2249-1708, Volume 3, Issue 1, pp. 147–152

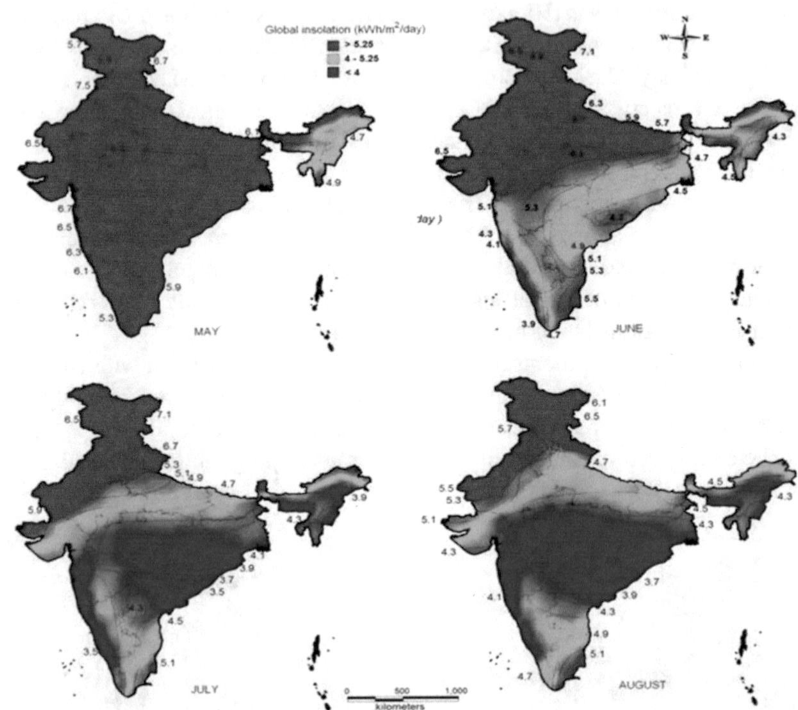

Source: Krishnan S. (2013) Solar Energy: A viable alternative for India. BIOINFO Renewable & Sustainable Energy, ISSN: 2249-1694 & EISSN: 2249-1708, Volume 3, Issue 1, pp.-147-152.

Fig. 8 Monthly average Insolation maps of India with isohels (May to Aug). Source: Krishnan S. (2013) Solar Energy: A viable alternative for India. BIOINFO Renewable & Sustainable Energy, ISSN: 2249-1694 & EISSN: 2249-1708, Volume 3, Issue 1, pp. 147–152

solar PVs, the central university of Punjab, Gian Sagar Medical College and others institutes generate 230 kilowatts of solar energy. Banerjee (2006) (cited by Krishnan, 2013) mentioned in his article, "Bank on solar, nuclear energy". There should be inter-disciplinary efforts from the scientists for better utilization of solar as well as nuclear energy and mentioned that solar energy would be a great source of energy in the future.

7 Review of Literature

Researchers have divided review of literature into three subheads they are (Tables 4, 5 and 6)

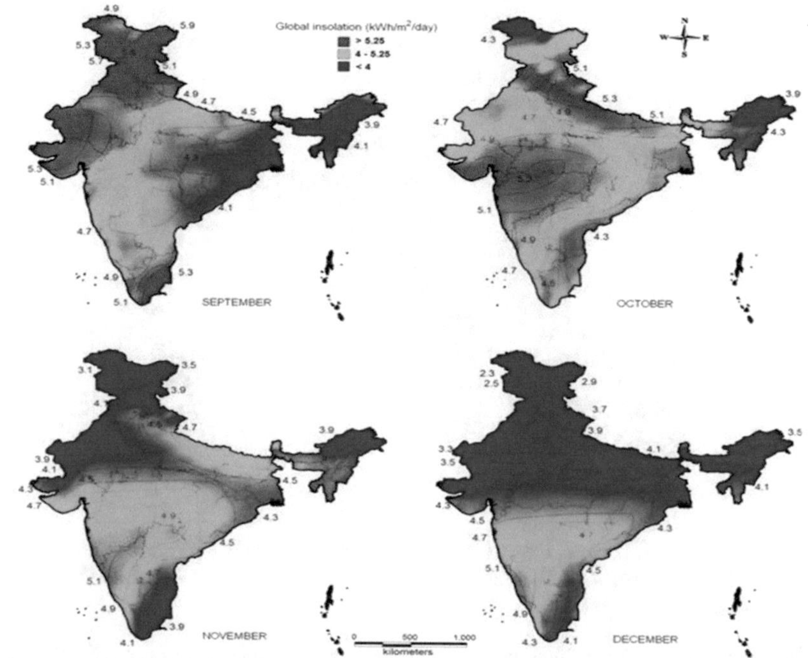

Source: Krishnan S. (2013) Solar Energy: A viable alternative for India. BIOINFO Renewable & Sustainable Energy, ISSN: 2249-1694 & EISSN: 2249-1708, Volume 3, Issue 1, pp.-147-152.

Fig. 9 Monthly average insolation map of India with Isohels (Sep to Dec). Source: Krishnan S. (2013) Solar Energy: A viable alternative for India. BIOINFO Renewable & Sustainable Energy, ISSN: 2249-1694 & EISSN: 2249-1708, Volume 3, Issue 1, pp. 147–152

1. Renewable energy (solar energy)
2. Solar products marketing (solar market)
3. Solar products

8 Findings

- The review of the literature on solar energy, solar products and solar products marketing in an Indian context illustrates the status of the need for solar energy, solar products and solar marketing.
- It has been noted from the literature that India consumed a large portion of commercial energy and the rural resident are consuming it lesser than the urban residents. Thus the consumption of energy will increase as the rural region of India is in need, and most of the rural residents are migrating to cities to avail themselves of the basic facilities and work, thus the urban area will be extended.

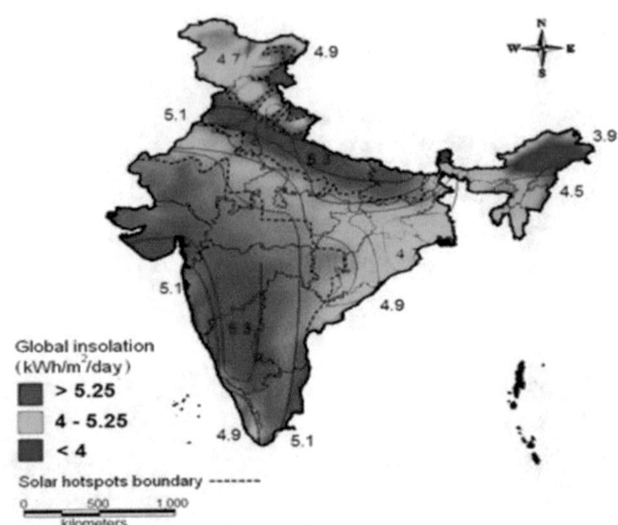

Source: Krishnan S. (2013) Solar Energy: A viable alternative for India. BIOINFO Renewable & Sustainable Energy, ISSN: 2249-1694 & EISSN: 2249-1708, Volume 3, Issue 1, pp.-147-152.

Fig. 10 Monthly average insolation map of India with Isohels with solar hotspots. Source: Krishnan S. (2013) Solar Energy: A viable alternative for India. BIOINFO Renewable & Sustainable Energy, ISSN: 2249-1694 & EISSN: 2249-1708, Volume 3, Issue 1, pp. 147–152

- The government of India has taken several initiatives and schemes such as Renewable Energy Certificates, and setting organizations like TERI, NGO to overcome the energy issue and other difficulties (pollution, climate change and sustainability) which is associated with the production of electrical energy through fossil fuels.
- It is also found that the government of India has also come up with different schemes for the rural and urban residents to fulfil the gap of energy by making solar energy a substitute to the non-renewable sources of energy.
- Few studies show that rural residents are in dire need of energy connectivity. Thus to fulfil the energy demand, alternate energy is required. The Indian government has come up with strategical solutions to produce 450 GW of energy through solar and wind energies. It is also found that many firms started to produce solar products.

9 Conclusion

The research work focuses on the three main components of solar products, solar energy and the solar market and tries to connect them to know the applicable part and the gap. It is found that the utilization of solar product energy can be easily tapped in

Table 4 Renewable energy (solar energy)

Sr. no	Author	Findings
1	Pachauri and Jiang (2008)	India's and China's rural households exceed urban households in terms of energy consumption and the factors for the energy transition in urban areas of both countries are income, energy prices, energy access and local fuel availability
2	Bhattacharya and Jana (2009)	Study found energy consumption in India is expected to continue in the future too. The installation and promotion of renewable energy devices are required for fulfilling the energy demand. This can reduce the damage caused to the environment
3	Ekholm et al. (2010)	The result of this study shows that biomass energy is the choice of rural consumers and it is an alternative to Kerosene
4	Kumar et al. (2010)	Results concluded that India can face the problem of energy shortage in the future on the basis of its growth; with the energy demand, the second problem is an environmental concern. There is no option for India to choose renewable energy systems as the future energy system. The renewable energy certificates (REC) can play a role in hindering the lacuna for renewable energy and thereby create a vibrant market. Renewable energy can benefit the national energy mix
5	Palit and Chaurey (2011)	The challenges for electrification in South Asian countries are technology, financial problems and government barrier
6	Khandker et al. (2012)	Energy poverty is not necessarily equal to income poverty and that energy policies matter a lot regarding eradicating energy poverty. Solid efforts are required by both governments (central as well as state) to ensure that the resources are allocated to the rural areas and to ensure that rural electrification and other energy sources are cost-effective
7	Millinger et al. (2012)	Researchers found solar power systems had a very little role as 400 million people live in India without the accessibility to electricity
8	Palit (2013)	Improved access to capital, development of effects after-sales services, customer-centric market development and stakeholder involvement assisted to scale-up policy implemented in Bangladesh, Sri Lanka, Nepal and India
9	Khare et al. (2013)	Development of many sectors in India requires solar energy as primary aspect
10	Urpelainen (2014)	Rural electrification can improve the economy and hence bring about an improvement in the education and health sector
11	Gupta (2015)	Growth in the economy is directly linked with the supply of power and studied noted that country made serious efforts to tap renewable sources of energy such as wind and solar energy
12	International Energy Agency (2015)	India's urbanization is a key driver of energy trends: an additional 315 million people are expected to live in India's cities by 2040. India's need for new infrastructure underlies strong demand for energy-intensive goods, while the rising level of vehicle ownership keeps transport demand on an even steeper upward curve

(continued)

Table 4 (continued)

Sr. no	Author	Findings
13	Razdan (2016)	Study highlights the slogan "Power for All" has been changed to "Power for all 2017" and hoping that it will be transferred soon into "Quality Power & Green Power"
14	Levin and Thomas (2016)	Study extending grid and implementing distributed solar home system (SHS). The outcome of this study is that solar home systems can fulfill the gap of energy in developing countries
15	Garud and Sharma (2016)	Electrification in rural India a sustainable way is required and the massive focus should be on the income generations steps which will uplift the financial condition of rural area
16	Alam and Khan (2016a)	The authors have come up with a logical idea on the basis of a review of the literature (secondary data) to solve this problem, that if the production of energy is done through renewable energy sources then the energy demand can easily be fulfilled and it will secure the rural population from the dangerous effects of the chullas' gases. The authors reported that electric energy would not sustain in the future, whereas renewable energy has the quality of sustaining. Thus renewable energy can be a tool for fulfilling the demand for energy and sustaining growth as well
17	Mathur (2016)	Various challenges which GOI will face by the demand for energy in India, for example (a) 23.6% of India's population still lives below $1.25(65.66 Indian Rupees) per day; about 75 million households are still not connected to grid electricity and 80% of the rural households use traditional biomass as primary sources for cooking. (b) The Human Development Index will change and also become very challenging to contain per capita power consumption. Population growth and energy growth are two main drivers of energy demand. (c) After 2011 population has increased from 1 billion to 1.2 and economic growth has increased at an average of 8 per cent a year. Primary energy demand grew by 5% a year—70% met through fossil fuels, 23% energy met from primary energy and the remaining 7 from the other natural gases. (d) Meeting the INDC target (Intended Nationally Determined Contribution (INDCs) agreement in Paris, submitted by all the countries indicates that the targets are clearly insufficient to keep the world within the safe limit of 2 degrees increase)
18	Sadath and Acharya's (2017)	Study found that income poverty and energy poverty are commensurate with each other predominantly in the Indian context
19	Alam and Khan (2017)	Study identified that economy of Uttar Pradesh (Indian state) can be improved by use of solar energy and its products, and also solar products awareness is required among Uttar Pradesh residents for the solar products to open new sphere industries in Uttar Pradesh
20	Abu-Elsamen et al. (2019)	Environmental awareness positively influences subjective norms and reduces perceived performance and financial risks, which are related to purchase intention

(continued)

Table 4 (continued)

Sr. no	Author	Findings
21	Alawiye and Ewulo (2019)	The study finds that all the respondents have good knowledge of what solar energy is Majority of the respondents have experienced solar energy as an alternative power source for ICT before. Respondents are aware of solar energy as an alternative power source in their libraries. They were also aware that solar energy can power their ICT equipment as they also believe that using solar energy, erratic power supply can be mitigated to a large extent
22	Graziano et al. (2019)	The results revealed that centralized, non-voluntary support policies may have larger effects if implemented beyond the town level and that SPEs change their determination power depending on the underlying built environment

India because of the geographical location whereas solar products are not commonly available in the market. This may be because of various reasons such as the quality of solar products, the technology of solar products and unready mindset of consumers and mass level unawareness.

10 Challenges and Opportunities for Solar Energy Conservation in India

Dependency The majority of solar modules are still made in China. India has to be capable of competing with China's robust industrial base, which poses a significant threat to Indian domestic producers. This is necessary to attain solar sector self-sufficiency.

Waste Management By 2050, India's solar waste is expected to reach 1.8 million tonnes. Solar cell makers, on the other hand, are not required by India's e-waste legislation to recycle or dispose of trash from this industry.

Space Cringe Many countries have struggled to deploy large-scale ground-mounted solar systems due to a shortage of space. In terms of installation, this allows for more R & D and innovation.

Low Tariffs Solar rates in India are among the lowest in the world, making them unsustainable for some developers and potentially compromising solar panel quality. Both in terms of tariff and profitability, it is necessary to preserve viability constraints.

Financing Mechanism A key issue is the lack of creative financing options that offer larger quantities at cheaper interest rates and for longer periods. The National Clean Energy and Environment Fund, Green Masala Bonds and other initiatives have been launched in this regard.

Table 5 Solar products marketing (solar market)

Sr. no	Author	Findings
1	Wiser (1998)	Green products have been attributed to the environment. Their consensus show that green marketing is significantly growing; the companies can make a profit by improving environmental performance & developing green products garnering the customer's interest. But obstacles are still there in selling green products
2	Velayudhan (2003)	Author recommended for the use of solar lanterns in India "(A) Product should be produced to meet the requirement of the end-user, (B) Promotion should focus on benefits than on subsidies, (C) Promotion should be extended beyond the government agencies and word of mouth communication should be encouraged, (D) Need and affordability should be the criteria for target group identification and (E) Make promotion to target group"
3	Markard and Truffer (2006)	The environmental characteristics depend on two factors the ecological quality of power generation and the promotional effects of the product design. The eco power depends on the power sources and the conversion technologies. This promotion is the underlying principle of products that diverts a certain fraction of electricity bills into a fund from which new RES (*Renewable Energy* Systems) capacity is financed
4	Solanki and Mudaliar (2010)	Companies can reach to the consumers through the Micro Finance Institutions (MFI) and Direct Marketing Process. TERI (The Energy and Renewable Institute) joined with NGOs for solar PV dissemination in rural areas
5	Bajpai (2012)	The authors have explained five types of strategies, "Strategy 1: multimedia usage, Strategy 2: integrate offline and online advertising; Strategy 3: message adaptation; Strategy 4: local social networks, Strategy 5: contests and discounts"
6	Suchi's (2013)	The obstacles which rural marketing get are as follows: "Understanding the Rural Consumer, Poor infrastructure, Physical distribution, Channel management, Promotion and marketing communication." The researcher has also suggested some marketing strategies for developing the rural market like the marketers should know the mindset of rural consumers; marketers should use Melas, Haats and Mandis for distribution of their product; delivery van should be carried out for delivering product and for sales promotion; audio-visual tools should be used for promoting the products; puppet show, folk dances, radio are other sources which attract rural consumers
7	Rai and Sharma (2014)	The critical points for promoting solar products are cost & awareness. If an individual is persuaded properly then there are chances of solar PV adoption, due to the inadequate knowledge, the rural consumers are not been motivated but they can be persuaded
8	Deshpande et al. (2014)	Solar rooftop agencies and solar power generation agencies have identified that solar power is unviable due to the cost and lower efficiencies of batteries for power backup, this can be reduced if grid-connected rooftop projects are made operational

(continued)

Table 5 (continued)

Sr. no	Author	Findings
9	Gupta (2015)	The best solution for this is to promote solar energy sources and other renewable sources of energy. The solar rooftop segment in India is still in its nascent stage but is growing fast and has great potential. Due to the high electricity bill, power shortage and the "go green" trend, the solar inverter is gaining great acceptance among the residents in India
10	Alam and Khan (2016b)	Authors identified the solution for women empowerment issue vis-à-vis the solar industry in India and come out with an output that the solar industry can boost women empowerment as well as the rural economy. With an example of two NGO's (1) Barefoot targeted to train the women of rural areas (Particularly those women, who are in the middle of their ages and who lives idly at their houses) by the training they are becoming engineers of solar products. (2) Frontier Marketers targeted solar products manufacturers and maintenance of solar products. Also created distribution channel for the supply of clean energy products and thus uplifting the rural people by making them entrepreneurs
11	Singh (2016)	The study reveals that a majority of enterprises provided solar-based energy in areas with grid access. If not limited by the grid, perhaps certain geographies play a role in a firm's ability to scale up. Statistical data analysis supported the claims by some experts that the market for off-grid solar technologies is determined by the seller. End users are not able to articulate what they need, particularly those users in areas without grid access who may need the technology options the most
12	Bashiri and Alizadeh (2018)	The empirical result shows a negative impact on adoption. Environmental concerns, knowledge of renewable energies, innovativeness and number of households, either of these factors positively increases the probability of adoption individually. The results of this study help policy-makers and renewable energy marketers to make energy-related decisions
13	Bondio et al. (2018)	PV is the technology of the middle class. This reasoning is made based on survey-stated concerns over rising electricity bills and survey data, which indicate that economic life events have a significant influence over perceptions of affordability

Solar Irradiance The power quality of PV systems is affected by changes in solar radiation and temperature. High distribution grid connection density and low irradiance can cause unfavourable variations in power supply quality, which must be resolved.

Balancing National Priorities with International Commitments India's domestic content requirement (DCR) clause has been challenged in court by the World Trade Organization (WTO), and the country must find a delicate balance between domestic ambitions and international goodwill.

Table 6 Solar products

Sr. no	Author	Findings
1	Murray (2012)	Study's result found that variables age was identified as a variable that also influenced the intent of solar panels
2	Venkatraman (2014)	This study suggested that the price of solar devices should be low so that low-income group consumers can also buy them; the size of the solar product should be small; awareness is highly needed peculiarly in rural areas as they are deprived of education and are unable to pay installation charges of solar energy.; the demonstration and exhibition can increase the popularity of the solar product in the market and there should be a variety of solar products
3	Kumar (2014)	The researcher found that people were using solar energy for small devices like emergency lights. Furthered authors stated that because of the higher cost of solar panels people are not much interested in installing solar panels
4	Aggarwal et al. (2019)	Study concluded that social beliefs followed by effort expectance concerns are key factors explaining approximately 20% of the purchase intent each, while the unit change in price value beliefs explains about 18% of the purchase intent
5	Kumar et al. (2019)	Solar pump users in Punjab, India, found cost, performance and government initiatives influencing the consumer buying behaviour
6	Kumar, Hundal, et al. (2020)	Study found perceived benefits, perceived compatibility and government initiatives were the reasons for farmers' intention to adopt solar-powered pumps and investment cost and lack of awareness were the reasons for farmers non-adoption of solar-powered pumps
7	Kumar, Syan, et al. (2020)	Study identified that increasing energy prices, product knowledge and experience, financial support and subsidies, perceived cost, have a positive influence on customers' purchase intention of the solar water heater
8	Parsad et al. (2020)	The study found government subsidies act as a primary motivator which helps in overcoming the initial risk of investment in the new technology. And the artificial neural network (ANN) model identified the technical barrier, knowledge and awareness factors that play a significant role in forecasting the investor investing decision

References

Abu-Elsamen, A. A., Akroush, M. N., Asfour, N. A., & Al Jabali, H. (2019). Understanding contextual factors affecting the adoption of energy-efficient household products in Jordan. *Sustainability Accounting, Management and Policy Journal, 10*(2), 314–332. https://doi.org/10.1108/SAMPJ-05-2018-0144

Aggarwal, A. K., Syed, A. A., & Garg, S. (2019). Factors driving Indian consumer's purchase intention of roof top solar. *International Journal of Energy Sector Management, 13*(3), 539–555.

Alam, S. A., & Khan, S. M. (2016a). Renewable energy as a tool for sustainable growth of rural India. In *Proceeding of rural development in Bhutan: Prospects & challenges*. Royal University of Bhutan.

Alam, S. A., & Khan, S. M. (2016b). Marketing strategies for rural women empowerment through solar industry in India: A View point. In *Proceeding of start-up India international conference on entrepreneurship and women empowerment*. NIESBUD (Ministry of Skill Development & Entrepreneurship, Govt. of India).

Alam, S. A., & Khan, S. M. (2017). Study of impact of solar energy based products on the economy of UP and the standard of life of its Residents. *International Journal of Society and Humanities, 11*, 71–79.

Alawiye, M. K., & Ewulo, O. R. (2019). Awareness and use of solar energy as alternative power source for ICT facilities in Nigerian university libraries and information centres. *Library Philosophy and Practice*, 1–18.

Bajpai, V. (2012). Social media marketing: strategies & its impact. *International Journal of Social Science & Interdisciplinary Research, 1*(7).

Banerjee, S. (2006). *Bank on solar, nuclear energy*. BARC.

Bashiri, A., & Alizadeh, S. H. (2018). The analysis of demographics, environmental and knowledge factors affecting prospective residential PV system adoption: A study in Tehran. *Renewable and Sustainable Energy Reviews, 81*, 3131–3139.

Bhattacharya, S. C., & Jana, C. (2009). Renewable energy in India: Historical developments and prospects. *Energy, 34*(8), 981–991. https://doi.org/10.1016/j.energy.2008.10.017

Bondio, S., Shahnazari, M., & McHugh, A. (2018). The technology of the middle class: Understanding the fulfilment of adoption intentions in Queensland's rapid uptake residential solar photovoltaics market. *Renewable and Sustainable Energy Reviews, 93*, 642–651.

Burt, Orris, Buchanan. (2013). Scientific evidence of health effects from coal use in energy generation. Environmental and Occupational Health Sciences_University of Illinois and Health Care Without Harm. https://noharm-uscanada.org/sites/default/files/documentsfiles/828/Health_Effects_Coal_Use_Energy_Generation.pdf

Deshpande, B. N., Priyadarshi, S., & Pandya, V. (2014). The Indian solar market: Updates, constraints and the road ahead.

Ekholm, T., Krey, V., Pachauri, S., & Riahi, K. (2010). Determinants of household energy consumption in India. *Energy Policy, 38*(10), 5696–5707. https://doi.org/10.1016/j.enpol.2010.05.017

Ernst & Young. 2013. Mapping India's RE growth potential: Status and outlooks—2013. Ernst & Young LLP.

Garud, S. S., & Sharma, P. (2016, August). Rural electrification: A development challenge. *Yojana A Development Monthly*, 23–26.

Graziano, M., Fiaschetti, M., & Atkinson-Palombo, C. (2019). Peer effects in the adoption of solar energy technologies in the United States: An urban case study. *Energy Research & Social Science, 48*, 75–84.

Gupta, A. (2015). Continuous government support and advent of micro solar inverter to foster future growth. *Solar Power,* 4–5.

Khandker, S. R., Barnes, D. F., & Samad, H. A. (2012). Are the energy poor also income poor? Evidence from India. *Energy Policy, 47*, 1–12. https://doi.org/10.1016/j.enpol.2012.02.028

Khare, V., Nema, S., & Baredar, P. (2013). Status of solar wind renewable energy in India. *Renewable and Sustainable Energy Reviews, 27*, 1–10. https://doi.org/10.1016/j.rser.2013.06.018

Krishnan, S. (2013). Solar Energy: A viable alternative for India. *BIOINFO Renewable & Sustainable Energy, 3*(1), 147–152. ISSN: 2249-1694 & EISSN: 2249-1708.

Kumar, A., Kumar, K., Kaushik, N., Sharma, S., & Mishra, S. (2010). Renewable energy in India: Current status and future potentials. *Renewable and Sustainable Energy Reviews, 14*(8), 2434–2442. https://doi.org/10.1016/j.rser.2010.04.003

Kumar, D. P. (2014). An overview of public perception about the suitability of solar power panels as an alternative energy source in Andhra Pradesh, 6.

Kumar, V., Hundal, B. S., & Kaur, K. (2019). Factors affecting consumer buying behaviour of solar water pumping system. *Smart and Sustainable Built Environment, 8*(4), 351–364.

Kumar, V., Hundal, B. S., & Syan, A. S. (2020). Factors affecting customers' attitude towards solar energy products. *International Journal of Business Innovation and Research, 21*(2), 271–293.

Kumar, V., Syan, A. S., Kaur, A., & Hundal, B. S. (2020). Determinants of farmers' decision to adopt solar powered pumps. *International Journal of Energy Sector Management, 14*(4).

Levin, T., & Thomas, V. M. (2016). Energy for sustainable development can developing countries leap frog the centralized electrification paradigm? *Energy for Sustainable Development, 31*, 97–107. https://doi.org/10.1016/j.esd.2015.12.005

Markard, J., & Truffer, B. (2006). The promotional impacts of green power products on renewable energy sources. *Direct and Indirect Eco-Effects, 34*, 306–321. https://doi.org/10.1016/j.enpol.2004.08.005

Mathur, R. (2016, August). India's energy challenges and sustainability development. *Yojana A Development Monthly*, 16–20

Millinger, M., Mårlind, T., & Ahlgren, E. O. (2012). Energy for sustainable development evaluation of Indian rural solar electrification: A case study in Chhattisgarh. *Energy for Sustainable Development, 16*(4), 486–492. https://doi.org/10.1016/j.esd.2012.08.005

Murray, G. (2012). *Exploring the intention of the south west of western Australian residents to purchase solar panels using the theory of planned behaviour approach.* Edith Cowan University.

Pachauri, S., & Jiang, L. (2008). The household energy transition in India and China. *Energy Policy, 36*, 4022–4035. https://doi.org/10.1016/j.enpol.2008.06.016

Palit, D. (2013). Energy for sustainable development solar energy programs for rural electrification: Experiences and lessons from south Asia. *Energy for Sustainable Development, 17*(3), 270–279. https://doi.org/10.1016/j.esd.2013.01.002

Palit, D., & Chaurey, A. (2011). Energy for sustainable development off-grid rural electrification experiences from south Asia: Status and best practices. *Energy for Sustainable Development, 15*(3), 266–276. https://doi.org/10.1016/j.esd.2011.07.004

Parsad, C., Mittal, S., & Krishnankutty, R. (2020). A study on the factors affecting household solar adoption in Kerala, India. *International Journal of Productivity and Performance Management, 69*(8), 1695–1720.

Rai, R., & Sharma, N. D. (2014). Marketing of solar products: An overview. *Science, Technology, and Management, 7*, 141–144.

Razdan, A. (2016, August). Energy sector: The challenge of power for all. *Yojana A Development Monthly*, 7–14.

Sadath, A. C., & Acharya, R. H. (2017). Assessing the extent and intensity of energy poverty using Multidimensional Energy Poverty Index: Empirical evidence from households in India. *Energy Policy, 102*(January), 540–550. https://doi.org/10.1016/j.enpol.2016.12.056

Singh, K. (2016). Business innovation and diffusion of off-grid solar technologies in India. *Energy for Sustainable Development, 30*, 1–13.

Solanki, C. S., & Mudaliar, S. (2010). Strategies to target rural PV market in developing countries—A perspective. *PVSC*, 2392–2396.

Suchi, K. P. (2013). The challenges and strategies of marketing in rural. *Asia Pacific Journal of Marketing & Management Review, 2*(7), 38–43. ISSN 2319-2836.

Tripathi, A. K. (2016, August). The national solar mission marching ahead in solar energy. *Yojana A Development Monthly*, 43–46.

Urpelainen, J. (2014). Energy for sustainable development grid and off-grid electrification: An integrated model with applications to India. *Energy for Sustainable Development, 19*, 66–71. https://doi.org/10.1016/j.esd.2013.12.008
Velayudhan, S. K. (2003). Dissemination of solar photovoltaic: A study on the government programme to promote solar lantern in India. *Energy Policy, 31*, 1509–1518.
Venkatraman, M. (2014). A study on customer's attitude towards solar energy devices. *International Research Journal of Business and Management, 5*(7), 53–57.
Wiser, R. H. (1998). Green power marketing: increasing customer demand for renewable energy. *Utilities Policy, 7*, 107–119.

Website/Others

Bridge to India. (2014). *Solar hand book*. Bridge to India.
Ernst & Young. (2013). *Mapping India's RE growth potential: Status and outlooks—2013*. Ernst & Young LLP.
International Energy Agency. (2015). *India energy outlook: World energy outlook special report*. International Energy Agency.
Mapping India's Renewable Energy growth potential (2013).
Oxforddictionaries.com. (n.d.). Energy. Retrieved from https://en.oxforddictionaries.com/definition/energy
Renewable Energy Global investor meet and Expo. (2015). MNRE. Retrieved from https://re-invest.in/
World Energy Council. (2016). World energy scenario I 2016. In collaboration with Accenture and Paul Scherrer Institute.

Precision Livestock Farming (PLF) Systems: Improving Sustainability and Efficiency of Animal Production

Christos Tzanidakis, Panagiotis Simitzis, and Panagiotis Panagakis

Abstract Precision Livestock Farming (PLF) plays a key role in the advancement of animal housing, since it is associated with the improvement of animals' health and welfare status, ensuring sustainability and efficiency of farms. The main objective of researchers is the development of systems for real-time continuous monitoring of the animals' everyday lives (i.e., animal-centric tools). Such systems based on both steady-state and dynamic models should have low installation costs, be precise, accurate, easy to use and environmentally friendly and provide the farmers with valuable information serving as decision support tools for the improvement of management practices. The data could be collected within the unit by simple sensors such as accelerometers, RFID sensors, etc., or more complex computer-based vision or sound and audio analysis systems. This chapter presents various PLF systems in basic livestock (i.e., dairy cows, sheep and goats, pigs, and poultry), indicating their benefits upon the production process.

Keywords Precision livestock farming · Real-time monitoring · Bio-responses · Ruminants · Pigs · Poultry

C. Tzanidakis
Department of Animal Science, North Carolina State University, Raleigh, NC, USA
e-mail: ctzanid@ncsu.edu

P. Simitzis (✉)
Department of Animal Science, Agricultural University of Athens, Athens, Greece
e-mail: pansimitzis@aua.gr

P. Panagakis
Department of Natural Resources Management and Agricultural Engineering, Agricultural University of Athens, Athens, Greece
e-mail: ppap@aua.gr

F. P. García Márquez, B. Lev (eds.), *Sustainability*, International Series in Operations Research & Management Science 333,
https://doi.org/10.1007/978-3-031-16620-4_15

1 Introduction

The global demand for meat and dairy products is continuously increasing due to rising economies such as Brazil, India, China and South Africa (OECD-FAO (Organisation for Economic Co-operation and Development/Food and Agriculture Organization of the United Nations), 2019). Farms become bigger and more intensified, while the number of farmers continuously declines (Berckmans, 2017; Andonovic et al., 2018; Tzanidakis et al., 2021). At the same time, public concerns on environmental impact of manure and animal welfare status are directing towards more environmentally friendly units that ensure welfare and, hence improved quality of the end product (OECD-FAO (Organisation for Economic Co-operation and Development/Food and Agriculture Organization of the United Nations), 2019). Use of technological advancements within the everyday animal production commonly, known as Precision Livestock Farming (PLF), has shown great potential in resolving such problems and increases the annual income of the unit, resulting in more sustainable farms. The main purpose of PLF is to detect, analyse and manage in real-time, any variable that influences the unit's production (Halachmi et al., 2019).

Technological evolution over the past two decades has induced the development of various methods for automatic monitoring of animals' behaviours (Andonovic et al., 2018), including measurements of animals' state and responses as a reply to external factors such as environmental and housing conditions and managerial techniques. The process is based on the principle that every single organism is a "Complex, Individually Different, Time-Varying and Dynamic (CITD) system" (Quanten et al., 2006) and that every living organism responds differently to various conditions and environmental changes (Berckmans, 2017). Furthermore, every organism will respond differently to an external stressor each time it occurs (Quanten et al., 2006; Berckmans, 2014a). Therefore, PLF systems are animal-centric support tools used in ideal conditions for monitoring and control processes and do not intend to replace the farmer, but rather provide valuable information and sometimes indicate solutions to various production and managerial problems (Berckmans, 2004).

The main purpose of PLF systems is to manage any variable that interferes with the production process in real-time and warn the farmer of every problem detected. Animal-based observations are the most direct method of welfare assessment (Temple et al., 2012), thus behavioural analysis is the main core of PLF systems. An animal exposed to discomfort conditions will express it using bio-responses (i.e., behavioural changes) that the system should detect. The development of a system that continuously monitors and detects in real-time, such behavioural changes is based on documentation and creation of datasets captured over a suitable time-period (Banhazi & Black, 2009). The resulting analysis will produce an automatic tool that classifies behavioural patterns thus, identifies behavioural differences caused by unsuitable conditions (Berckmans, 2009; Statham et al., 2009). Therefore, in a sense the animals' bio-responses are used as a sensor, providing data for the system's developed mathematical equations and algorithms that translate them into welfare, performance indicators and production sustainability elements (Vranken & Berckmans, 2017). The second step is the development of mathematical models based on the behavioural pattern analysis and the classification of specific behavioural indicators for these behaviours (Berckmans, 2014b). The final step is the development of a model matching bio-responses such as time spent resting, or feeding and drinking, ruminating, floor occupation area, aggressive events and general activity to a parameter of interest such as growth parameters like body weight, feed intake and conversion rate, or environmental conditions like temperature, relative humidity, etc. (Tzanidakis et al., 2021). It is evident that this task is not simple and knowledge of animal responses to different conditions and variations is a prerequisite for this type of research and application development. The responses may include both steady-state and dynamic components thus, mechanistic, empirical and biological models are necessary leading to complex modelling methods (Smouse et al., 2010). The resulting model should:

1. Include the relationship between the parameter of interest, for example, space availability and the behaviour and
2. Predict current behaviour from past information.

The comparison between the predicted and the real measured behaviour will reveal if the animal status has remained the same or changed. This change is determined by the prediction error (i.e., the difference between the predicted and the real measured value), or by the change in the model parameters defined by the model structure. This information is used as input for the controller, adjusting the necessary parameter settings in effort to bring the animal to its "normal state" (i.e., comfort behaviour; Werkheiser, 2018). The monitoring and the control algorithms are based on experimental data collected for a suitable time-period. An extra dataset not used for the development of the algorithms will then validate the sensitivity and specificity of the model. It is proposed that a minimum of 85% for sensitivity and 89% for specificity should be achieved for the development of a commercial application (Berckmans, 2014b; Petrie & Watson, 2006; Oczak et al., 2013, 2014; Viazzi et al., 2014). It is evident that PLF applications utilise animals' basic behaviours to monitor, manage and control certain variables/parameters thus, the animals are unconsciously controlling these parameters themselves. Such applications lead to real-time problem solutions, improve feed efficiency, diet provision, health and welfare, housing conditions, and help the producers identify animals in need of special attention (Dominiak et al., 2019), while at the same time, improve the economic viability of the unit (Berckmans, 2014b) (Fig. 1).

In this chapter, an effort is made to illustrate the majority of PLF applications in livestock. At first, applications in dairy cattle sector are presented, including milking robots, sensors for automatic oestrus detection and health/welfare disorders (lameness and mastitis) identification. Similar innovative PLF technologies implemented in small ruminants are then discussed. In the following part, precision pig farming technologies are reviewed and are mainly focused on camera-based monitoring, weight estimation, thermal comfort assessment and sound surveillance. Finally, PLF

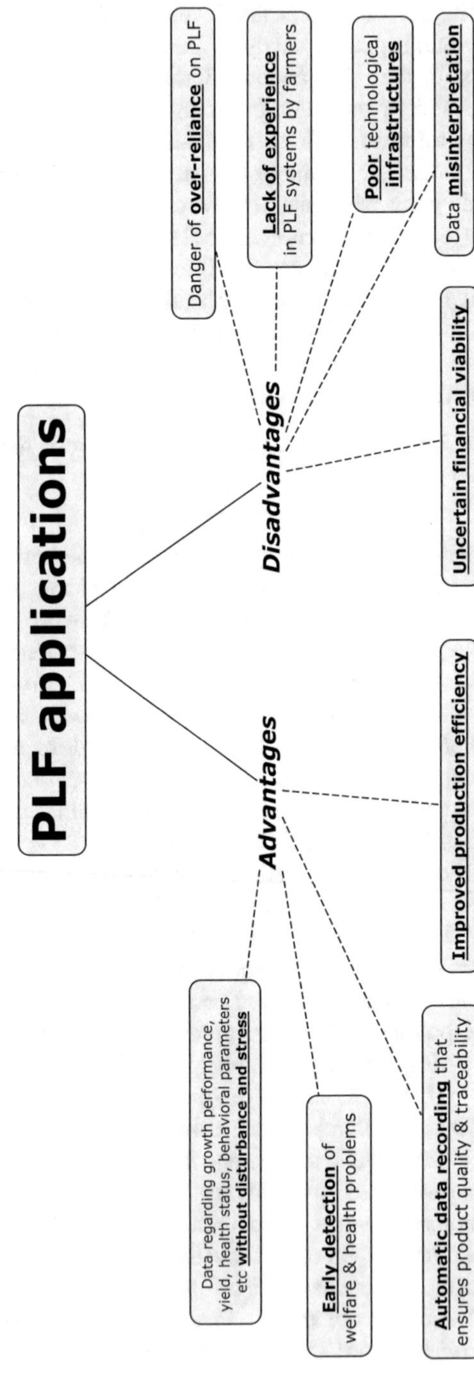

Fig. 1 Potential advantages and disadvantages of PLF applications [adapted by Wathes et al. (2008)]

applications in poultry sector are reported, such as RFID-based technologies and several sensors for sound, video and image processing.

2 PLF Advancements in Dairy Sector

It is expected that by 2050, the average consumption of milk and dairy products will increase by approximately 50% (Thorton, 2010). In addition, milk yield and quality requirements are constantly rising, resulting in higher energy demands per unit and in less time for the farmer to spend with individual animals (Berckmans, 2014a; Mottram, 2016; Norton et al., 2019). These changes directly affect the relationship between the animals and humans, animal health and welfare status, and the consumers' perception of food and feed safety and the units' sustainability (Zucali et al., 2020). PLF may be the answer in minimising or even solving these problems.

PLF aims at the fully automated continuous monitoring of individual animals by utilising technological and computer power as part of the production process (Berckmans, 2014b). Each individual animal is considered as a main gear of the system providing all the information needed for the system's processes (Berckmans, 2014b). PLF systems' function is mainly based on monitoring the animals' behaviours (e.g., feeding, drinking, lying, etc.) and behavioural changes due to external factors such as housing conditions (e.g., temperature and relative humidity variations, air flow, solar radiation), or biological changes (e.g., oestrus, calving, diseases) that greatly affect the animals' health and well-being. When behavioural changes as such are detected, the system triggers a warning signal enabling the farmer to take immediate action, leading to early problem solution or immediate housing practices assessment (Berckmans, 2017). At the same time, the farmer can monitor the animals' everyday lives despite the size of the herd (Norton et al., 2019). Therefore, it can potentially improve animal health and welfare, quality and quantity of the end product and enhance the economic viability of the unit.

Various PLF systems have been introduced in the past decades, from collecting and analysing data either at group level or monitoring individual animals using static, moving or animal-mounted sensors (Rutter, 2012). Stygar et al. (2021) reported that 129 commercial applications from 67 different providers are available in the dairy industry today. These technologies include:

1. Milking robots, accelerometer-based for monitoring health status
2. Load cells systems combined with RFID technologies (i.e., collars, leg, ear, and halter mounted sensors) for movement tracking, individual identification, and lameness detection
3. Boluses used for body temperature monitoring, pH analysis, rumen activity and individual identification
4. Camera-based systems for body temperature monitoring (i.e., thermal cameras), automatic body condition scoring and monitoring feeding and drinking behaviour
5. GPS sensors for activity monitoring

6. Sound analysis systems
7. Mobile applications for body condition scoring and weight estimation

However, they reported that only 14% of the commercially available systems have been validated externally, undermining their credibility.

2.1 Milking Robot: Automatic Milking Parlor

PLF revolutionised the milking process with the introduction of automatic milking robots. These automatic milking machines utilise the higher robotic technological achievements achieving better quality and increased quantity of milk, while at the same time improve the welfare of the farms (John et al., 2016). The whole management process has been changed since the introduction of the milking robots. The milking process is no longer performed within specific time intervals during the day, but each individual cow can choose the preferable time of being milked (Kuipers & Rossing, 1996).

In the early years of milking, robot machines were mostly used for indoor farming. However, automatic milking systems have been developed for semi-grazing and pasture-grazing farms. Ketelaar-de Lauwere (1999) introduced the pasture-based automatic milking robots and Greenall et al. (2004) and Woolford et al. (2004) developed similar systems for fully grazing farming units. Clark et al. (2015) reported that the interest for such systems will be increased within the next decade as existing management principles are followed, achieving high levels of pasture utilisation. However, Gargiulo et al. (2020) reported that the adoption rates for pasture-based systems have been considerably lower than expected. This trend could be attributed to the fact that these systems are still either under development or in need of modification regarding the maintenance and safety processes, improved pasture and robot utilisation, and thus increased profitability (John et al., 2016; Gargiulo et al., 2020).

2.2 Automatic Oestrus Detection

Early oestrus detection is a labor demanding and time-consuming process in today's dairy farms. Resumption of oestrus cycle after calving is essential for cows' fertility (Mayo et al., 2019). However, the difficulty of the process is constantly increasing (Dobson et al., 2007; Fricke et al., 2014) as cows demonstrate less intense symptoms and for shorter time (i.e., as short as 3 h; Roelofs et al., 2005). Boyd et al. (2004) reported that in 1992 cows showed active ovaries at a rate of 95%, whereas Lopez et al. (2005) reported a decreased rate of 71% in 2005. Mottram (2016) reported that 32% of the fertility cycles were not recognised, and a 5–21% of the cows were not inseminated properly as a result of both the introduction of high milk yielded

Holstein breeds and the limited time and tools for the farmers to detect oestrus. Holman et al. (2011) suggested that a preferable practice should be to follow a five 30-min observation period programme resulting in 86% oestrus identification rate. These demands add human labor, which is in shortage in today's units (Thorton, 2010; Van Hertem et al., 2017), resulting in an increase of the total cost of the unit. PLF applications could provide solutions as they have been successfully tested in the past with promising results concerning the fertility cycle detection rates instead of the visual observation methods (Mayo et al., 2019; Michaelis et al., 2014; Stevenson et al., 2014).

PLF oestrus detection systems found in the literature include heat mount detectors (Stevenson et al., 1996; Hempstalk et al., 2013), pedometers (Roelofs et al., 2005; Peter & Bosu, 1986; Koelsch et al., 1994; Kamphuis et al., 2013) and animal-mounted detectors (Lopez et al., 2005; Michaelis et al., 2014; Stevenson et al., 1996, 2014; Hempstalk et al., 2013; Kamphuis et al., 2012; Haladjian et al., 2018), infrared sensors (Marquez et al., 2019, 2021), camera-based systems combined with a variety of motion tracking, temperature and pump pressure sensors and biosensors (Mayo et al., 2019; Hempstalk et al., 2013). However, these systems are still under research and only a limited number of commercial applications for automatic oestrus detection are available. The Herd Navigator (HN) is one of the most common applications used in today's units and combines automatic sampling of milk and complex five-point sensing systems for more holistic analysis. This system has shown great potential in oestrus detection (Samsonova et al., 2015; Gaillard et al., 2016; Yu & Maeda, 2017) and successful pregnancy rates (Gaillard et al., 2016). However, there are only a few published scientific reports validating the system and further research is needed for more concrete and wholistic conclusion extraction about the system's economic beneficial impact. Furthermore, a considerable amount is required for the installation costs, thus this PLF application is not feasible for small units (Yu & Maeda, 2017).

2.3 *Health/Welfare Disorders Identification*

Lameness and mastitis are the main health issues of the dairy cattle industry (Booth et al., 2004). They negatively affect the health and welfare of the animals and produce added labor on the unit's everyday processes thus, degrading the economic viability of the unit (Cha et al., 2010). The economic loss can be traced in decreased milk yield (Hernandez et al., 2002; Onyiro et al., 2008; Van Hertem et al., 2013), reduced reproductive performance (Morris et al., 2011; Peake et al., 2011), and increased culling risk (Booth et al., 2004). PLF monitoring and early detection systems could reduce the impact or even eliminate these problems (Mottram, 2016).

2.3.1 Lameness

Lameness is among the top three health-related causes of economic loss in dairy industry. Lameness results in reduced mobility, milk yield, reproductive ability, loss of body condition and intense pain (Morris et al., 2011). To date, visual observation is the most common technique the farmers use to detect the disease. However, it is a time-consuming, labor-demanding process, sensitive to variations between the observers, and on many occasions, it is omitted resulting in economic degradation of the units due to production losses and increased treatment costs (Thomsen et al., 2008). Therefore, early detection of the disease is essential for effective treatment and disease progress prevention.

A variety of promising PLF systems have been reported for assessing the problem including mount detectors (Van Hertem et al., 2013; Kokin et al., 2014; Thorup et al., 2016; Taneja et al., 2020), convolutional neural network (CNN) models (Pastel & Kujala, 2007), camera-based systems (Poursaberi et al., 2010; Viazzi et al., 2013; Romanini et al., 2013; Jabbar et al., 2017; Zhao et al., 2018; Kang et al., 2020; Wu et al., 2020) and IoT sensors (Taneja et al., 2020; Byabazaire et al., 2019). Pastel and Kujala (2007) developed a 4-balance probabilistic Neural Network model based on weekly measurements of leg load (i.e., leg weight pressure) for lameness detection. Viazzi et al. (2013) developed a camera-based model using the population approach. However, they reported that individual monitoring is the most effective way to detect lameness compared with population threshold. In addition, Romanini et al. (2013) developed a model, based on 3D video data analysis. It should be noted that the level of misclassification of the systems leads to multiple false alarm signals thus, further research is needed for the development of a commercial application.

2.3.2 Mastitis

The existence of bovine mastitis can be traced back to at least 3100 BC, the same era as the milking process was discovered (Ruegg, 2017). However, only with the invention of the microscope which allowed the classification of microorganisms was the disease completely analysed. Breed and Brew (1917) first reported the existence of streptococci bacteria in milk. Since then, mastitis has become one of the most important research areas emphasising on its effects on public health (Ruegg, 2017), deterioration of milk quality characteristics (Kitchen, 1981), impairment of animal welfare (Kemp et al., 2008) and negative impact on the economy of the units (Huijps et al., 2008).

During the twenty-first century scientists were able to develop various models and PLF applications including health status evaluation and behaviour monitoring using biometric sensors (Park et al., 2015; Neethirajan & Kemp, 2021) and numerous animal-mounted wireless sensors (Jegadeesan & Venkatesan, 2017), individual identification using camera (Kumar et al., 2015; Kumar & Singh, 2019; Okura et al., 2019; Martins et al., 2020; Pezzuolo et al., 2018a) and thermal camera analysis

(Bewley et al., 2008; Halachmi et al., 2008, 2013) and drones (Andrew et al., 2020), and body weight and body condition score estimation using camera-based monitoring and image processing techniques (Kumar et al., 2015; Kumar & Singh, 2019; Okura et al., 2019; Martins et al., 2020; Pezzuolo et al., 2018a) and ultrasonic sensors (Halachmi et al., 2008), promoting animals' health and welfare and at the same time, allow the farmer to monitor in real-time individual animals (Rosa, 2021). The milking robot introduction to the managerial process of the farms has resulted in the development of a variety of mastitis early detection automatic systems (Jensen et al., 2016). These systems' operation is based on two elements: (a) sensors data collection and (b) development of models for data analysis and alert production combined with decision support/making systems (Rutten et al., 2013). A series of variables is detected, documented, analysed and evaluated (Kamphuis et al., 2013) including milk's electrical conductivity (Cavero et al., 2006; Khatun et al., 2017), milk colour (Song & van der Tol, 2010), lactate dehydrogenase (Friggens et al., 2007), milk yield, body weight, lactose, fat and protein percentages, blood percentage (i.e., the volume of blood in 1 mL of milk) and Somatic Cell Counts (SCC) (Woolford et al., 2004; Jensen et al., 2016; Højsgaard & Friggens, 2010; Marino et al., 2021). Furthermore, various PLF technologies are under development including infrared and thermal cameras (Zaninelli et al., 2018; Zhang et al., 2020), biometric sensors (either invasive or non-invasive) for real-time individual health and behaviour monitoring and blockchain technology (Neethirajan & Kemp, 2021). Farmers today can choose between a variety of systems fitting their individual needs to simplify their everyday processes, cut down the workload hence minimise the labor costs, improve the welfare of the animals, therefore, increase the sustainability of the units. Figure 2 presents the majority of PLF technologies applied in dairy cows farming.

Small ruminants such as sheep and goats are often managed as a herd, allowing only group welfare status to be considered. Innovative technologies are a unique opportunity to monitor and improve welfare management from the farm-level manual to automated or semi-automated assessment and management at an individual level, along the value chain leading to a reduction of on-farm labour and veterinary care costs (Morgan-Davies et al., 2018). Various PLF applications have been developed over the past decades including flock management using the commercially available virtual fence (Fay et al., 1989; Jouven et al., 2011; Brunberg et al., 2015, 2017; Marini et al., 2018a, b; Kearton et al., 2019), flock monitoring using drones and image analysis techniques (Xu et al., 2020; Al-Thani et al., 2020), weight monitoring using complex automatic weight crates (Wishart, 2019), individual tracking using animal-mounted GPS sensors (Beker et al., 2010; Betteridge et al., 2010; Meunier et al., 2015; Virgilio et al., 2018) and tri-axial accelerometer loggers (Burgunder et al., 2018), health status evaluation by analysing data such as heart rate and body temperature collected from implanted sensors (Fuchs et al., 2019), monitoring sexual behaviour using animal-mounted accelerometers (Mozo et al., 2019), grazing and ruminating behaviour monitoring using animal-mounted accelerometers/gyroscopes sensors (Mansbridge et al., 2018) and tri-axial loggers (Barwick et al., 2018), monitoring resting, grazing and explorative behaviours using animal-

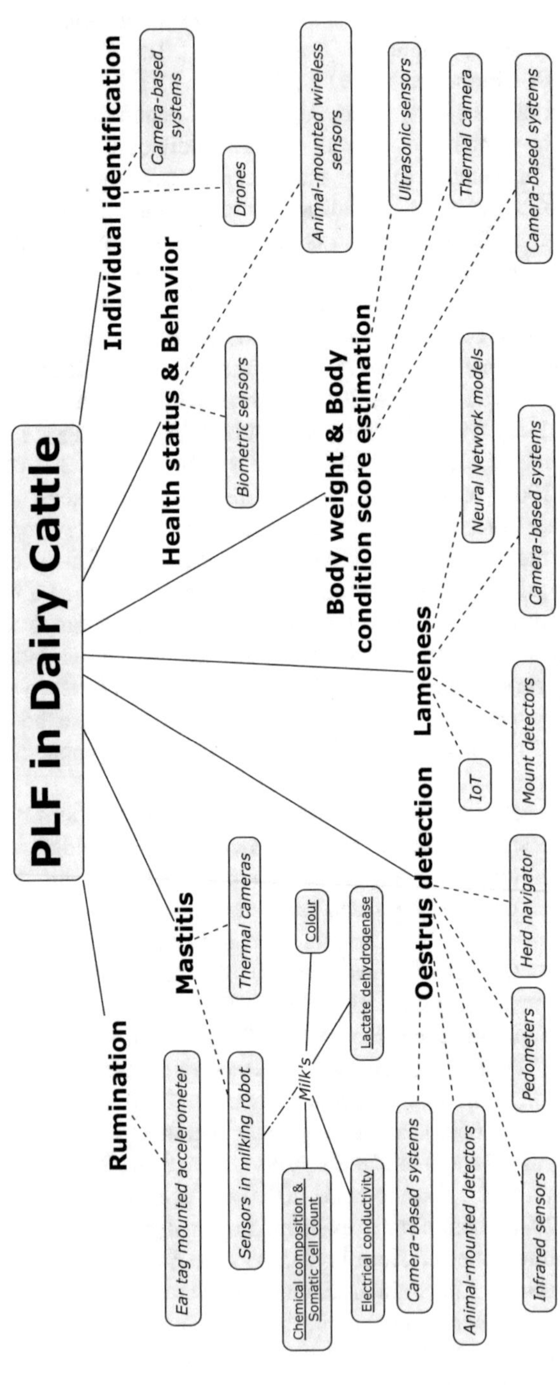

Fig. 2 Application of PLF technologies in Dairy Cattle

mounted tri-axial accelerometer loggers (Barwick et al., 2018), individual identification using injectable transponders (Ribó et al., 2001; Bortolotti et al., 2013), RFID sensors (Morgan-Davies et al., 2018; Voulodimos et al., 2010; Cappai et al., 2018), endoruminal boluses (Cappai et al., 2018; Hentz et al., 2014) and drones—image analysis (Xu et al., 2020), monitoring feeding behaviour using cameras, microphones and animal-mounted gyroscopes (Meunier et al., 2015), animal-mounted accelerometers (Meunier et al., 2015; Alvarenga et al., 2016; Cronin et al., 2016), oestrus detection using the commercially available Alpha-D detector system (Alhamada et al., 2016, 2017), lambing behaviour detection using animal-mounted temperature sensors (Abecia et al., 2019), age identification using sound analysis (Bishop et al., 2017), and monitoring standing and lying behaviour using camera-based analysis and ultrawideband real-time location systems (Ren et al., 2020) and animal-mounted accelerometers (Zobel et al., 2015). However, the fact that small ruminants farming companies are characterised as small and family farms mostly, along with the hesitation in acceptance and adaptation of the new technologies by the farmers, PLF applications are not common and are mainly used for research purposes (Vaintrub et al., 2021). Table 1 presents a variety of research applications in small ruminants.

3 PLF Advancements in the Pig Sector

The considerable increase of raw computer power over the past two decades has allowed researchers to expand their studies in the area of PLF research and increase their perception in problem definition, detection and assessment. It is a fact that PLF area is still under research and commercial PLF systems and decision support control tools are limited. However, new farmers and big companies are interested in adopting technologies as such and pig farms are oriented towards PLF automatisation utilisation (Berckmans, 2017; Tzanidakis et al., 2021).

PLF systems mainly consist of a powerful computer that performs all the calculations, one or more complex dynamic control algorithms based on behavioural analysis and a system of sensors that captures all the data needed. These sensors include cameras, microphones and sound analysis systems, RFID and accelerometers, weight estimation systems, recently introduced Convolutional Neural Network and IoT (i.e., Internet of Things) systems, and any other sensor that can collect data of any parameter of interest within the production process (Berckmans, 2014b, 2017). Various PLF applications have been developed over the past decades including individual monitoring (Lu et al., 2016; Ju et al., 2018; Hansen et al., 2018; Tian et al., 2019; Ahn et al., 2021; Van der Zande et al., 2021), welfare monitoring (Islam et al., 2015; Ni et al., 2017; Zhang et al., 2019), disease detection (Islam et al., 2015; Ni et al., 2017; Zhang et al., 2019; Cornou et al., 2008; Scheel et al., 2015; Cross et al., 2018; Manteuffel, 2009; Maselyne et al., 2016, 2018; Aerts et al., 2005; Wang et al., 2019a; Ferrari et al., 2008), feeding (Cornou et al., 2008; Scheel et al., 2015; Cross et al., 2018; Manteuffel, 2009; Maselyne et al., 2016, 2018; Adrion et al.,

Table 1 Application of PLF technologies in Small Ruminants

Parameter under surveillance	Sensors for data collection	References
Flock management	Virtual fence (i.e., animal-mounted collars embed with electromagnetic transmitters, and ground-installed receivers and sound speakers)	Fay et al. (1989), Jouven et al. (2011), Brunberg et al. (2015, 2017), Marini et al. (2018a, b), Kearton et al. (2019)[a]
Flock monitoring	Drones—Image analysis	Xu et al. (2020), Al-Thani et al. (2020)
Weight monitoring	Automatic weigh-crates	Wishart (2019)
Animal tracking	Animal-mounted GPS sensors	Beker et al. (2010), Betteridge et al. (2010), Meunier et al. (2015), Virgilio et al. (2018)
	Animal-mounted tri-axial accelerometer loggers	Burgunder et al. (2018)
Health status detection	Implanted sensors (heart rate and body temperature)	Fuchs et al. (2019)
Sexual behaviour of rams	Animal-mounted accelerometers	Mozo et al. (2019)
Grazing and ruminating behaviour monitoring	Animal-mounted accelerometer/gyroscope sensor	Mansbridge et al. (2018)
	Animal-mounted tri-axial accelerometer loggers	Barwick et al. (2018)
Resting, grazing and searching behaviours	Animal-mounted tri-axial accelerometer loggers	Barwick et al. (2018)
	GIS systems	Virgilio et al. (2018)
	Animal-mounted GPS sensors	Beker et al. (2010), Betteridge et al. (2010), Meunier et al. (2015), Virgilio et al. (2018)
Individual Identification	Injectable transponders	Ribó et al. (2001), Bortolotti et al. (2013)
	RFID sensors	Morgan-Davies et al. (2018), Voulodimos et al. (2010), Cappai et al. (2018)
	Endoruminal bolus	Cappai et al. (2018), Hentz et al. (2014)
	Drones—Image analysis	Xu et al. (2020)
Feeding behaviour	Camera-based analysis	Meunier et al. (2015)
	Microphones	Meunier et al. (2015)
	Animal-mounted gyroscopes	Meunier et al. (2015)
	Animal-mounted accelerometers	Meunier et al. (2015), Alvarenga et al. (2016), Cronin et al. (2016)
	GPS sensors	Yiakoulaki et al. (2018)
Oestrus detection	Alpha-D detector	Alhamada et al. (2016, 2017)
Lambing detection	Animal-mounted temperature sensors	Abecia et al. (2019)

(continued)

Table 1 (continued)

Parameter under surveillance	Sensors for data collection	References
Age identification	Sound recorders analysis	Bishop et al. (2017)
Standing/Lying behaviour monitoring	Camera-based analysis	Ren et al. (2020)
	Ultrawideband real-time location	Ren et al. (2020)
	Animal-mounted accelerometers	Zobel et al. (2015)[a]

[a]These PLF systems are applied in goats, the rest in sheep

2018; Kashiha et al., 2014a; Stygar et al., 2017; Nir et al., 2018) and drinking behaviour monitoring (Dominiak et al., 2019; Kashiha et al., 2013a; Maselyne et al., 2015; Chen et al., 2020a), weight estimation (Wang et al., 2008; Banhazi et al., 2011; Jun et al., 2018; Lu et al., 2018; Pezzuolo et al., 2018b; Balontong et al., 2020; Meckbach et al., 2021; Zhang et al., 2021), aggressive behaviour detection (Viazzi et al., 2014; Lee et al., 2016; Chen et al., 2019, 2020b; Pandey et al., 2021) and control (Ismayilova et al., 2013; Sonoda et al., 2013), mounting (Nasirahmadi et al., 2016) and resting behaviour monitoring (Shao & Xin, 2008; Kashiha et al., 2013b), activity monitoring and movement detection (Nasirahmadi et al., 2016; Kashiha et al., 2014b; Soerensen & Pedersen, 2015; Xiao et al., 2019a; Li et al., 2019a; Seo et al., 2019; Shi et al., 2019; Oczak et al., 2022), health status monitoring (Soerensen & Pedersen, 2015), monitoring thermoregulatory behaviour (Hrupka et al., 2000) and thermal comfort assessment (Nilsson et al., 2015; Ramirez et al., 2018), growth detection (Lee et al., 2019), inflammation and lesions detection (Ruminski et al., 2007), body temperature monitoring (Xin, 1999; Xin & Shao, 2002; Warriss et al., 2006), lying (Huynh et al., 2005) and agonistic behaviours monitoring (Boileau et al., 2019), pain (Marx et al., 2003; Von Borell et al., 2009; Diana et al., 2019) and stress detection (Schön et al., 2004; Vandermeulen et al., 2015; Da Silva et al., 2019), age and sex identification (Cordeiro et al., 2018), tail biting detection (Wallenbeck & Keeling, 2013; Domun et al., 2019; Liu et al., 2020; Ollagnier et al., 2021; D'Eath et al., 2021), individual liveweight trajectory estimation (Revilla et al., 2019) and reproductive and respiratory syndrome detection (Süli et al., 2017). Table 2 demonstrates various published papers and applications in Precision Pig Farming, addressing a variety of production process aspects and problems.

3.1 *Camera-Based Monitoring*

Camera-based analysis has been the main topic of multiple PLF applications since their introduction into production process. Bio-responses are recorded and analysed (a) in slow-motion (Stygar et al., 2017) or frame-by-frame (Oczak et al., 2013) for individual behavioural analysis (McGlone, 1986) or (b) a faster speed for activity analysis over a specific population of animals (e.g., pen level). PLF researchers and

Table 2 Various applications and used technologies in Precision Pig Farming

Parameter under surveillance	Sensors for data collection	References
Individual monitoring	Cameras	Lu et al. (2016), Ju et al. (2018), Hansen et al. (2018), Tian et al. (2019), Ahn et al. (2021), Van der Zande et al. (2021)
Welfare monitoring	Thermal cameras	Islam et al. (2015), Ni et al. (2017), Zhang et al. (2019)
Disease detection	Thermal cameras	Islam et al. (2015), Ni et al. (2017), Zhang et al. (2019)
	Electronic feeders and various embedded sensors	Cornou et al. (2008), Scheel et al. (2015), Cross et al. (2018), Manteuffel (2009), Maselyne et al. (2016, 2018)
	Microphones and Sound data collection and analysis systems	Aerts et al. (2005), Wang et al. (2019a), Ferrari et al. (2008)
Feeding behaviour monitoring	Electronic feeders and various embedded sensors	Cornou et al. (2008), Scheel et al. (2015), Cross et al. (2018), Manteuffel (2009), Maselyne et al. (2016, 2018)
	Ultra-high-Frequency (UHF) RFID sytem	Adrion et al. (2018)
	Cameras	Adrion et al. (2018), Kashiha et al. (2014a), Stygar et al. (2017), Nir et al. (2018)
Weight estimation	Cameras	Wang et al. (2008), Banhazi et al. (2011), Jun et al. (2018), Lu et al. (2018), Balontong et al. (2020), Meckbach et al. (2021), Zhang et al. (2021)
	Depth sensors	Pezzuolo et al. (2018b)
Drinking behaviour monitoring	Cameras	Kashiha et al. (2013a), Chen et al. (2020a)
	Flow meters	Dominiak et al. (2019)
		Maselyne et al. (2015)
	HF RFID system	Maselyne et al. (2015)
	Cameras	Maselyne et al. (2015)
Aggressive behaviour control	Cameras	Ismayilova et al. (2013)
	Electronic feeders	Sonoda et al. (2013)
Aggressive behaviour detection	Infrared Measurement Equipment (IRME)	Pandey et al. (2021)
	Accelerometers	
	Bluetooth sensors	
	Microphones and Sound data collection and analysis systems	
	Cameras	Viazzi et al. (2014), Lee et al. (2016), Chen et al. (2019, 2020b)
Mounting behaviour monitoring	Cameras	Nasirahmadi et al. (2016)

(continued)

Table 2 (continued)

Parameter under surveillance	Sensors for data collection	References
Resting behaviour monitoring	Cameras	Shao and Xin (2008), Kashiha et al. (2013b)
Activity and movement detection	Cameras	Nasirahmadi et al. (2016), Kashiha et al. (2014b), Xiao et al. (2019a), Li et al. (2019a), Seo et al. (2019), Shi et al. (2019), Oczak et al. (2022)
	Accelerometers	Oczak et al. (2022)
	Infrared Measurement Equipment (IRME)	Soerensen and Pedersen, (2015)
Health status monitoring	Infrared Measurement Equipment (IRME)	Soerensen and Pedersen (2015)
Thermoregulatory behaviour monitoring	Pressure sensors	Hrupka et al. (2000)
	Climate control system	
Thermal comfort assessment	Cameras	Nilsson et al. (2015)
	Thermal cameras	Ramirez et al. (2018)
Growth detection	Cameras and IoT embedded sensors	Lee et al. (2019)
Inflammation and lesions detection	Thermal cameras	Ruminski et al. (2007)
Body temperature monitoring		Xin (1999), Xin and Shao (2002), Warriss et al. (2006)
Lying behaviour monitoring		Huynh et al. (2005)
Agonistic behaviours monitoring		Boileau et al. (2019)
Pain detection	Microphones and Sound data collection and analysis systems	Marx et al. (2003), Von Borell et al. (2009), Diana et al. (2019)
Stress detection	Microphones and Sound data collection and analysis systems	Schön et al. (2004), Vandermeulen et al. (2015)
	Artificial Neural Network	Schön et al. (2004), Vandermeulen et al. (2015)
	Microphones and Sound data collection and analysis systems	Da Silva et al. (2019)
Age and sex definition	Microphones and Sound data collection and analysis systems	Cordeiro et al. (2018)
Tail biting detection	Cameras	Liu et al. (2020)
	3D camera system	D'Eath et al. (2021)
	Flow meters	Larsen et al. (2019)
	Climate control system	
	Photoelectric sensors and artificial environment systems	Domun et al. (2019)
	Electronic feeders and various embedded sensors	Wallenbeck and Keeling (2013), Ollagnier et al. (2021)

(continued)

Table 2 (continued)

Parameter under surveillance	Sensors for data collection	References
Individual live weight trajectory estimation	Electronic feeders and various embedded sensors	Revilla et al. (2019)
Estimation of nest building timing	Accelerometers	Oczak et al. (2019)
Reproductive and respiratory syndrome detection	Implanted microchips	Süli et al. (2017)
	Accelerometers	
	RFID tags	

system manufacturers should bear in mind the internal characteristics of cameras such as resolution quality, effective distance from the lens, focal length, lens type (e.g., fisheye, etc.) to determine the exact number of them that needs to be installed. This is an essential during the experimental design as it greatly affects the economy of the experiment and the data analysis process (i.e., video and image editing and processing). More specifically, even a single camera above the pigpen can capture the full field of view (FoV) depending on pen size and camera specifications (i.e., zoom lens and image stabilisation, focus analysis, frames per second, etc.), and could be efficient enough for all data collection needed and at relatively low installation costs (Berckmans, 2014b). Real-time camera-based monitoring systems have been already applied for investigating and controlling the feeding behaviour and weight estimation (Kashiha et al., 2014a; Stygar et al., 2017; Nir et al., 2018) and weight monitoring (Balontong et al., 2020), detecting aggressive behaviours (Lee et al., 2016; Chen et al., 2019, 2020b; Pandey et al., 2021; Ismayilova et al., 2013), mounting behaviour detection (Boileau et al., 2019), monitoring resting behaviours (Shao & Xin, 2008; Kashiha et al., 2013b), monitoring drinking behaviour (Kashiha et al., 2013a; Chen et al., 2020a), automatic detection and monitoring of individual pigs (Lu et al., 2016; Ju et al., 2018; Hansen et al., 2018; Tian et al., 2019; Ahn et al., 2021; Van der Zande et al., 2021), monitoring the tail posture for tail injuries, health status evaluation and tail biting outbreaks prediction (D'Eath et al., 2021), tail biting behaviours detection (Liu et al., 2020), disease detection (Zhang et al., 2019), monitoring movement and general activity at pen level (Oczak et al., 2022), detection of lesions and inflammation (Ruminski et al., 2007), welfare monitoring (Islam et al., 2015), body temperature monitoring (Xin, 1999; Warriss et al., 2006), and thermal comfort classification (Nilsson et al., 2015; Ramirez et al., 2018). Although there is a growing interest towards the development of PLF applications, it should be noted that the limited datasets used during the behavioural analysis (i.e., usually one or two pens, pigs of a certain hybrid, housed in a single environment, similar age and weight, etc.) and the low recording duration (i.e., 8 hours per day or less, less than ten days per recording session) make the developed systems highly uncertain and too limited for commercial applications, as the real pig housing parameters vary from farm to farm.

3.1.1 Camera-Based Systems Problems

Blob merging is the most common problem in today's camera-based tracking systems (Viazzi et al., 2014; Van der Zande et al., 2021; Lee et al., 2016; Cowton et al., 2019). Illumination variances, environmental changes such as vapour condensation or dust accumulation on the camera lens during the data collection, similar body appearances or body shape deformations, partial or semi-partial occlusion and overlapping are the most common parameters interfering with the video's clarity and increase the noise of the recordings (Lee et al., 2016; Gangsei & Kongsro, 2016; Guo et al., 2017; Zhang et al., 2018; Brünger et al., 2018; Sa et al., 2019). Furthermore, the top-view angle of the camera data limits the ability of the built-in algorithm/software to distinguish individual blobs. This problem especially appears when the pigs get too close to each other, or when they sleep in piles (i.e., one on top of the other)—a behavioural response to cold environments. The software merges two or more individual blobs into one misinterpreting two or more pigs as one and eventually cannot identify which pig is which, failing the main purpose of PLF that is individual monitoring. Thus, today's commercially available tracking software lacks the ability of complete detection of behaviours and the problem of real-time monitoring and complete automatic continuous behaviour detection is yet to be overcome (Tzanidakis et al., 2021).

Another problem in utilisation of PLF systems as support tools is that data presentation varies from system to system in terms of frequency, duration and formats. Farmers rarely have the knowledge to combine and analyse the information and, on many occasions, the already adopted systems are not utilised properly if not at all (Van Hertem et al., 2017; Hartung et al., 2017). It is suggested that PLF manufacturers should simplify their data presentation models and provide short-time educational seminars for the farmers to enhance their ability to use this information.

In recent years, Convolutional Neural Network (CNN) approaches were introduced to assess blob merging problems. Various systems have been published in the concept area such as these of Zhang et al. (Zhang et al., 2018) that describe a software based on three stages of analysis: (a) object detection, (b) multiple objects tracking (MOT) and (c) data association. Moreover, Seo et al. (Seo et al., 2019) and Lee et al. (Lee et al., 2019) models based on combination of infrared and depth informative sensors and a fast CNN object detection technique named YOLO (i.e., "You Only Look Once"). Xiao et al. (2019a) developed a tool to track pigs' movement housed under commercial environmental conditions including no-light, sudden illumination changes, adhesion, and occlusion scenes. On the other hand, an algorithm developed by Li et al. (2019a) was based on analysing the captured data as the combination of two datasets, comprising a) dominant orientation templates and (b) brightness ratio templates. Furthermore, a camera-based system on LabVIEW created by Shi et al. (2019) acquires real-time data for pigs' body components measurement such as body length, body width, body height, hip width and hip height. There are also other systems, i.e., by Hansen et al. (2018) that classifies pigs' face characteristics to identify individual animals based on human face recognition techniques found in the literature, by Shao and Xin (2008) that is a real-time camera-based system for tracking pigs at rest (i.e., as indicator of thermal comfort) and classify their thermal bio-responses in cold, comfortable, and/or warm/hot conditions, by Nasirahmadi et al. (2019) that is a model that detects and classifies the lateral and sternal lying posture of pigs under commercial conditions using two-dimensional images, by Chen et al. (2019) automatically detect aggressive behaviours based on data collected using a camera depth sensor and, finally by Lee et al. (2019) that is a camera-based system using an IoT embedded device to detect individual undergrown pigs. It should be noted that the above applications are still under research and further research is needed under various environmental and housing conditions and different pig hybrids for the development of a robust commercial application.

3.2 *Weight Estimation*

Pig weight trajectory is one of the most important indicators of feed efficiency and feed conversion rate, growth and health status of the animals (Wang et al., 2008). The most common method of weighing the pigs is with the use of individual ground scales and measuring sticks (Zhang et al., 2021). It is generally, a labor intense and time-consuming process that induces stress for the animals and increases the possibility of injuries for both the workers and the pigs and thus, negatively affects the welfare of animals and the efficiency of the unit. Over the past three decades, researchers in the PLF area are focused in developing automatic systems that measure weight without interfering with the pigs' everyday lives while, at the same time minimise the workload for the workers. According to Zhang et al. (2021), there are four methods of weight measuring automatically:

(a) The projection method, where a slide with grinds is projected on the back of the animal and weight is estimated by calculating the shoulder height and area using the principle of stereo projection. It is stated that this method is not automated.
(b) The two-dimensional image method, where top-view 2D images captured are analysed and weight is estimated by multiplying the measured body size with constants related to the camera's specifications and physiology of the animals with an average error of 3.38%–5.3% (Jun et al., 2018; Lu et al., 2018; Schofield et al., 1999; Yang et al., 2006; Kollis et al., 2007; Liu et al., 2013).
(c) The three-dimensional method, where a depth sensor captures 3D images and body size and shape is extracted. Then the same process as in (b) method is used to estimate the animal's weight with an average absolute error of 1.44–5.81% (Pezzuolo et al., 2018a; Kongsro, 2014; Wang et al., 2018; Fernandes et al., 2019).
(d) The ellipse fitting method, where an ellipse is fitted on a back image of the pig and body weight is estimated based on the parameters of the fitted ellipse with an

average relative error of 3–3.8% (Kashiha et al., 2014a; Wang et al., 2008; Banhazi et al., 2011).

All methods use the same PLF principles of image processing and data analysis including background subtraction, enhancement, binarization, noise filtering and head and tail removal (Zhang et al., 2021). This process is time-consuming and, in many cases, fails to comply and provide concrete results posing a great challenge for the development of a commercial application. It should be noted that an alternate method was introduced by Psota et al. (2019), where a camera-based system detects several body parts of the pigs (i.e., both ears, the back right between the shoulders and the tail) in different pen locations. However, although this method demonstrated an improved ability to detect individual animals compared with other systems, the images used to create the dataset were not a continuous timeline but fixed.

3.3 *Thermal Analysis Technologies and Thermal Comfort Assessment*

Pigs tend to change their postural behaviour by regulating heat loss to achieve the minimum deviation from their thermal comfort zone (Xin, 1999; Xin & Shao, 2002; Ye & Xin, 2000). Thermal comfort zone is the effective environmental temperature within which pigs' optimal health status and genetic potential performance are guaranteed (Baker, 2004; Mitchell, 2006; Andersen et al., 2008). Pigs do not pant effectively, and their sweating rate is very low, namely 30 g/m^2 h (Ingram, 1965). Therefore, they lack the ability to adapt to extreme environmental variations (Huynh et al., 2005). Continuous exposure to such climate discomfort negatively affects feed efficiency and growth performance (Renaudeau et al., 2012), increases the frequency of aggression events and undesirable abnormal behaviours such as belly nosing, ear and tail biting (Geers et al., 1989) and stress levels (Morgan & Tromborg, 2006), induces huddling behaviour (Hillmann et al., 2004a), vocalization (Hillmann et al., 2004b) and undermines the stability of hierarchy within the pig pen (Edwards, 2008). Therefore, thermal camera-based PLF systems integrating thermal (dis)-comfort behavioural responses as part of their climate control protocol, could improve the welfare and productivity of pigs, optimise energy/feed resources consumption, and therefore, enhance the economy and sustainability of pig barns (Berckmans, 2014b; Brown-Brandl et al., 2013).

The measurement of rectal temperature is the method commonly used to measure pigs' core body temperature under commercial conditions (Dewulf et al., 2003). This is a stress and labor intense method for both pigs and workers (Godyn & Herbut, 2017), increasing labor costs and compromising the welfare thus, negatively affecting the unit's economic stability (Zhang et al., 2019). Researchers in this area are working towards the development of systems that can efficiently and automatically regulate climate conditions near to thermal comfort zone without interfering with pigs' everyday lives. Various systems and methods have been published including the use of floor pressure sensors to detect pigs' thermoregulatory behaviour such as dispersion within the pig pen (Hrupka et al., 2000), the use of CCTV cameras (Nilsson et al., 2015) and thermal cameras (Ramirez et al., 2018) for thermal comfort assessment, the use of thermal cameras for body temperature monitoring (Xin, 1999; Xin & Shao, 2002; Warriss et al., 2006), detection of lesions and inflammation (Ruminski et al., 2007), lying (Huynh et al., 2005) and agonistic behaviours (Boileau et al., 2019) monitoring, welfare monitoring (Islam et al., 2015; Ni et al., 2017), disease detection (Zhang et al., 2019), the use of infrared sensors to detect aggressive behaviours (Pandey et al., 2021) and health status evaluation (Soerensen & Pedersen, 2015). It is evident that a computer-based system able to accurately analyse pig bio-responses or any parameter of interest, in real-time and use them as input for optimal environmental control would improve the welfare, performance and the sustainability of today's units. For a process that data collection is performed by thermal cameras, the angle of view of the camera is of great importance in an effort to minimise temperature measurement errors (Jiao et al., 2016). However, usually there is inadequate equipment in field research and limited knowledge of its functions (Soerensen & Pedersen, 2015), although this research area has already demonstrated very promising results suggesting that soon could be in the centre of PLF research and commercial on-farm applications.

3.4 Sound Surveillance Technologies

Pigs tend to coordinate and synchronise their behaviour in space and time and even one bark alarm may make the whole group or even all the groups in a chamber to freeze and attend towards the sound source (Berckmans, 2014b; Aerts et al., 2005; Talling et al., 1998; Marchant-Forde et al., 2001; Špinka, 2009). Furthermore, vocalizations and screams are behavioural expressions among pigs that conveying information about their current health and welfare status (Vandermeulen et al., 2015; Hillmann et al., 2004b) and allowing the farmers in early detection of problems such as disease outbreaks or aggressiveness escalation. PLF applications in the area of sound surveillance and analysis vary including cough and screams classification (Chung et al., 2013), cough analysis for disease detection (Aerts et al., 2005; Ferrari et al., 2008; Van Hirtum & Berckmans, 2002; Silva et al., 2009) and housing environment quality evaluation (Wang et al., 2019a), aggressive behaviour detection (Pandey et al., 2021), pain (Marx et al., 2003; Von Borell et al., 2009; Diana et al., 2019) and stress detection (Schön et al., 2004; Vandermeulen et al., 2015; Da Silva et al., 2019; Cordeiro et al., 2018), and age and sex identification (Cordeiro et al., 2018). Very promising results were reported by Spensley et al. (1995) and Talling et al. (1996) indicating that pigs' defense mechanisms were activated using a series of sounds varying in nominal intensity and frequency. In addition, Li et al. (2019a) reported that pigs showed different bio-responses in various types of music preferring mostly either music played by string instruments at a slow tempo (i.e., 65 bpm) or from wind instruments at a fast tempo (i.e., 200 bpm). Von Borell et al. (2009)

developed a tool based on the classification of three different classes of piglet vocalizations (i.e., grunting, squealing and screaming). They found that vocalization analysis in pigs can help identify both pain and behavioural changes indicating that a system that combines both camera-based and audio analysis could potentially improve contemporary PLF systems. Therefore, a PLF tool that analyses pigs' vocal behaviours and automatically detects pain could potentially be developed, providing the farmer with an early warning signal and/or helping for an early solution, prior to escalation.

The main problem with sound surveillance and analysis systems is the multi-factorial origin of pigs' bio-responses in different sounds along with the high statistical errors produced by high noise within the collected datasets. For example, Aerts et al. (2005) reported that the labelling sound of the system intrigued the pigs to start coughing voluntarily to listen to this particular sound, leading to experimental errors. Furthermore, Wegner et al. (2019) reported that age, sex, floor type (e.g., fully slatted or partly slatted), feed quantity and quality, feed type (e.g., liquid or dry feed), ventilation systems and the use of bedding materials provide different sound datasets. Therefore, each housing system is unique and should be treated exclusively (Tzanidakis et al., 2021).

As in all PLF research experiments, a reference set of data is a prerequisite for the development an automatic scream classifier (Vandermeulen et al., 2015; Hemeryck & Berckmans, 2015). Audio labelling the captured data should be carried out by a single observer (Chung et al., 2013; Guarino et al., 2008) and then used to build the datasets. It should be noted that even a low-cost microphone can be used to collect the data further reducing the system cost (Chung et al., 2013). However, although research results provide evidence that utilisation of certain pigs' behaviour elements through sound analysis is feasible, and that it will be an essential part of future PLF systems, more experimentation is needed under different commercial housing conditions and hybrids to improve the PLF evaluation parameters and minimise false alarms.

3.5 RFID Technologies, Electronic Feeders and Drinkers, and Various Multi-sensor Combined Systems

Numerous sensors have been used to monitor a variety of parameters of interest within a pig barn such as radio-frequency identification (RFID) chips (Adrion et al., 2017) and depth sensors (Kim et al., 2017) along with more complex systems including neural networks analysis and, electronic feeders and drinkers (Berckmans, 2017). The series of related studies and research conducted in this area is discussed in the following text.

3.5.1 Electronic Feeders and Drinkers Combined with Various Systems

Electronic feeders and drinkers are currently being used by breeding companies around the globe to test feeding behaviour, growth, and performance of their breeds as data collected from these systems is usually very accurate. Eissen et al. (1998) suggested that frequent chacking and correction of feeding stations function for certain recording periods and frequent maintenance, may improve the system's viability by reducing the errors occurred when a feeding station is not functioning optimally. Cornou et al. (2008) reported that data collected from electronic sow feeders could be used in oestrus, lameness, and health disorders detection. However, this method demonstrated a high number of false alarms and further research is needed for more concrete results. Furthermore, when combined with other sensors, they have shown great potential in monitoring pigs' social behaviour (Hoy et al., 2012). Systems as such have been adopted already within the production process of many international companies around the globe and are being used for a variety of purposes such as disease detection, feed intake documentation and feed control and meat quality estimation (Wallenbeck & Keeling, 2013).

Manteuffel (2009) used a novel approach for active feeding control. The pigs were trained to respond to individual jingles produced by Call Feeding Stations (CFS) installed in each pen. The experimental groups showed improved meat quality, lower frequency of appearance of abnormal behaviours (i.e., belly nosing and healing of skin lesions) compared with the control groups, while growth performance was not affected. Furthermore, fighting events frequency was generally minimised, and the pigs demonstrated improved health and welfare and reduced stress levels. Wallenbeck and Keeling (2013) developed a model that predicts future tail biting victims 9 weeks before the first visual tail injuries, by analysing their feeding behaviours (i.e., daily frequency of feeder visits and daily feed intake) from data collected from electronic feeders. However, data analysis in both studies was based on a single farm and thus, further research is needed for the development of accurate commercial applications.

Revilla et al. (2019) monitored piglets during the post-weaning period (i.e., initial age of 28 days). They developed a dynamic model that estimates the individual live weight trajectory based on data collected from electronic feeders and provides information on individual amplitude and length of perturbation caused by weaning and the dynamic of animal recovery. This model could be used to develop a PLF application that will provide the farmer information on each pig's weight dynamic traits hence, which pig requires special feeding treatment. However, hybrid, housing conditions and management practices should be available as selection buttons for the producers to insert.

Maselyne et al. (2016) used High-Frequency RFID tags on each ear of individual pigs and successfully detected pigs' individual visits at the feeder. In addition, Maselyne et al. (2018) used the same method and developed a system that detects the feeding patterns of individual pigs by analysing their variables such as the number of trough visits, the duration, and the average interval between the visits.

However, the main question asked by the producers is the units' gain from this method. Therefore, further research is necessary to improve its practicality. Cross et al. (2018) developed a model based on finishing pigs' feeding behaviour to achieve automated disease detection, like pneumonia outbreaks, at an early stage. Scheel et al. (2015) aiming to develop an on-farm application with an early warning system, proposed a method to automatically detect sow lameness at early stages by combining data collected from accelerometers placed in ear tags and electronic feeders. It should be noted that the RFID ear tags used in the industry need extra time for data collection, have limited range for efficient measurement (i.e., a maximum of 120 cm) and at the same time may negatively affect the welfare serving as an extra stressor for the pigs (Maselyne et al., 2014). A combination of the RFID tag methods (Jun et al., 2018; Lu et al., 2018; Pezzuolo et al., 2018b) could possibly provide indication for feed conversion estimation in growing-fattening pigs.

3.5.2 Other Systems and Sensors

Under commercial housing conditions, sows lack the ability to express the natural behaviour of nest building as they are confined in farrowing crates preventing them from crushing the piglets resulting in welfare degradation. Oczak et al. (2019) used accelerometers placed in ear tags and successfully predicted the starting and the ending of nest building behaviour. However, the model was based on recorded data and was not applied in real-time thus, further research is needed for the development of a commercial application.

Maselyne et al. (2015) developed a system to monitor drinking behaviour of individual pigs using high-frequency RFID technology. HF RFID sensors were installed around the nipple drinkers and on ear tags, whereas flow meters were installed before each nipple drinker. The number and duration of the drinking bouts were overestimated by 10 and 19%, respectively, therefore further research is needed for a PLF signal warning application as an early problem detection system. Süli et al. (2017) used implanted transponder microchips and accelerometers placed on individual leather collars to detect reproductive and respiratory syndrome virus disease. RFID tags were used to collect measurements of the transponders. However, further research is needed to test whether this combination of sensors affects pigs' welfare, performance, and quality of the end product. Adrion et al. (2018) tested an ultra-high frequency (UHF) RFID system for its efficiency in tracking pigs' trough visits using cameras, but the ear tissue negatively affected the performance of system and led to inaccuracies in their results. Finally, Domun et al. (2019) developed three dynamic models and used a neural network to detect behavioural changes such as tail biting, diarrhoea and fouling under various environmental conditions. They reported that the results were promising, but further research is needed to improve the model's sensitivity and specificity. It is worth noting that the data were collected by educated staff and this may be a problem if applied in commercial units as the staff, although experienced in everyday practice, lacks skills in conducting experiments. A combination of camera and a sound-based system would be more appropriate for continuous monitoring.

4 PLF Advancements in the Poultry Sector

The poultry sector represents the most highly selected livestock production sector mainly due to the birds' short production cycle (i.e., within 40 days or less from hatch to slaughter weight approximately) and their worldwide popularity (Knowles et al., 2008; Wilhelmsson et al., 2019). Furthermore, broilers represent the greatest population of farmed animals worldwide in total numbers with a standing population of 22.7 billion (Bennett et al., 2018). OECD (2021) report on worldwide meat consumption clearly demonstrates a constant increase in poultry meat consumption of up to 70.5% between 1990 and 2017 and it is expected that by 2050 the global demand will be over double compared with 2005 and the chicken eggs demand will be 40% greater (Smith et al., 2015). However, due to the low profit per animal for the farmers and the massive numbers of animals per flock, high stocking densities ranging from 30 to 40 kg of live weight per m^2 (Meluzzi et al., 2008), the welfare of individual animals is compromised (Rowe et al., 2019). Furthermore, this constant increase in demand, the public concern about poultry welfare (Li et al., 2019b), along with the more and more aggressive competitiveness of the agriculture-related sectors (Godfray et al., 2010) potentially affects the profitability and sustainability of the poultry production units (Astill et al., 2020). The intensification of the production process resulted in a considerable increase in frequency of appearance of health-related problems such as sudden death syndrome, ascites, lameness and contact dermatitis (Bessei, 2006; De Jong et al., 2016) up to 400% since the 1960s (Zuidhof et al., 2014).

PLF focuses at increasing quality and production through the improvement of animal health and welfare (Rowe et al., 2019) thus it is invaluable for the poultry sector as it can be applied in various parts of the production process such as improving the quality of the environment, the feed conversion rate and control feed intake, along with enhancing the welfare of the units (Astill et al., 2020). The multi-factorial origin of parameters influencing animal welfare such as stocking density, climate conditions and thermal stress, unsuitable social environments, and difficulty in accessing resources, are causes of stress leading to performance and welfare deterioration (Meluzzi & Sirri, 2009; Tactacan et al., 2009). However, the common commercial practices of housing such as vast numbers of housing hens and high stocking densities combined with inadequate farm equipment, resulting in great difficulty for continuous monitoring and management of single animal welfare (Sassi et al., 2016). Furthermore, there are many indicators for comprehensive welfare assessment such as behavioural responses and stress and performance indicators. These indexes interact together on many occasions making the evaluation difficult and labor and time consuming (Li et al., 2019b). It is though possible to monitor a proportion of birds and collect information that will be used as inputs for assessing flock health and welfare (Astill et al., 2020).

The past decade there has been a great increase of interest in research and farmers seem to be eager to invest and adopt technologies to improve performance, quality, and welfare of today's units (Rowe et al., 2019). A variety of sensors has been implemented in PLF poultry research area including cameras and vision and image analysis systems (Reiter & Bessei, 2010; Dawkins et al., 2012; De Montis et al., 2013; Fernández et al., 2016; Mortensen et al., 2016; Aydin, 2017a, b; Zhuang & Zhang, 2019), microphones and sound analysis systems, temperature and humidity monitoring sensors such as thermometers, infrared sensors and thermal cameras (Pereira et al., 2006; Nääs et al., 2010; Nascimento et al., 2011; Giloh et al., 2012; Masey-O'Neil et al., 2014; Shen et al., 2016; Abreu et al., 2017; Knížková et al., 2018; Xiong et al., 2019; Bloch et al., 2020) and RFID technologies (Pereira & Nääs, 2008; Zhang et al., 2016; Taylor et al., 2017; Wang et al., 2019b; Van der Sluis et al., 2020; Zukiwsky et al., 2020, 2021; You et al., 2021a, b). Figure 3 presents a series of published papers describing relevant experiments conducted over the past decades in the poultry sector.

4.1 *Sound Analysis Systems*

Animal vocalizations analysis can be a useful pool of information concerning their welfare as social interactions, alarm signals, health status and behavioural needs data can be extracted and used as protocols for automatic management and monitoring systems (Manteuffel et al., 2004). Furthermore, systems used in this area can collect all the data needed such as pecking sounds, vocalizations and environmental noises, without meddling with the birds' everyday activities and interactions (Li et al., 2019b). Various systems can be found in the literature assessing a variety of important production parameters including monitoring growth rate and feeding behaviour (Aydin et al., 2014, 2015; Aydin & Berckmans, 2016; Huang et al., 2021; Fontana et al., 2015, 2016, 2017), disease detection and health status monitoring (Sadeghi et al., 2015; Banakar et al., 2016; Raj & Jayanthi, 2018; Huang et al., 2019; Carpentier et al., 2019; Mahdavian et al., 2021), activity monitoring (Mahdavian et al., 2020), sex and genetic strains identification (Pereira et al., 2015), and different vocal traits detection for special feed rewards (McGrath et al., 2017). Furthermore, Lee et al. (2015) developed a model that detects chickens under heat or fear stress automatically under commercial conditions based on sound analysis. The average stress detection and classification accuracy were reported to be greater than 96% indicating that this system can be adopted as an environmental support control tool in today's units, improving welfare and thus, performance and profitability of the units. It is evident that sound analysis is a very useful tool when implemented in PLF control and early warning systems. However, although there are a few commercial applications, to date it is impossible to achieve the basic concept of PLF that is individual welfare assessment.

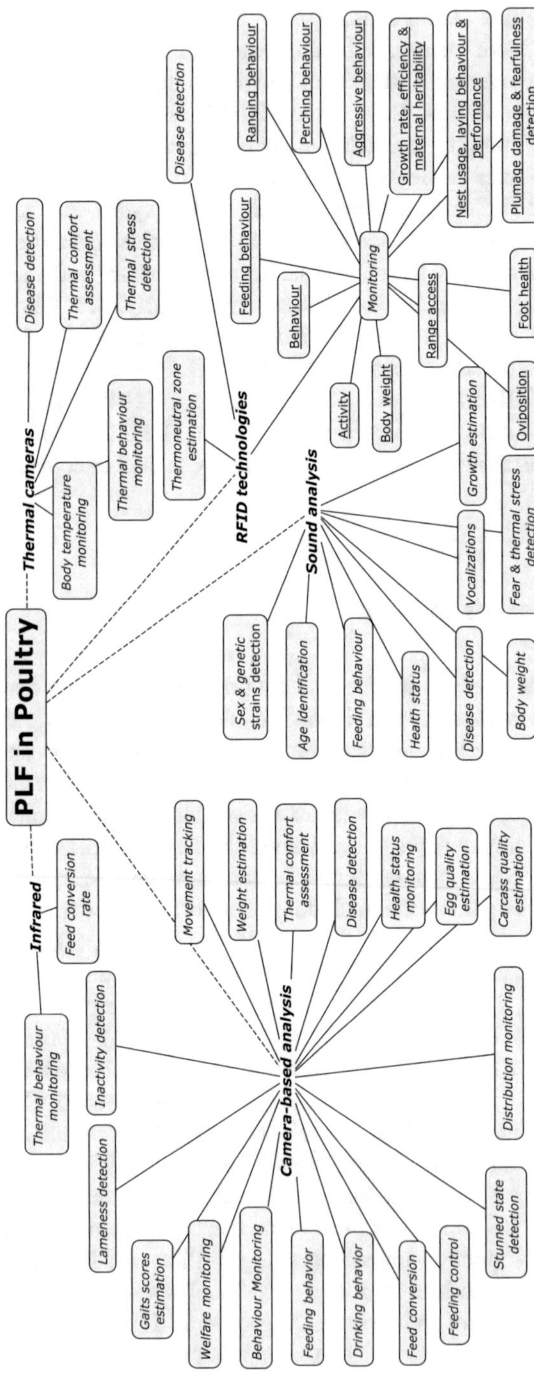

Fig. 3 PLF applications in the poultry sector

4.2 RFID-Based Technologies

The main area of focus in RFID technologies is to detect individuals' location and activity remotely (Li et al., 2019b). Furthermore, an RFID microchip can be injected or mounted on individual birds and with the collected data nesting, feeding and drinking, perching and ranging behaviours can be monitored continuously (Yu & Huang, 2018). This information can be analysed and utilised for health and welfare status monitoring and evaluation (Siegford et al., 2016). In addition, Wiltschko et al. (2015) and Stadig et al. (2018) reported that this technology has minor, non-persistent effects on growth, behaviour and performance in chickens implying that RFID technology wearables are animal friendly and can be used for behavioural research studies focused on welfare measures and production parameters monitoring. In concept, RFID technology has found applications in various PLF research areas including real-time nest usage, laying behaviour and performance monitoring of individual hens (Chien & Chen, 2018; Ringgenberg et al., 2015), individual activity monitoring (Van der Sluis et al., 2020, 2021) and early disease detection (Zhang et al., 2016; Raj & Jayanthi, 2018; Dogra et al., 2010; Ammad-Uddin et al., 2014; Roy & Sarkar, 2016; Ahmed et al., 2021), environmental preference for laying hens monitoring (Sales et al., 2015), feeding behaviours monitoring (Li et al., 2017, 2018, 2021; Oliveira et al., 2019), growth monitoring and maternal heritability of body weight estimation (Zukiwsky et al., 2021), ranging behaviour monitoring as welfare indicator (Taylor et al., 2017; Durali et al., 2014; Campbell et al., 2017, 2018a, b; Taylor et al., 2018, 2020), individual perching monitoring as health status indicator (Wang et al., 2019b), predicting daily oviposition events for individual egg laying (You et al., 2021b), hen's egg-laying timing, position and egg production status monitoring (Lin et al., 2019), monitoring range access as indicator of plumage damage and fearfulness (Hartcher et al., 2016), foot health and spatial cognition (Campbell et al., 2016; Larsen et al., 2018) and range use monitoring as related with weather conditions (Sztandarski et al., 2021). However, although there is great interest in research for the utilisation of RFID technologies, only a few RFID applications are available commercially compared with dairy and pig farming. This is probably due to the physical characteristics of birds (i.e., small size and light weight) which require large amounts of on-bird mounted microchips that greatly increase the installation costs (Li et al., 2019b).

4.3 Video and Image Processing: Camera-Based Analysis Systems

Video and image analysis has been widely used in PLF research since the beginning of PLF for animal behaviour monitoring. As in the other animals, this technology combines the raw power of a computer, behavioural models (i.e., software and algorithms) and all the information gathered from the monitoring system for data analysis. The developmental philosophy of the system is exactly the same as described in the third paragraph of the Introduction, where the sensors collect the data, and the computer analyses the parameters of interest based on a previously designed model/software, providing signals or managerial suggestions for the producer to act and prevent any problem prior its appearance or at an early stage, promoting welfare, quality and performance. Furthermore, the same problems in video and image analysis as described in Sect. 3.1.1. chapter are to be detected during data collection.

Various sensors have been used in the past including common CCTV cameras, IR cameras and infrared sensors and depth sensors (Li et al., 2019b). It is a non-invasive, continuous, and automatic process used for health and welfare monitoring, tracking individual animals and weight estimation, and adjusting environmental conditions according to the preferences of the animals, at a considerably low purchase and installation cost. More specifically, technologies as such include precision feeding (Mehdizadeh et al., 2015), monitoring breeder behaviour for welfare assessment (Fernández et al., 2016; Pereira et al., 2013), monitoring feeding behaviour as a result of the type of the feeder (Neves et al., 2015) and feed composition (Neves et al., 2019), movement and activity tracking (Fujii et al., 2009; Chao et al., 2014; González et al., 2017; Fang et al., 2020), continuous monitoring for early detection and prediction of diseases (Zhuang et al., 2018; Okinda et al., 2019) and health status identification (Raj & Jayanthi, 2018; Xiao et al., 2019b), monitoring activity and special density for thermal comfort assessment (So-In et al., 2014), monitoring activity for broiler gait scores prediction (Van Hertem et al., 2018), detect stunned stated of broilers (Ye et al., 2020a, b), monitoring distribution of the animals for problems detection (Kashiha et al., 2014c; Novas & Usberti, 2017) and using a 3D camera-based system for lameness assessment (Aydin, 2017a, b) and weight estimation (Mortensen et al., 2016; Amraei et al., 2018). It should be noted that a considerable amount of research has been published in the area of carcass' (Chmiel et al., 2011; Geronimo et al., 2019; Kaswati et al., 2020) and eggs' (Omid et al., 2013; Firouz & Omid, 2015; Sun et al., 2017; Narin et al., 2018; Alon et al., 2019; Mota-Grajales et al., 2019; Guanjun et al., 2019; Narushin et al., 2020; Okinda et al., 2020; Nasiri et al., 2020) quality estimation. However, although it is evident that there is a considerable amount of research conducted in the area, the main issues of PLF technologies that prevent commercial application development (i.e., non-continuous individual identification) are still present.

5 Conclusions

In conclusion, PLF techniques and technologies have been commercially developed in the dairy sector with the introduction of automatic cow milking systems. A variety of sensors has been used on the milking machine ever since, providing information concerning the animals' health status, oestrus and disease detection (e.g., mastitis and lameness), quantity and quality of milk, individual identification, weight, etc. For small ruminants, although commercial applications are already available from the previous decade for sheep and goat farms including flock management (e.g., drones and virtual fence), resting, grazing and feeding behaviour monitoring using accelerometers, camera-based and sound analysis systems, GPS and GIS systems, detection and identification of health-related problems such as lameness and mastitis, individual monitoring etc., their use within the production process is almost absent due to economic and cultural constraints, poor technological infrastructures (e.g., electricity, telephone and internet networks) and lack of information and competence. In parallel, ongoing research on commercial pig barns has already shown a great potential in minimising or even solving various problems such as abnormal behaviours, namely tail-biting and belly nosing, aggression escalation, but also health-related problems such as lameness and disease spreading. Finally, PLF poultry research has deepened more than any other field representing publications from multiple different variables of the production process being monitored and controlled. However, although research applications in the area are vast, compared with the other competitive sectors, adoption of the technology is almost absent, probably due to its installation costs or the farmers' lack of information for its beneficial impact on the production process. It is proposed that future PLF applications should include within their manual economic analyses of the installation and usage costs along with the estimated benefits for the farmers. Furthermore, PLF companies should present the usage of the applications data utilisation in more simple ways that the farmers can understand and benefit from them.

References

Abecia, J. A., Maria, G. A., Estévez-Moreno, L. X., & Miranda-De La Lama, G. C. (2019). Daily rhythms of body temperature around lambing in sheep measured non-invasively. *Biological Rhythm Research, 51*(6), 988–993. https://doi.org/10.1080/09291016.2019.1592352

Abreu, L. H. P., Yanagi, T., Jr., Campos, A. T., Bahuti, M., & Fassani, É. (2017). Cloacal and surface temperatures of broilers subject to thermal stress. Agricultural. *Building and Environment, 37*(05). https://doi.org/10.1590/1809-4430-Eng.Agric.v37n5p877-886/2017

Adrion, F., Kapun, A., Holland, E.-M., Staiger, M., Löb, P., & Gallmann, E. (2017). Novel approach to determine the influence of pig and cattle ears on the performance of passive UHF-RFID ear tags. *Computers and Electronics in Agriculture, 140*(1), 168–179. https://doi.org/10.1016/j.compag.2017.06.004

Adrion, F., Kapuri, A., Eckert, F., Holland, E.-M., Staiger, M., Götz, S., & Gallmann, E. (2018). Monitoring trough visits of growing-finishing pigs with UHF-RFID. *Computers and Electronics in Agriculture, 144*, 144–153. https://doi.org/10.1016/j.compag.2017.11.036

Aerts, J.-M., Jans, P., Halloy, D., Gustin, P., & Berckmans, D. (2005). Labeling of cough from pigs for on-line disease monitoring by sound analysis. *Transactions of the American Society of Agricultural Engineers (ASAE), 48*(1), 351–354. https://doi.org/10.13031/2013.17948

Ahmed, G., Malick, R. A. S., Akhunzada, A., Zahid, S., Sagri, M. R., & Gani, A. (2021). An approach towards IoT-based predictive service for early detection of diseases in poultry chickens. *Sustainability, 13*(23), 13396. https://doi.org/10.3390/su132313396

Ahn, H., Son, S., Kim, H., Lee, S., Chung, Y., & Park, D. (2021). EnsemblePigDet: Ensemble Deep Learning for accurate pig detection. *Applied Sciences, 11*, 5577. https://doi.org/10.3390/app11125577

Alhamada, M., Debus, N., Lurette, A., & Bocquier, F. (2016). Validation of automated electronic oestrus detection in sheep as an alternative to visual observation. *Small Ruminant Research, 134*, 97–104. https://doi.org/10.1016/j.smallrumres.2015.12.032

Alhamada, M., Debus, N., Lurette, A., & Bocquier, F. (2017). Automatic oestrus detection system enables monitoring of sexual behaviour in sheep. *Small Ruminant Research, 149*, 105–111. https://doi.org/10.1016/j.smallrumres.2017.02.003

Alon, A. S., Marasigan, R. I., Jr., Nicolas-Mindoro, J. G., & Casuat, C. D. (2019). An image processing approach of multiple eggs' quality inspection. *International Journal of Advanced Trends in Computer Science and Engineering, 8*(6), 2794–2799. https://doi.org/10.30534/ijatcse/2019/18862019

Al-Thani, N., Albuainain, A., Alnaimi, F., & Zorba, N. (2020). Drones for sheep livestock monitoring. *2020 IEEE 20th Mediterranean Electrotechnical Conference (MELECON)*. 16–18 June 2020. doi: https://doi.org/10.1109/MELECON48756.2020.9140588

Alvarenga, F. A. P., Borges, I., Palkovič, L., Rodina, J., Oddy, V. H., & Dobos, R. C. (2016). Using a three-axis accelerometer to identify and classify sheep behaviour at pasture. *Applied Animal Behaviour Science, 181*, 91–99. https://doi.org/10.1016/j.applanim.2016.05.026

Ammad-Uddin, M., Ayaz, M., Aggoune, E.-H., & Sajjad, M. (2014). Wireless sensor network: A complete solution for poultry farming. *IEEE 2nd International Sympoium on Telecommunication Technologies (ISTT)*. 24–26 November 2014. IEEE. 15435656. doi: https://doi.org/10.1109/ISTT.2014.7238228

Amraei, S., Mehdizadeh, S. A., & Nääs, I. A. (2018). Development of a transfer function for weight prediction of live broiler chicken using machine vision. *Engenharia Agricola, 38*(5), 776–782. https://doi.org/10.1590/1890-4430-eng.agric.v38n5p776-782/2018

Andersen, H. M.-L., Jørgensen, E., Dybkjær, L., & Jørgensen, B. (2008). The ear skin temperature as an indicator of the thermal comfort of pigs. *Applied Animal Behaviour Science, 113*(1–3), 43–56. https://doi.org/10.1016/j.applanim.2007.11.003

Andonovic, I., Michie, C., Cousin, P., Janati, A., Pham, C., & Diop, M. (2018). Precision livestock farming technologies. *Proceedings of the Global Internet of Things Summit*. 4–7 June 2018, Bilbao. doi: https://doi.org/10.1109/GOITS.2018.8534572

Andrew, W., Greatwood, C., & Burghardt, T. (2020). Fusing animal biometrics with autonomous robotics: Drone-based search and individual ID of Friesian cattle. *Proceedings of the IEEE/CVF Winter Conference on Applications of Computer Vision (WACV) Workshops*. 1–5 March 2020. doi: https://doi.org/10.1109/WACVW50321.2020.9096949

Astill, J., Dara, R. A., Fraser, E. D. G., Roberts, B., & Sharif, S. (2020). Smart poultry management: Smart sensors, big data, and the internet of things. *Computers and Electronics in Agriculture, 170*, 105291. https://doi.org/10.1016/j.compag.2020.10591

Aydin, A. (2017a). Development of an early detection system for lameness of broilers using computer vision. *Computers and Electronics in Agriculture, 136*(C), 140–146. https://doi.org/10.1016/j.compag.2017.02.019

Aydin, A. (2017b). Using 3D vision camera system to automatically assess the level of inactivity in broiler chickens. *Computers and Electronics in Agriculture, 135*, 4–10. https://doi.org/10.1016/j.compag.2017.01.024

Aydin, A., & Berckmans, D. (2016). Using sound technology to automatically detect the short-term feeding behaviours of broiler chickens. *Computers and Electronics in Agriculture, 121*, 25–31. https://doi.org/10.1016/j.compag.2015.11.010

Aydin, A., Bahr, C., Viazzi, S., Exadactylos, V., Buyse, J., & Berckmans, D. (2014). A novel method to automatically measure the feed intake of broiler chickens by sound technology. *Computers and Electronics in Agriculture, 101*, 17–23. https://doi.org/10.1016/j.compag.2013.11.012

Aydin, A., Bahr, C., & Berckmans, D. (2015). A real-time monitoring tool to automatically measure the feed intakes of multiple broiler chickens by sound analysis. *Computers and Electronics in Agriculture, 114*, 1–6. https://doi.org/10.1016/j.compag.2015.03.010

Baker, J. E. (2004). Effective environmental temperature. *Journal of Swine Health Production, 12*, 140–143. https://www.aasv.org/shap/issues/v12n3/v12n3ptip.html

Balontong, A. J., Gerardo, B., & Medina, R. P. (2020). Swine management system in PLF integrating image processing for weight monitoring. *International Journal of Advanced Trends in Computer Science and Engineering, 9*(1), 1. https://doi.org/10.30534/ijatcse/2020/0291.12020

Banakar, A., Sadeghi, M., & Shushtari, A. (2016). An intelligent device for diagnosing avian diseases: Newcastle, infectious bronchitis, avian influenza. *Computers and Electronics in Agriculture, 127*, 744–753. https://doi.org/10.1016/j.compag.2016.08.006

Banhazi, T. M., & Black, J. L. (2009). Livestock farming: A suite of electronic systems to ensure the application of the best practice management on livestock farms. *Australian Journal of Multi-Discipinary Engineering, 7*(1), 1–14. https://doi.org/10.1080/14488388.2009.11464794

Banhazi, T. M., Tscharke, M., Ferdous, W. M., Saunders, C., & Lee, S. H. (2011). Improved image analysis based system to reliably predict the liveweight of pigs on farm: Preliminary results. *Australian Journal of Multi-Disciplinary Engineering, 8*, 107–119. https://doi.org/10.1080/14488388.2011.11464830

Barwick, J., Lamb, D. W., Dobos, R., Welch, M., & Trotter, M. (2018). Categorising sheep activity using a tri-axial accelerometer. *Computers and Electronics in Agriculture, 145*, 289–297. https://doi.org/10.1016/j.compag.2018.01.007

Beker, A., Gipson, T. A., Puchala, R., Askar, A. R., Tesfai, K., Detweiler, G. D., Asmare, A., & Goetsch, A. L. (2010). Energy expenditure and activity of different types of small ruminants grazing varying pastures in the summer. *Journal of Applied Animal Research, 37*(1), 1–14. https://doi.org/10.1080/09712119.2010.9707086

Bennett, C. E., Thomas, R., Williams, M., Zalasiewicz, J., Edgeworth, M., Miller, H., Coles, B. F., Foster, A., Burton, E. J., & Marume, U. (2018). The broiler chicken as a signal of a human reconfigured biosphere. *Royal Society Open Science, 5*(12), 180325. doi: https://doi.org/10.1098/esos.180325 [Online]. Available from: https://royalsocietypublishing.org/doi/10.1098/rsos.180325.

Berckmans, D. (2004). Automatic on-line monitoring of animals by precision livestock farming. In G. Clement & F. Madec (Eds.), *2006. Livestock production and society* (pp. 27–30). Wageningen Academic Publishers. https://doi.org/10.3920/978-90-8686-567-3

Berckmans, D. (2009). Automatic on-line monitoring of animal health and welfare by precision livestock farming. *Proceedings of the European Forum Livestock Housing for the Future*. 22—23 October. Lille, pp. 155–165. [Online]. Available from: http://www.rmt-batiments.org/IMG/pdf/Proceedings_Presentation-26.pdf

Berckmans, D. (2014a). Precision livestock farming technologies for welfare management in intensive livestock systems. *Revue Scientifique et Technique, 33*(1), 189–196. https://doi.org/10.20506/rst.33.1.2273

Berckmans, D. (2014b). My vision of precision livestock farming in 2020. In: *BPEX Innovation Conference 2014*. 2014. 24 June 2014. Warwickshire. [Online]. Available form: https://youtube.com/watch?v=jEUQytFAHwo

Berckmans, D. (2017). General introduction to Precision Livestock Farming. *Animal Frontiers, 7*, 6–11. https://doi.org/10.2527/af.2017.0102

Bessei, W. (2006). Welfare of broilers: a review. *World's Poultry Science Journal, 62*(3), 455–466. https://doi.org/10.1017/S0043933906001085

Betteridge, K., Hoogendoorn, C., Costall, D., Carter, M., & Griffiths, W. (2010). Sensors for detecting and logging spatial distribution of urine patches of grazing female sheep and cattle. *Computers and Electronics in Agriculture, 73*(1), 66–73. https://doi.org/10.1016/j.compag.2010.04.005

Bewley, J. M., Peacock, A. M., Lewis, O., Boyce, R. E., Roberts, D. J., Coffey, M. P., Kenyon, S. J., & Schutz, M. M. (2008). Potential for estimation of body condition scores in dairy cattle from digital images. *Journal of Dairy Science, 91*(9), 3439–3453. https://doi.org/10.3168/jds.2007-0836

Bishop, J. C., Falzon, G., Trotter, M., Kwan, P., & Meek, P. D. (2017). Sound analysis and detection, and the potential for precision livestock farming – a sheep vocalization case study. *Zenodo*. https://doi.org/10.5281/zenodo.897209

Bloch, V., Barchilon, N., Halachmi, I., & Druyan, S. (2020). Automatic broiler temperature measuring by thermal camera. *Biosystems Engineering, 199*, 127–134. https://doi.org/10.1016/j.biosystemseng.2019.08.011

Boileau, A., Farish, M., Turner, S. P., & Camerlink, I. (2019). Infrared thermography of agonistic behaviour in pigs. *Physiology and Behavior, 210*, 112637. https://doi.org/10.1016/j.physbeh.2019.112637

Booth, C. J., Warnick, L. D., Grohn, Y. T., Maizon, D. O., Guard, C. L., & Jansen, D. (2004). Effect of lameness on culling in dairy cows. *Journal of Dairy Science, 87*(12), 4115–4122. https://doi.org/10.3168/jds.S0022-0302(04)73554-7

Bortolotti, L., Zampieri, A., & Miatto, A. (2013). Practical experience on the use of injectable transponders in flock of Quessant sheep. *Large Animal Review, 19*(5), 219–223.

Boyd, H., Barrett, D. C., & Mihm, M. (2004). Problems associated with oestrus cyclicity. In A. H. Andrews, R. W. Blowey, H. Boyd, & R. G. Eddy (Eds.), *Bovine medicine. Diseases and husbandry for cattle* (2nd ed., pp. 530–551). Blackwell.

Breed, R. S., & Brew, J. D. (1917). The control of public milk supplies by the use of the microscopic method. *Journal of Dairy Science, 1*(3), 259–271. https://doi.org/10.3168/jds.S0022-0302(17)94379-6

Brown-Brandl, T. M., Eigenberg, R. A., & Purswell, J. L. (2013). Using thermal imaging as a method of investigating thermal thresholds in finishing. *Biosystems Engineering, 114*, 327–333. https://doi.org/10.1016/j.biosystemseng.2012.11.015

Brunberg, E. I., Bøe, K. E., & Sørheim, K. M. (2015). Testing a new virtual fencing system on sheep. *Acta Agriculturae Scandinavica, Section A—Animal Science, 65*(3–4), 168–175. https://doi.org/10.1080/09064702.2015.1128478

Brunberg, E. I., Bergslid, I. K., Bøe, K. E., & Sørheim, K. M. (2017). The ability of ewes with lambs to learn a virtual fencing system. *Animal, 11*(11), 2045–2050. https://doi.org/10.1017/S1751731117000891

Brünger, J., Traulsen, I., & Koch, R. (2018). Model-based detection of pigs in images under sub-optimal conditions. *Computers and Electronics in Agriculture., 152*, 59–63. https://doi.org/10.1016/j.compag.2018.06.043

Burgunder, J., Petrželková, K., Modrý, D., Kato, A., & MacIntosh, A. J. J. (2018). Fractal measurements in activity patterns: Do gastrointestinal parasites affect the complexity of sheep behaviour? *Applied Animal Behaviour Science, 205*, 44–53. https://doi.org/10.1016/j.applanim.2018.05.014

Byabazaire, J., Olariu, C., Taneja, M., & Davy, A. (2019). Lameness detection as a service: Application of machine learning to an internet of cattle. *Proceedings of the 16th IEEE Annual Consumer Communications and Networking Conference*. 11–14 January 2019. doi: https://doi.org/10.1109/CCNC.2019.8651681

Campbell, D. L. M., Hinch, G. N., Dowing, J. A., & Lee, C. (2016). Fear and coping styles of outdoor-preferring, moderate-outdoor and indoor-preferring free-range laying hens. *Applied Animal Behaviour Science, 185*, 73–77. https://doi.org/10.1016/j.applanim.2016.09.004

Campbell, D. L. M., Hinch, G. N., Dyall, T. R., Warin, L., Little, B. A., & Lee, C. (2017). Outdoor stocking density in free-range laying hens: Radio-frequency identification of impacts on range use. *Animal, 11*(1), 121–130. https://doi.org/10.1017/S1751731116001154

Campbell, D. L. M., Hinch, G. N., Dowing, J. A., & Lee, C. (2018a). Early enrichment in free-range laying hens: Effects on ranging behaviour, welfare and response to stressors. *Animal, 12*(3), 575–584. https://doi.org/10.1017/S1751731117001859

Campbell, D. L. M., Talk, A. C., Loh, Z. A., Dyall, T. R., & Lee, C. (2018b). Spatial cognition and range use in free-range hens. *Animals, 8*(2), 26. https://doi.org/10.3390/ani8020026

Cappai, M. G., Rubiu, N. G., Nieddu, G., Bitti, M. P. L., & Pinna, W. (2018). Analysis of fieldwork activities during milk production recording in dairy ewes by means of individual ear tag (ET) alone or plus RFID based electronic identification (EID). *Computers and Electronics in Agriculture, 144*, 324–328. https://doi.org/10.1016/j.compag.2017.11.002

Carpentier, L., Vranken, E., Berckmans, D., Paeshuyse, J., & Norton, T. (2019). Development of sound-based poultry health monitoring tool for automated sneeze detection. *Computers and Electronics in Agriculture, 162*, 573–581. https://doi.org/10.1016/j.compag.2019.05.013

Cavero, D., Tolle, K. H., Buxade, C., & Krieter, J. (2006). Mastitis detection in dairy cows by application of fuzzy logic. *Livestock Science, 105*(1–3), 207–213. https://doi.org/10.1016/j.livsci.2006.06.006

Cha, E., Hertl, J. A., Bar, D., & Gröhn, Y. T. (2010). The cost of different types of lameness in dairy cows calculated by dynamic programming. *Preventive Veterinary Medicine, 97*(1), 1–8. https://doi.org/10.1016/j.prevetmed.2010.07.011

Chao, K., Kim, M. S., & Chan, D. E. (2014). Control interface and tracking control system for automated poultry inspection. *Computer Standards and Interfaces., 36*(2), 271–277. https://doi.org/10.1016/j.csi.2011.03.006

Chen, C., Zhu, W., Liu, D., Steibel, J., Siegford, J., Wurtz, K., Han, J., & Norton, T. (2019). Detection of aggressive behaviours in pigs using a RealSense depth sensor. *Computers and Electronics in Agriculture, 166*, 105003. https://doi.org/10.1016/j.compag.2019.105003

Chen, C., Zhu, W., Steibel, J., Siegford, J., Han, J., & Norton, T. (2020a). Classification of drinking and drinker-playing in pigs by a video-based deep learning method. *Biosystems Engineering, 196*, 1–14.

Chen, C., Zhu, W., Liu, D., Steibel, J., Wurtz, K., Han, J., & Norton, T. (2020b). Recognitions of aggressive episodes of pigs based on convolutional neural network and long short-term memory. *Computers and Electronics in Agriculture, 169*, 105166. https://doi.org/10.1016/j.compag.2019.105166

Chien, Y.-R., & Chen, Y.-X. (2018). An RFID-based smart nest box: An experimental study of laying performance and behavior of individual hens. *Sensors (Basel), 18*(3), 859. https://doi.org/10.3390/s18030859

Chmiel, M., Slowinski, M., & Dasiewicz. (2011). Application of computer vision systems for estimation of fat content in poultry meat. *Food Control, 22*(8), 1424–2427. https://doi.org/10.1016/j.foodcont.2011.03.002

Chung, Y., Oh, S., Lee, J., & Park, D. (2013). Automatic detection and recognition of pig wasting diseases using sound data in audio surveillance systems. *Sensors, 13*(10), 12929–12942. https://doi.org/10.3390/s131012929

Clark, C. E. F., Farina, S. R., Gracia, S. C., Islam, M. R., Kerrisk, K. L., & Fulkerson, W. J. (2015). A comparison of conventional and automatic milking system pasture utilization and pre- and post-grazing pasture mass. *Grass and Forage Science, 71*(1), 153–159. https://doi.org/10.1111/gfs.12171

Cordeiro, A. F. S., Nääs, L. A., Leitäo, F. S., de Almeida, A. C. M., & de Moura, D. J. (2018). Use of vocalization to identify sex, age, and distress in pig production. *Biosystems Engineering, 173*, 57–63. https://doi.org/10.1016/j.biosystemseng.2018.03.007

Cornou, C., Vinther, J., & Kristensen, A. R. (2008). Automatic detection of oestrus and health disorders using data from electronic feeders. *Livestock Science, 118*, 262–271. https://doi.org/10.1016/j.livsci.2008.02.004

Cowton, J., Kyriazakis, I., & Bacardit, J. (2019). Automated individual pig localization, tracking and behaviour metric extraction using deep learning. *IEEE, 7*, 108049–108060. https://doi.org/10.1109/ACCESS.2019.2933060

Cronin, G. M., Beganovic, D. F., Sutton, A. L., Palmer, D. J., Thomson, P. C., & Tammen, I. (2016). Manifestation of neuronal ceroid lipofuscinosis in Australian Merino sheep:

Observations on altered behaviour and growth. *Applied Animal Behaviour Science, 175*, 32–40. https://doi.org/10.1016/j.applanim.2015.11.012

Cross, A. J., Rohrer, G. A., Brown-Brandl, T. M., Cassady, J. P., & Keel, B. N. (2018). Feed-forward and generalized regression neural networks in modelling feeding behaviour of pigs in the grow-finish phase. *Biosystems Engineering, 173*, 124–133. https://doi.org/10.1016/j.biosystemseng.2018.02.005

D'Eath, R. B., Foister, S., Jack, M., Bowers, N., Zhu, Q., Barclay, D., & Baxter, E. M. (2021). Changes in tail posture detected by a 3D machine vision system are associated with injury from damaging behaviours and ill health on commercial pig farms. *PLoS One, 16*(10), e0258895. https://doi.org/10.1371/journal.pone.0258895

Da Silva, P. J., Nääs, I. A., Abe, J. M., & Cordeiro, A. F. S. (2019). Classification of piglet (Sus Scrofa) stress conditions using vocalization pattern and applying paraconsistent logic Eτ. *Computers and Electronics in Agriculture, 166*, 105020. https://doi.org/10.1016/j.compag.2019.105020

Dawkins, M. S., Cain, R., & Roberts, S. J. (2012). Optical flow, flock behaviour and chicken welfare. *Animal Behaviour, 84*(1), 219–223. https://doi.org/10.1016/j.anbehav.2012.04.036

De Jong, I. C., Hindle, V. A., Butterworth, A., Engel, B., Ferrari, P., Gunnink, H., Moya, T. P., Tuyttens, F. A. M., & van Reenen, C. G. (2016). Simplifying the welfare quality® assessment protocol for broiler chicken welfare. *Animal, 10*(1), 117–127. https://doi.org/10.1017/S1751731115001706

De Montis, A., Pinna, A., Barra, M., & Vranken, E. (2013). Analysis of poultry eating and drinking behavior by software eYeNamic. *Proceedings of the 10th Conference of the Italian Society of Agricultural Engineering*. 44, S2. doi: https://doi.org/10.4081/jae.2013.275

Dewulf, J., Koenen, F., Laevens, H., & de Kruif, A. (2003). Infrared thermography is not suitable for detection of fever in pigs. *Vlaams Dierfeneeskundig Tijdschrift, 72*, 373–379.

Diana, A., Carpentier, L., Piette, D., Boyle, L. A., Berckmans, D., & Norton, T. (2019). An ethogram of biter and bitten pigs during an ear biting event: First step in development of a Precision Livestock Farming tool. *Applied Animal Behaviour Science, 215*, 26–36. https://doi.org/10.1016/j.applanim.2019.03.011

Dobson, H., Smith, R. F., Royal, M. D., Knight, C. H., & Sheldon, I. M. (2007). The high producting dairy cow ant its reproductive performance. *Reproduction in Domestic Animals, 2*(2), 17–23. https://doi.org/10.1111/j.1439-0531.2007.00906.x

Dogra, S., Chatterjee, S., Ray, R., Ghosh, S., Bhattacharya, D., & Sarkar, S. K. (2010). A novel proposal for the detection of Avian Influenza and managing poultry in a cost efficient way implementing RFID. *International Conference on Advances in Recent Technologies in Communication and Computing*. 16–17 October 2010. IEEE. 11688401. doi: https://doi.org/10.1109/ARTCom.2010.48

Dominiak, K., Pedersen, L. J., & Kristensen, A. R. (2019). Spatial modeling of pigs' drinking patterns as an alarm reducing method I. Developing a multivariate dynamic linear model. *Computers and Electronics in Agriculture, 161*, 79–91. https://doi.org/10.1016/j.compag.2018.06.032

Domun, Y., Pedersen, L. J., White, D., Adeyemi, O., & Norton, T. (2019). Learning patterns for time-serries data to discriminate predictions of tail-biting, fouling and diarrhoea in pigs. *Computers and Electronics in Agriculture, 163*, 104878. https://doi.org/10.1016/j.compag.2019.104878

Durali, T., Groves, P., Cowieson, A. J., & Singh, M. (2014). Evaluating range usage of commercial free range broilers and its effect on bird performance using radio frequency identification (RFID) technology. *25th Annual Australian Poultry Science Symposium*. February 2014. Sydney, New South Wales, Australia. pp. 103–106. [Online]. Available from: https://d1wqtxts1xzle7.cloudfront.net/33391116/APSS_Proceedings_2014-with-cover-page-v2.pdf?Expires=1640626827&Signature=IYoqiQHQYrby92zGt12eErf51JeDYKXAn14WkZwB0eL-wm2Kr-ZitjJKQm4YPO8go2Gc2yv0Nw-XbH7UuYG4nNul4wL2bUjQ3uuhkr56vpTGcKaaahf3Gc6wrxRQwLUnOYCkpJIgg-0ILXWHmqTX~88MTXNdONT5rqWZ894PMnIuO1

zVNDhG-WdB5WIpmSef9mLQlla8qG-5wLl-ErFO6f93yLpuro06tZ94dPB4sc6uy9h~6nzXDeVPWlkwp6ejreuH4ciK1CJFeJI0~lGPxt5aPinA9~Hi-741rvEcCI47vmxZfbHJMrI82cKKmjuojx7SSs2GxNetql5zSh1~ww__&Key-Pair-Id=APKAJLOHF5GGSLRBV4ZA#page=123

Edwards, S. A. (2008). Tail biting in pigs: Understanding the intractable problem. *The Veterinary Journal, 171*(2), 198–199. https://doi.org/10.1016/j.tvjl.2005.04.010

Eissen, J., Kanis, E., & Merks, J. W. M. (1998). Algorithms for identifying errors in individual feed intake data of growing pigs in group-housing. *Applied Engineering in Agriculture, 14*(6), 667–673.

Fang, C., Huang, J., Cuan, K., Zhuang, X., & Zhang, T. (2020). Comparative study on poultry target tracking algorithms based on deep regression network. *Biosystems Engineering, 190*, 176–183. https://doi.org/10.1016/j.biosystemseng.2019.12.002

Fay, P. K., McElligot, V. T., & Havstad, K. M. (1989). Containment of free-ranging goats using pulsed-radio-wave-activates shock collars. *Applied Animal Behaviour Science, 23*(1–2), 165–171. https://doi.org/10.1016/0168-159(89)90016-6

Fernandes, A. F. A., Dórea, J. R. R., Fitzgerald, R., Herring, W., & Rosa, G. J. (2019). A novel automated system to acquire biometric and morphological measurements and predict body weight of pigs via 3D computer vision. *Journal of Animal Science, 97*(1), 496–508. https://doi.org/10.1093/jas/sky418

Fernández, P. A., Norton, T., Exadactylos, V., Vranken, E., & Berckmans, D. (2016). Analysis of behavioural patterns in broilers using camera-based technology. *International Conference in Agricultural Engineering CIGR AgEng 2016*, 26–29 June 2016. Aarhus, Denmark.

Ferrari, S., Silva, M., Guarino, M., Aerts, J.-M., & Berckmans, D. (2008). Cough sound analysis to identify respiratory infection in pigs. *Computers and Electronics in Agriculture, 64*, 318–325. https://doi.org/10.1016/j.compag.2008.07.003

Firouz, M. S., & Omid, M. (2015). Detection of poultry egg freshness by dielectric spectroscopy and machine learning techniques. *Labensmittel-Wissenschaft und Technologie, 62*(2), 1034–1042. https://doi.org/10.1016/j.lwt.2015.02.019

Fontana, I., Tullo, E., Butterworth, A., & Guarino, M. (2015). An innovative approach to predict the growth in intensive poultry farming. *Computers and Electronics in Agriculture, 119*, 178–183. https://doi.org/10.1016/j.compag.2015.10.001

Fontana, I., Tullo, E., Scrase, A., & Butterworth, A. (2016). Vocalisation sound pattern identification in young broiler chickens. *Animal, 10*(9), 1567–1574. https://doi.org/10.1017/S1751731115001408

Fontana, I., Tullo, E., Carpetnier, L., Berckmans, D., Butterworth, A., Vranken, E., Borton, T., Berckmans, D., & Guarino, M. (2017). Sound analysis to model weight of broiler chickens. *Poultry Science, 96*(11), 3938–3943. https://doi.org/10.3382/ps/pex215

Fricke, P. M., Giordano, J. O., Valenza, A., Lopes, G., Jr., Amundson, M. C., & Carvalho, P. D. (2014). Reproductive performance of lactating dairy cows managed for first service using timed artificial insemination with or without detection of estrus using activity monitoring system. *Journal of Dairy Science, 97*(5), 2771–2781. https://doi.org/10.3168/jds.2013-7366

Friggens, N. C., Chagunda, M. G. G., Bjerring, M., Ridder, C., Hojsgaard, S., & Larsen, T. (2007). Estimating degree of mastitis from time series measurements in milk: A test model based on lactate dehydrogenase measurements. *Journal of Dairy Science, 90*(12), 5415–5427. https://doi.org/10.3168/jds.2007-0148

Fuchs, B., Sørheim, K. M., Chincarini, M., Brunberg, E., Stubsjøen, S. M., Bratbergsengen, K., Hvasshovd, S. O., Zimmermann, B., Lande, U. S., & Grøva, L. (2019). Heart rate sensor validation and seasonal and diurnal variation of body temperature and heart rate in domestic sheep. *Veterinary and Animal Science, 8*, 100075. https://doi.org/10.1016/j.vas.2019.100075

Fujii, T., Yokoi, H., Tada, T., Suzuki, K., & Tsukamoto, K. (2009). Poultry tracking system with camera using particle filters. *IEEE International Conference on Robotics and Biometrics*. 22–25 February 2009. doi: https://doi.org/10.1109/ROBIO.2009.4913289

Gaillard, C., Barbu, H., Sørensen, M. T., Sehested, J., Callesen, H., & Vestergaard, M. (2016). Milk yield and estrous behavior during eight consecutive estruses in Holstein cows fed standardized or high energy diets and grouped according to live weight changes in early lactation. *Journal of Dairy Science, 99*(4), 3134–3143. https://doi.org/10.3168/jds.2015-10023

Gangsei, L. E., & Kongsro, J. (2016). Automatic segmentation of Computed Tomography (CT) images of domestic pig skeleton using 3D expansion of Dijkstra's algorithm. *Computers and Electronics in Agriculture, 121*, 191–194. https://doi.org/10.1016/j.compag.2015.12.002

Gargiulo, J. I., Lyons, N. A., Kempton, K., Armstrong, D. A., & Garcia, S. C. (2020). Physical and economic comparison of pasture-based automatic and conventional milking systems. *Journal of Dairy Science, 103*(9), 8231–8240. https://doi.org/10.3168/jds.2020-18317

Geers, R., Dellaert, B., Goedseels, V., Hoogerbrugge, A., Vranken, E., Maes, F., & Berckmans, D. (1989). An assessment of optimal air temperatures in pig houses by the quantification of behavioural and health related problems. *Animal Production, 48*(3), 17–22. https://doi.org/10.1017/S0003356100004098

Geronimo, B. C., Mastelini, S. M., Carvalho, R. H., Júnior, S. B., Barbin, D. F., Shimokomaki, M., & Ida, E. L. (2019). Computer vision system and near-infrared spectroscopy for identification and classification of chicken with wooden breast, and physiochemical and technological characterization. *Infrared Physics and Technology, 96*, 303–310. https://doi.org/10.1016/j.infrared.2018.11.036

Giloh, M., Shinder, D., & Yahav, S. (2012). Skin surface temperature of broiler chickens is correlated to body core temperature and is indicative of their thermoregulatory status. *Poultry Science, 91*(1), 175–188. https://doi.org/10.3382/ps.2011-01497

Godfray, H. C. J., Beddington, J. R., Crute, I. R., Haddad, L., Lawrence, D., Muir, J. F., Pretty, J., Robinson, S., Thomas, S. M., & Toulmin, C. (2010). Food Security: The challenge of feeding 9 billion people. *Science, 327*(5967), 812–818. https://doi.org/10.1126/science.1185383

Godyn, D., & Herbut, P. (2017). Application of continuous body temperature measurements in pigs – a review. *Animal Science, 56*, 209–220. https://doi.org/10.22630/AAS.2017.56.2.22

González, C., Pardo, R., Fariña, J., Valdés, M. D., Rodriguez-Andina, J. J., & Portela, M. (2017). Real-time monitoring og poultry activity in breeding farms. *43rd Annual Conference of the IEEE Industrial Electronics Society*. 29 October–1 November 2017. doi: https://doi.org/10.1109/IECON.2017.8216605

Greenall, R., Warren, E., Warren, M., Meijering, A., Hogeveen, H., & de Koning, C. (2004). Integrading automatic milking installations (AMIS) into grazing systems – Lessons from Australia. In A. Meijering, H. Hogeveen, & C. De-Koning (Eds.), *Automatic milking: A better understanding* (pp. 273–279). Wageningen Academic Publishers.

Guanjun, B., Mimi, J., Yi, X., Shibo, C., & Qinghua, Y. (2019). Cracked egg recognition based on machine vision. *Computers and Electronics in Agriculture, 158*, 159–166. https://doi.org/10.1016/j.compag.2019.01.005

Guarino, M., Jans, P., Costa, A., Aerts, J.-M., & Berckmans, D. (2008). Field test of algorithm for automatic cough detection in pig houses. *Computers and electronics in Agriculture, 62*(1), 22–28. https://doi.org/10.1016/j.compag.2007.08.016

Guo, H., Ma, X., Ma, Q., Wang, K., Su, W., & Zhu, D. (2017). LSSA_CAU: An interactive 3D point clouds analysis software for body measurement of livestock with similar forms of cows or pigs. *Computers and Electronics in Agriculture, 138*, 60–68. https://doi.org/10.1016/j.compag.2017.04.014

Halachmi, I., Polak, P., Roberts, D. J., & klopčič, M. (2008). Cow body shape and automation of condition scoring. *Journal of Dairy Science, 91*(11), 4444–4451. https://doi.org/10.3168/jds.2007-0785

Halachmi, I., Klopčič, M., Polak, P., Roberts, D. J., & Bewley, J. M. (2013). Automatic assessment of dairy cattle body condition score using thermal imaging. *Computers and Electronics in Agriculture, 99*, 35–40. https://doi.org/10.1016/j.compag.2013.08.012

Halachmi, I., Guarino, M., Bewley, J., & Pastell, M. (2019). Smart animal agriculture: Application of real-time sensors to improve animal well-being and production. *Annual Review of Animal Biosciences, 7*, 403–425. https://doi.org/10.1146/annurev-animal-020518-114851

Haladjian, J., Haug, J., Nüske, S., & Bruegge, B. (2018). A wearable sensor system for lameness detection in dairy cattle. *Multimodal Technologies and Interaction, 2*(2), 27. https://doi.org/10.3390/mti2020027

Hansen, M. F., Smith, M. L., Salter, M. G., Baxter, E. M., Farish, M., & Grieve, B. (2018). Towards on-farm pig face recognition using convolutional neural networks. *Computers in Industry, 98*, 145–152. https://doi.org/10.1016/j.comind.2018.02.016

Hartcher, K. M., Hickey, K. A., Hemsworth, P. H., Cronin, G. M., Wilkinson, S. J., & Singh, M. (2016). Relationships between range access as monitored by radio frequency identification technology, fearfulness, and plumage damage in free-range laying hens. *Animal, 10*(5), 847–853. https://doi.org/10.1017/S1751731115002463

Hartung, J., Banhazi, T., Vranken, E., & Guarino, M. (2017). European farmers' experiences with Precision Livestock Farming systems. *Animal Frontiers, 7*(1), 38–44. https://doi.org/10.2527/af.2017.0107

Hemeryck, M., & Berckmans, D. (2015). Pig cough monitoring in the EU-PLF project: first results. In I. Halachmi (Ed.), *Precision livestock farming applications. Making sense of sensors to support farm management* (pp. 197–209). Wageningen Academic Publishers.

Hempstalk, K., Burke, C. R., & Kamphuis, C. (2013). *Verification of an automated camera-based system of oestrus detection in dairy cows. New Zeeland Society of Animal Production* (Vol. 73, pp. 26–28). Hamilton. http://www.nzsap.org/proceedings/2013/verification-automated-camera-based-system-oestrus-detection-dairy-cows

Hentz, F., Umstätter, C., Gilaverte, S., Prado, O. R., Silva, S. J. A., & Monteiro, A. L. G. (2014). Electronic bolus design impacts on administration. *Journal of Animal Science, 92*(6), 2686–2692. https://doi.org/10.2527/jas.2013-7183

Hernandez, J., Shearer, J. K., & Webb, D. W. (2002). Effect of lameness on milk yield in dairy cows. *Journal of the American Veterinary Medical Association, 220*(5), 640–644. https://doi.org/10.2460/javma.2002.220.640

Hillmann, E., Mayer, C., & Schrader, L. (2004a). Lying behaviour and adrenocortical response as indicators of the thermal tolerance of pigs of different weights. *Animal Welfare, 13*(3), 229–335.

Hillmann, E., Mayer, C., Scön, P. C., Puppe, B., & Schrader, L. (2004b). Vocalisation of domestic pigs (Sus scrofa domestica) as an indicator for their adaptation towards ambient temperatures. *Applied Animal Behaviour Science, 89*(3–4), 195–206. https://doi.org/10.1016/j.applanim.2004.06.008

Højsgaard, S., & Friggens, N. C. (2010). Quantifying degree if mastitis from common trends in panel of indicators for mastitis in dairy cows. *Journal of Dairy Science, 93*, 582–592. https://doi.org/10.3168/jds.2009-2445

Holman, A., Thompson, J., Routly, J. E., Cameron, J., Jones, D. N., Grove-White, D., Smith, R. F., & Dobson, H. (2011). Comparison of oestrus detection methods in dairy cattle. *The Veterinary Record, 169*(2), 47. https://doi.org/10.1136/vr.d2344

Hoy, S., Schamun, S., & Weirich, C. (2012). Investigations on feed intake and social behaviour of fattening pigs fed at an electronic feeding station. *Applied Animal Behaviour Science, 139*(1–2), 58–64. https://doi.org/10.1016/j.applanim.2012.03.010

Hrupka, B. J., Leibbrandt, V. D., Crenshaw, T. D., & Benevenga, N. J. (2000). The effect of thermal environment and age on neonatal pig behaviour. *Journal of Animal Science, 78*(3), 583–591. https://doi.org/10.2527/2000.783583x

Huang, J., Wang, W., & Zhang, T. (2019). Method for detecting avian influenza disease of chickens based on sound analysis. *Biosystems Engineering*. https://doi.org/10.1016/j.biosystemseng.2019.01.015

Huang, J., Zhang, T., Cuan, K., & Fang, C. (2021). An intelligent method for detecting poultry eating behaviour based on vocalization signals. *Computers and Electronics in Agriculture, 180*, 105884. https://doi.org/10.1016/j.compag.2020.105884

Huijps, K., Lam, T. J., & Hogeveen, H. (2008). Costs of mastitis: Facts and perception. *Journal of Dairy Research, 75*(1), 113–120. https://doi.org/10.1017/S0022029907002932

Huynh, T. T. T., Aarnik, A. J. A., Gerrits, W. J. J., Heetkamp, M. J. H., Canh, T. T., Spoolder, H. A. M., Kemp, B., & Verstegen, M. W. A. (2005). Thermal behaviour of growing pigs in response to high temperature and humidity. *Applied Animal Behaviour Science, 91*, 1–16.

Ingram, D. L. (1965). Evaporative cooling in the pig. *Nature, 207*, 415–416.

Islam, M. M., Ahmed, S. T., Mun, H. S., Bostami, A. B. M. R., Kim, Y. J., & Yang, C. J. (2015). Use of thermal imaging for early detection of signs of disease in pigs challenged orally with Salmonella typhimurium and Escherichia coli. *African Journal of Microbiology Research, 9*, 1667–1674. https://doi.org/10.5897/AJMR2015.7580

Ismayilova, G., Sonoda, L., Fels, M., Rizzi, R., Oczak, M., Viazzi, S., Vranken, E., Hartung, J., Berckmans, D., & Guarino, M. (2013). Acoustic-reward learning as a method to reduce the incidence of aggressive and abnormal behaviours among newly mixed piglets. *Animal Production Science, 54*(8), 1084–1090. https://doi.org/10.1071/AN13202

Jabbar, K. A., Hansen, M. F., Smith, M. L., & Smith, L. N. (2017). Early and non-intrusive lameness detection in dairy cows using 3-dimensional video. *Biosystems Engineering, 153*, 63–69. https://doi.org/10.1016/j.biosystemseng.2016.09.017

Jegadeesan, S., & Venkatesan, G. K. D. P. Distant biometry in cattle farm using wireless sensor networks. *International Conference on Communication and Electronics Systems (ICCES)*. 21—22 October 2017. doi: https://doi.org/10.1109/CESYS.2016.7889964

Jensen, D. B., Hogeveen, H., & De Vries, A. (2016). Bayesian integration of sensor information and a multivariate dynamic linear model for prediction of dairy cow mastitis. *Journal of Dairy Science, 99*(9), 7344–7361.

Jiao, L., Dong, D., Zhao, X., & Han, P. (2016). Compensation method for the influence of angle of view on animal temperature measurement using thermal imaging camera combined with depth image. *Journal of Thermal Biology, 62*(A), 15–19. https://doi.org/10.1016/j.jtherbio.2016.07.021

John, A. J., Clark, C. E. F., Freeman, M. J., Kerrisk, K. L., Garcia, S. C., & Halachmi, I. (2016). Review: Milking robot utilization, a successful precision livestock farming evolution. *Animal, 10*(9), 1484–1492. https://doi.org/10.1017/S1751731116000495

Jouven, M., Leroy, H., Ickowicz, A., & Lapeyronie, P. (2011). Can virtual fences be used to control grazing sheep? *Rangelands, 34*(1), 111–123. https://doi.org/10.1071/RJ11044

Ju, M., Choi, Y., Seo, J., Sa, J., Lee, S., Chung, Y., & Park, D. (2018). A Kinect-based segmentation of touching-pigs for real-time monitoring. *Sensors (Basel), 18*(6), 1746. https://doi.org/10.3390/s18061746

Jun, K., Kim, S. J., & Ji, H. W. (2018). Estimating pig weights from images without constraint on posture and illumination. *Computers and Electronics in Agriculture, 153*, 169–176.

Kamphuis, C., DelaRue, B., Burke, C. R., & Jago, J. (2012). Field evaluation of 2 collar-mounted activity meters for detecting cows in estrus on a large pasture-grazed dairy farm. *Journal of Dairy Science, 95*(6), 3045–3056. https://doi.org/10.3168/jds.2011-4934

Kamphuis, C., Dela Rue, B., Mein, G., & Jago, J. (2013). Development of protocols to evaluate in-line mastitis-detection systems. *Journal of Dairy Science, 96*(6), 4047–4058. https://doi.org/10.3168/jds.2012-6190

Kang, X., Zhang, X. D., & Liu, G. (2020). Accurate detection of lameness in dairy cattle with computer vision: A new and individualized detection strategy based on the analysis of the supporting phase. *Journal of Dairy Science, 103*(11), 10628–10638. https://doi.org/10.3168/jds.2020-18288

Kashiha, M., Bahr, C., Haredasht, S. A., Ott, S., Moons, C. R. H., Niewld, T. A., Ödberg, F., & Berckmans, D. (2013a). The automatic monitoring of pigs water use by cameras. *Computers and Electronics in Agriculture, 90*, 164–169. https://doi.org/10.1016/j.compag.2012.09.015

Kashiha, M., Bahr, C., Ott, S., Moons, C. P. H., Niewold, T. A., Ödberg, F. O., & Berckmans, D. (2013b). Automatic identification of marked pigs in a pen using image pattern recognition.

Computers and Electronics in Agriculture, 93, 111–120. https://doi.org/10.1016/j.compag.2013.01.013

Kashiha, M., Bahr, C., Ott, S., Moons, C. P. H., Niewold, T. A., Ödberg, F. O., & Berckmans, D. (2014a). Automatic weight estimation of individual pigs using image analysis. *Computers and Electronics in Agriculture, 107*, 38–44. https://doi.org/10.1016/j.compag.2014.06.003

Kashiha, M., Bahr, C., Ott, S., Moons, C. P. H., Niewold, T. A., Tuyttens, F., & Berckmans, D. (2014b). Automatic monitoring of pig locomotion using image analysis. *Livestock Science, 159*, 141–148. https://doi.org/10.1016/j.livsci.2013.11.007

Kashiha, M., Bahr, C., Vranken, E., Hong, S.-W., & Berckmans, D. (2014c). Monitoring system to detect problems in broiler houses based on image processing. *Proceedings of the International Conference of Agricultural Engineering*. 6–10 July 2014. C0403. [Online]. Available from: http://www.geyseco.es/geystiona/adjs/comunicaciones/304/C04030001.pdf

Kaswati, E. L. N., Saputro, A. H., & Imawan, C. (2020). Examination of chicken meat quality based on hyperspectral imaging. *Journal of Physics: Conference Series. 4th International Seminar on Sensors, Instrumentation, Measurement and Metrology*. 14 November 2019. doi: https://doi.org/10.1088/1742-6596/1528/1/012045

Kearton, T., Marini, D., Cowley, F., Belson, S., & Lee, C. (2019). The effect of virtual fencing stimuli on stress responses and behaviour in sheep. *Animals (Basel), 9*(1), 30. https://doi.org/10.3390/ani9010030

Kemp, M., Nolan, A., Cripps, P., & Fitzpatrick, J. (2008). Animal-based measurements on the severity of mastitis in dairy cows. *The Veterinary Record, 163*(6), 175–179. https://doi.org/10.1136/vr.163.6.175

Ketelaar-de Lauwere, C. C. (1999). *Cow behaviour and managerial aspects of fully automatic milking loose housing systems*. PhD Thesis, Wageningen University.

Khatun, M., Clark, C. E. F., Lyons, N. A., Thomson, P. C., Kerrisk, K. L., & García, S. C. (2017). Early detection of clinical mastitis from electrical conductivity data in automatic milking system. *Animal Production Science, 57*(7), 1226–1232. https://doi.org/10.1071/AN16707

Kim, J., Chung, Y., Choi, Y., Sa, J., Kim, H., Chung, Y., Park, D., & Kim, H. (2017). Depth-based detection of standing pigs in moving noise environments. *Sensors, 17*(12), 2757. https://doi.org/10.3390/s17122757

Kitchen, B. J. (1981). Review of the progress of dairy science: Bovine mastitis: Milk composition changes and related diagnostic tests. *Journal of Dairy Research, 48*(1), 167–188. https://doi.org/10.1017/s0022029900021580

Knížková, I., Kunc, P., Langrová, I., Vadlejch, J., & Jankovská, I. (2018). Thermal profile of broilers infected by Eimeria tenella. *Proceedings of the 14th Quantitative InfraRed Thermography Conference*. 25–29 June 2018. https://www.qirt2018.de/portals/qirt18/doc/P3.pdf

Knowles, T. G., Kestin, S. C., Haslam, S. M., Brown, S. N., Green, L. E., Butterworth, A., Pope, S. J., Pfeiffer, D., & Nicol, C. J. (2008). Leg disorders in broiler chickens: Prevalence, risk factors prevention. *PLoS One, 3*(2), e1545. https://doi.org/10.1371/journal.pone.0001545

Koelsch, R. K., Aneshansley, D. J., & Buttler, W. R. (1994). Analysis of activity measurement for accurate estrus detection in dairy-cattle. *Journal of Agricultural Engineering Research, 58*(2–3), 107–114. https://doi.org/10.1006/jaer.1994.1040

Kokin, E., Praks, J., Veermäe, I., Poikalainen, V., & Vallas, M. (2014). IceTag3DTM accelerometric device in cattle lameness detection. *Agronomy Research, 12*(1), 223–230. Accessed July 5, 2021, from https://www.researchgate.net/profile/V-Poikalainen/publication/287290567_IceTag3D_accelerometric_device_in_cattle_lameness_detection/links/57a86c8708aed76703f55455/IceTag3D-accelerometric-device-in-cattle-lameness-detection.pdf

Kollis, K., Phang, C. S., Banhazi, T. M., & Searle, S. J. (2007). Weight estimation using image analysis and statistical modelling: A preliminary study. *Applied Engineering in Agriculture, 23*(1), 91–96. https://doi.org/10.13031/2013.22332

Kongsro, J. (2014). Estimation of pig weight using a Microsoft Kinect prototype imaging system. *Computers and Electronics in Agriculture, 109*, 32–35. https://doi.org/10.1016/j.compag.2014.08.008

Kuipers, A., & Rossing, W. (1996). Robotic milking of dairy cows. In C. J. C. Philips (Ed.), *Progress in dairy science* (pp. 263–280). CABI publishing.
Kumar, S., & Singh, S. K. (2019). Cattle recognition: A new frontier in visual animal biometrics research. Proceedings of the National Academy of Sciences. *India Section A: Physical Sciences, 90*, 689–708. https://doi.org/10.1007/s40010-019-00610-x
Kumar, M., Veeraraghavan, A., & Sbharwal, A. (2015). DistancePPG: Robust non-contact vital signs monitoring using a camera. *Biomedical Optics Express., 6*(5), 1565–1588. https://doi.org/10.1364/BOE.6.001565
Larsen, H., Hemsworth, P. H., Cronin, G. M., Gebhardt-Henrich, S. G., Smith, C. L., & Rault, J. L. (2018). Relationship between welfare and individual ranging behaviour in commercial free-range laying hens. *Animal, 12*(11), 1–9. https://doi.org/10.1017/S1751731118000022
Larsen, M. L. V., Pedersen, L. J., & Jensen, D. B. (2019). Prediction of tail biting events in finisher pigs from automatically recorded sensor data. *Animals, 9*(7), 458. https://doi.org/10.3390/ani9070458
Lee, J., Noh, B., Jang, S., Park, D., Chung, Y., & Chang, H.-H. (2015). Stress detection and classification of laying hens by sound analysis. *Asian-Australian Journal of Animal Sciences, 28*(4), 592–598. https://doi.org/10.5713/ajas.14.0654
Lee, J., Jin, L., Park, D., & Chung, Y. (2016). Automatic recognition of aggressive behaviour in pigs using a depth sensor. *Sensors, 16*, 631–642. https://doi.org/10.3390/s16050631
Lee, H., Sa, J., Chung, Y., Park, D., & Kim, H. (2019). Deep learning-based overlapping-pig separation by balancing accuracy and execution time. In: *WSCG 2019: full papers proceedings. 27th International Conference in Central Europe in Computer Graphics, Visualization and Computer Vision. Computer Science Research Notes*. 27–31 May, 2019. Pilsen/Prague, Czech Repiblic, pp. 17–25. doi: https://doi.org/10.24132/CSRN.2019.2901.1.3.
Li, L., Zhao, Y., Oliveira, J., Verhoijsen, W., Liu, K., & Xin, H. (2017). A UHF RFID system for studying individual feeding and nesting behaviors of group-housed laying hens. *Transactions of the ASABE, 60*(4), 1337–1347. https://doi.org/10.13031/trans.12202
Li, G., Zhao, Y., Hailey, R., Zhang, N., Liang, Y., & Purswell, J. L. (2018). Radio-frequency identification (RFID) system for monitoring specific behaviors of group housed broilers. *10th International Livestock Environment Symposium (ILES X)*. 25–27 September 2018. ASABE. ILES18-051. doi: https://doi.org/10.13031/iles.ILES18-051
Li, B., Liu, L., Shen, M., Sun, Y., & Lu, M. (2019a). Group-housed pig detection in video surveillance of overhead views using multi-feature template matching. *Biosystems Engineering, 181*, 28–39. https://doi.org/10.1016/j.biosystemseng.2019.02.018
Li, N., Ren, Z., Li, D., & Zeng, L. (2019b). Review: Automated techniques for monitoring the behaviour and welfare of broilers and laying hens: Towards the goal of precision livestock farming. *Animal, 14*(3), 617–625. https://doi.org/10.1017/S1751731119002155
Li, G., Zhao, Y., Purswell, J. L., & Magee, C. (2021). Effects of feeder space on broiler feeding behaviors. *Poultry Science, 100*(4), 101016. https://doi.org/10.1016/j.psj.2021.01.038
Lin, D.-Y., Wu, M.-C., Tzeng, S.-J., & Lai, Y.-Y. (2019). *Egg production recording system of Taiwan native chicken. FFTC Agricultural Policy Platform (FFTC-AP)*. https://ap.fftc.org.tw/article/1622
Liu, T., Li, Z., Teng, G., & Luo, C. (2013). Predicition of pig weight based on radical basis function neural network. *Trabsactions of the Chinese Society of Agricultural Machinery, 44*(8), 245–249. https://doi.org/10.6041/j.issn.1000-1298.2013.08.042
Liu, D., Oczak, M., Maschat, K., Baumgartner, J., Pletzer, B., He, D., & Norton, T. (2020). A computer vision-based method for spatial-temporal action recognition of tail-biting behaviour in group-housed pigs. *Biosystems Engineering, 195*, 27–41. https://doi.org/10.1016/j.biosystemseng.2020.04.007
Lopez, H., Caraviello, D. Z., Satter, L. D., Fricke, P. M., & Wiltbank, M. C. (2005). Relationship between level of milk production and multiple ovulations in lactating cows. *Journal of Dairy Science, 88*(8), 2783–2793. https://doi.org/10.3168/jds.S0022-0302(05)72958-1

Lu, M., Xiong, Y., Li, K., Liu, L., Yan, L., Ding, Y., Lin, X., Yang, X., & Shen, M. (2016). An automatic splitting method for the adhesive piglets' gray scale image based on the ellipse shape feature. *Computers and Electronics in Agriculture, 120*, 53–62. https://doi.org/10.1016/j.compag.2015.11.008

Lu, M., Norton, T., Youssef, A., Radojkovic, N., Fernández, A. P., & Berckmans, D. (2018). Extracting body surface dimensions from top-view images of pigs. *International Journal of Agricultural and Biological Engineering, 11*, 182–191. https://doi.org/10.25165/j.ijbe.20181105.4054

Mahdavian, A., Minaei, S., Yang, C., Almasganj, F., Rahimi, S., & Marchetto, P. M. (2020). Ability evaluation of the voice activity detection algorithm in bioacoustics: A case study on poultry calls. *Computers and Electronics in Agriculture, 168*, 105100. https://doi.org/10.1016/j.compag.2019.105100

Mahdavian, A., Minaei, S., Marchetto, P. M., Almasganj, F., Rahimi, S., & Yang, C. (2021). Acoustic features of vocalization signal in poultry health monitoring. *Applied Acoustics, 175*, 107756. https://doi.org/10.1016/j.apacoust.2020.107756

Mansbridge, N., Mitsch, J., Bollard, N., Ellis, K. A., Miguel-Pacheco, G., Dottorini, T., & Kaler, J. (2018). Feature selection and comparison of machine learning algorithms in classification of grazing and rumination behaviour in sheep. *Sensors, 18*(10), 3532. https://doi.org/10.3390/s18103532

Manteuffel, G. (2009). Active feeding control and environmental enrichment with call-feeding-stations. In: Lokhorst, C., Koerkamp, P. W. G. G. (Eds.), Precision livestock farming '09. Papers presented at the *4th European Conference on Precision Livestock Farming*. 6–8 July, 2009 (pp. 283–288). Wageningen Academic Publishers.

Manteuffel, G., Puppe, B., & Schon, P. C. (2004). Vocalization of farm animals as a measure of welfare. *Applied Animal Behaviour Science, 88*(1–2), 163–182. https://doi.org/10.1016/j.applanim.2004.02.012

Marchant-Forde, J. N., Whittaker, X., & Broom, D. M. (2001). Vocalisations of the adult female domestic pig during a standard human test and their relationships with behavioural and heart measures. *Applied Animal Science, 72*, 23–39. https://doi.org/10.1016/S0168-159(00)00190-8

Marini, D., Llewellyn, R., Belson, S., & Lee, C. (2018a). Controlling within-field sheep movement using virtual fencing. *Animals (Basel), 8*(3), 31. https://doi.org/10.3390/ani8030031

Marini, D., Meuleman, M. D., Belson, S., Rodenburg, T. B., Llewellyn, R., & Lee, C. (2018b). Developing an ethically virtual fencing system for sheep. *Animals (Basel), 8*(3), 33. https://doi.org/10.3390/ani8030033

Marino, R., Petrera, F., Speroni, M., Rutigliano, T., Gali, A., & Abeni, F. (2021). Unraveling the relationship between milk yield and quality at the test day with rumination time recorded by a PLF technology. *Animals, 11*(6), 1583. https://doi.org/10.3390/ani11061583

Marquez, H. J. P., Ambrose, D. J., Schaefer, A. L., Cook, N. J., & Bench, C. J. (2019). Infrared thermography and behavioral biometrics associated with estrus indicators and ovulation in estrus-synchronized dairy cows housed in tiestalls. *Journal of Dairy Science, 102*(5), 4427–4440. https://doi.org/10.3168/jds.2018-15221

Marquez, H. J. P., Ambrose, D. J., Schaefer, A. L., Cook, N. J., & Bench, C. J. (2021). Evaluation of infrared thermography combined with behavioral biometrics for estrus detection in naturally cycling dairy cows. *Animal, 15*(7), 100205. https://doi.org/10.1016/j.animal.2021.100205

Martins, B. M., Mendes, A. L. C., Silva, L. F., Moreira, T. R., Costa, J. H. C., Rotta, P. P., Chizzotti, M. L., & Marcondes, M. I. (2020). Estimating body weight, body condition score, and type traits in dairy cows using three dimensional cameras and manual body measurements. *Livestock Science, 236*, 104054. https://doi.org/10.1016/j.livsci.2020.104054

Marx, G., Horn, T., Thielebein, J., Knubel, B., & von Borell, E. (2003). Analysis of pain-related vocalization in young pigs. *Journal of Sound and Vibration, 266*(3), 687–698. https://doi.org/10.1016/S0022-460X(03)00594-7

Maselyne, J., Sayes, W., de Ketelaere, B., Mertens, K., Vangeyte, J., Hessel, E. F., Millet, S., & van Nuffel, A. (2014). Validation of a high frequency radio frequency identification (HF RFID) system for registering feeding patterns of growing-finishing pigs. *Computers and Electronics in Agriculture, 102*, 10–18. https://doi.org/10.1016/j.compag.2013.12.015

Maselyne, J., Adriaens, I., Huybrechts, T., de Ketelaere, B., Millet, S., Vangeyte, J., van Nuffel, A., & Saeys, W. (2015). Assessing the drinking behaviour of individual pigs using RFID registrations. In I. Halachmi (Ed.), *Precision livestock farming applications. Making sense of the sensors to support farm management* (pp. 209–215). Wageningen Academic Publishers.

Maselyne, J., Sayes, W., Briene, P., Mertens, K., Vangeyete, J., de Katelaere, B., Hessel, E. F., Sonck, B., & van Neuffel, A. (2016). Methods to construct feeding visits from RFID registrations of growing-finishing pigs at the feed trough. *Computers and Electronics in Agriculture, 128*, 9–19. https://doi.org/10.1016/j.compag.2016.08.010

Maselyne, J., van Nuffel, A., Briene, P., Vangeyte, J., de Ketelaere, B., Miller, S., van den Hof, J., Maes, D., & Sayes, W. (2018). Online warning systems for individual fattening pigs based on their feeding pattern. *Biosystems Engineering, 173*, 143–156. https://doi.org/10.1016/j.biosystemseng.2017.08.006

Masey-O'Neil, H. V., Singh, M., & Cowieson, A. J. (2014). Effects of exogenous xylanase on performance, nutrient digestibility, volatile fatty acid production and digestive tract thermal profiles of broilers fed on wheat- or maize-based diet. *British Poultry Science, 55*(3), 351–359. https://doi.org/10.1080/00071668.2014.898836

Mayo, L. M., Silvia, W. J., Ray, D. L., Jones, B. W., Stone, A. E., Tsai, I. C., Clark, J. D., Bewley, J. M., & Heersche, G., Jr. (2019). Automated estrous detection using multiple commercial precision dairy monitoring technologies in synchronized dairy cows. *Journal of Dairy Science, 102*(3), 2645–2656. https://doi.org/10.3168/jds.2018-14738

McGlone, J. J. (1986). Agonistic behaviour in food animals: review of research and techniques. *Journal of Animal Science, 62*(4), 1130–1139. https://doi.org/10.2527/jas1986.6241130x

McGrath, N., Dunlop, R., Dwyer, C., Burnman, O., & Phillips, C. J. C. (2017). Hens vary their vocal repertoire and structure when anticipating different types of reward. *Animal Behaviour, 130*, 79–96. https://doi.org/10.1016/j.anbehav.2017.05.025

Meckbach, C., Tiesmeyer, V., & Traulsen, I. (2021). A promishing approach towards precise animal weight monitoring using neural networks. *Computers and Electronics in Agriculture, 183*, 106056. https://doi.org/10.1016/j.compag.2021.106056

Mehdizadeh, S. A., Neves, D. P., Tscharke, M., Nääs, I. A., & Banhazi, T. M. (2015). Image analysis method to evaluate beak and head motion of broiler chickens during feeding. *Computers and Electronics in Agriculture, 114*, 88–95. https://doi.org/10.1016/j.compag.2015.03.017

Meluzzi, A., & Sirri, F. (2009). Welfare of broiler chicken. *Italian Journal of Animal Science., 8*, 161–173. https://doi.org/10.4081/ijas.2009.s1.161

Meluzzi, A., Fabbri, C., Folegatti, E., & Sirri, F. (2008). Survey of chicken rearing conditions in Italy: effects of litter quality and stocking density on productivity, foot dermatitis and carcase injuries. *British Poultry Science, 49*(3), 257–264. https://doi.org/10.10180/00071660802094156

Meunier, B., Giname, C., Houdebine, M., Fleurance, G., Mialon, M.-M., Siberberg, M., & Boisy, A. (2015). Development of a multi-sensor and multi-application device for monitoring indoor and outdoor sheep behaviour. In the *Proceeding of the 7th European Conference of Precision Livestock Farming (EC-PLF)*. 15 September 2015, p. 12.

Michaelis, I., Burfeind, O., & Heuwieser, W. (2014). Evaluation of oestrus detection in dairy cattle comparing an automated activity monitoring system to visual observation. *Reproduction in Domestic Animals, 49*(4), 621–628. https://doi.org/10.1111/rda.12337

Mitchell, M. A. (2006). Using physiological models to define environmental control strategies. In R. Gous, T. Morris, & C. Fisher (Eds.), *Mechanistic modelling in pig and poultry production* (pp. 209–228). CABI.

Morgan, K. N., & Tromborg, C. T. (2006). Sources of stress in captivity. *Applied Animal Behaviour Science, 102*(3–4), 262–302. https://doi.org/10.1016/j.applanim.2006.05.032

Morgan-Davies, C., Lambe, N., Wishart, H., Waterhouse, T., Kenyon, F., McBean, D., & McCracken, D. (2018). Impacts of using a precision livestock system targeted approach in mountain flocks. *Livestock Science, 208*, 67–76. https://doi.org/10.1016/j.livsci.2017.12.002

Morris, M. J., Kaneko, K., Walker, S. L., Jones, D. N., Routly, J. E., Smith, R. F., & Dobson, H. (2011). Influence of lameness on follicular growth, ovulation, reproductive hormone concentrations and estrus behavior in dairy cows. *Theriogenology, 76*(4), 658–668. https://doi.org/10.1016/j.theriogenology.2011.03.019

Mortensen, A. K., Lisouski, P., & Ahredt, P. (2016). Weight prediction of broiler chickens using 3D computer vision. *Computers and Electronics in Agriculture, 123*, 319–326. https://doi.org/10.1016/j.compag.2016.03.011

Mota-Grajales, R., Torres-Peña, J. C., Camas-Anzueto, J. L., Pérez-Patricio, M., Coutiño, R. G., López-Estrada, F. R., Escobar-Gómez, E. N., & Guerra-Crespo, H. (2019). Defect detection in eggshell using a vision system to ensure the incubation in poultry production. *Measurement, 135*, 39–46. https://doi.org/10.1016/j.measurement.2018.09.059

Mottram, T. (2016). Animal broad invited review: Precision livestock farming for dairy cows with a focus on oestrus detection. *Animal, 10*(10), 1575–1584. https://doi.org/10.1017/S1751731115002517

Mozo, R., Alabart, J. L., Rivas, E., & Folch, J. (2019). New method to automatically evaluate the sexual activity of the ram based on accelerometer records. *Small Ruminant Research, 172*, 16–22. https://doi.org/10.1016/j.smallrumres.2019.01.009

Nääs, I. A., Romanini, C. E. B., Neves, D. P., Nascimento, G. R., & Vercelino, R. A. (2010). Broiler surface temperature of 42 day old chickens. *Scientia Agricola, 67*(5), 497–502. https://doi.org/10.1590/S0103-90162010000500001

Narin, B., Buntan, S., Chumuang, N., & Ketcham, M. (2018). Crack on eggshell detection system based on image processing technique. *18th International Symposium on Communications and Information Technologies (ISCIT)*. 26–29 September 2018. https://doi.org/10.1109/ISCIT.2018.8587980

Narushin, V. G., Lu, G., Cugley, J., Romanov, M. N., & Griffin, D. K. (2020). A 2-D imaging-assisted geometrical transformation method for non-destructive evaluation of the volume and surface of avian eggs. *Food Control, 112*, 107112. https://doi.org/10.1016/j.foodcont.2020.107112

Nascimento, G. R., Nääs, I. A., Pereira, D. F., Baracho, M. S., & Garcia, R. (2011). Assessment of broiler surface temperature variation when exposed to different air temperatures. Brazilian. *Journal of Poultry Science, 13*(4). https://doi.org/10.1590/S1516-635X2011000400007

Nasirahmadi, A., Hensel, O., Edwards, S. A., & Strum, B. (2016). Automatic detection of mounting behaviours among pigs using image analysis. *Computers and Electronics in Agriculture, 124*, 295–302.

Nasirahmadi, A., Strum, B., Edwards, S., Jeppsson, K.-H., Olsson, A.-C., Müller, S., & Hensel, O. (2019). Deep learning and machine vision approaches for posture detection of individual pigs. *Sensors, 19*(17). https://doi.org/10.3390/s19173738

Nasiri, A., Omid, M., & Taheri-Garavand, A. (2020). An automatic sorting system for unwashed eggs using deep learning. *Journal of Food Engineering, 283*, 11036. https://doi.org/10.1016/j.foodeng.2020.110036

Neethirajan, S., & Kemp, B. (2021). Digital livestock farming. *Sensing and Bio-Sensing Research, 32*, 100408. https://doi.org/10.1016/j.sbsr.2021.100408

Neves, D. P., Mhdizadeh, S. A., Tscharke, M., Nääs, I. A., & Banhazi, T. M. (2015). Detection of flock movement and behaviour of broiler chickens at different feeders using image analysis. *Information Processing in Agriculture, 2*(3–4), 177–182. https://doi.org/10.1016/j.inpa.2015.08.002

Neves, D. P., Mehdizadeh, S. A., Santana, M. R., Amadori, M. S., Banhazi, T. M., & Nääs, I. A. (2019). Young broiler feeding kinematic analysis as a function of the feed type. *Animals (Basel), 9*(12), 1149. https://doi.org/10.3390/ani9121149

Ni, J.-Q., Liu, S., Radcliffe, J. S., & Vonderohe, C. (2017). Evaluation and characterization of Passive Infrared Detectors to monitor pig activities in an environmental research building. *Biosystems Engineering*. https://doi.org/10.1016/j.biosystemseng.2017.03.014

Nilsson, M., Herlin, A. H., Guzhva, O., Åström, K., Ardö, H., & Bergsten, C. (2015). Continuous surveillance of pigs in a pen using learning-based segmentation in computer vision. In I. Halachmi (Ed.), *Precision livestock farming applications. Making sense of sensors to support farm management* (pp. 217–223). Wageningen Academic Publishers.

Nir, O., Parmet, Y., Werner, D., Adin, G., & Halachmi, I. (2018). 3D Computer-vision system for automatically estimating heifer height and body mass. *Biosystems Engineering, 173*, 4–10. https://doi.org/10.1016/j.biosystemeng.2017.11.014

Norton, T., Chen, C., Larsen, M. L. V., & Berckamns, D. (2019). Review: Precision livestock farming: building 'digital representations' to bring the animals closer to the farmer. *Animal, 13*(12), 3009–3017. https://doi.org/10.1017/S175173111900199X

Novas, R. V., & Usberti, F. (2017). Live monitoring in poultry houses: A broiler detection approach. *30th SIBGRAPI Conference on Graphics, Patterns and Images*. 17–20 October 2017 (pp. 216–222). doi: https://doi.org/10.1109/SIBGRAPI.2017.35

Oczak, M., Ismayilova, G., Costa, A., Viazzi, S., Sonoda, L. T., Fels, M., Bahr, C., Hartung, J., Guarino, M., Berckmans, D., & Vranken, E. (2013). Analysis of aggressive behaviours of pigs by automatic video recordings. *Computers and Electronics in Agriculture, 99*(1), 209–217. https://doi.org/10.1016/j.compag.2013.09.015

Oczak, M., Viazzi, S., Ismayilova, G., Sonoda, L. T., Roulston, N., Fels, M., Bahr, C., Hartung, J., Guarrino, M., Berckmans, D., & Vranken, E. (2014). Classification of aggressive behaviour in pigs by activity index and multilayer feed forward neural network. *Biosystems Engineering, 119*(1), 89–97. https://doi.org/10.1016/j.biosystemseng.2014.01.005

Oczak, M., Maschat, K., & Baumgartner, J. (2019). Dynamics of sows' activity housed in farrowing pens with possibility of temporary carting might indicate the time when sows should be confined in a crate before onset of farrowing. *Animals, 10*, E6. https://doi.org/10.3390/ani10010006

Oczak, M., Bayer, F., Vetter, S., Maschat, K., & Baumgartner, J. (2022). Comparison of the automated monitoring of the sow activity in farrowing pens using video and accelerometer data. *Computers and Electronics in Agriculture, 192*, 106517. https://doi.org/10.1016/j.compag.2021.106517

OECD. (2021). *Meat consumption (indicator)*. https://doi.org/10.1787/fa290fd0-en. Accessed: 21 December 2021.

OECD-FAO (Organisation for Economic Co-operation and Development/Food and Agriculture Organization of the United Nations). (2019). *Agricultural Outlook 2019–2028. Special Focus: Latin America*. OECD Publishing.

Okinda, C., Lu, M., Liu, L., Nyalala, I., Muneri, C., Wang, J., Zhang, H., & Shen, M. (2019). A machine vision system for early detection and prediction of sick birds: A broiler chicken model. *Biosystems Engineering, 188*©, 229–242. https://doi.org/10.1016/j.biosystemseng.2019.09.015

Okinda, C., Sun, Y., Nyalala, I., Korohou, T., Opiyo, S., Wang, J., & Shen, M. (2020). Egg volume estimation based on image processing and computer vision. *Journal of Food Engineering, 283*, 110041. https://doi.org/10.1016/j.foodeng.2020.110041

Okura, F., Ikuma, S., Makihara, Y., Muramatsu, D., Nakada, K., & Yagi, Y. (2019). RGB-D video-based individual identification of dairy cows using gait and texture analyses. *Computers and Electronics in Agriculture, 165*, 104944. https://doi.org/10.1016/j.compag.2019.104944

Oliveira, J. L., Xin, H., & Wu, H. (2019). Impact of feeder space on laying hen feeding behavior and production performance in enriched colony housing. *Animal, 13*(2), 374–383. https://doi.org/10.1017/S1751731118001106

Ollagnier, C., Kasper, C., Wallenbeck, A., Keeling, L., Bee, G., & Bigdeli, S. A. (2021). Machine learning algorithms can predict tail biting outbreaks in pigs using feeding behaviour records. *bioRxiv*. https://doi.org/10.1101/2021.05.11.443554

Omid, M., Firouz, M. S., Dehrouyeh, M. H., Mohtasebi, S. S., & Ahmadi, H. (2013). An expert egg grading system based on machine vision and artificial intelligence techniques. *Journal of Food Engineering, 118*(1), 70–77. https://doi.org/10.1016/j.foodeng.2013.03.019

Onyiro, O. M., Offer, J., & Brotherstone, S. (2008). Risk farctors and milk yield losses associated with lameness in Holstein-Friesian dairy cattle. *Animal, 2*(8), 1230–1237. https://doi.org/10.1017/S1751731108002279

Pandey, S., Kalwa, U., Kong, T., Guo, B., Gauger, P. C., Peters, D. J., & Yoon, K.-J. (2021). Behavioural monitoring tool for pig farmers: Ear tag sensors, machine intelligence, and technology adoption roadmap. *Animals, 11*, 2665. https://doi.org/10.3390/ani11092665

Park, M.-C., Jung, H.-C., Kim, T.-K., & Ha, O.-K. (2015). Design of cattle health monitoring system using wireless bio-sensor networks. In A. Hussain & M. Ivanovic (Eds.), *Electronics, communications and networks IV* (pp. 225–228).

Pastel, M. E., & Kujala, M. (2007). A probabilistic Neural Network Model for lameness detection. *Journal of Dairy Science, 90*(5), 2283–2292. https://doi.org/10.3168/jds.2006-267

Peake, K. A., Biggs, A. M., Smith, R. F., Christley, R. M., Routly, J. E., & Dobson, H. (2011). Effects of lameness, subclinical mastitis and loss of body condition on the reproductive performance of dairy cows. *Veterinary Record, 168*(11), 301. https://doi.org/10.1136/vr.c6180

Pereira, D. F., & Nääs, I. A. (2008). Estimating the thermoneutral zone for broiler breeders using behavioral analysis. *Computers and Electronics in Agriculture, 62*(1), 2–7. https://doi.org/10.1016/j.compag.2007.09.001

Pereira, D. F., de Alencar Nääs, I. A., & de Moura, D. J. (2006). Digital monitoring of broiler breeder behavior for assessment of thermal welfare. *Computers in Agriculture and Natural Resources, Proceedings of the 4th World Congress Conference*, 24–26 July 2006. doi: https://doi.org/10.13031/2013.21955

Pereira, D. F., Miyamoto, B. C. B., Maia, G. D. N., Sales, G. T., Magalhães, M. M., & Gates, R. S. (2013). Machine vision to identify boiler breeder behavior. *Computers and Electronics in Agriculture, 99*, 194–199. https://doi.org/10.1016/j.compag.2013.09.012

Pereira, E. M., Nääs, I. D., & Garcia, R. G. (2015). Vocalization of broilers can be used to identify their sex and genetic strain. *Journal of the Brazilian Association of Agricultural Engineering*. ISSN: 1809-4430. doi: https://doi.org/10.1590/1809-4430-Eng.Agric.v35n2p192-196/2015

Peter, A. T., & Bosu, W. T. K. (1986). Postpartum ovarian activity in dairy-cows – correlation between behavioural oestrus, pedometer measurement and ovulations. *Theriogenology, 26*(1), 111–115. https://doi.org/10.1016/0093-691x(86)90117-2

Petrie, A., & Watson, P. (2006). *Statistics for veterinary and animal science* (2nd ed.). Blackwell Publishing.

Pezzuolo, A., Guarino, M., Sartori, L., & Marinello, F. (2018a). A feasibility on the use of a structured light depth-camera for three-dimensional body measurements of dairy cows in free-stall barns. *Sensors (Basel), 18*(2), 673. https://doi.org/10.3390/s18020673

Pezzuolo, A., Guarino, M., Sartori, L., González, L. A., & Marinello, F. (2018b). On-barn pig weight estimation based on body measurements by a Kinect V1 depth camera. *Computers and Electronics in Agriculture, 148*, 29–36. https://doi.org/10.1016/j.compag.2018.03.003

Poursaberi, A., Bahr, C., Pluk, A., Van Neuffel, A., & Berckmans, D. (2010). Real-time automatic lameness detection based on back posture extraction in dairy cattle: Shape analysis of cow with image processing techniques. *Computers and Electronics in Agriculture, 74*(1), 110–119. https://doi.org/10.1016/j.compag.2010.07.004

Psota, E. T., Mittek, M., Pérez, L. C., Schmidt, T., & Mote, B. (2019). Multi-pig part detection and association with a fully convolutional network. *Sensors (Basel), 19*(4), 852. https://doi.org/10.3390/s19040852

Quanten, S., Valck, E., Cluydts, R., Aerts, J.-M., & Berckmans, D. (2006). Individualized and time-variant model for the functional link between thermoregulation and sleep onset. *Journal of Sleep Research, 15*(2), 183–198. https://doi.org/10.1111/j.1365-2869.2006.00519.x

Raj, A. A. G., & Jayanthi, J. G. (2018). IoT-based real-time poultry monitoring and health status identification. *IEEE. 11th International Symposium on Mechatronics and its Applications (ISMA)*, 4–6 March 2018. 17669265. doi: https://doi.org/10.1109/ISMA.2018.8330139

Ramirez, B. C., Hoff, S. J., & Harmon, J. D. (2018). Thermal environment sensor array: Part 2 applying the data to assess grow-finish pig housing. *Biosystems Engineering, 174*, 341–351. https://doi.org/10.1016/j.biosystemseng.2018.08.003

Reiter, K., & Bessei, W. (2010). Gait analysis in laying hens and broilers with and without leg disorders. *Equine Veterinary Journal, 29*(S23), 110–112. https://doi.org/10.1111/j.2042-3306.1997.tb0567.x

Ren, K., Karlsson, J., Liuska, M., Hartikainen, M., Hansen, I., & Jørgensen, G. H. M. (2020). A sensor-fusion-system for tracking sheep location and behaviour. *International Journal of Distributed Sensor Networks, 16*(5), 155014772092177. https://doi.org/10.1177/155014772092177

Renaudeau, D., Gilbert, H., & Noblet, J. (2012). Effect of climatic environment on feed efficiency in swine. In A. M. Gaines, B. A. Peterson, O. F. Mendoza, & J. F. Patience (Eds.), *Feed efficiency in swine* (pp. 183–210). Wageningen Academic Publishers.

Revilla, M., Friggens, N. C., Broudiscou, L. P., Lemonnier, G., Blanc, F., Ravon, L., Mercat, M. J., Billon, Y., Rogel-Gaillard, C., Le Floch, N., Estellé, J., & Muñoz-Tamayo, R. (2019). Towards the quantitative characterization of piglets' robustness to weaning: a modelling approach. *Animal, 16*, 1–11. https://doi.org/10.1017/S1751731119000843

Ribó, O., Korn, C., Meloni, U., Cropper, M., De Winne, P., & Cuypers, M. (2001). IDEA: A large-scale project on electronic identification of livestock. *Revue Scientifique et technique (International Office of Epizootics), 20*(2), 426–436. https://doi.org/10.20506/rst.20.2.1281

Ringgenberg, N., Fröhlich, E. K. F., Harlander-Matauschek, A., Toscano, M. J., Würbel, H., & Roth, B. A. (2015). Effects of variation in nest curtain design on pre-laying behaviour of domestic hens. *Applied Animal Behaviour Science, 170*, 34–43. https://doi.org/10.1016/j.applanim.2015.06.008

Roelofs, J. B., Van Eerdenburg, E. J., Soede, N. M., & Kemp, B. (2005). Pedometer readings for oestrus detection and as a predictor for time of ovulation in airy cattle. *Theriogenology, 64*(8), 1690–1703. https://doi.org/10.1016/j.theriogenology.2005.04.004

Romanini, C. E. B., Bahr, C., Viazzi, S., Van Hertem, T., Schlageter-Tello, A., Halachmi, I., Lokhorst, K., & Berckmans, D. (2013). Application of image based filtering to improve the performance of an automated lameness detection system for dairy cows. *Proceedings of Annual International Meeting of ASABE*. July 21–23, 2013. doi: https://doi.org/10.13031/aim.20131620675.

Rosa, G. J. M. (2021). Grand challenge in precision livestock farming. *Frontiers in Animal Science*. https://doi.org/10.3389/fanim/2021.650324

Rowe, E., Dawkins, M. S., & Gebhardt-Henrich, S. G. (2019). A systematic review of Precision Livestock Farming in the poultry sector: I technology focused on improving bird welfare? *Animals, 9*, 614. https://doi.org/10.3390/ani9090614

Roy, S., & Sarkar, S. K. (2016). RFID real time system for early detection of Avian Influenza for poultry based industry. *International Conference of Microelectronics, Computing and Communications (MicroCom)*. 23–25 January 2016. IEEE. 16177762. doi: https://doi.org/10.1109/MicroCom.2016.7522460

Ruegg, P. L. (2017). A 100-year review: Mastitis detection, management, and prevention. *Journal of Dairy Science, 100*(12), 10381–10397. https://doi.org/10.3168/jds.2017-13023

Ruminski, J., Kaczmarek, M., Renkielska, A., & Nowakowski, A. (2007). Thermal parametric imaging in the evaluation of skin burn depth. Institute of Electrical and Electronics Engineers Professional Technical Group of Bio-Medical Engineering (IEEE) Transactions on. *Biomedical Engineering, 54*, 303–312. https://doi.org/10.1109/TBME.2006.86607

Rutten, D. J., Velthuis, G. J., Steeneveld, W., & Hogeveen, H. (2013). Invited review: Sensors to support health management of dairy farms. *Journal of Dairy Science, 96*(4), 1928–1952. https://doi.org/10.3168/jds.2012-6107

Rutter, S. M. (2012). A "smart" future for ruminant livestock production? *Cattle Practice, 20*(3), 186–193. Rec. No. 20133027451.

Sa, J., Choi, Y., Lee, H., Chung, Y., Park, D., & Cho, J. (2019). Fast pig detection with a top-view camera under various illumination conditions. *Symmetry, 11*, 266. https://doi.org/10.3390/sym11020266

Sadeghi, M., Banakar, A., Khazaee, M., & Soleimani, M. R. (2015). An intelligent procedure for the detection and classification of chickens infected by clostridium perfringens based on their vocalization. *Brazilian Journal of Poultry Science, 17*(4), 537–544. https://doi.org/10.1590/1516-635X1704537-544

Sales, G. T., Green, A. R., Gates, R. S., Brown-Brandl, T. M., & Eigenberg, R. A. (2015). Quantifying detection performance of a passive low-frequency RFID system in an environmental preference chamber for laying hens. *Computers and Electronics in Agriculture, 114*, 261–268. https://doi.org/10.1016/j.compag.2015.03.008

Samsonova, J. V., Safronova, V. A., & Osipov, A. P. (2015). Pretreatment-free lateral flow enzyme immunoassay for progesterone detection in whole cows' milk. *Talanta, 132*, 685–689. https://doi.org/10.1016/j.talanta.2014.10.043

Sassi, N. B., Averós, X., & Estevez, I. (2016). Technology and poultry welfare. *Animals, 6*, 62. https://doi.org/10.3390/ani6100062

Scheel, C., Traulsen, I., & Krieter, J. (2015). Detecting lameness in sows using acceleration data from ear tags. In I. Halachmi (Ed.), *Precision livestock farming applications. Making sense of sensors to support farm management* (pp. 39–44). Wageningen Academic Publishers.

Schofield, C. P., Marchant, J. A., White, R. P., Brandl, N., & Wilson, M. (1999). Monitoring pig growth using a prototype imaging system. *Journal of Agricultural Engineering Research, 72*(3), 205–210. https://doi.org/10.1006/jaer.1998.0365

Schön, P. C., Puppe, B., & Manteuffel, G. (2004). Automated recording of stress vocalisations as a tool to document impaired welfare in pigs. *Animal Welfare, 13*, 105–110.

Seo, J., Sa, J., Choi, Y., Chung, Y., Park, D., & Kim, H. (2019). A YOLO-based separation of touching-pigs for smart pig farm applications. In: *21st International Conference on Advanced Communication Technology (ICACT)*. 17–20 February, 2019. PyeongChang Kwangwoon_do, Korea (South). IEEE, 18636911. doi: https://doi.org/10.23919/ICACT.2019.8701968.

Shao, B., & Xin, H. (2008). Areal-time computer vision assessment and control of thermal comfort for group housed pigs. *Computers and Electronics in Agriculture, 62*, 15–21. https://doi.org/10.1016/j.compag.2007.09.006

Shen, R.-N., Lei, P.-K., Liu, Y.-C., Haung, Y.-J., & Lin, J.-L. (2016). Development of temperature measurement system for broiler flock with thermal imaging. *Engineering in Agriculture, Environment and Food, 9*(3), 291–295. https://doi.org/10.1016/j.eaef.2016.03.001

Shi, C., Zhang, J., & Teng, G. (2019). Mobile measuring system based on LabVIEW for pig body components estimation in a large-scale farm. *Computers and Electronics in Agriculture, 156*, 399–405. https://doi.org/10.1016/j.compag.2018.11042

Siegford, J. M., Berezowski, J., Biswas, S. K., Daigle, C. L., Gebhardt-Henrich, S. G., Hernandez, C. E., Thurner, S., & Toscano, M. J. (2016). Assessing activity and location of individual hens in large groups using modern technology. *Animals, 6*(2), 10. https://doi.org/10.3390/ani6020010

Silva, M., Exadactylos, V., Ferrari, S., Guarino, M., Aerts, J.-M., & Berckmans, D. (2009). The influence of respiratory disease on the energy envelope dynamics of pig cough sounds. *Computers and Electronics in Agriculture, 69*, 80–85. https://doi.org/10.1016/j.compag.2009.07.002

Smith, D., Lyle, S., Berry, A., Manning, N., Zaki, M., & Neely, A. (2015). *Data and Analytics: Internet of animal health things opportunities and challenges. Technical Report*. University of Cambridge. https://doi.org/10.13140/RG.2.1.1113.8409

Smouse, E. P., Focardi, S., Moorcroft, R. P., Kie, G. J., Forester, D. J., & Morales, M. J. (2010). Stochastic modelling of animal movement. Philosophical transactions of the Royal Society of

London. *Series B, Biological Sciences, 365*(1550), 2201–2211. https://doi.org/10.1098/rstb.2010.0078

Soerensen, D. D., & Pedersen, L. J. (2015). Infrared skin temperature measurements for monitoring health in pigs: a review. *Acta Veterinaria Scandinavica, 57*(1), 5. https://doi.org/10.1186/s13028-015-0094-2

So-In, C., Poolsanguan, S., & Rujirakul, K. (2014). A hybrid mobile environmental and population density management system for smart poultry farms. *Computers and Electronics in Agriculture., 109*, 287–301. https://doi.org/10.1016/j.compag.2014.10.004

Song, X., & van der Tol, P. P. J. (2010). Automatic detection of clinical mastitis in Astronaut A3 TM milking robot. *Proceeding of the 1st North American Conference on Precision Dairy Management*. March 2010. Accessed June 13, 2021, from http://www.precisiondairy.com/proceedings/s8vandertol.pdf

Sonoda, L. T., Fels, M., Rauterberg, S., Viazzi, S., Ismayilova, G., Oczak, M., Bahr, C., Guarino, M., Vranken, E., Berckmans, D., & Hartung, J. (2013). Cognitive enrichment in piglet rearing: an approach to enhance animal welfare and to reduce aggressive behaviour. *ISRN Veterinary Science, 1*, 389186. https://doi.org/10.1155/2013/389186

Spensley, J. C., Wathes, C. M., Waran, N. K., & Lines, J. A. (1995). Behavioural and physiological responses of piglets to naturally occurring sounds. *Applied Animal Behaviour Science, 44*(2–4), 277. https://doi.org/10.1016/0168-1591(95)92367-3

Špinka, M. (2009). Behaviour of pigs. In P. Jensen (Ed.), *The ethology of domestic animals: An introductory text* (2nd ed., pp. 177–191). CABI Publishing.

Stadig, L. M., Rodenburg, T. B., Ampe, B., Reubens, B., & Tuyttens, F. A. M. (2018). An automated positioning system for monitoring chickens' location: Effects of wearing a backpack on behaviour, leg health and production. *Applied Animal Behaviour Science, 198*, 83–88. https://doi.org/10.1016/j.applanim.2017.09.016

Statham, P., Green, L., Bichard, M., & Mendl, M. (2009). Predicting tail biting from behavior of pigs prior to outbreaks. *Applied Animal Behaviour Science, 121*(3–4), 157–164. https://doi.org/10.1016/j.applanim.2009.09.011

Stevenson, J. S., Smith, M. W., Jaeger, J. R., Corah, L. R., & Lefever, D. G. (1996). Detection of estrus by visual observation and radiotelemetry in peripubertal, estrus-synchronized beef heifers. *Journal of Animal Science, 74*(4), 729–735. https://doi.org/10.2527/1996.744729x

Stevenson, J. S., Hill, S., Nebel, R., & DeJarnette, J. (2014). Ovulation timing and conception risk after automated activity monitoring in lactating dairy cows. *Journal of Dairy Science, 97*(7), 4296–4308. https://doi.org/10.3168/jds.2013-7873

Stygar, A. H., Dolechck, K. A., & Kristensen, A. R. (2017). Analyses of body weight patterns in growing pigs: A new view on body weight in pigs for frequent monitoring. *Animal, 12*, 295–302. https://doi.org/10.1017/S1751731117001690

Stygar, A. H., Gómez, Y., Bertesell, G. V., Costa, E. D., Canall, E., Niemi, J. K., Llonch, P., & Pastell, M. (2021). A systematic review on commercially available and validated sensor technologies for welfare assessment for dairy cattle. *Frontiers in Veterinary Science, 8*, 634338. https://doi.org/10.3389/fvets.2021.634338

Süli, T., Halas, M., Benyeda, Z., Boda, R., Belák, S., Martínez-Avilés, M., Fernández-Carrión, E., & Sánchez-Vizcaíno, J. M. (2017). Body temperature and motion: Evaluation of an online monitoring system in pigs challenged with Porcine Reproductive & Respiratory Syndrome Virus. *Research in Veterinary Science, 114*, 482–488. https://doi.org/10.1016/j.rvsc.2017.09.021

Sun, K., Ma, L., Pan, L., & Tu, K. (2017). Sequenced wave signal extraction and classification algorithm for duck egg crack on-line detection. *Computers and Electronics in Agriculture, 142*(A), 429–439. https://doi.org/10.1016/j.compag.2017.09.034

Sztandarski, P., Marchewka, J., Wojciechowski, F., Riber, A. B., Gunnarsson, S., & Horbańczuk, J. O. (2021). Associations between weather conditions and individual range use by commercial and heritage chickens. *Poultry Science, 100*(8), 101265. https://doi.org/10.1016/j.psj.2021.101265

Tactacan, G. B., Guenter, W., Lewis, N. J., Rodriguez-Lecompte, J. C., & House, J. D. (2009). Performance and welfare of laying hens in conventional and enriched cages. *Poultry Science, 88*(4), 698–707. https://doi.org/10.3382/ps.2008-00369

Talling, J. C., Waran, N. K., Wathes, C. M., & Lines, J. A. (1996). Behavioural and physiological responses of pigs to sound. *Applied Animal Behaviour Science, 48*(3–4), 187–201. https://doi.org/10.1016/0168-1591(96)01029-5

Talling, J. C., Waran, N. K., Wathes, C. M., & Lines, J. A. (1998). Sound avoidance by domestic pigs depends upon the characteristics of the signal. *Applied Animal Behaviour Science, 58*, 255–266. https://doi.org/10.1016/S0168-159(97)00142-1

Taneja, M., Byabazaire, J., Jalodia, N., Davy, A., Olariu, C., & Malone, P. (2020). Machine learning based fog computing assisted data-driven approach for early lameness detection in dairy cattle. *Computers and Electronics in Agriculture, 171*, 105286. https://doi.org/10.1016/j.compag.2020.105286

Taylor, P. S., Hemsworth, P. H., Groves, P. J., Gebhardt-Henrich, S. G., & Rault, J.-L. (2017). Ranging behaviour of commercial free-range broiler chickens 2: Individual variation. *Animals, 7*(7), 55. https://doi.org/10.3390/ani7070055

Taylor, P. S., Hemsworth, P. H., Groves, P. J., Gebhardt-Henrich, S. G., & Rault, J.-L. (2018). Ranging behavior relates to welfare indicators pre- and post-range access in commercial broilers. *Poultry Science, 97*(6), 1861–1871. https://doi.org/10.3382/ps/pey060

Taylor, P. S., Hemsworth, P. H., Groves, P. J., Gebhardt-Henrich, S. G., & Rault, J.-L. (2020). Frequent range visits further from the shed relate positively to free-range broiler chicken welfare. *Animal, 14*(1), 138–149. https://doi.org/10.1017/S1751731119001514

Temple, D., Courboulav, V., Manteca, X., Velarde, A., & Dalmau, A. (2012). The welfare of growing pigs in five different production systems: Assessment of feeding and housing. *Animal, 6*(4), 656–667. https://doi.org/10.1017/S1751731111001868

Thomsen, P. T., Munksgaard, L., & Tøgersen, F. A. (2008). Evaluation of a lameness system for dairy cows. *Journal of Dairy Science, 91*(1), 119–126. https://doi.org/10.3168/jds.2007-0496

Thorton, P. K. (2010). Livestock production: Recent trends, future prospects. *Philosophical Transactions of the Royal Society of London B: Biological Sciences, 365*(1554), 2853–2867. https://doi.org/10.1098/rstb.2010.0134

Thorup, V. M., Nielsen, B. L., Robert, P.-E., Giger-Reverdin, S., Konka, J., Michie, C., & Friggens, N. C. (2016). Lameness affects cow feeding but not rumination behavior as characterized from sensor data. *Frontiers of Veterinary Science, 10*(3), 37. https://doi.org/10.3389/fvets.2016.00037

Tian, M., Guo, H., Chen, H., Wang, Q., Long, C., & Ma, Y. (2019). Automated pig counting using deep learning. *Computers and Electronics in Agriculture, 163*, 104840. https://doi.org/10.1016/j.compag.2019.05.049

Tzanidakis, C., Simitzis, P., Arvanitis, K., & Panagakis, P. (2021). An overview of the current trends in Precision Pig Farming technologies. *Livestock Science, 249*, 104530. https://doi.org/10.1016/j.livsci.2021.104530

Vaintrub, M. O., Levit, H., Chincarini, M., Fusaro, I., Giammarco, M., & Vignola, G. (2021). Review: Precision livestock farming automats and new technologies: possible applications in extensive dairy sheep farming. *Animal, 15*, 100143. https://doi.org/10.1016/j.animal.2020.100143

Van der Sluis, M., de Haas, Y., de Klerk, B., Rodenburg, T. B., & Ellen, E. D. (2020). Assessing the activity of individual group-housed broilers through life using a passive radio frequency identification system – A validation study. *Sensors, 20*(13), 3612. https://doi.org/10.3390/s20133612

Van der Sluis, M., Ellen, E. D., de Haas, Y., de Klerk, B., & Rodenburg, T. B. (2021). Automated activity recordings throughout life in broilers: Heritability of activity and the relationship with body weight. *54th Congress of the International Society for Applied Ethology*. 26–6 August 2021. Bangalore, India. Online]. Available from: https://edepot.wur.nl/556936

Van der Zande, L. E., Guzhva, O., & Rodenburg, T. B. (2021). Individual detection and tracking of group housed pigs in their home pen using computer vision. *Frontiers of Animal Science, 2*, 669312. https://doi.org/10.3389/fanim.2021.669312

Van Hertem, T., Maltz, E., Antler, A., Romanini, C. E. B., Viazzi, S., Bahr, C., Schlageter-Tello, A., Lokhorst, C., Berckmans, D., & Halachmi, I. (2013). Lameness detection based on multivariate continuous sensing of milk yield, rumination, and neck activity. *Journal of Dairy Science, 96*(7), 4286–4298. https://doi.org/10.3168/jds.2012-6188

Van Hertem, T., Rooijakkers, L., Berckmans, D., Fernández, A. P., Norton, T., & Vranken, E. (2017). Appropriate data visualisation is key to precision livestock farming acceptance. *Computers and Electronics in Agriculture, 138*, 1–10.

Van Hertem, T., Norton, T., Berckmans, D., & Vranken, E. (2018). Predicting broiler gait scores from activity monitoring and flock data. *Biosystems Engineering, 173*, 93–102. https://doi.org/10.1016/j.biosystemseng.2018.07.002

Van Hirtum, A., & Berckmans, D. (2002). Automated recognition of spontaneous versus voluntary cough. *Medical Engineering and Physics, 24*(7–8), 541–545. https://doi.org/10.1016/S1350-4533(02)00056-5

Vandermeulen, J., Bahr, C., Tullo, E., Fontana, I., Ott, S., Kashiha, M., Guarino, M., Moons, C. P. H., Tuyttens, F. A. M., Niewold, T. A., & Berckmans, D. (2015). Discerning pig screams in production environments. *PLoS One, 10*, e0123111. https://doi.org/10.1371/journal.pone.0123111

Viazzi, S., Bahr, C., Schlageter-Tello, A., Van Hertem, T., Romanini, C. E. B., Pluk, A., Halachmi, I., Lokhorst, C., & Berckmans, D. (2013). Analysis of individual classification of lameness using automatic measurement of back posture in dairy cattle. *Journal of Dairy Science, 96*(1), 257–266. https://doi.org/10.3168/jds.2012-5806

Viazzi, S., Ismayilova, G., Oczak, M., Sonoda, L. T., Fels, M., Guarino, M., Vranken, E., Hartung, J., Bahr, C., & Berckmans, D. (2014). Image feature extraction for classification of aggressive interactions among pigs. *Computers and Electronics in Agriculture, 104*, 57–62. https://doi.org/10.1016/j.compag.2014.03.010

Virgilio, A. D., Morales, J. M., Lambertucci, S. A., Shepard, E. L. C., & Wilson, R. P. (2018). Multi-dimensional Precision Livestock Farming: a potential toolbox for sustainable rangeland management. *PeerJ, 6*, e4867. https://doi.org/10.7717/peerj.4867

Von Borell, E., Bünger, B., Schmidt, T., & Horn, T. (2009). Vocal-type classification as a tool to identify stress in piglets under on-farm conditions. *Animal Welfare, 18*(4), 407–416.

Voulodimos, A. S., Patrikakis, C. Z., Sideridis, A. B., Ntafis, V. A., & Xylouri, E. M. (2010). A complete farm management system based on animal identification using RFID technology. *Computers and Electronics in Agriculture, 70*(2), 380–388. https://doi.org/10.1016/j.compag.2009.07.009

Vranken, E., & Berckmans, D. (2017). Precision Livestock Farming for pigs. *Animal Frontiers, 7*, 32–37. https://doi.org/10.2527/af.2017.0106

Wallenbeck, A., & Keeling, L. J. (2013). Using data from electronic feeders on visit frequency and feed consumption to indicate tail biting outbreaks in commercial pig production. *Journal of Animal Science, 91*, 2879–2884. https://doi.org/10.2527/jas.2012-5848

Wang, Y., Yang, W., Winter, P., & Walker, L. (2008). Walk-through weighing of pigs using a machine vision and an artificial neural network. *Biosystems Engineering, 100*, 117–125. https://doi.org/10.1016/j.biosystemseng.2007.08.008

Wang, K., Guo, H., Ma, Q., Su, W., & Zhu, D. (2018). A portable and automatic Xtion-based measurement system for pig body size. *Computers and Electronics in Agriculture, 148*, 291–298. https://doi.org/10.1016/j.compag.2018.03.018

Wang, X., Zhao, X., He, Y., & Wang, K. (2019a). Cough sound analysis to assess air quality in commercial weaner barns. *Computers and Electronics in Agriculture, 160*, 8–13. https://doi.org/10.1016/j.compag.2019.03.001

Wang, K., Liu, K., Xin, H., Chai, L., Wang, Y., Fei, T., Oliveira, J., Pan, J., & Ying, Y. (2019b). An RFID-based automated individual perching monitoring system for group housed poultry. *Transactions of the ASABE, 62*(3), 695–704. https://doi.org/10.13031/trans.12105

Warriss, P. D., Pope, S. J., Brown, S. N., Wilkins, L. J., & Knowles, T. G. (2006). Estimating the body temperature of groups of pigs by thermal imaging. *The Veterinary Record, 158*(1), 331–334. https://doi.org/10.1136/vr.158.10.331

Wathes, C. M., Kristensen, H. H., Aerts, J. M., & Berckmans, D. (2008). Is precision livestock farming an engineer's daydream or nightmare, an animal's friend or foe, and a farmer's panacea or pitfall? *Computers and Electronics in Agriculture, 64*(1), 2–10. https://doi.org/10.1016/j.compag.2008.05.005

Wegner, B., Spiekermeier, I., Nienhoff, H., Große-Kleimann, J., Rohn, K., Meyer, H., Pite, H., Gerhardy, H., Kreienbrock, L., Beilage, E. G., Kemper, N., & Fels, M. (2019). Status quo analysis of noise levels in pig fattening units in Germany. *Livestock Science, 230*, 103847. https://doi.org/10.1016/j.livsci.2019.103847

Werkheiser, I. (2018). Precision Livestock Farming and farmers' duties to livestock. *Journal of Agricultural and Environmental Ethics, 31*(2), 181–195. https://doi.org/10.1007/s10806-018-9720-0

Wilhelmsson, S., Yngvesson, J., Jönsson, L., Gunnarsson, S., & Wallenbeck, A. (2019). Welfare Quality® assessment of a fast-growing and a slower-growing broiler hybrid, reared until 10 weeks and fed a low-protein, high protein or mussel-meal diet. *Livestock Science, 219*, 71–79. https://doi.org/10.1016/j.livsci.2018.11.010

Wiltschko, R., Thalau, P., Gehring, D., Nießner, C., Ritz, T., & Wiltschko, W. (2015). Magnetoreception in birds: the effect of radio-frequency fields. *Journal of the Royal Society Interface, 12*(103), 20141103. https://doi.org/10.1098/rsif.2014.1103

Wishart, H. M. (2019). *Precision livestock farming: potential application for sheep systems in harsh environments*. Thesis submitted in fulfilment of the degree of PhD: University of Edinburgh. 2019. [Online]. Available from: https://era.ed.ac.uk/handle/1842/36196

Woolford, M., Claycomb, R., Jago, J., Davis, K., Ohnstad, I., Wieliczko, R., Copeman, P., & Bright, K. (2004). Automatic dairy farming in New Zealand using extensive grazing systems. In A. Meijering, H. Hogeveen, & C. De-Koning (Eds.), *Automatic milking: A better understanding* (pp. 280–285). Wageningen Academic Publishers.

Wu, D., Wu, Q., Yin, X., Jiang, B., Wang, H., He, D., & Song, H. (2020). Lameness detection of dairy cows based on the YOLOv3 deep learning algorithm and a relative step size characteristic vector. *Biosystems Engineering, 189*, 150–163. https://doi.org/10.1016/j.biosystemseng.2019.11.017

Xiao, D., Feng, A., & Liu, J. (2019a). Detection and tracking of pigs in natural environments based on video analysis. *International Journal of Agricultural and Biological Engineering, 12*, 116–126. https://doi.org/10.25165/j.ijabe.20191204.4591

Xiao, L., Ding, K., Gao, Y., & Rao, X. (2019b). Behavior-induced health condition monitoring of caged chickens using binocular vision. *Computers and Electronics in Agriculture, 156*, 254–262. https://doi.org/10.1016/j.compag.2018.11.022

Xin, H. (1999). Assessing swine thermal comfort by image analysis of postural behaviours. *Journal of Animal Science, 77*, 1–9. https://doi.org/10.2527/1999.77suppl_21x

Xin, H., & Shao, B. (2002). Real-time assessment of swine thermal comfort by computer vision. In: *Proceedings of the world congress of computers in agriculture and natural resources*. 13–15 March 2002. Iguacu Falls (pp. 362–369). doi: https://doi.org/10.13031/2013.8353.

Xiong, X., Lu, M., Yang, W., Duan, G., Yuan, Q., Shen, M., Norton, T., & Berckmans, D. (2019). An automatic head surface temperature extraction method for top-view thermal image with individual broiler. *Sensors, 19*(23), 5286. https://doi.org/10.3390/s19235286

Xu, B., Wang, W., Falzon, G., Kwan, P., Guo, L., Sun, Z., & Li, C. (2020). Livestock classification and counting in quadcopter aerial images using Mask R-CNN. *International Journal of Remote Sensing, 41*(21), 8121–8142. https://doi.org/10.1080/01431161.2020.1734245

Yang, Y., Teng, G., Li, B., & Shi, Z. (2006). Measurement of pig weight based on computer vision. *Transactions of the Chinese Society of Agricultural Engineering, 22*, 127–131.

Ye, W., & Xin, H. (2000). Thermographical quantification of physiological and behavioural responses of group-housed young pigs. *Transactions of the ASAE, 43*, 1843–1851. https://doi.org/10.13031/2013.3089

Ye, C.-W., Yousaf, K., Qi, C., Liu, C., & Chen, K.-J. (2020a). Broiler stunned state detection based on an improved fast region-based convolutional neural network algorithm. *Poultry Science, 99*(1), 637–646. https://doi.org/10.3382/ps/pez564

Ye, C.-W., Yu, Z.-W., Kang, R., Yousaf, K., Qi, C., Chen, K., & Huang, Y.-P. (2020b). An experimental study of stunned state detection for broiler chickens using an improved convolution neural network algorithm. *Computers and Electronics in Agriculture, 170*, 105284. https://doi.org/10.1016/j.compag.2020.105284

Yiakoulaki, M. D., Hasanagas, N. D., Michelaki, E., Tsiobani, E. T., & Antoniou, I. (2018). Social network analysis of sheep grazing different plant functional groups. *Grass and Forage Science, 74*(1), 129–140. https://doi.org/10.1111/gfs.12398

You, J., Lou, E., Afrouziyeh, M., Zukiwsky, N. M., & Zuidhof, M. J. (2021a). A supervised machine learning method to detect anomalous real-time broiler breeder body weight data recorded by a precision feeding system. *Computers and Electronics in Agriculture, 185*, 106171. https://doi.org/10.1016/j.compag.2021.106171

You, J., Lou, E., Afrouziyeh, M., Zukiwsky, N. M., & Zuidhof, M. J. (2021b). Using an artificial neural network to predict the probability of oviposition events of precision-fed broiler breeder hens. *Poultry Science, 100*(8), 101187. https://doi.org/10.1016/j.psj.2021.101187

Yu, W., & Huang, S. (2018). Traceability of food safety based on block chain and RFID technology. *11th International Symposium on Computational Intelligence and Design (ISCID)*. 8–9 December 2018. doi: https://doi.org/10.1109/ISCID.2018.00083

Yu, G.-M., & Maeda, T. (2017). Inline progesterone monitoring in the dairy industry. *Trends in Biotechnology, 35*(7), 579–582. https://doi.org/10.1016/j.tibtech.2017.02.007

Zaninelli, M., Redaelli, V., Luzi, F., Bronzo, V., Mitchell, M., Dell'Orto, V., Bontempo, V., Cattaneo, D., & Savoini, G. (2018). First evaluation of infrared thermography as a tool for the monitoring of udder health status in farms of dairy cows. *Sensors (Basel), 18*(3), 862. https://doi.org/10.3390/s18030862

Zhang, F. Y., Hu, Y. M., Chen, L. C., Guo, L. H., Duan, W. J., & Wang, L. (2016). Monitoring behavior of poultry based on RFID radio frequency network. *International Journal of Agricultural and Biological Engineering, 9*(6), 139–147. https://doi.org/10.3965/ijabe.20160906.1568

Zhang, L., Gray, H., Ye, X., Collins, L., & Allison, N. (2018). Automatic individual pig detection and tracking in pig farms. *Sensors, 19*, 1188. https://doi.org/10.3390/s19051188

Zhang, Z., Zhang, H., & Liu and T. (2019). Study on body temperature detection of pig based on infrared technology: A review. *AI in Agriculture, 1*, 14–26.

Zhang, X., Kang, X., Feng, N., & Liu, G. (2020). Automatic recognition of dairy cow mastitis from thermal images by deep learning detector. *Computers and Electronics in Agriculture, 178*, 105754. https://doi.org/10.1016/j.compag.2020.105754

Zhang, J., Zhuang, Y., Ji, H., & Teng, G. (2021). Pig weight and body size estimation using a multiple output regression Convolutional Neural Network: A fast and fully automatic method. *Sensors, 21*, 3218. https://doi.org/10.3390/s21093218

Zhao, K., Bewley, J. M., He, D., & Jin, X. (2018). Automatic lameness detection in dairy cattle based on leg swing analysis with image processing technique. *Computers and Electronics in Agriculture, 148*, 226–236. https://doi.org/10.1016/j.compag.2018.03.014

Zhuang, X., & Zhang, T. (2019). Detection of sick broilers by digital image processing and deep learning. *Biosystems Engineering, 179*, 106–116. https://doi.org/10.1016/j.boisystemseng.2019.01.003

Zhuang, X., Bi, M., Guo, J., Wu, S., & Zhang, T. (2018). Development of an early warning algorithm to detect sick broilers. *Computers and Electronics in Agriculture, 144*, 102–113. https://doi.org/10.1016/j.compag.2017.11.032

Zobel, G., Weary, D. M., Leslie, K., Chapinal, N., & von Keyserlingk, M. A. G. (2015). Technical note: Validation of data loggers for recording behavior in dairy goats. *Journal of Dairy Science, 98*(2), 1082–1089. https://doi.org/10.3168/jds.2014-8635

Zucali, M., Lovareli, D., Celozzi, S., Bacenetti, J., Sandrucci, A., & Bava, L. (2020). Management options to reduce the environmental impact of dairy goat milk production. *Livestock Science, 231*, 103888. https://doi.org/10.1016/j.livsci.2019.103888

Zuidhof, M. J., Schneider, B. L., Carney, V. L., Korver, D. R., & Robinson, F. E. (2014). Growth, efficiency, and yield of commercial broilers from 1957, 1978, and 2005. *Poultry Science, 93*(12), 2970–2982. https://doi.org/10.3382/ps.2014-04291

Zukiwsky, N. M., Girard, T., & Zuidhof, M. (2020). Effect of an automated marking system on aggressive behavior of precision-fed broiler breeder chicks. *The Journal of Applied Poultry Research, 29*(4), 786–797. https://doi.org/10.1016/j.japr.2020.06.005

Zukiwsky, N. M., Afrouziyeh, M., Robinson, F. E., & Zuidhof, M. J. (2021). Broiler growth and efficiency in response to relaxed maternal feed restriction. *Poultry Science, 100*(4), 100993. https://doi.org/10.1016/j.psj.2021.01.016

Environmental Assessment of Rural Road Construction in India

Yash Aryan, Anil Kumar Dikshit, and Amar Mohan Shinde

Abstract Construction of road pavement consumes significant quantities of raw materials and energy and involves use of heavy machineries. In India, total length of road network is 6.38 million km, out of which highest share (71%) is that of the rural roads. The main aim of this chapter is to present the development of detailed life cycle inventory for material phase and construction phase of rural road pavement, followed by evaluation of their environmental impacts. The scope of the study included the raw material extraction, the transportation of raw materials to the site and the construction of rural road using heavy machineries for a typical rural road section in Bihar, India, using GaBi software version 10.5. Maintenance and end-of-life phases were beyond the scope. The functional unit adopted was 1 km of road pavement constructed. A total of seven impact categories were considered. The pavement was constructed in three layers viz. granular sub-base layer, water-bound macadam layer and top bituminous layer. It was found that the granular sub-base layer construction had the highest contribution towards the impact on environment for all impact categories due to the high consumption of aggerates.

Keywords Environmental impacts · Life cycle assessment · Rural road

1 Introduction

Roads are essential for economic development of country as they provide connectivity and access to resources, consumer goods, jobs and services. The rate of road construction and road density is increasing with increase in population. As per an

Y. Aryan · A. K. Dikshit (✉)
Environmental Science and Engineering Department, Indian Institute of Technology Bombay, Mumbai, Maharashtra, India
e-mail: dikshit@iitb.ac.in

A. M. Shinde
Department of Civil Engineering, Manipal Institute of Technology, Manipal Academy of Higher Education, Manipal, Karnataka, India

F. P. García Márquez, B. Lev (eds.), *Sustainability*, International Series in Operations Research & Management Science 333,
https://doi.org/10.1007/978-3-031-16620-4_16

estimate, a total of 43 million km of paved roads existed worldwide in 2010, while additional 14 million km shall be constructed by 2030 (Grael et al., 2021). India has about 6.386 million km of road network, the second largest in the world. Out of the total road network, National Highways (NH)/Expressway are 132,500 km, State Highways (SH) 186,528 km, District Roads 632,154 km, Rural Roads 4,535,511 km, Urban Roads 544,683 km and Project Roads 354,921 km (MoRTH, 2021). The total road length in India has been increasing year by year and there has been around 47% increase in the total road length in last 20 years. The total road length has been presented in Fig. 1. The majority length road pavements in India are surfaced while some are non-surfaced. The surfaced and total road length in India has been presented in Fig. 2.

Certainly, development of transportation infrastructure is necessary and has an important role in the rapid economic growth of country but not without its impact on the environment. Road networks which include road pavements and associated infrastructure have significant environmental impacts on the environment. Commonplaces like this need no reference. The life cycle assessment (LCA) provides a comprehensive approach for assessing the environmental impacts of any product/ process or system. LCA tools has been already widely in use to evaluate the environmental impacts of various process/product/system all over the world. The procedures of LCA have been standardized as a part of the ISO 14000 Environmental Management Standards (EMS) (Guinée & Lindeijer, 2002). The methodological framework is given in ISO 14044 that defines four inter-related stages of LCA (1) goal and scope definition, (2) inventory assessment, (3) impact assessment and (4) improvement assessment (AzariJafari et al., 2016). Over the last two decades, several researchers have used LCA methods to assess the environmental impacts of road pavements (Jiang & Wu, 2019). In the present chapter, the environmental impacts due to the construction of rural roads were assessed. As the present study

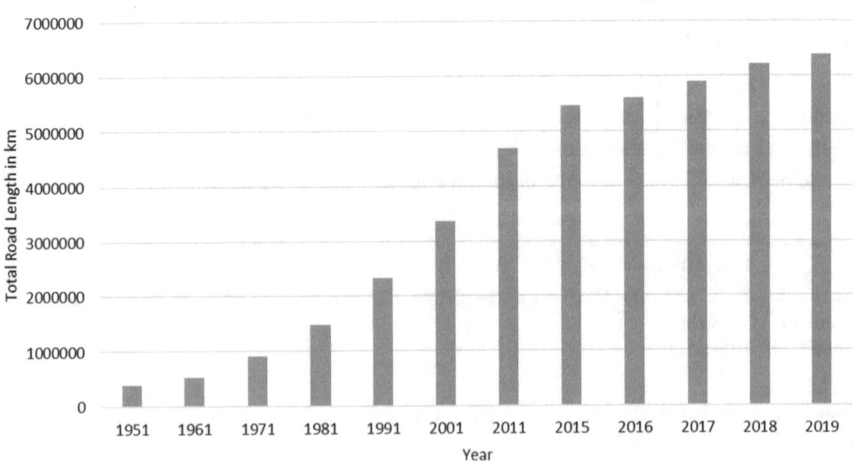

Fig. 1 Total road length in India (MoRTH, 2018)

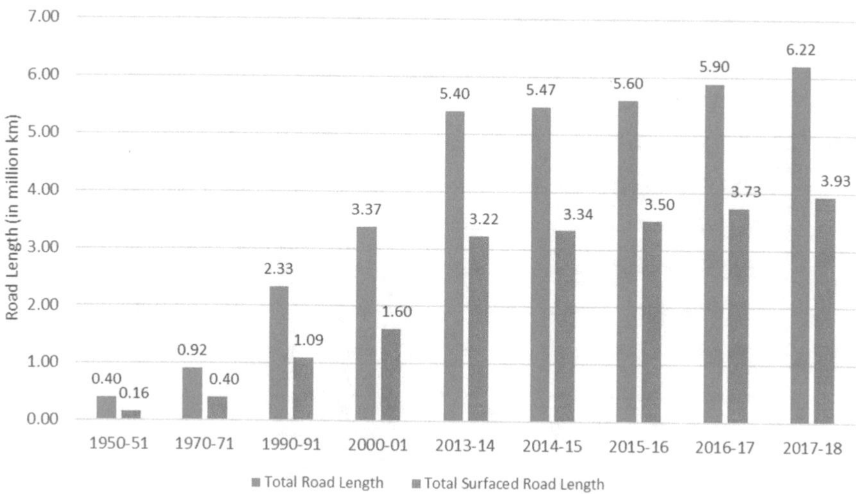

Fig. 2 Surfaced and total road length in India (in million km) (MoRTH, 2018)

considered only material and construction phase, the rain water effect, end-of-life, recycling or renewal considerations were out of scope of the present study.

1.1 Different Classes of Roads in India

Roads are classified into six classes in India. These are discussed as follows:

Expressway: Expressways are the highest class of roads in India with superior highway infrastructure and higher access specifications permitting only fast-moving vehicles. These are initially constructed as eight lanes and have the provision to be expanded to four or more number of lanes further. The expressway may be owned by Central or State Government.

National Highways: National highways connect various states, state capitals and major parts of India and run through the length and breadth of the country. These highways are owned by the Central Government.

State Highways: State Highways (SH) are owned by the State Government. These highways traverse the length and width of a state connecting the state capital, district headquarters and important towns and cities and link up with the National Highways and adjacent State Highways.

Major District Roads: Major District Roads (MDR) are important roads within the district connecting important places in the district and link up with SH or NH. These roads are owned by the State Government.

Project Roads: Project roads are special category roads. These roads come under the jurisdiction of different organizations such as Public Sector Units (PSUs), the

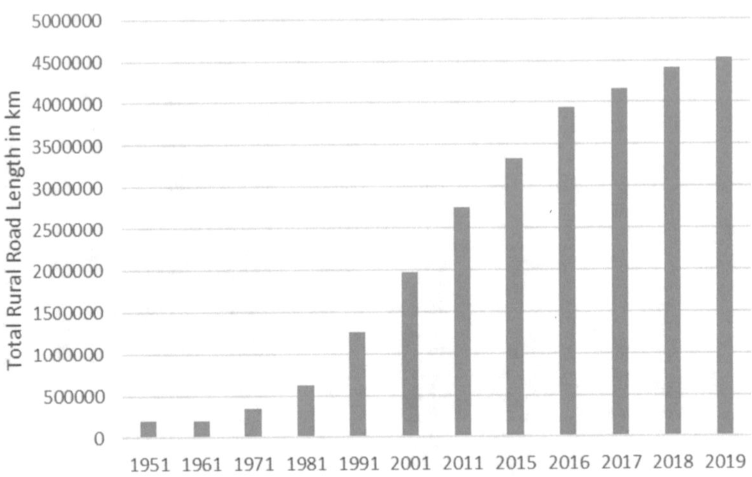

Fig. 3 Total rural road length in India (MoRTH, 2018)

government-owned enterprises, Railways, Border Road Organization, state departments such as forest, power etc.

Rural Roads: Rural roads are low-volume traffic roads but are very important for rural development as rural roads provide access to economic and social services. These roads connect the village to various districts through major district roads, state highways or national highways. The rural roads in India are constructed by the Panchayati Raj Institutions such as Zilla Parishad, Panchayat Samiti, Gram Panchayat, State Rural Works Departments and State Public Works Departments. The year-wise growth of rural road length in India has been shown in Fig. 3.

The Panchayati Raj Institutions had constructed the highest length of rural roads (57%) followed by the State Rural Works Departments/Public Works Departments (22%) and the Prime Minister Rural Road Scheme (21%).

As mentioned earlier, rural roads constitute 71% of the total road network having the highest growth rate compared to other classes of roads in India. The category-wise growth of different classes of roads in India is shown in Fig. 4.

1.2 Types of Pavements

There are two types of pavements all over the world (1) flexible pavement and (2) rigid pavement. These are discussed in following sections:

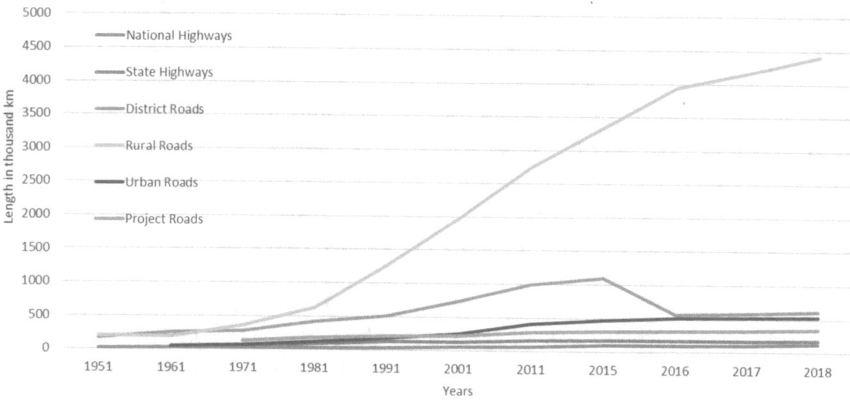

Fig. 4 Category-wise growth of different classes of roads in India till 2018 (MoRTH, 2018)

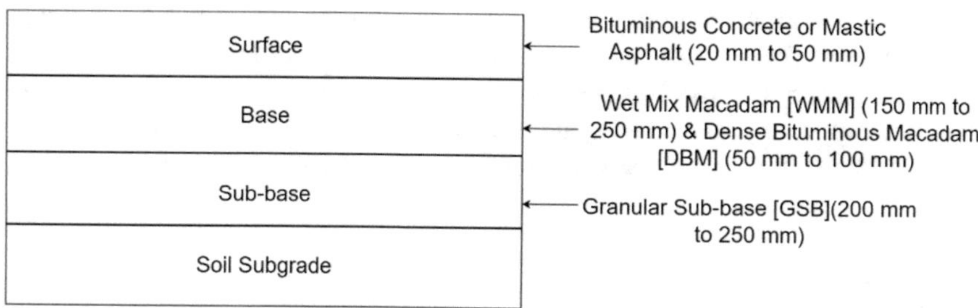

Fig. 5 Typical structure of flexible pavement

1.2.1 Flexible Pavement

Flexible pavement consists of four components or layers (1) soil subgrade, (2) sub-base course, (3) base course and (4) surface course (Fig. 5)

The subgrade (SG) consists of natural soil, which is compacted to increase the carrying capacity. The sub-base consists of roller compacted aggregates and is called as granular sub-base (GSB). Typical depth of this layer varies from 200 to 250 mm. The aggregates size for this layer ranges from 4.75 to 53 mm as per the grading depending on the pavement design. The next layer is called base layer. It consists of wet mix macadam (WMM) of 150–250 mm and dense mix macadam (DBM) of 50–100 mm thickness. The aggregates size for WMM layer ranges from 74 μm to 45 mm and for DBM layer ranges from 4.75 to 37.5 mm. Finally, the wearing course (WC) consists of 20–50 mm thick bituminous concrete or mastic asphalt having aggregates size varying from 5 to 20 mm. This type of pavement is flexible in its structural action under load because it has low or negligible flexural strength. The flexible pavement layers transfer the load to the lower layers by grain-to-grain transfer. Generally, the initial cost of construction of flexible pavement is less as compared to rigid pavement, but flexible pavement has less life span in comparison with rigid pavement. The depth of each layer of flexible pavement depends on the carrying capacity of the soil and traffic volume. The flexible pavement does not require any joint to take care of the contraction during the heat in summer months or expansion during the cold in winter months as bitumen is a plastic material and accommodates these minor changes due to temperature.

1.2.2 Technologies for Construction of Flexible Pavement

In India, currently there are two technologies for the construction of flexible pavement (1) Hot mix asphalt technology and (2) Cold mix asphalt technology.

Hot mix technology is a quite old and simple technology. In this technology, the bitumen is heated to 160–170 °C to make it workable and aggregates are heated to 150–160 °C consuming enormous amount of heat. This method or technology is most extensively used in India for the construction of flexible pavement.

In *cold mix technology*, asphalt mix is produced by mixing unheated mineral aggregate with foamed or emulsified bitumen. Cold mix technology is economical during operation and also less polluting compared to the hot mix technology as it does not require any heating of aggregate. Cold asphalt mixes are typically suitable for light to medium traffic roads when used in base and surface courses.

1.2.3 Rigid Pavement

Rigid pavements consist of three components or layers, (1) soil subgrade, (2) base course and (3) cement concrete slab (Fig. 6).

The subgrade constitutes compacted natural soil and the base layer consists of granular sub-base (GSB) and dry lean concrete (DLC). The top surface layer consists of pavement quality concrete (PQC) on which the wheel of the vehicle interacts. Similar to the flexible pavement, the thickness of each layer of the rigid pavement depends on the soil strength and traffic volume. In general, the depth of GSB layers varies from 150 to 200 mm, DLC layers varies from 50 to 100 mm and the surface PQC layer varies from 200 to 300 mm. The aggregates size in DLC layer should not exceed 25 mm and aggregates cement ratio must not exceed 15:1 having cement

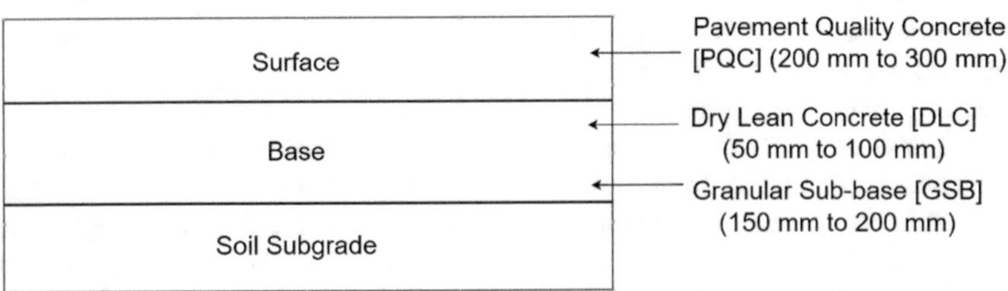

Fig. 6 Typical structure of rigid pavement

content not less than 150 kg/m^3. Similarly, the aggregates size in PQC layer should not exceed 25 mm having cement content about 400 kg/m^3. These pavements have high flexural rigidity and transfer the wheel load by its slab action. The rigid pavement has lesser number of layers and suitable for soil with low bearing capacity as the concrete slab can distribute the traffic load into a large area with small depth thus reducing stress on the soil strata. Rigid pavement consists of dowel bars and tie bars. Dowel bars (longitudinal joints) are short steel bars that provide a mechanical connection between slabs without restricting horizontal joint movement. Tie bars (transverse joints) are either deformed steel bars or connectors used to hold the adjacent faces of slab edge at the same level. The rigid pavement also has provision for accommodating the expansion and contraction due to temperature change. These joints are called as contraction joints and are placed transverse at regular intervals.

In India, roads consist of wearing course built from bitumen (flexible pavement) or concrete (rigid pavement) or WBM (no wearing course). The surfaced rural roads (having either bituminous or cement concrete wearing course) in India are around 65%, while the remaining 35% of the rural roads in India are WBM roads (having no wearing course) (MoRTH, 2018).

The total rural road length and surfaced rural road length in different states in India are shown in Fig. 7.

2 Methodology

The four steps involved in assessing environmental impacts associated with construction of rural roads are discussed in forthcoming sections:

2.1 *Goal and Scope Definition*

In goal and definition phase, LCA practitioner defines the purpose of the proposed study, assumptions of the study and the target audience of the study. The goal definition influences the choice of modelling depending upon the decision context. Based on goal, scope of the study shall elaborate on what to study, how to study, where the assessment is framed and outlined. Functional unit is defined in goal and scope phase which defines the function or service quantitatively for which the assessment is performed (Hauschild et al., 2018). Functional unit also enables different systems to be treated as functionally equivalent for comparison (Institute for Environment and Sustainability, 2010). The goal of the present study is to develop life cycle inventory (LCI) for the rural road pavement construction and to evaluate the environmental impacts due to the construction of the rural road pavement. The functional unit adopted for the present study is construction of 1 km of rural road. The scope includes the material phase and the construction phase of the pavement life cycle. LCA was promised above however, maintenance is omitted?

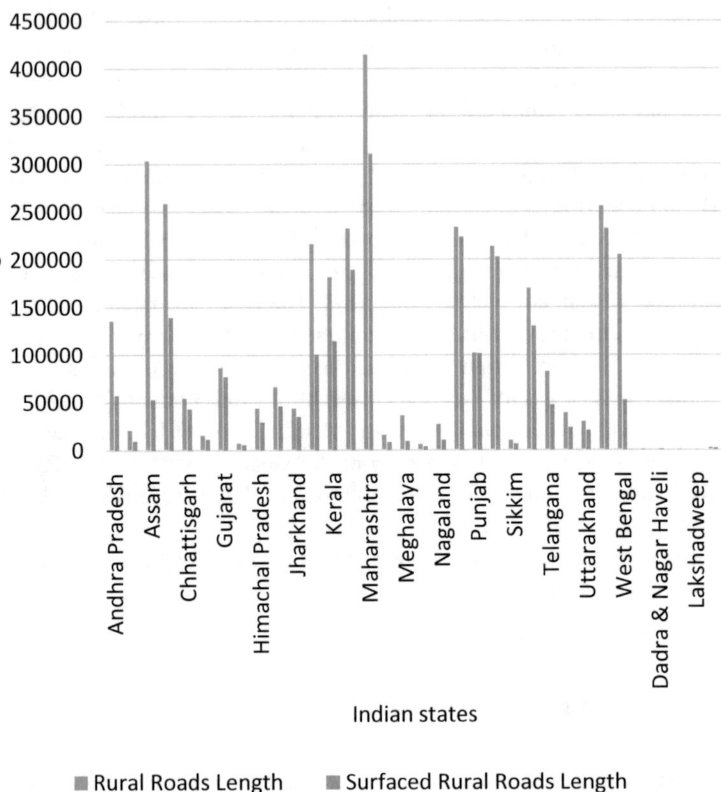

Fig. 7 State/UT-wise details of total and surfaced rural road's length as of March 2018 (MoRTH, 2018)

System boundary is also defined in this phase depicting the inputs and outputs that are to be considered in the study. The system boundary for the present study is shown in Fig. 8.

The material phase includes the extraction, processing and production of raw materials used during the construction phase of the rural road pavement. The transportation of raw materials to the construction site and machineries used during the construction and have also been considered for impact assessment.

2.2 *Life Cycle Inventory (LCI)*

LCI is an important stage of any LCA study as the quality of LCA work depends upon the quality of data collected. The development of LCI is one of the most time and effort consuming as it involves collection, compilation and interpretation based upon goal and scope of the study. Collecting and compiling the best possible data for

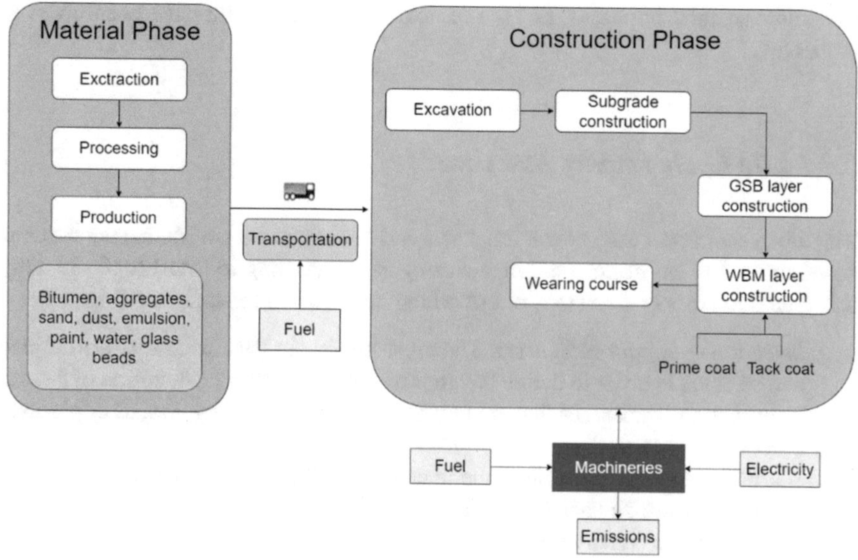

Fig. 8 System boundary for the present study

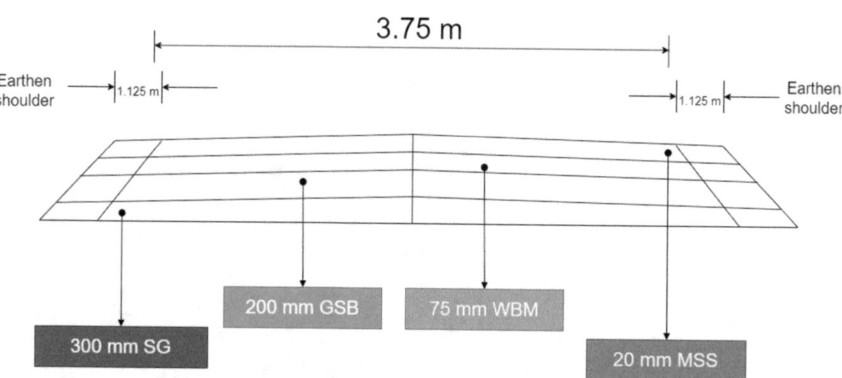

Fig. 9 Cross-section of the rural road

the extensive system boundary was one of the major challenges in the present study. The data required to develop LCI was collected from Rural Works Department, road contractors and site engineers in the state of Bihar in India. The rural road considered for this study is a single lane having width 3.75 m and its cross-section has been shown in Fig. 9. The present study used various India specific data from the professional database of GaBi 10.5 LCA software for better and accurate impact assessment. The transportation distance of crushed aggregates, moorum, fine aggregates, bitumen and emulsion to the road construction site were 27 km, 24 km, 19 km, 33 km and 30 km, respectively. The impacts due to the production and maintenance of trucks were not considered in the system boundary (however this is a major part of the environmental impact of LCI). The detailed collected data have been given in Appendix.

2.3 Life Cycle Impact Assessment

The life cycle impact assessment translates and quantifies the environmental burdens based on LCI to environmental impact category. According to ISO 14040 standard, this phase consists of five steps out of which first three are mandatory (Fig. 10):

(i) *Selection of impact categories:* The most suitable impact categories are defined and selected in order to define the impacts caused by the consumption of natural resources and emissions that are induced during the life cycle stages of product, process or system.
(ii) *Classification:* In this step, the elementary flows from the inventory are assigned to the impact categories according to their ability to contribute to different environmental problems.
(iii) *Characterization:* After the impact categories are selected and defined, the quantification of contributions to the different impact categories is performed

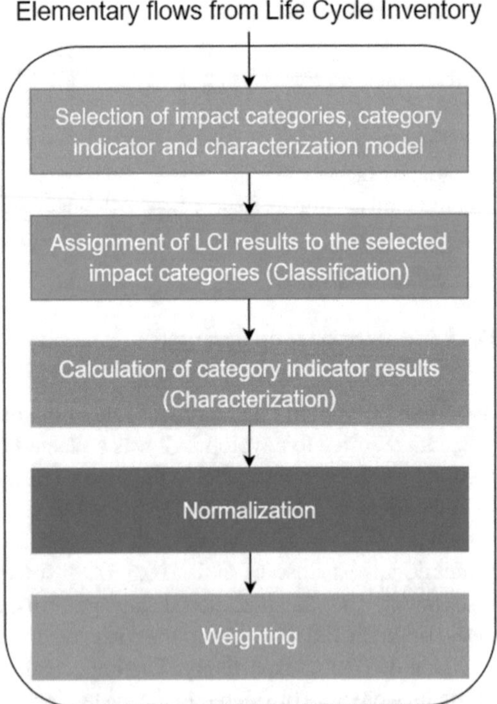

Fig. 10 Five steps of life cycle impact assessment

by assigning the relative contribution of each input and output within the product system to the environmental load and translating them into indicators.

(iv) *Normalization:* It is an optional step in LCA and is used to simplify the interpretation of the results. Normalization solves the incompatibility of units and shows to what extent an impact category indicator result has a relatively high or a relatively low value compared to a reference.

(v) *Weighting:* This step is also optional like Normalization and is difficult to perform for midpoint impact categories. Weighting expresses the relative importance of the impact categories and involves assigning distinct quantitative weights to all impact categories. Weighting is useful when decisions are to be made on the basis of the LCA results with other information like economic costs of the alternatives.

There are two approaches for characterization (1) midpoint approach and (2) end-point approach. Midpoint approach is also called problem-oriented approach as the cause-effect chain starts with a specific process or an activity which led to emissions and consequently, primary changes in the environment appear (Menoufi, 2011). End point approach is also known as damage-oriented methods as this method gives the results in terms of damage caused by the change in environment.

The present study used GaBi 10.5 commercial LCA software to perform life cycle impact assessment. GaBi is an object-oriented and process-based proprietary LCA software with inbuilt database. It uses the processes, flows and plans to assess the environmental impacts. The plan is at the top level in hierarchy, representing the actual processes connected by flows. A total of 7 impact categories were considered for consistent and meaningful understanding of environmental impacts viz. five impact categories [Abiotic Depletion (ADP fossil), Acidification Potential (AP), Eutrophication Potential (EP), Global Warming Potential (GWP) and Human Toxicity Potential (HTP)] from CML 2001 (University of Leiden 2001) method, 1 impact category [Fine Particulate Matter Formation (PMF)] from ReCiPe 2016 midpoint method and 1 impact category [Smog Air (SA)] from Tool for Reduction and Assessment of Chemicals and Other Environmental Impacts (TRACI) 2.1 method.

The LCI was prepared for the study based on all input data provided by Rural Works Department, Government of Bihar. The preparation of GaBi plans requires the creation of flows and processes and then plans for each road pavement layers were interconnected to obtain top-level plan for the impact assessment. The GaBi plan for a typical layer, say WBM layer of rural road pavement is presented in Fig. 11. This plan shows the flows and the processes including the raw material and machinery required for construction of the WBM layer. The plan also includes the prime coat and tack coat which is a part of the WBM layer. Similarly, plans for GSB and WC layers were prepared.

The top-level plan for construction of rural road has been shown in Fig. 12. The top-level plan of the rural road pavement includes the plan for all the layers of the rural road pavement and interconnected into a top plan for the evaluation of environmental impacts.

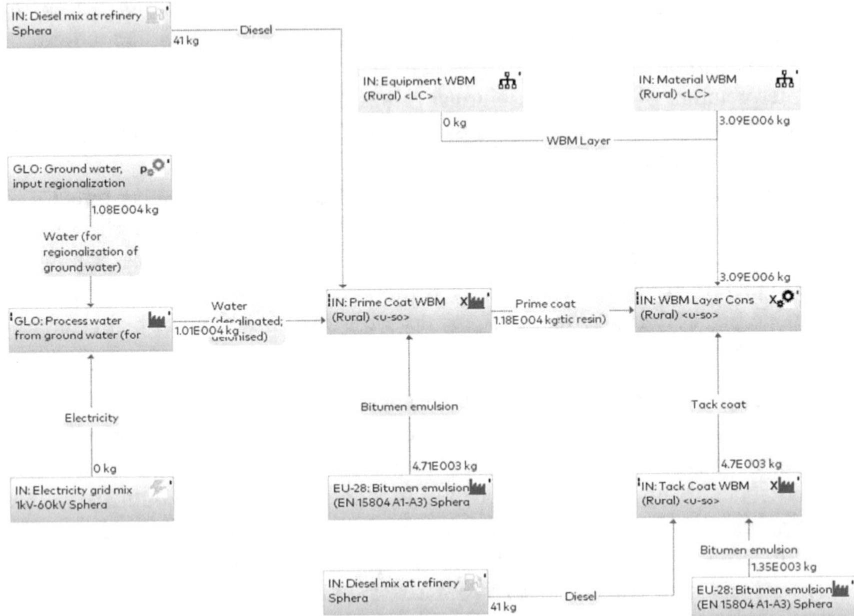

Fig. 11 GaBi plan for WBM layer to calculate the environmental impacts

2.4 *Interpretation*

The fourth and last step is interpretation also called as improvement analysis. In this step, the results of the impact assessment step are presented, discussed and interpreted. The interpretation step is presented here in next section as a part of results and discussions.

3 Results and Discussions

The model was run with data collected from site engineers and contractors, Rural Works Department, Bihar, and the relative contribution of different layers on each impact category for material and construction phase of rural road pavement has been illustrated in Fig. 13. The construction of WC layer had the highest contribution to ADP fossil impact category (36.4%), followed by construction of WBM layer (25.1%). The above result can be attributed to the use of significant amount of bitumen in WC layer. The AP impact category was dominated by the sub-base GSB layer (38.6%) due to the use of the coarse aggregates or gravel in significant amount. Grael et al. (2021) also observed similar higher contribution from gravel on AP impact category. The sub-base GSB layer had the highest contribution to EP impact

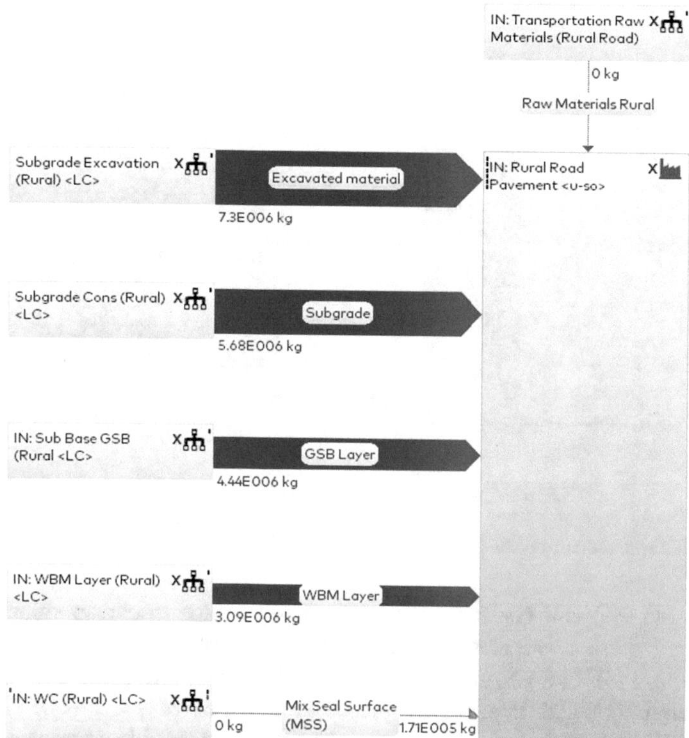

Fig. 12 Top-level plan for rural road pavement to calculate the environmental impacts

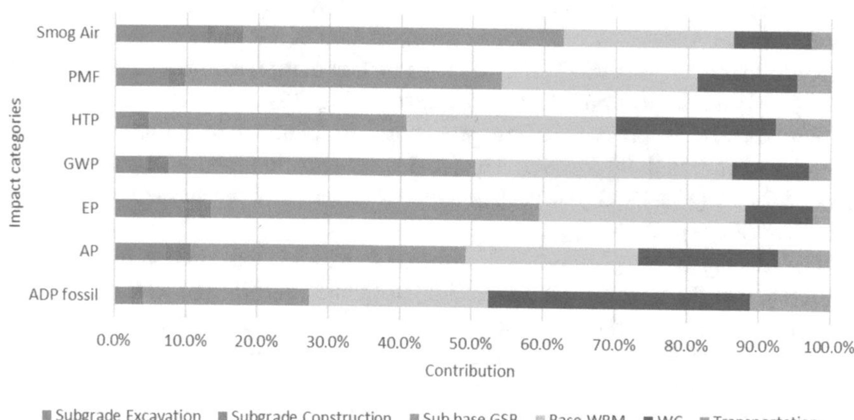

Fig. 13 The relative contribution to each impact category from the construction of rural road pavement. *PMF* particulate matter formation, *HTP* human toxicity potential, *GWP* global warming potential, *EP* eutrophication potential, *AP* acidification potential, *ADP* abiotic depletion, *GSB* granular sub-base, *WBM* water-bound macadam, *WC* wearing course

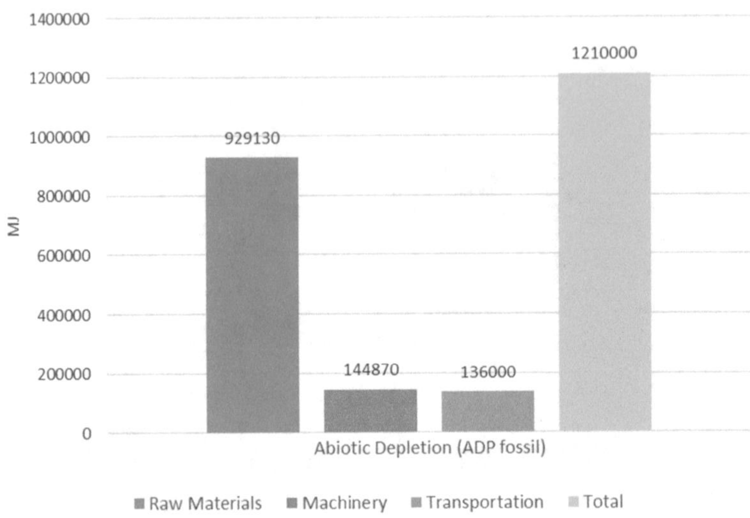

Fig. 14 Environmental impacts in MJ/km on ADP fossil impact category

category (45.9%) and GWP impact category (43%) due to consumption of large quantity of coarse aggregates. Similarly, the GSB layer had highest contribution to HTP (36%), PMF (44.4%) and Smog Air (44.9%) impact categories.

The environmental impacts are presented in terms of per km of rural road constructed. The rural road pavement construction had significant impact on ADP fossil impact category due to the use of bitumen and crushed aggregated in significant amount (Fig. 14). The total energy consumption impacts were 121,0000 MJ/km, out of which raw material contributed 929,130 MJ (77%), construction machinery contributed 144,870 MJ (12%) and transportation of raw materials to the site was responsible for 136,000 MJ (11%). The construction of rural road pavement had total 132 kg SO_2 eq. impact on AP impact category (Fig. 15) due to use of bitumen as sulphur dioxide is emitted during production of bitumen in refinery. The raw material contributed 99 kg SO_2 eq. (75%), construction machinery contributed 24 kg SO_2 eq. (18%) and transportation of raw materials to the site was responsible for 10 kg SO_2 eq. (7%) (Fig. 15).

The EP impact category was less affected due to the construction of rural road pavement and the total impact was 21 kg Phosphate eq. (Fig. 16). Out of this total impact, the raw material contributed 17 kg Phosphate eq. (80%), construction machinery contributed 4 kg Phosphate eq. (19%) and transportation of raw materials to the site contributed 1 kg Phosphate eq. (1%). The construction of rural road pavement had significantly high environmental impacts on GWP impact category (Fig. 17). The total GWP impact was 50,100 kg CO_2 eq., out of which raw material contributed 42,238 kg CO_2 eq. (84%), construction machinery contributed 6302 kg CO_2 eq. (12%) and transportation of raw materials to the site was responsible for 1560 kg CO_2 eq. (4%). The less impacts on GWP due to the transportation of raw

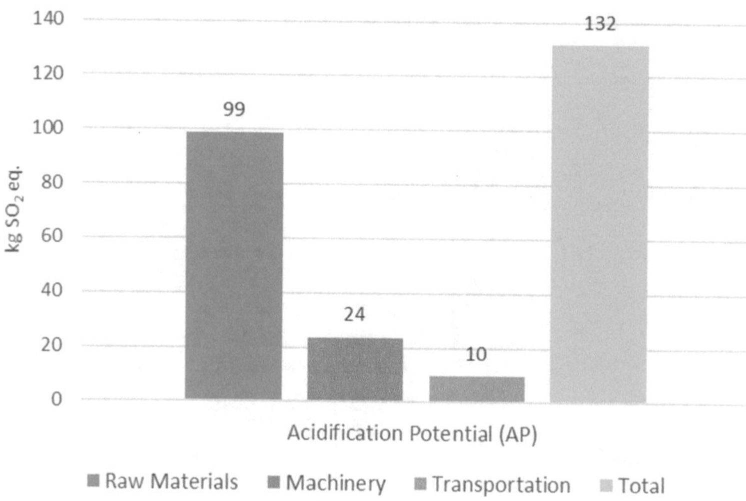

Fig. 15 Environmental impacts in SO_2 eq. on AP impact category

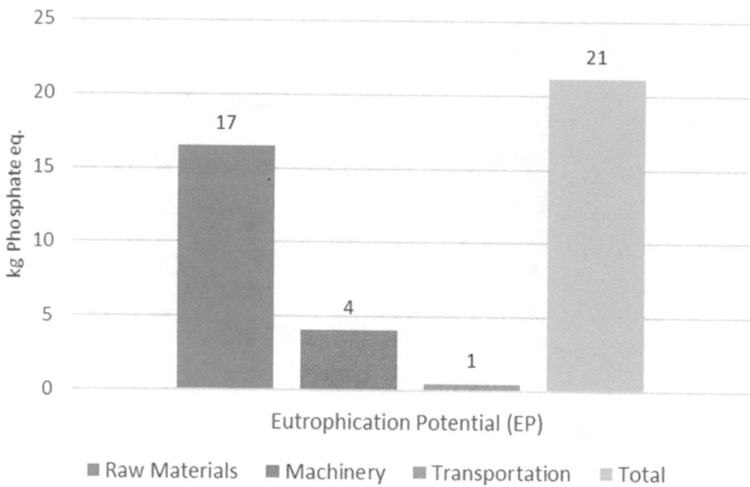

Fig. 16 Environmental impacts in kg phosphate eq. on EP impact category

materials to the site can be attributed to the implementation of BSVI engines and norms in India.

The rural road construction had significant impact on HTP impact category (2510 kg DCB eq.) as shown in Fig. 18. The high impacts on HTP can be attributed to the extraction and processing of the coarse aggregates and production of the bitumen. Out of this total impact, the raw material contributed 2072 kg DCB eq. (82%), construction machinery contributed 243 kg DCB eq. (10%) and transportation of raw materials to the site contributed 195 kg DCB eq. (8%).

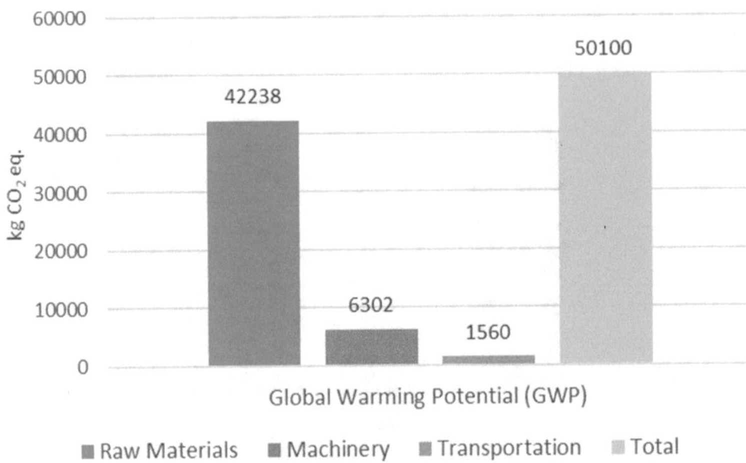

Fig. 17 Environmental impacts in kg CO_2 eq. on GWP impact category

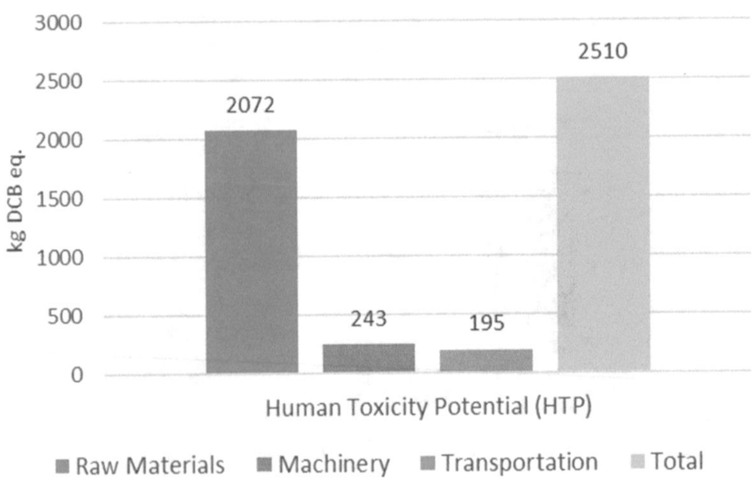

Fig. 18 Environmental impacts in kg DCB eq. on HTP impact category

The construction of rural road had very less impacts on PMF impact category (49 kg PM2.5 eq.) (Fig. 19). The raw material contributed 40 kg PM2.5 eq. (82%), construction machinery contributed 7 kg PM2.5 eq. (14%) and transportation of raw materials to the site contributed 2 kg PM2.5 eq. (4%). Smog Air impact category was significantly affected due to the construction of rural road pavement construction (Fig. 20). The total impact on Smog Air impact category was 3260 kg O_3 eq. out of which raw material contributed 2297 kg O_3 eq. (70%), construction machinery contributed 871 kg O_3 eq. (27%) and transportation of raw materials to the site was responsible for 93 kg O_3 eq. (3%).

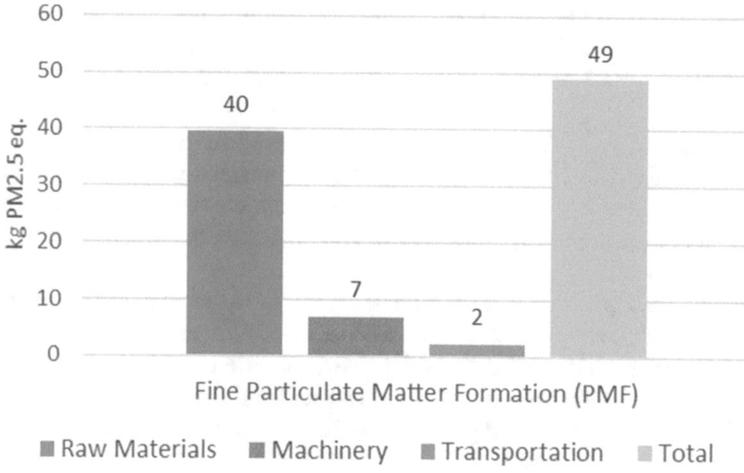

Fig. 19 Environmental impacts in kg PM2.5 eq. on PMF impact category

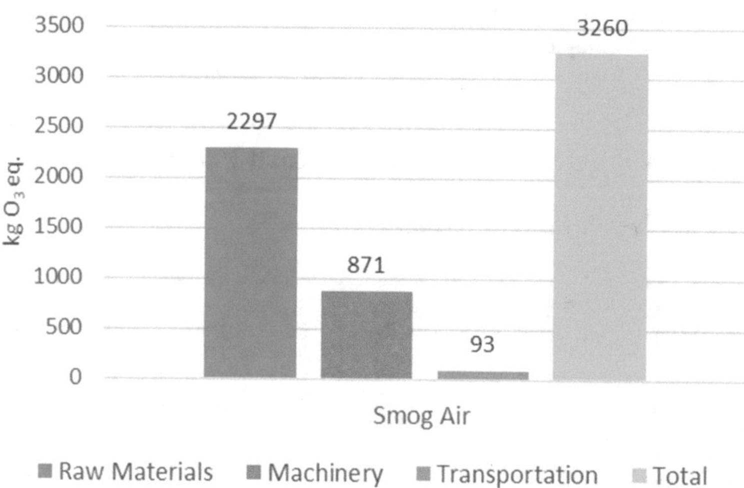

Fig. 20 Environmental impacts in O_3 eq. on smog air impact category

4 Conclusions

The present study evaluated environmental impacts due to the construction of 1 km rural road pavement in India. The study adopted LCA approach and considered total seven impact categories. The study found that the construction of sub-base Granular Sub-Base layer of rural road pavement had the highest contribution on six impact categories except ADP fossil impact category. The rural road pavement construction had significant high impacts on Abiotic Depletion fossil (1,210,000 MJ), Global Warming Potential (50,100 kg CO_2 eq.), Human Toxicity Potential (2510 kg DCB eq.) and Smog Air (2297 kg O_3 eq.) impact categories while very less impacts on Acidification Potential (132 kg SO_2 eq.), Eutrophication Potential (21 kg Phosphate eq.) and Particulate Matter Formation (49 kg PM2.5 eq.) impact categories. The extraction, processing and consumption of raw materials were responsible for the highest impacts on all the impact category, followed by the use of construction machinery and at last, the transportation of raw materials to the site. The raw materials contributed to around 80% impacts, while construction machinery contributed about 15% and transportation of raw materials contributed about 5% to overall impacts.

The rural road growth rate is the highest compared to the other classes of roads in India. The rural roads in India are being constructed at rapid rate about average of 357,141 km every year, so that the village areas in different states get connected to the nearby markets, districts and other important places within the proper rural road network. The environmental impacts due to the construction of significant length of rural roads in India will be highly significant considering the impact due to 1 km rural road construction, which the present study evaluated. These environmental impacts can be reduced by using the recycled aggregates for the construction of rural road pavement, which is possible as rural roads carry low traffic volume and design speed is also less compared to other superior classes of roads in India.

Appendix: Collected Data for Rural Road Project

Rural Flexible Pavement (Bihar)

Construction equipment emissions					
Construction equipment vehicle (CEV III)	CO_2 (kg)	CO (kg)	HC (kg)	NO_x (kg)	PM (kg)
Excavation					
Hydraulic Excavator	2012	25.10	5.33	14.85	1.51
Subgrade construction					
Road Grader/Motor Grader	442	4.68	2.53	1.87	0.38
Vibratory Roller (12 tonne)	510	6.35	2.46	2.62	0.40
GSB construction					
Road Grader/Motor Grader	540	5.71	2.32	3.05	0.46
Vibratory Roller (12 tonne)	498	6.20	3.12	1.84	0.38
WBM construction					
Road Grader	172	1.83	1.04	0.68	0.15
Vibratory Roller (12 tonne)	160	2.0	0.9	0.7	0.12
WC construction					
Paver	100	1.0	0.58	0.44	0.05
Smooth Wheeled Tandem Roller	200	2.66	1.11	1.0	0.16

Construction materials and fuel consumption	
Excavation (per km)	
Earthwork	2754.68 m^3
Number of hours worked	39.54 h
Diesel consumed by Hydraulic Excavator (1.1 cum bucket capacity)	711.72 l
Subgrade construction	
Volume required	2144.63 m^3
Road Grader/Motor Grader number of hours worked	13.96 h
Road Grader/Motor Grader diesel consumed	181.47 l
Vibratory Roller (12 tonne) number of hours worked	10.41 h
Vibratory Roller (12 tonne) diesel consumed	156.15 l
Water required	187.65 Kl

GSB (Plant Mix Method)	Length 1000 m	Width 7.86 m	Depth 0.20 m
Wet Mix Plant 80 hp @ 100 tonne per hour working time run by 82.5 KVA generator	44 h		
Diesel fuel consumed to run WMP plant	827.2 l		
Road Grader/Motor Grader number of hours worked	17.05 h		
Diesel consumed by Road Grader/Motor Grader	221.65 l		
Vibratory Roller (12 tonne) number of hours worked	10.17 h		
Diesel consumed by Vibratory Roller (12 tonne)	152.55 l		
Materials	**Volume**	**Mass**	
Grading I	2116.15 cum	4443.92 tonne	
53–26.5 mm @ 27.5%	581.94 cum	1222.07 tonne	
26.5–9.5 mm @ 22.5%	476.13 cum	999.87 tonne	
9.5–4.75 mm @ 10%	211.62 cum	444.40 tonne	
4.75 mm below @ 40%	846.46 cum	1777.56 tonne	
Water required	355.52 Kl		

WBM (water-bound macadam)	Length 1000 m	Width 6.72 m	Depth 0.075 m
Road Grader/Motor Grader number of hours worked	5.46 h		
Diesel consumed by Road Grader/Motor Grader	71.0 l		
Vibratory Roller (12 tonne) number of hours worked	3.26 h		
Diesel consumed by Vibratory Roller (12 tonne)	48.93 l		
Materials	**Volume**		**Mass**
Gravel	1478.4 cum		2365.44 tonne
Moorum	403.20 cum		725.76 tonne

(continued)

WBM (water-bound macadam)	Length 1000 m	Width 6.72 m	Depth 0.075 m
Water required	201.6 Kl		

Prime coat over WBM		
Prime Coat	6720 sqm	11.76 tonne
Mechanical broom working hours	2.0 h	
Mechanical broom fuel consumption	18.0 l	
Air compressor 250 cfm working hours	2.0 h	

Tack coat over WBM	
Mechanical broom working hours	2.0 h
Mechanical broom fuel consumption	18.0 l
Air compressor 250 cfm working hours	2.0 h
Air compressor 250 cfm fuel consumed	22.0 l
Bitumen pressure distributor working hours	1.86 h
Bitumen pressure distributor fuel consumed	9.30 l
Materials	
RS1 grade bitumen emulsion @ 0.20 kg/sqm	1.35 tonne
Bitumen pressure distributor fuel consumed	9.30 l
Materials	
SS1 grade bitumen emulsion @ 0.70 kg/sqm	4.71 tonne
Water	10.1 Kl
Air compressor 250 cfm fuel consumed	22.0 l
Bitumen pressure distributor working hours	1.86 h

Wearing course (WC)	Length	Width	Depth
	1000 m	3.75 m	0.02 m
HMP 160 TPH working hours	1.48 h		
HMP 160 TPH fuel consumed diesel	959 l		
Air compressor 250 cfm working hours	0.45 h		
Air compressor 250 cfm fuel consumed	4.95 l		
Paver 175 HP (300 TPH) working hours	1.97 h		
Paver 175 HP (300 TPH) diesel fuel consumption	30.0 l		
Smooth wheeled tandem roller working hours	6.64 h		
Smooth wheeled tandem roller diesel fuel consumed	60.0 l		
Materials			
Bitumen	8.25 tonne		
Crushed aggregate	162.0 tonne		

References

AzariJafari, H., Yahia, A., & Amor, M. B. (2016). Life cycle assessment of pavements: Reviewing research challenges and opportunities. *Journal of Cleaner Production, 112*, 2187–2197.

Grael, P. F., Oliveira, L. S., Oliveira, D. S., & Bezerra, B. S. (2021). Life cycle inventory and impact assessment for an asphalt pavement road construction—A case study in Brazil. *The International Journal of Life Cycle Assessment, 26*(2), 402–416.

Guinée, J. B., & Lindeijer, E. (Eds.). (2002). *Handbook on life cycle assessment: Operational guide to the ISO standards* (Vol. 7). Springer.

Hauschild, M., Rosenbaum, R., & Olsen, S. (2018). Introduction to LCA methodology. In: *LCA: Theory and practice* (p. 12). Springer.

Institute for Environment and Sustainability. (2010). *International reference life cycle data system handbook-general guide for LCA*. Institute for Environment and Sustainability, European Commission, Luxembourg Publications, Europe.

Jiang, R., & Wu, P. (2019). Estimation of environmental impacts of roads through life cycle assessment: a critical review and future directions. *Transportation Research Part D: Transport and Environment, 77*, 148–163.

Menoufi, K. A. I. (2011). *Life cycle analysis and life cycle impact assessment methodologies: A state of the art*. Universitat de Lleida. Available at: https://repositori.udl.cat/handle/10459.1/4583

MoRTH. (2018). Basic road statistics of India 2017-2018. Available at: https://morth.nic.in/sites/default/files/BRS_Final.pdf.

MoRTH. (2021). Annual report 2020–2021. Available at: https://morth.nic.in/sites/default/files/Annual%20Report%20%202021%20%28-English%29_compressed.pdf

Perspectives on the Sustainable Steel Production Process: A Critical Review of the Carbon Dioxide (CO_2) to Methane (CH_4) Conversion Process

Wandercleiton Cardoso, Renzo Di Felice, and Raphael Colombo Baptista

Abstract Synthetic natural gas (SNG) can be obtained by methanation. Many thermodynamic reaction details involved in this process are not yet fully known. In this chapter, a comprehensive thermodynamic analysis of the reactions involved in the methanation of carbon oxides (CO and CO_2) is carried out using the Gibbs free energy minimization method. The equilibrium constants of eight reactions involved in the methanation reactions were calculated at different temperatures. The effects of temperature, pressure, H_2/CO (and H_2/CO_2) ration, and addition of other compounds (H_2O, O_2, and CH_4) in the feed gas on the conversion of CO and CO_2, CH_4 selectivity and yield, and carbon capture were carefully studied. It was found that low temperatures, high pressure, and a large H_2/CO (and H_2/CO_2) ratio are favorable for the methanation reactions. Concluding, the conversion of carbon dioxide (CO_2) into methane could be a solution for a new technology.

Keywords Carbon dioxide · Methanation · Circular economy · Blast furnace

1 Introduction

Carbon dioxide (CO_2) is a very important gas for human survival, but the excessive emissions of this gas, resulting from the burning of fossil fuels, increase the concentration of CO_2 in the atmosphere, modifying the natural phenomenon of the greenhouse effect and becoming a major environmental problem (Arens et al., 2014; Garbarino et al., 2014; Quader et al., 2016).

These carbon dioxide emissions will contribute to this effect for a long time to come, due to the increased use of fossil fuels in developed and emerging countries.

W. Cardoso (✉) · R. Di Felice
University of Genoa, Liguria, Italy
e-mail: wandercleiton.cardoso@dicca.unige.it; renzo.difelice@unige.it

R. C. Baptista
Ternium Group, Rio de Janeiro, Brazil

F. P. García Márquez, B. Lev (eds.), *Sustainability*, International Series in Operations Research & Management Science 333,
https://doi.org/10.1007/978-3-031-16620-4_17

In this context, it is evident the need to reduce the impacts caused by high concentrations of CO_2 in the environment and an alternative for the reduction of carbon dioxide would be the reuse of it as raw material in processes that can generate attractive products to the market (Arens et al., 2014; Wang et al., 2017).

Fossil resources have long been the great driving force of economies around the world, providing for industrialization. However, the carbon dioxide (CO_2) released from the burning of these fuels is partly responsible for the greenhouse effect, which causes hundreds of environmental problems, such as increasing the temperature of the Earth's surface, raising sea levels, changing the rainfall system, and more. Therefore, measures to curb environmental degradation have become a global necessity (Gazzani et al., 2013; Hussain et al., 2019).

The use of renewable energy sources such as wind, solar, and biomass may make an important contribution to a diversified and efficient energy matrix. It is a fact that countries dependent on fossil fuels such as coal are looking for ways to reduce their dependence on this form of energy and are looking for renewable alternatives (Schildhauer & Biollaz, 2015).

However, it is important to point out that the mining of non-renewable minerals in Latin America has increased six fold between 1970 and 2017, putting a huge strain on the natural environment. Environmental sustainability became part of the competitive strategy of industrial companies, especially in the steel, energy, and chemical sectors (Lee et al., 2020).

Nowadays, innovations are always needed to produce "cleaner," using less polluting and more recyclable materials and processes. Meeting environmental requirements challenges the innovation and creativity of companies, making them more competitive.

Environmental innovation has become essential for competitive survival. But the problem is complex and requires multiple solutions shared by all stakeholders, from changing production processes to prevent and minimize waste to developing new processes for recycling and/or reusing waste and end-of-life products.

The steel plant integrated through the blast furnace is the most widely used steel production route in the world and is continuously upgraded to meet new production, operational and environmental requirements. Blast furnaces operate on a large scale and generate about 600 kg of waste per ton of steel produced, which may be reused and sold as a by-product. Continuous improvement of integrated facilities would be one of the ways to achieve a higher level of sustainability in the world, considering that a blast furnace cannot be replaced overnight by an alternative process (Cardoso et al., 2021a; Cardoso & Di Felice, 2022).

In short, in the globalized and competitive world, industries need to make efforts to improve both their production methods and their end products and services if they want to remain competitive and survive. Investing in more sustainable technologies and approaches encourages the emergence of innovative products that may revolutionize their markets.

In this sense, the main objective of the chapter of this book is to describe the actions taken by the steel industry to develop the circular economy, leading to broad benefits for society, including the supply of sustainable products and the launch of less polluting substances (carbon dioxide) in order to preserve natural resources for future generations.

2 The Steel Industry

At the present stage of social development, the world is inconceivable without the use of steel. Steel production is a strong indicator of a country's level of economic development. However, the production of steel products requires technology that needs to be renewed cyclically. Therefore, steel manufacturers are constantly investing in research (Cardoso & Di Felice, 2022).

In the twenty-first century, environmental sustainability became part of the competitive strategy of industrial companies, especially in the steel, energy, and chemical sectors. Concepts such as industrial ecology, circular economy, and recycling became part of the scenario of regulations and technical standards that require constant and growing technological research and development.

Nowadays, innovations are always needed to produce "cleaner," using less polluting and more recyclable materials and processes. Meeting environmental requirements is a challenge for the innovation and creativity of companies, making them more competitive (Cardoso & Di Felice, 2022).

Innovation in materials, driven by environmental concerns, has become essential for competitive survival. However, the problem is complex and requires multiple solutions shared by all stakeholders, from changing production processes to prevent and minimize waste, to developing new processes for recycling and/or reusing waste and end-of-life products, and spreading the circular economy.

The circular economy, which is used to be addressed only in the realm of industry, has become the agenda of society as a whole due to growing concerns about the sustainability of the planet. It is a new model of production and consumption that goes beyond recycling and decouples economic development from the recurrent use of natural resources that characterizes the linear economy. In this new economic model, almost nothing is thrown away. Materials that are used or generated in the production process may be reused and recycled over and over again in the manufacture of new products, reducing the need for natural resources.

Every year, the steel industry drives the circular economy through technological and process innovations. Some examples of the steel industry's actions to develop the circular economy, enabling the reuse of input materials, the reduction of raw material consumption, and the disposal of materials, are the use of by-products in road construction and the introduction of technologies enabling the reuse of gases from steel production.

Continuous improvement of integrated plants would be one of the ways to increase sustainability, since a blast furnace cannot be replaced by an alternative process overnight. Blast furnaces, although efficient, no longer meet new environmental and market requirements such as: low gas emissions, greater environmental sustainability, greater efficiency in the treatment and recycling of the waste produced, greater flexibility in the use of materials and raw materials, low fixed costs, low costs of installation, maintenance and expansion, low operating costs, and low flexibility in changing the scale of production.

The new needs of the steel market have meant that this type of plant (blast furnace plants) may no longer remain profitable on the market. Therefore, there are alternative and complementary processes to the blast furnace. Good examples of these processes, which are already in the maturity and consolidation phase, are the direct reduction processes, such as Midrex® and HYL. Due to the use of natural gas as a reducing agent, the Midrex® and HYL processes are not economically feasible in some countries due to a lack energy policy that provides a fixed quota of natural gas to the steel industry (direct reduction processes), as is already the case for other industries. However, alternative processes for biogas production and the conversion of carbon dioxide (CO_2) into methane may be a solution for these new technologies (Cardoso et al., 2021b).

Steel mills produce about 600 kg of waste per ton of cast iron. Minimizing waste production means increasing production efficiency and reducing environmental impact. More than 98% of these residues may be sold in the form of various products with good profit expectations.

Currently, alternative and complementary technologies to the blast furnace have proven to be fundamental to the process of renewal of the global steel industry, especially for the recycling of steel residues. Iron ore is chemically reduced by two types of technologies characterized by the type of reduction reaction or the type of steelmaking process, known as bath smelting and self-reduction.

Bath smelting: iron oxides dissolved in metal baths and/or liquid slag are rapidly reduced by carbon, producing liquid metal.

Self-reduction: solid–solid reaction occurs via gaseous intermediates inside agglomerates (fine ore and carbonaceous materials), producing liquid metal or sponge iron. The close contact between the particles and the absence of nitrogen lead to very fast reactions.

The steel industry is one of the most energy-intensive segments of the industry, and its specific energy consumption may be influenced by several factors, among which the following stand out: the technological path, the processes used, and the type and quality of fuels used.

Coal, in the form of coke and pulverized coal, enables the transformation of iron ore into pig iron in blast furnaces. Nowadays, blast furnace is the most important way of steel production in the world. The blast furnace is a metallurgical-chemical reactor whose function is to reduce iron ore. Cast iron, the main product of steel production, consists of 90–95% iron, 3.0–4.5% carbon, alloying elements such as silicon (Si) and manganese (Mn), and impurities such as phosphorus (P) and sulfur (S) (Cusumano et al., 1976).

In the blast furnace pig iron is produced in a liquid state at a temperature of about 1500 °C. After this process, it enters the melt shop, where it is cleaned and treated in order to obtain certain properties and be transformed into steel. In addition to pig iron, the blast furnace also produces by-products such as slag and blast furnace gas

(Cardoso, Barros, et al., 2021; Cardoso, Di Felice, & Baptista, 2022; Cardoso et al., 2021c).

Slag is obtained by melting and separating the gangue from the raw materials and fluxes. It consists mainly of thermodynamically stable oxides such as MgO, CaO, Al_2O_3, and SiO_2, which make up to 95% by weight of the slag. Blast furnace gas is a gas mixture containing CO, H_2, N_2, and CO_2 in its composition. Blast furnace gas contains about 3–6% H_2, 45–54% N_2, 21–27% CO, and 22–27% CO_2 (Gottschlich et al., 1989).

In this sense, the steel industry is one of the most energy-intensive industries in the world, and the use of coal as the primary fuel to make iron and steel produces higher carbon dioxide emissions than any other industry.

Annual global demand for steel will continue to grow over time, with a rapid projection that it will reach more than 2300 million metric tons (Mt) of crude steel by 2050. Developing countries, especially China and India, will account for most of this growth.

According to the **International Energy Agency (IEA)**, the steel industry is responsible for the largest share of CO_2 emissions from the global manufacturing sector, accounting for about 27%. Steel production is an energy-intensive process and is responsible for about 7% of global carbon dioxide emissions. Regardless of the technology used, the production of 1 ton of steel releases about 1.8 tons of CO_2 into the atmosphere, a remarkable and very alarming fact (Lee et al., 2020).

Moreover, environmental legislation has become increasingly stringent in recent years. In particular, since 2008, with the entry into force of the **Kyoto Protocol**, the European Union proposes in its roadmap a low carbon economy to reduce greenhouse gas emissions by 80% by 2050 compared to 1990 (Su et al., 2016).

In 2015, the **Paris Agreement** was signed, the first legally binding agreement between nations to reduce emissions worldwide. Therefore, in this context, the steel industry is under pressure to limit its CO_2 emissions. There are two solutions to this problem: (1) Indirect measures (**ETS emissions trading scheme**) and (2) direct measures (industrial technological innovation).

The **ETS** is the world's largest emissions trading market and accounts for over 75% of international carbon trading. The world's major industries, including the steel industry, use this type of flexible mechanism to comply with the conceded cap on CO_2 emissions. The industries that are defined as "short," i.e. exceeding the imposed cap, buy up the emission allowances to come back within their own limit (Li et al., 2015; Liu et al., 2011).

However, these indirect measures are gradually being penalized precisely to encourage technological innovation. The total number of allowances issued has been reduced by 1.74% per year over the period 2013–2020. The target is to reduce the total number of allowances issued by 2.2% per year from 2021.

Looking at the new European benchmark for CO_2 emissions for the iron/steel industry, it cannot be met with the currently applied procedures, even using the indirect measures (**ETS**) just described. In fact, the new benchmarks set by the **European Union** are below 10% of the minimum achievable emissions for pig iron production.

Due to the new benchmarks and the lower volume of quotas granted, the steel industry is therefore forced to resort to major technological innovations that revolutionize the previously established process schemes.

In this sense, the development of new technologies to properly treat the carbon dioxide emitted by the steel industry is a viable alternative to solve this problem. Among these new technologies, the one consisting in the conversion of carbon monoxide (CO) and/or carbon dioxide (CO_2) into methane (CH_4) has attracted the attention of the main researchers of this new technological process.

3 Methanation Process

CO and CO_2 methanation processes were first discovered by **Sabatier** and **Senderens** in 1902 and have since been researched and developed for more than 100 years. Methanation of CO gained greater importance for syngas production during the oil crisis in the late 1970s. The aim was to use syngas from coal gasification to produce a substitute for natural gas. At that time, various concepts of methanation were developed. The developments of CO_2 methanation processes were primarily based on methanation research from CO, but they were driven by the desire to use alternative feedstock gases. However, few of these concepts reached commercial scale due to the high cost required to purify these gases (Koytsoumpa & Karellas, 2018; Moller, 2014).

Due to growing environmental awareness and the urgent task of reducing anthropogenic greenhouse gas emissions, research into CO_x methanation processes was revived at the beginning of the twenty-first century.

Methane is an energy carrier of great importance for industry, the energy sector and the transport sector worldwide. Methane is a constitutive element of the modern economy. In fact, it may be transported efficiently and cheaply using existing natural gas pipelines and the wide distribution network. In recent years, the increase in the price of natural gas, the desire to reduce dependence on natural gas imports, and the replacement of petroleum products have led to the production of synthetic or substitute natural gas attracting increasing attention in various countries.

In recent years, the debate on the finite nature of fossil resources and climate change has led to an increase in research spending on catalytic and biological methane production from carbon-oxide-rich gases (methanation). In this chapter, the focus is on the former type of methanation. Research in catalytic methanation processes focuses on two options, CO methanation and CO_2 methanation (Boll et al., 2006; Van Herwijnen et al. 1973).

CO methanation processes (Eq. 1) use carbon monoxide and hydrogen as reactants for the catalytic production of methane and water, and CO2 methanation processes (Eq. 2) use carbon dioxide and hydrogen as reactants.

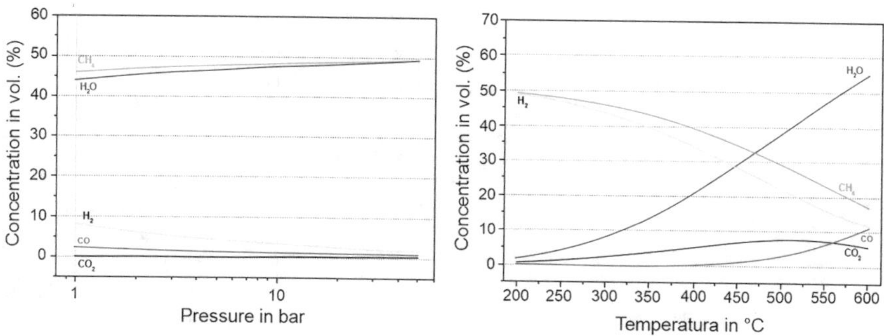

Fig. 1 Pressure and temperature influence on the equilibrium composition of CO methanation and water–gas shift reaction (left: 300 °C, right: 1 bar). Educts gas composition: $y_{CO} = 0.25$, $y_{H2} = 0.75$

$$CO + 3H_2 \leftrightarrow CH_4 + H_2O(g) \quad - 206\,kJ\,mol^{-1}(at\,298\,K) \tag{1}$$

$$CO_2 + 4H_2 \leftrightarrow CH_4 + 2H_2O(g) \quad - 165\,kJ\,mol^{-1}(at\,298\,K) \tag{2}$$

Both reactions are basically exothermic. While the stoichiometric methanation of carbon monoxide releases 206 kJ of heat per mole (Eq. 1), the conversion of carbon dioxide releases 165 kJ per mole. CO_2 methanation (Eq. 2) is a linear combination of CO methanation and the reverse water–gas shift reaction (RWGS) (Eq. 3), which is an endothermic reaction and in practical operation is always accompanied by the CO methanation reaction using nickel catalysts. However, the conversion of CO_2 is inhibited when the CO concentration exceeds a certain threshold.

$$CO_2 + H_2 \leftrightarrow CO + H_2O(g) \quad 41\,kJ\,mol^{-1}(at\,298\,K) \tag{3}$$

The equilibrium of the two reactions is affected by pressure and temperature. In thermodynamic equilibrium, high pressures favor the production of methane. High temperatures, on the other hand, limit methane formation. The influence of pressure and temperature on the chemical equilibrium of CO methanation and the water–gas shift reaction is exemplified in Fig. 1. The equilibrium composition was calculated based on a simplified equilibrium model. Instead, the influence of pressure and temperature on the equilibrium composition of CO_2 methanation and the water–gas shift reaction is shown in Fig. 2.

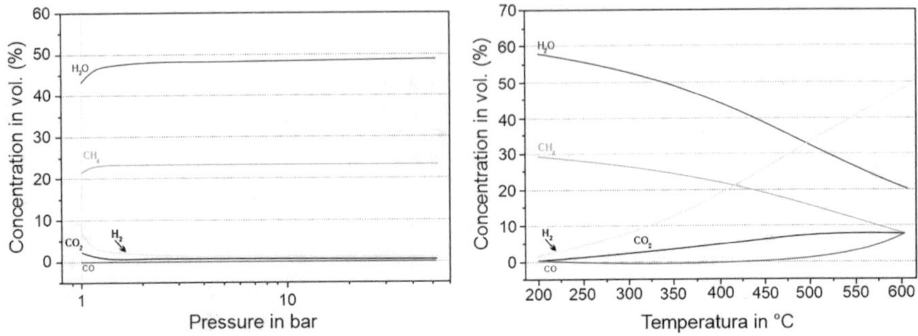

Fig. 2 Pressure and temperature influence on equilibrium composition of CO_2 methanation and water–gas shift reaction (left: 300 °C, right: 1 bar). Educts gas composition: $y_{H2} = 0.8$, $yCO_2 = 0.2$

4 Thermodynamic Analysis

In order to analyze CO_x methanation, the starting point must be thermodynamics, which is useful in evaluating the processes involved in such a complex reaction system. Thermodynamic equilibrium calculations of complex chemical systems based on the minimization of Gibbs free energy may provide answers to numerous questions, including the nature of the thermodynamically stable reaction products produced and their selectivity and yield, whether a chemical reaction is endothermic or exothermic, the influence of reaction parameters such as temperature, pressure, or reactant ratios, and whether the observed reaction temperatures are thermodynamically constrained (Boon et al., 2015; Nahar & Madhani, 2010).

The total **Gibbs free energy** of the system is obtained from Eq. (4) and the introduction of **Lagrange multipliers** for species i, subject to the mass balance constraints in Eq. (5).

$$G_T = \sum_{i=1}^{m} n_i \mu_i = \sum_{i=1}^{m} n_i \mu_i^{\Theta} + RT \sum n_i \ln \frac{f_i}{P^{\Theta}} \tag{4}$$

$$\mu_i + \sum_{j=0}^{k} \lambda_j a_{ji} = 0 \tag{5}$$

where λ is the Lagrange multiplier, a_{ji} is the number of atoms of element j in species i. Therefore, the combination of Eqs. (4) and (5) gives Eqs. (6) and (7).

$$\mu_i^{\Theta} + RT \ \ln \left(\frac{f_i}{P^{\Theta}} \right) + RT \sum_{j=0}^{k} \lambda_j a_{ji} = 0 \tag{6}$$

Table 1 Possible reactions involved in the methanation of carbon oxides

Equation	Equation formula	ΔH_{298K} (kJ mol^{-1})	Reaction description
(8)	$CO + 3H_2 \leftrightarrow CH_4 + H_2O$	–206.1	CO methane reaction
(9)	$CO_2 + 4H_2 \leftrightarrow CH_4 + 2H_2O$	–165.0	CO_2 methane reaction
(10)	$2CO + 2H_2 \leftrightarrow CH_4 + CO_2$	–247.3	Inversed reforming
(11)	$2CO \leftrightarrow C + CO_2$	–2172.4	Boudouard reaction
(12)	$CO + H_2O \leftrightarrow CO_2 + H_2$	–41.2	Water–gas reaction
(13)	$CH_4 \leftrightarrow 2H_2 + C$	–74.8	Methane cracking
(14)	$CO + H_2 \leftrightarrow C + H_2O$	–131.3	Carbon monoxide reduction
(15)	$CO_2 + 2H_2 \leftrightarrow C + 2H_2O$	–90.1	Carbon dioxide reduction
(16)	$nCO + (2n{+}1)H_2 \leftrightarrow C_nH_{(2n+2)} + nH_2O$	–	Complex reactions
(17)	$nCO + 2nH_2 \leftrightarrow C_nH_{2n} + nH_2O$	–	Complex reactions

$$\mu_i^{\Theta} + RT\sum_{j=0}^{k}\lambda_j a_{ji} = 0 \tag{7}$$

During methanation, some side reactions may occur that affect the purity of the gas product. Table 1 lists the main possible reactions during methanation. In addition to the normal methanation reactions (Eqs. 8 and 9), a carbon monoxide methanation reaction may also occur at a lower H_2/CO ratio (Eq. 10).

The carbon monoxide disproportionation reaction (Eq. 11), also known as the **Boudouard reaction**, is of great importance because carbon on the catalyst surface is considered a necessary intermediate during the methanation reaction. In addition, water plays an important role through the water–gas shift reaction (Eq. 12), which would change the surface and catalytic chemistry of methanation catalysts. In these reactions, it should be noted that Eqs. (8), (9), and (11) may be considered as three independent reactions. The other reactions may be described as a linear combination of these three reactions (Gao et al. 2011, 2015).

Figure 3 shows the equilibrium constants K of the first eight possible reactions (see Table 1) that occur during methanation as a function of temperature. K is calculated below using the **Van't Hoff equation (18)**.

$$\frac{d}{dT}\ln K_{\mathrm{eq}} = \frac{\Delta_r H^{\Theta}}{R} \tag{18}$$

where ln is the natural logarithm, K_{eq} is the thermodynamic equilibrium constant, and R is the ideal gas constant. This equation is exact at any temperature and pressure. It is derived from the requirement that the Gibbs free energy of the reaction must be stationary in a state of chemical equilibrium.

In practice, the equation is often integrated between two temperatures, assuming that the standard enthalpy of reaction $\Delta_r H^{\Theta}$ is constant (and furthermore, it is often

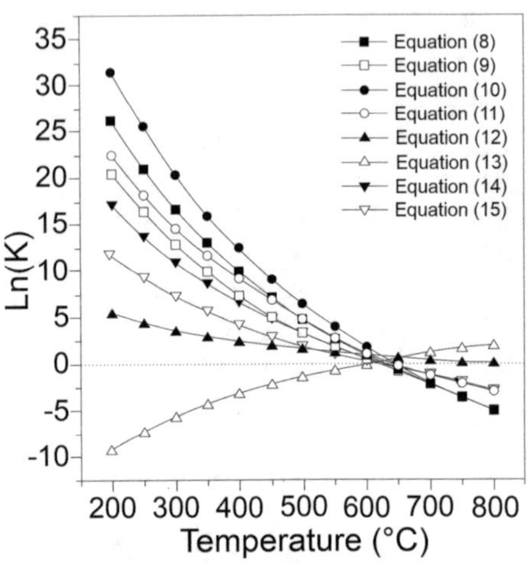

Fig. 3 Calculated K values of the reactions involved in the methanation

assumed to be equal to the value at standard temperature). Since, in reality, $\Delta_r H^\Theta$ and the standard reaction entropy $\Delta_r S^\Theta$ vary with temperature in most processes, the integrated equation is only an approximation. In practice, approximations are also made to the activity coefficients within the equilibrium constant.

An important application of the integrated equation is to estimate a new equilibrium constant at a new absolute temperature, assuming a constant standard enthalpy change over the temperature range. In order to obtain the integrated equation, it is convenient to first rewrite the **Van't Hoff equation** (19) and the definite integral between temperatures T_1 and T_2 as in Eq. (20).

$$\frac{d \ln K_{\text{eq}}}{d \frac{1}{T}} = \frac{\Delta_r H^\Theta}{R} \tag{19}$$

$$\ln \frac{K_2}{K_1} = \frac{\Delta_r H^\Theta}{R} \times \left(\frac{1}{T_1} - \frac{1}{T_2} \right) \tag{20}$$

It may be seen that all exothermic reactions are suppressed with increasing temperature, except for the methane cracking reaction, according to Eq. (13). Equations (8)–(11) and (14) play a key role in the methanation reaction system due to their high equilibrium constants in the temperature range of 200–500 °C.

Consumption of CO via Eqs. (8), (10), (11), and (12) may result in almost complete conversion of CO at low temperature due to the high K values of these reactions. In contrast, it is difficult to completely convert CO_2 because the reactions show in Eqs. (10)–(12) produce it.

Moreover, it may be deduced that carbon deposition mainly comes from the Boudouard reaction in Eq. (11), because its K value is much larger than that of the reactions in Eqs. (13)–(15).

4.1 Performance of Methanation Processes

In order for a better understanding of the parts discussed below, the expressions for the calculation of conversion (Eq. 21), selectivity (Eq. 22), and yield (Eq. 23) related to the methanation reaction starting from CO or CO_2 are given below. For methane synthesis, the following main performance parameters are defined, assuming that no methane is present in the output stream. The three parameters **(conversion, selectivity, and yield)** are also expressed with Eq. (24). **Conversion:** The carbon conversion (C) is defined here in molar basis and specifies the percentage of a specific reactant component, which is reacted based on the initial feed (Czekai et al., 2007).

Selectivity: Although the conversion characterizes how much of the initial feed has reacted, the selectivity characterizes to which product, as different products may occur by multiple reactions. The selectivity (S) is defined as how much of the desired product is formed in relation to the converted reactant and incorporates in its definition the stoichiometric coefficient of the specific reactant and product. **Yield:** The yield (Y) is defined as the ratio of produced moles of the specific product and the input of the specific reactant.

$$C_{CO/CO_2}(\%) = \frac{\dot{n}_{CO/CO_{2,reacted}}}{\dot{n}_{CO/CO_{2,in}}} = \frac{\dot{n}_{CO/CO_{2,in}} - \dot{n}_{CO/CO_{2,out}}}{\dot{n}_{CO/CO_{2,in}}} \times 100 \tag{21}$$

$$\frac{S_{CO}^{CH_4}}{CO_{2,reacted}}(\%) = \frac{\dot{n}_{CH_4}}{\dot{n}_{\frac{CO}{CO_{2,reacted}}}} \times \frac{\nu_{\frac{CO}{CO_2}}}{\nu_{CH_4}} = \frac{\dot{n}_{CH_4}}{\dot{n}_{CO_{2,in}} - \dot{n}_{CO_{2,out}}} \times 100 \tag{22}$$

$$Y_{CO/CO_{2,in}}^{CH_4}(\%) = \frac{\dot{n}_{CH_4}}{\dot{n}_{CO/CO_{2,in}}} \times \frac{\nu_{\frac{CO}{CO_2}}}{\nu_{CH_4}} \times 100 = \frac{\dot{n}_{CH_4}}{\dot{n}_{CO/CO_{2,in}}} \times 100 \tag{23}$$

$$Y = C \times S \tag{24}$$

4.2 CO Methanation

The conventional synthesis of synthetic natural gas (SNG) is based on the catalytic methanation of carbon monoxide and hydrogen with a stoichiometric H_2/CO ratio. Looking first at the reaction at atmospheric pressure, we find that the products below low temperatures (200–300 °C) contain mainly CH_4 and H_2O with CO_2 as a by-product, without carbon deposition. As the temperature increases, the mole fraction of CH_4 decreases while the unreacted fraction of CO, H_2, CO_2 and deposited carbon increases simultaneously (Ensell & Stroud, 1983; Kopyscinski et al., 2010; Sushil & Sandun, 2006).

Carbon deposition occurs when the temperature is higher than 450 °C. This is because the Boudouard equation (11), due to its higher equilibrium constant,

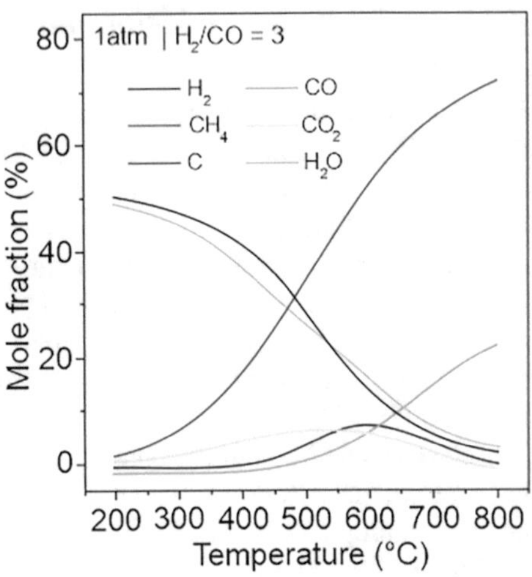

Fig. 4 Product composition of CO methanation at equilibrium at 1 atm

proceeds more easily above this temperature than CO methanation (Eq. 8), as shown in the previous Fig. 3. Figure 4 instead shows the typical product fraction at equilibrium calculated using the Gibbs free energy minimization method at 1 atm.

For better illustration, Fig. 4 shows the effects of different pressure and temperature values on the methanation performance of CO (H_2/ CO = 3). In Fig. 5a, you can see that a higher pressure leads to a higher conversion of CO at the same reaction temperature because the methanation of CO is a volume reduction reaction. On the other hand, a much higher pressure (100 atm) below 700 °C does not lead to a much higher CO conversion than at 30 atm. At the same reaction pressure, a lower temperature leads to a higher CO conversion due to the exothermic reaction of CO methanation (Eq. 8).

Figure 5b shows how the selectivity of CH_4 varies at different values of temperature and pressure. It should be noted that high selectivity of CH_4 may be achieved at a pressure above 1 atm and a relatively low temperature.

This is because CO is partially converted to by-products such as CO_2 and solid carbon by a reverse reforming reaction of methane-CO_2 (Eq. 10) and a Boudouard equation (11), which occur at low pressure and high temperature, respectively (Table 1).

Therefore, methane selectivity may be greatly improved by increasing the pressure from 1 to 30 atm, but further increasing the pressure (100 atm) would not result in significant improvement. This could be the reason why the lower operating pressure is usually 20–30 atm for most industrial technologies.

Figure 5c shows the variation of CH_4 yield as a function of temperature at different pressures. A CH_4 yield close to 100% was observed at temperatures below 250% C at 1 atm or higher pressure.

Figure 5d shows the variation of the yield of deposited carbon as a function of temperature at different pressures. It may be seen that the carbon is produced at 1 atm in the temperature range of 400–800 °C and reaches a maximum at about 600 °C.

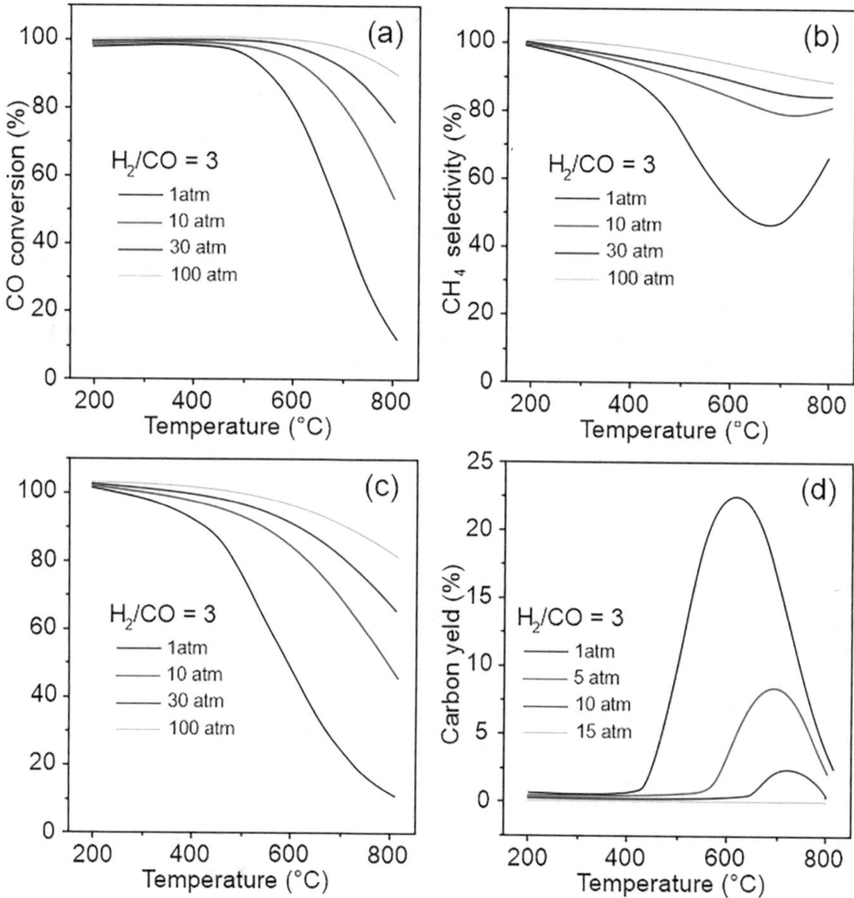

Fig. 5 Effects of temperature and pressure on CO conversion (**a**), CH_4 selectivity (**b**), CH4 yield (**c**), and carbon yield (**d**)

When the pressure is further increased to 10 atm, a smaller amount of carbon is found in the temperature range between 650 and 800 °C and the maximum is reached at about 725 °C.

When the pressure is higher than 15 atm, the deposited carbon is not observed. So, you may see that high pressure may greatly facilitate the deposition of carbon. Carbon deposition is known to result from several reactions, including the Boudouard equation (11), the cracking of methane (Eq. 13), the reduction of CO (Eq. 11), and the reduction of CO_2 (Eq. 15).

Since the equilibrium constant K of Eq. (11) is higher than that of Eq. (8) in the temperature range of 450–600 °C at 1 atm (see Figs. 4 and 5), Eq. (11) takes place more readily than Eq. (8), resulting in carbon deposition, which occurs mainly by CO disproportionation.

However, when the reaction temperature exceeds about 600 °C, at which the maximum amount of carbon yield is obtained, carbon deposition may occur due to Eq. (13). Considering the relatively low CH_4 concentration above 600 °C, Eq. (13) may result in less carbon deposition at this stage. However, the reverse reactions of Eqs. (14) and (15) may become dominant and consume much of the deposited carbon above 600 °C (Rostrup-Nielsen et al., 2007).

As a result, the total deposited carbon decreases when the temperature exceeds 600 °C. Therefore, the carbon yield curves in Fig. 5d first show an increase and then a decrease in the value and at the end it decreases to a negligible value around 800 °C. According to the analysis of all these series of curves, low temperature and high-pressure values are highly recommended in CO methanation in accordance with the exothermic nature of the reaction (Lefebvre et al., 2015).

4.3 *Effect of H_2/CO Ratio*

As mentioned earlier, the H_2/CO ratio must be about 3, which is normally controlled by the water–gas displacement reaction (8). However, it is very difficult to adjust this value so that it is exactly 3 in the methanation region. Therefore, it is necessary to know the influence that this ratio may have on the process of methanation.

Figure 6a shows the effect of different H_2/CO ratios at different pressures on the CO conversion. The CO conversion does not change at different H_2/CO ratios (1, 3 and 5) at atmospheric pressure. The situation changes at 30 atm, where a higher ratio seems to be beneficial for CO conversion. Figure 6b shows the CH_4 selectivity at different ratios H_2/CO. It may be seen that a higher H_2/CO ratio leads to higher CH_4 selectivity at both 1 and 30 atm. The high-pressure reaction also leads to high CH_4 selectivity at different ratios, especially at high reaction temperatures. Figure 6c shows the CH_4 yield at different H_2/CO ratios and different pressures. It has been shown that high ratio, high pressure, and low temperature are useful to obtain higher yield of CH_4.

Finally, Fig. 6d shows the carbon yield deposited at different ratios of H_2/CO. It may be observed that the low hydrogen to carbon monoxide ratio of 1 leads to high carbon yield at both 1 and 30 atm. Conversely, the deposited carbon is not observed at high ratios of 5 (1 atm) and 3 (30 atm) because CO is mainly converted to CH_4 via reaction (8) (see Table 1) and to CO_2 via reaction (10). From all these considerations, it may be concluded that the ratio H_2/CO must not be lower than 3 to obtain a relatively high yield of CH_4, and this is the ratio we usually work with in technologies for methanation from CO.

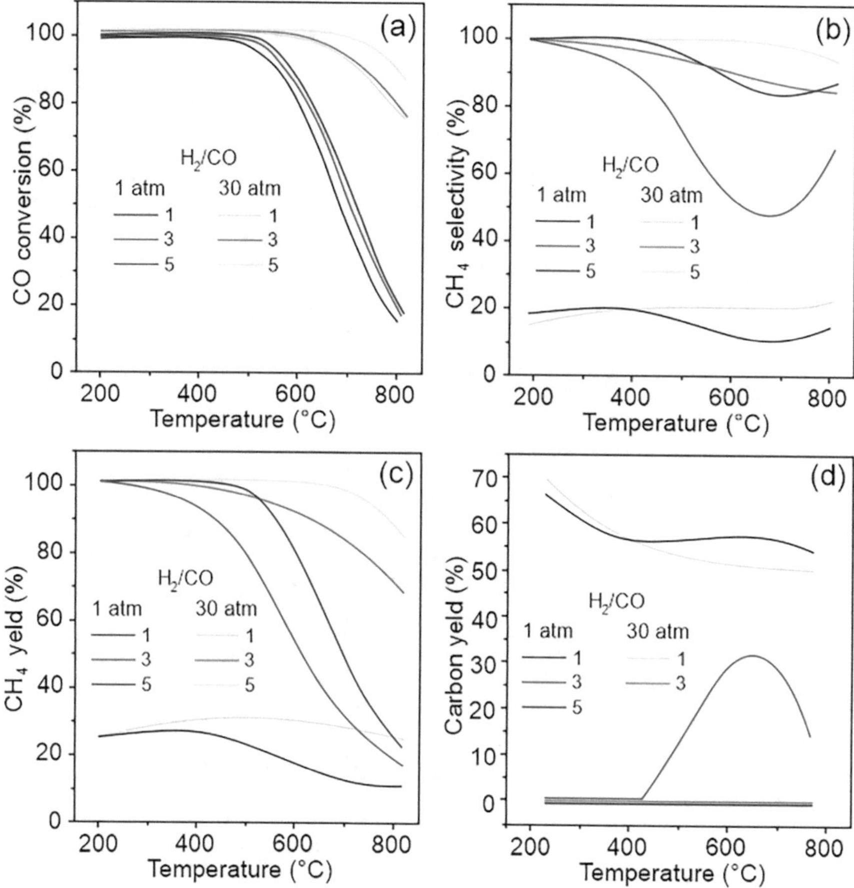

Fig. 6 Effect of H_2/CO ratio on CO conversion (**a**), CH_4 selectivity (**b**), CH_4 yield (**c**), and carbon yield (**d**)

4.4 *Effect of H_2O*

For the sake of completeness, may refer to the study of the influence which the presence of water vapor may have on the methanation of carbon monoxide. Industrial experience has shown that the addition of water vapor to reagents may largely prevent the formation of carbon on methanation catalysts, which is an aspect that should not be underestimated. Figure 7a shows that additional water vapor at different ratios leads to a slight decrease in the conversion of CO at high temperature (1 and 30 atm), since water vapor is one of the reaction products that inhibit the methanation of CO (reaction 8) to some extent according to Le Chatelier's principle (Kuznecova & Gusca, 2017).

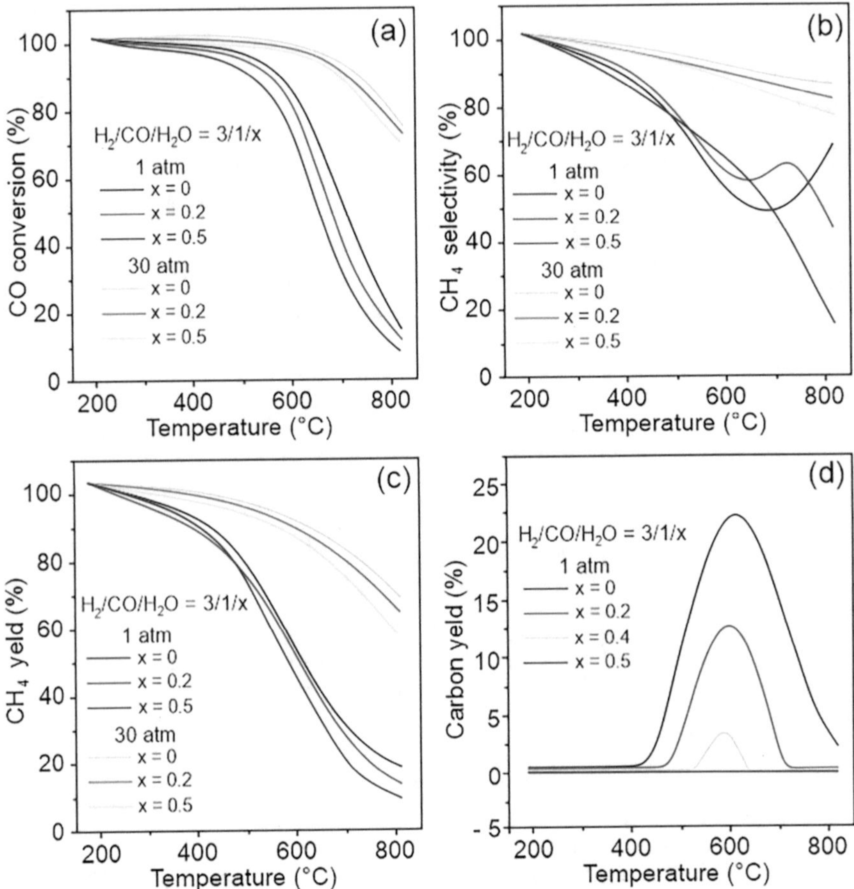

Fig. 7 Effect of water steam added into the feed on CO conversion (**a**), CH_4 selectivity (**b**), CH_4 yield (**c**), and carbon yield (**d**)

In addition, Fig. 7b shows that the introduction of water vapor causes a small difference in CH_4 selectivity, with small decreases as the vapor ratio increases at 30 atm and at a temperature above 500 °C. Figure 7c shows the effect of adding steam on CH_4 yield. The trend is similar to that of the two previously studied plots. Finally, Fig. 7d shows the effect of steam on carbon yield at 1 atm. Additional steam significantly reduces carbon capture at atmospheric pressure, especially at high temperatures (Zhu et al., 2015).

In particular, it may be seen that no carbon is formed at a steam ratio of 0.5, which is due to the fact that the addition of water would significantly inhibit reactions (14) and (15). Thus, adding a small amount of water to the CO methanation system could reduce carbon deposition (Bailera et al., 2017).

This is consistent with current methanation processes where excess steam is required for the first methanation reactor to avoid carbon deposition. In addition, the carbon deposition diagram is not given at 30 atm, since at this pressure there is no deposition of this chemical element even without the addition of steam.

4.5 Effect of O_2

It is known that traces of O_2 in the gas are inevitable in the gasification process. At present, there is no report on the effect of O_2 on methanation. Figure 8 shows the effect of O_2 introduced into the feed gas on CO methanation.

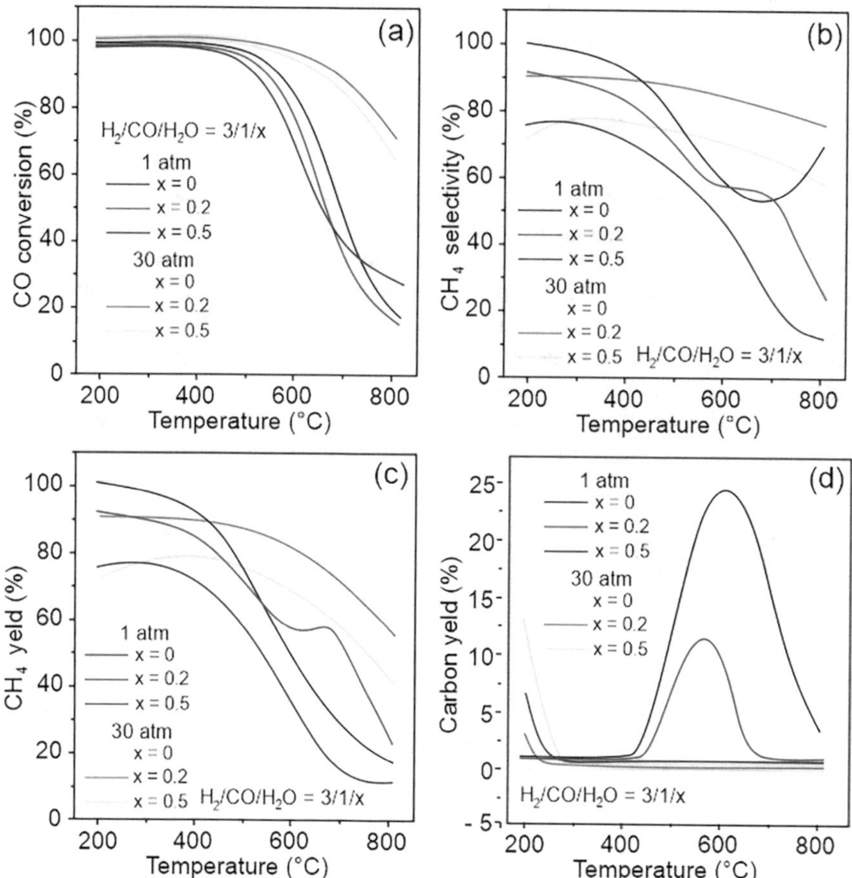

Fig. 8 Effect of O_2 contained in the feed gas on CO conversion (**a**), CH_4 selectivity (**b**), CH_4 yield (**c**), and carbon yield (**d**)

Figure 8a shows the effect of O_2 on the conversion of CO. From this, it may be seen that the conversion of CO decreases slightly with an increase in O_2 at both 1 and 30 atm, which is due to the reactions of H_2 and CO with O_2, especially at high temperatures.

Figure 8b shows the effect of O_2 on CH_4 selectivity, suggesting that O_2 could lead to a significant decrease in CH_4 selectivity throughout the temperature range as CO is partially oxygenated to CO_2.

As a result, the CH_4 yield decreases with an increase in O_2 of the reactants, as shown in Fig. 8c. However, as shown in Fig. 8d, the introduction of O_2 may lead to a large decrease in the deposited carbon at 1 atm, since O_2 reacts with carbon to form CO_2, especially at a ratio of H_2/ CO /O_2 = 3/1/ 0.5.

Similarly, at 30 atm, there is no carbon deposition at high temperature. Therefore, considering the CH_4 yield, it is necessary to completely remove the O_2 in the feed gas for the methanation process.

4.6 CO_2 Methanation

It is possible to consider the typical product fraction of CO_2 methanation at equilibrium calculated at 1 atm (Fig. 9), as in CO methanation. When the feed gas contains H_2 and CO_2 with a stoichiometric H_2/CO_2 molar ratio of 4, the products consist mainly of CH_4 and H_2O at a relatively low temperature (200–250 °C). However, if the temperature is increased to above 450 °C, the by-product CO increases due to the reverse water gas shift reaction, and in the meantime, the unreacted CO_2 and H_2 also increase while the CH_4 product decreases.

Since CO_2 methanation is also a highly exothermic reaction, increasing the temperature is unfavorable for this reaction. However, when the temperature exceeds

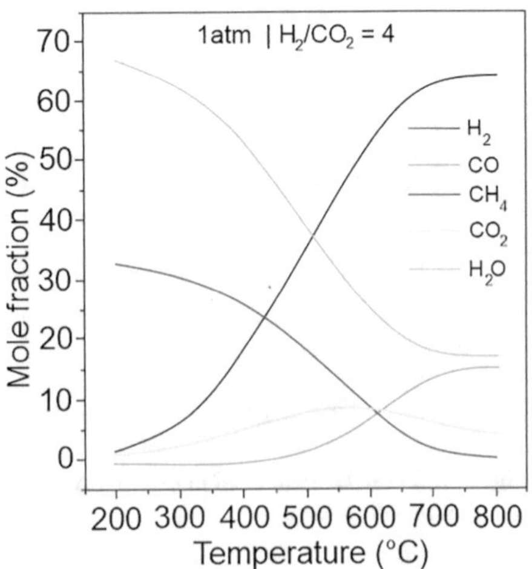

Fig. 9 Product composition of CO_2 methanation at equilibrium at 1 atm

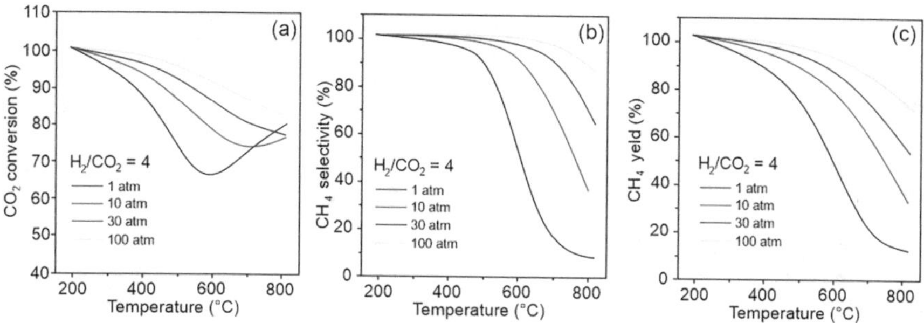

Fig. 10 Effects of temperature and pressure on CO_2 conversion (**a**), CH_4 selectivity (**b**) and CH_4 yield (**c**)

about 550 °C, the mole fraction of CO_2 reaches its maximum and then decreases because the reverse water gas shift reaction dominates. The calculation also shows that no significant carbon deposition occurs under these conditions.

Figure 10 shows the effects of pressure and temperature on CO_2 methanation on CO_2 conversion, CH_4 selectivity, and CH_4 yield. In Fig. 10a, it may be seen that CO_2 conversion decreases with increasing temperature at temperatures below 600 °C and increases with pressure. However, when CO_2 methanation occurs above 600 °C (at 1 atm), CO_2 conversion gradually increases. This is mainly because the reverse water gas shift dominates above 600 °C and consumes CO_2. It is observed that the trend with variations in the pressure and temperature is similar to CO methanation, but CO_2 is more difficult than CO to be methanized at the same temperature and pressure although the CH_4 selectivity below 500 °C in the CO_2 methanation system is relatively higher than that in CO methanation.

4.7 *Effect of H_2/CO_2 Ratio*

It may be seen that CO_2 conversion in Fig. 11a and CH_4 selectivity in Fig. 11b are influenced to a significant extent by the H_2/CO_2 ratio. A high H_2/CO_2 ratio generally results in high CO_2 conversion and CH_4 selectivity at both 1 and 30 atm. When the H_2/CO_2 ratio is equal to 2, CO_2 conversion of only about 50–70% may be achieved at 1 atm or 30 atm, and maximum CH_4 selectivity of 73% and 88% is achieved at 1 atm and 30 atm, respectively.

Figure 11c shows the effects on CH_4 yield. When the H_2/CO_2 ratio is 2, the CH_4 yield is only about 40% at 1 atm and 45% at 30 atm. The CH_4 yield may be significantly increased by increasing the H_2/CO_2 ratio under the same conditions. Figure 11d shows the carbon yield. When the H_2/CO_2 ratio is 2, abundant carbon is deposited (up to 50%) at both 1 and 30 atm below 500 °C. When the H_2/CO_2 ratio is equal to or greater than 4, no carbon deposition occurs. Thus, to achieve high CH_4

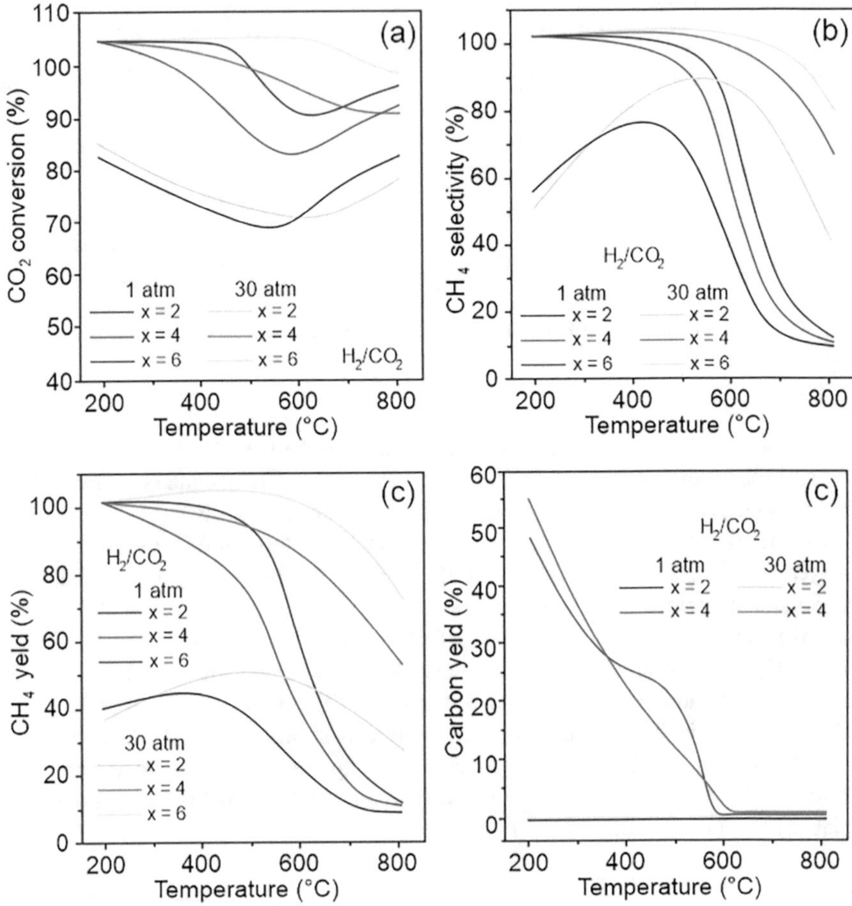

Fig. 11 Effect of H_2/CO_2 ratio on CO_2 conversion (**a**), CH_4 selectivity (**b**), CH_4 yield (**c**), and carbon yield (**d**)

yield and avoid carbon deposition, the H_2/CO_2 ratio should not be less than 4 even at 30 atm.

4.8 *Effect of H_2O*

Figure 12 gives the effect of adding steam into the feed gas ($H_2/CO_2/H_2O = 3/1/x$, $x = 0$, 0.2, and 0.5, here $x = 0$ means that no additional H_2O is added). It is seen that at 1 and 30 atm, the addition of steam leads to a little decrease in CO_2 conversion (Fig. 12a), and no significant difference in CH_4 selectivity (Fig. 12b) and CH_4 yield

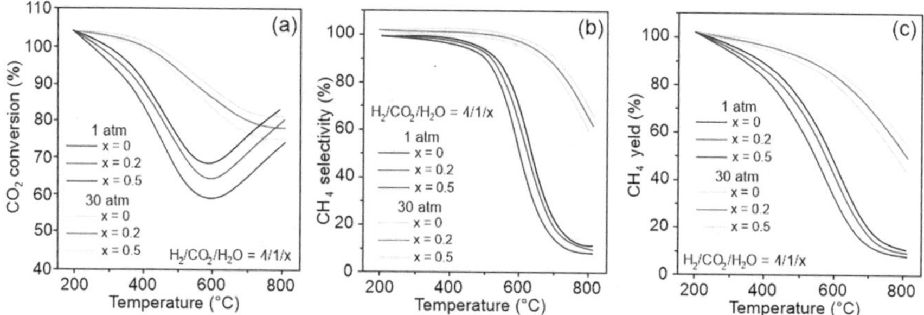

Fig. 12 Effect of water steam added into the feed on CO_2 conversion (**a**), CH_4 selectivity (**b**), CH_4 yield (**c**)

(Fig. 12c). This is mainly because H_2O is one of the products in CO_2 methanation reaction and slightly inhibits the production of CH_4.

5 Catalysts for Methanation

Catalysts are essential for methanation processes. Not only the design of the reactor is clearly influenced by the catalyst used, its activity and selectivity, but also the upstream and downstream processes. Methanation catalysts usually consist of active metal particles dispersed on metal oxide supports. So far, various active metals such as Ni, Fe, Co, Ru, Rh, Pt, Pd, W, and Mo have been carefully studied in the methanation reaction of carbon monoxide and carbon dioxide to produce SNG. The following is an overview of methanation catalysts (Argyle & Bartholomew, 2015; Gao et al., 2015; Gotz et al., 2016; Schildhauer & Biollaz, 2015).

Nowadays, the constant challenge is to develop thermally stable catalysts with high activity at low temperatures. Many metals, mainly from groups 8 to 10, are active for the methanation reaction, as shown in Fig. 13.

Because of the kinetic limitations of the methanation reaction, it is necessary to use a catalyst for the reaction to occur. However, due to the process conditions applicable to methanation, such catalysts must have a number of properties. The catalysts must have high activity, good selectivity with respect to CH_4 formation, high thermal stability and resistance to deactivation by sintering or carbon deposition, and low cost (Gruber et al., 2015).

The main methods for preparing supported catalysts for methanation are impregnation—wet and initial—and precipitation or co-precipitation. However, other methods have also shown promise, such as precipitation–precipitation (DP), sol-gel, or even the use of auxiliary techniques such as dielectric barrier discharge (DBD), luminescent discharge, and ultrasound (Silva et al., 2016).

In impregnation, metals are deposited on the surface and in the pores of a porous material that serves as a carrier for the active phase. Wet impregnation, or

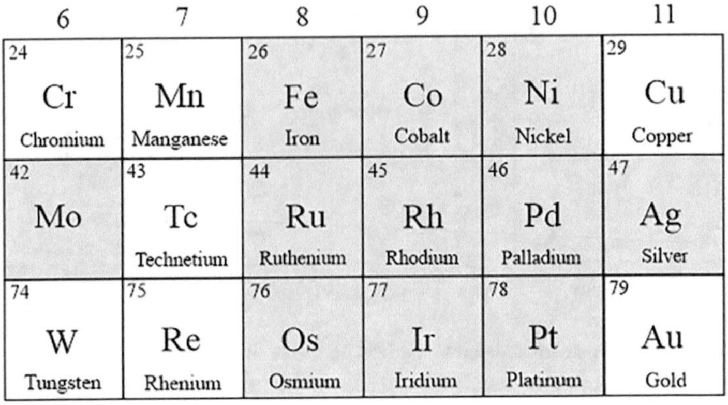

6	7	8	9	10	11
24 Cr Chromium	25 Mn Manganese	26 Fe Iron	27 Co Cobalt	28 Ni Nickel	29 Cu Copper
42 Mo	43 Tc Technetium	44 Ru Ruthenium	45 Rh Rhodium	46 Pd Palladium	47 Ag Silver
74 W Tungsten	75 Re Rhenium	76 Os Osmium	77 Ir Iridium	78 Pt Platinum	79 Au Gold

Fig. 13 Active metals for methanation are highlighted in gray

impregnation with an excess of solvent, produces a solution containing metallic precursors whose volume is greater than the pore volume of the support, thereby "soaking" the material, whereas, in incipient or dry impregnation, the volume of the precursor solution is equal to or less than the pore volume of the support, thereby avoiding excess liquid.

The impregnation methods are easy to handle, with few steps, it is not necessary to perform filtration, drying being an important step. While wet impregnation allows greater homogeneity in the distribution of metals on the substrate, incipient impregnation ensures greater control over the metal content on the surface.

Precipitation involves the co-precipitation of a solution of metal ions with a support by the action of an alkaline solution; when the metallic solution contains two or more cations to be precipitated, it is referred to as co-precipitation. The presence of several metals allows the formation of mixed oxides of lamellar double hydroxides or hydrotalcite-like compounds that form in the precipitate. Control of pH, temperature, and constant stirring are crucial parameters in this method, which is easily reproducible while maintaining the synthesis conditions.

Among the elements highlighted, elements such as Ru, Fe, Ni, Co, and Mo seem to play a more important role in the process of methanation and a more detailed analysis will be carried out on these active compounds. Ruthenium catalyst is considered to be the most chemically active and stable. A comparison of Ru and Ni catalysts shows that a methane yield of 96% was obtained with Ru catalysts at an operating temperature of 300 °C, while a maximum yield of 80% was obtained with Ni catalysts at 400 °C. One obstacle to the use of Ru catalysts is its high price.

The Nickel catalyst is the most studied catalyst as it offers a reasonable combination of good properties and price. However, one of the disadvantages of Ni catalysts is its short life due to carbon deposits blocking the pores and thus deactivating the catalyst. Molybdenum-based catalysts have the advantage of longer life and resistance to sulfur poisoning. This aspect should not be underestimated, as deactivation due to the formation of sulfites and sulfates is crucial in choosing the

Table 2 Catalyst scores based on properties

Catalyst	Catalyst properties			
	Selectivity	Activity	Stability	Costs
Ru	3	5	4	1
Fe	2	4	5	5
Ni	4	3	3	4
Co	1	2	1	3
Mo	5	1	2	2

optimal catalyst for the process. Finally, as far as the cobalt-based catalyst is concerned, it seems to be the most expensive, shows similar activity to Ni in CO and CO_2 methane, and does not require an induction phase.

In order to determine the most suitable catalyst, it is also better to rely on additional specific properties such as selectivity: higher selectivity means greater methane yield, activity: interaction of the catalyst with COx, stability: thermal and chemical resistance, cost: price of the catalyst. The catalysts are rated on a scale of one to five (where 1—the least preferred option, 5—the most preferred option) according to four main criteria.

As may be seen in Table 2, the preferred option is characterized by a higher value (e.g., when the selectivity is higher than the methane yield), while for the cost criterion, the preferred option is related to a lower catalyst price.

The results show that although Ru is the most active catalyst, it receives a lower overall score than Fe and Ni catalysts due to its high price and low use for industrial processes.

Co and Mo catalysts received the lowest score, although these catalysts are suitable for CO and CO_2 methanation, there are better options than these catalysts due to their main properties. Therefore, it is possible to say that Ni-based catalysts are the leaders on the industrial market for its good properties and reasonable price (Ngo et al., 2020).

5.1 Active Phases

The activity and selectivity of catalysts is strongly influenced by the choice of active phase, with the transition metals of the VIII group in the periodic table widely considered to be active for methanation, as they favor chemisorption, dissociation, and activation of CO_2.

Metals used as active phases include noble metals such as Rh, Ru, Pd, and Pt, as well as Ni, Fe, Co, and Cu, to name a few. Although noble metals have high activity, stability, and selectivity in the formation of CH_4, they are hindered by high cost and low availability, which significantly limits their use in industrial applications

Rhodium (Rh) is one of the most active metals, although selectivity is largely influenced by the presence of other metals or by process parameters such as temperature and pressure. Ruthenium (Ru) is considered the most active and stable metal for methanation, even at low temperatures (<300 °C), although selectivity depends on the dispersion of the metal on a support, as well as the choice of support and the addition of other modifiers. Palladium (Pd) has good catalytic activity and also promotes dissociation of H_2 molecules, providing hydrogen atoms for the reaction (Rieke, 2015).

According to literature, there are catalysts prepared with Pd, Rh, and Ru promoted with Mn and supported on Cu-Al_2O_3 used in CO_2 methanation. The catalyst with 5% Ru content exhibits the highest CO_2 conversions, reaching 100% at 300 °C. The catalyst with 10% Ru content shows no signs of significant deactivation in a stability test at 220 °C for 7 h with conversions above 94% (Rönsch et al., 2016)

Iron (Fe) is a highly active metal, though not very selective for methanation. Not only can Fe be used over a wide temperature range, but it is also very inexpensive and very durable. Cobalt (Co) has high activity and selectivity, but is more expensive. Nickel (Ni) is usually the metal most commonly used as the active phase, as it is highly active and selective for the formation of CH_4 and is also inexpensive and widely available. However, nickel catalysts suffer from deactivation, mainly due to carbon deposits—especially at high temperatures—that block the pores and the active metal, as well as low stability, which significantly reduces their durability (Rönsch et al., 2014).

Several researchers have prepared Ni/γ-Al_2O_3 catalysts by initial impregnation, varying the Ni content between 2.5 and 50%, and tested them during CO_2 methanation. Increasing the Ni content resulted in a decrease in pore size and specific area, as well as an increase in the number of medium- and strong-base sites, although it did not significantly affect the size of Ni crystallites due to the strong interaction between metal and support (Zhang et al., 2013).

Low Ni contents were not active for methanation but favored the RWGS reaction, especially with the increase of the reaction temperature. At contents above 10%, the highest yield of CH4 shows that methanation is favored over the RWGS reaction, with a yield of about 70% and a CO_2 conversion of 60% at 400 °C at a content of 25%, which remains stable between 400 and 500 °C at higher contents.

In stability tests, the catalysts maintained their activity for 7 h at 450 °C, although the catalysts with 15 and 25% Ni showed a decrease in CH_4 yield and an increase in CO yield. Although there was no evidence of sintering, carbon formed especially in catalysts with contents below 15%.

In 2016, a team of researchers investigated co-precipitated Co-Cu-ZrO_2 catalysts with different Co/Cu ratios. The 20Co-40Cu-40ZrO_2 catalyst exhibited the largest surface area and pore volume and was also the most active catalyst, achieving CO_2 conversion of about 70% and CH_4 yield of 60% at 300 °C, although the binary 60Co-40ZrO_2 catalyst showed greater selectivity in the formation of CH_4, over 90% between 240 and 300 °C. The 20Co-40Cu-40ZrO_2 catalyst showed high resistance to deactivation as it maintained the same level of CO_2 conversion in a stability test at 280 °C for 50 h.

Subsequently, a team of researchers in 2017 investigated CO_2 methanation over catalysts with 3% Ni, Rh, and Pd on SiO_2, Al_2O_3, CeO_2, and ZSM-5 prepared by initial impregnation (with the exception of Ni/ZSM catalyst-5, which was prepared by ion exchange), where catalysts on SiO_2 exhibited a larger surface area. Catalysts supported on CeO_2 exhibited the highest conversion of CO_2 and the highest selectivity for CH_4, followed by catalysts supported on Al_2O_3; the Rh/Al_2O_3 catalyst exhibited the lowest CO formation and the highest selectivity for CH_4 at 350 °C, while among the Ni catalysts, the Ni/CeO_2 catalyst exhibited the highest conversion and the catalysts supported on SiO_2 exhibited the lowest selectivities in the formation of CH_4. The Rh/CeO_2 catalyst achieved the highest conversion of CO_2, 44%, combined with a high selectivity for CH_4 at 350 °C.

5.2 *Promoters*

The addition of promoters is done to modify the properties of monometallic catalysts, changing the physical and chemical properties through the interaction of the components and alloying, generally resulting in a synergistic effect. Structural properties such as metal dispersion, surface area, and porosity or chemical properties such as alkalinity and acidity, activity and selectivity may be modified by the addition of promoters, allowing a change in reducibility and increased resistance of the catalyst to a particular type of deactivation (Campbell & Goodman, 1982).

Transition metals, such as alkali and alkaline earth metals or metal oxides used as the active phase, are commonly used as promoters in catalysts for methanation. Fe, Co, and Ru may increase the activity and selectivity of catalysts, while copper (Cu) may be used to suppress carbon formation and increase reducibility, although it is selective for the formation of CO. Molybdenum (Mo) may increase resistance to poisoning in systems using reagents contaminated with sulfur compounds and increase selectivity for CH_4.

Liu et al. (2014b) investigated Ni/Al_2O_3 catalysts obtained by impregnation promoted with Fe, Co, and Cu for application in CO_2 methanation, with a total metal content of 15% and a Ni:M ratio = 3:1. The alloy formed between Ni and Fe in the 75Ni25Fe/A12O3 catalyst was found to be the most active, while the 75Ni25Cu/Al_2O_3 catalyst exhibited a relatively inactive Ni-Cu alloy compared to the other catalysts. Garbarino et al. (2014) also prepared Ni catalysts promoted with Fe, Co, and Cu by impregnation, but on a ZrO_2 support, with 30% Ni content and 0–7% promoters. The additions of Fe and Co increased the CO_2 conversion between 210 and 250 °C, reaching 100% at 270 °C, and the Fe-containing catalyst increased the selectivity for CH_4 to over 90% at the same temperatures compared to the Ni-only catalyst. The Cu-containing catalyst decreased the conversion of CO_2 by about 20% and the selectivity for CH_4 by 5% compared to the Ni catalyst.

A group of researchers investigated a range of Ni/Al_2O_3 catalysts with contents between 5 and 25%, as well as promoted Ni-Co/Al_2O_3 catalysts with contents of 10–20% Ni and 3–10% Co, used in the methanation of CO_2. The addition of Co resulted in a slight decrease in surface area compared to catalysts with only Ni, while the pore volume did not change significantly. The addition of Co increased the activity of the catalysts, with the 20Ni-8Co/Al_2O_3 catalyst achieving 90% CO_2

conversion between 300 and 350 °C, close to equilibrium conversion, with 100% selectivity for CH_4. In stability tests, the 10Ni-10Co/Al_2O_3 catalyst showed no signs of deactivation or decrease in selectivity for CH_4 in 200 h of reaction at 350 °C.

In 2019, Guilera et al. investigated Ni/λ-Al_2O_3 catalysts promoted by a series of metal oxides (CeO_2, La2O3, Sm_2O_3, Y_2O_3, and ZrO_2) by wet impregnation, with 25% Ni content and 15% oxide. The addition of oxides resulted in a decrease in surface area and pore volume, as well as increased Ni dispersion and reducibility. The catalysts containing La_2O_3 and CeO_2 had moderately basic sites that promoted an increase in Ni activity and achieved CO_2 conversions of over 90% at 300 °C.

5.3 *Supports*

The choice of support is of great importance in the preparation of catalysts for methanation. The supports are responsible for a number of properties of the catalysts, such as surface area, porosity, acid/alkaline sites, dispersion of the metallic phase, thermal stability, and the interaction between the active phase and the support, which may increase the catalytic activity. Inert materials that do not form compounds with the active phase are desirable. Widely used are metal oxides such as Al_2O_3, SiO_2, ZrO_2, CeO_2, TiO_2, and ZnO, although materials such as zeolites and hydrotalcites are also used (Gottschlich et al., 1989; Hochgesand et al., 2006).

Alumina (Al_2O_3) is one of the most commonly used supports because it has a large surface area, readily accessible pores, is relatively inert and inexpensive. The most commonly used alumina transition phase in catalysis is γ-Al_2O_3, and its acidity may affect the properties of the catalyst. Silica (SiO_2) is also commonly used because it is relatively inert, has a large surface area, and has no phase transitions (Liu et al., 2014a; Suopajärvi et al., 2018).

Zhang et al. (2013) compared Ni-Al_2O_3-HT catalysts obtained from Ni-Al hydrotalcites prepared by precipitation and impregnated with K (K-Ni-Al_2O_3-HT) with catalysts impregnated with Ni/γ-Al_2O_3 (20 and 78% Ni) in the methanation of CO_2. The catalyst Ni-Al_2O_3-HT exhibited high metallic dispersion and small crystallite diameter compared to the impregnated catalysts. It also exhibited strong interaction between Ni and Al_2O_3 and was the most active catalyst with CO_2 conversion of 82.5% and selectivity of 99.4% CH_4 at 350 °C, maintaining activity for 20 h in stability tests. The addition of K increased the amount of strongly basic sites, increasing the conversion to 84% and the selectivity for CH_4 to 99.8% at 350 °C.

In 2017, a group of researchers used wet impregnation to prepare catalysts with 10% Ni on SiO_2, TiO_2, γ-Al_2O_3, CeO_2, and ZrO_2, as well as co-precipitation, to obtain Ni-CeO_2 catalysts for use in the methanation of CO and CO_2. The Ni/SiO_2 catalyst had the largest surface area and pore volume, while the Ni/TiO_2 catalyst had the largest pore diameter. The smaller diameter of the crystallites and the presence of strong basic sites were attributed to the high activity of the Ni/CeO_2 catalyst, which achieved 100% CO and CO_2 conversion and 100% CH_4 yield at about 230 °C. Among the co-precipitated catalysts, Ni0.8Ce0.2Ox had the smallest crystallite diameter, was the most active, and achieved 100% CO_2 conversion and CH_4 yield at 220 °C, albeit at H_2/CO_2 ratio = 50 nos catalytic tests.

In 2018, CO_2 methanation tests were performed with Ni catalysts (10%) on ZSM-5, SBA-15, MCM-41, Al_2O_3, and SiO_2 prepared by wet impregnation. The Ni/SBA-15 catalyst had the largest surface area and pore volume, while the Ni/SiO_2 catalyst had a much larger pore diameter than the other catalysts. The Ni/ZSM-5 catalyst had a larger amount of small Ni particles, which had a strong interaction with the support and were more active in methanation, with a larger amount of weak and intermediate basic sites, achieving the highest CO_2 conversion, 76% and 99% selectivity for CH_4 at 400 °C, respectively, with the order of activity being Ni/ZSM-5 > Ni/SBA-15 > Ni/Al_2O_3 > Ni/SiO_2 > Ni/MCM-41. The Ni/ZSM-5 catalyst also showed high stability when tested at 400 °C.

Tada et al. (2014) investigated Ru catalysts (3%) prepared by deposition on CeO_2 aggregates and SiO_2-CeO_2 and TiO_2-CeO_2 nanocomposites obtained by a solvothermal method for CO_2 methanation. The Ni/CeO_2 catalyst exhibited the largest metallic dispersion, with the largest surface area attributed to the Ni/SiO_2-CeO_2 catalyst. However, the Ni/TiO_2-CeO_2 catalyst was the most active catalyst, achieving a CH_4 yield of 80% at 300% and remaining stable in the test at 400 °C for 24 h.

In 2019, the research group coordinated by scientist Gabriella Garbarino investigated the γ-Al_2O_3 support with La_2O_3 in different proportions (4, 14, and 37%) using Ni (13.6%) as the active phase, and the preparation was carried out by incipient impregnation. Although the addition of La_2O_3 decreased the surface area, it changes the acidic sites of the alumina, resulting in the catalyst having more basic sites, which increase the CO_2 adsorption and make the catalyst more active. The catalyst with an La_2O_3 content of 14% achieved a CH_4 yield of 90% at 350 °C, close to thermodynamic equilibrium, making it the most active catalyst (Garbarino et al., 2014).

6 Discussion and Conclusion

The proliferation of alternative processes to the blast furnace using consolidated direct reduction and reduction smelting technologies is a worldwide trend. However, the main aspect delaying the introduction of new reduction technologies is the long life of blast furnaces. Even if the new technologies would achieve better performance from a technological point of view, it would still take several decades before blast furnaces would be completely eliminated worldwide (Cardoso & Di Felice, 2021; Cardoso, Di Felice, Santos, et al., 2022).

In this sense, the steel industry is a global industry with an extensive chain of consumption sectors. Today, about 1.6 billion tons of steel are consumed annually in the world. Steel is used for a variety of purposes, such as construction, machinery and equipment, and vehicles. It is an important product for human life in everyday life, even if it is not always visible, because it is behind a painting or the facade of a building, or even underground, where it transports water for the population or drills wells for oil production. The steel industry is vital to maintaining the sustainability of modern society, as the world's population will continue to grow and countries will need to continue their development process, for which they will need more steel to accommodate this growth (Itman et al., 2013, 2014).

The targets proposed by the steel industry are linked to the introduction of new standards for clean technologies, greater energy efficiency, and the development of low-carbon infrastructure. This requires the adoption of coherent measures and policies related to climate change. Public policies must be created that promote the efficiency of technological processes and innovation, as well as financing instruments.

In the steel industry, carbon is present in the fuels used for energy production. In the case of integrated plants, it plays its role as a reducing agent for iron ore. Subsequently, part of this carbon is incorporated into the products and the other part is emitted in the form of CO_2 after combustion.

The emission of greenhouse gasses is directly related to energy and resource conservation due to its potential impact on global climate change, as emissions represent a loss of matter and energy that could be reduced. Since it is not yet possible to avoid the generation of greenhouse gasses in steel production, they need to be mitigated to minimize their potential impact on climate change. The CO_2 emission rate currently ranges from 1510 to 1950 kg/gross in coke-integrated plants.

Carbon capture and storage (CCS) technology, which is an alternative for sequestering CO_2 produced in blast furnaces, will tend to become more widespread from 2030 onwards, once studies confirm its techno-economic feasibility, and may be integrated into both emerging primary iron-making processes and blast furnaces (under construction or renovation). In this sense, the blast furnace tends to maintain its supremacy in primary iron production, as CCS technology will optimize its energy and environmental performance.

Measures that have the potential to reduce the sector's CO_2 emissions are divided into mitigation measures in production processes and investments in low-carbon technologies and infrastructure. Mitigation actions in production processes may result from energy efficiency improvements, operational control, substitution of raw materials used, and process changes that include the introduction of new incremental or disruptive technologies. In this regard, investments in CO_2 methanation technologies are particularly promising (Wang et al., 2014, 2017).

Synthetic methane may be produced at a range of pressures and temperatures using the Sabatier reaction. The best temperature range for reactor operation is between 300°C and 400°C (according Fig. 9) at a pressure of 1 bar. In order to operate at higher temperatures, it is necessary to increase the pressure of the reagents. The Sabatier reaction may be used as a mechanism to capture carbon dioxide and allows the production of sustainable fuels such as synthetic methane (Yang & Jackson, 2013).

References

Arens, M., Hasanbeigi, A., & Pricea, L. (2014). Alternative emerging ironmaking technologies for energy-efficiency and carbon dioxide emissions reduction: A technical review. *Renewable and Sustainable Energy Reviews, 33*, 645–685.

Argyle, M. D., & Bartholomew, C. H. (2015). Heterogeneous catalyst deactivation and regeneration: A review. *Catalysts, 5*, 145–269.

Bailera, M., Lisbona, P., Romeo, L. M., & Espatolero, S. (2017). Power to gas projects review: Lab, pilot and demo plants for storing renewable energy and CO_2. *Renewable and Sustainable Energy Reviews, 69*, 292–312.

Boll, W., Hochgesand, G., & Müller, W. -D. (2006). Methanation and methane synthesis. In *Ullmann's encyclopedia of industrial chemistry* (p. 85). Wiley-VCH Verlag GmbH &Co. KGaA.

Boon, J., Cobden, P., van Dijk, H., & van Sint Annaland, M. (2015). High-temperature pressure swing adsorption cycle design for sorption-enhanced water-gas shift. *Chemical Engineering Science, 122*, 219–231.

Campbell, C. T., & Goodman, D. W. (1982). A surface science investigation of the role of potassium promoters in nickel catalysts for CO hydrogenation. *Surface Science, 123*(2–3), 413–426.

Cardoso, W., Barros, D., Baptista, R., & Di Felice, R. (2021). Mathematical modelling to control the chemical composition of blast furnace slag using artificial neural networks and empirical correlation. *IOP Conference Series: Materials Science and Engineering, 1203*, 032096.

Cardoso, W., & Di Felice, R. (2021). Prediction of silicon content in the hot metal using Bayesian networks and probabilistic reasoning. *International Journal of Advances in Intelligent Informatics, 7*(03), 268–281.

Cardoso, W., & Di Felice, R. (2022). A novel committee machine to predict the quantity of impurities in hot metal produced in blast furnace. *Computers & Chemical Engineering*, 107814.

Cardoso, W., Di Felice, R., & Baptista, R. (2021a). Artificial neural networks based on committee machine to predict the amount of sulfur and phosphorus in the hot metal of a blast furnace. In *XLII Ibero-Latin-American congress on computational methods in engineering CILAMCE-2021* (pp. 01–08).

Cardoso, W., Di Felice, R., & Baptista, R. (2021b). Mathematical modeling of a solid oxide fuel cell operating on biogas. *Bulletin of Electrical Engineering and Informatics, 10*(06), 2929–2942.

Cardoso, W., Di Felice, R., & Baptista, R. (2021c). Artificial neural networks for modelling and controlling the variables of a blast furnace. In *2021 IEEE 6th International Forum on Research and Technology for Society and Industry (RTSI)* (pp. 148–152).

Cardoso, W., Di Felice, R., & Baptista, R. (2022). Mathematical modelling to predict fuel consumption in a blast furnace using artificial neural networks. In F. P. García Márquez (Ed.), *International conference on intelligent emerging methods of artificial intelligence & cloud computing. IEMAICLOUD 2021. Smart innovation, systems and technologies* (Vol. 273, pp. 01–10). Springer.

Cardoso, W., Di Felice, R., Santos, B. N., Schitine, A. N., Machado, T. A. P., Galdino, A. G. S., & Dixini, P. V. M. (2022). Modeling of artificial neural networks for silicon prediction in the cast iron production process. *International Journal of Artificial Intelligence, 11*(02), 530–538.

Cusumano, J. A., Dalla Betta, R. A., & Levy, R. B. (1976). Scientific resources relevant to the catalytic problems in the conversion of coal. *Energy Research and Development Administration,* 583.

Czekai, I., Loviat, F., Raimondi, F., Wambach, J., Biollaz, S., & Wokaun, A. (2007). Characterization of surface processes at the Ni-based catalyst during the methanation of biomass-derived synthesis gas: X-ray photoelectron spectroscopy. *Applied Catalysis, 68*, 329.

Ensell, R., & Stroud, H. (1983). The British gas HICOM methanation process for SNG production. In *Proceedings of the international gas research conference* (pp. 472–481). British Corporation UK.

Gao, J., Liu, Q., Gu, F., Liu, B., Zhong, Z., & Su, F. (2015). Recent advances in methanation catalysts for the production of synthetic natural gas. *RSC Advances, 5*, 22759–22776.

Gao, J., Wang, Y., Ping, Y., Hu, D., Xu, G., Gu, F., & Su, F. (2011). A thermodynamic analysis of methanation reactions of carbon oxides for the production of synthetic natural gas. *RSC Advances, 2*, 2358–2368.

Garbarino, G., Riani, P., Magistri, L., & Busca, G. (2014). A study of the methanation of carbon dioxide on Ni/Al2O3 catalysts at atmospheric pressure. *International Journal of Hydrogen Energy, 39*(22), 11557–11565.

Gazzani, M., Macchi, E., & Manzolini, G. (2013). CO_2 capture in natural gas combined cycle with SEWGS. Part A: Thermodynamic performances. *International Journal of Greenhouse Gas Control, 12*, 493–501.

Gottschlich, D., Roberts, D., Wijmans, J., Bell, C.-M., & Baker, R. (1989). Economic comparison of several membrane configurations for H2/N2 separation. *Gas Separation & Purification, 3*(4), 170–179.

Gotz, M., Lefebvre, J., Mors, F., Koch, A. M., Graf, F., Bajohr, S., Reimert, R., & Kolb, T. (2016). Renewable power-to-gas: A technological and economic review. *Renewable Energy, 85*, 1371–1390.

Gruber, M., Harth, S., Trimis, D., Bajohr, S., Posdziech, O., Brabandt, J., & Koppel, W. (2015). Integrated high-temperature electrolysis and methanation for effective power to gas conversion (HELMETH). In *Gasfachliche Aussprachetagung, Essen-Germany*.

Hochgesand, G., Kriebel, M., & Schlichting, H. (2006). Absorption processes. In *Ullmann's encyclopedia of industrial chemistry* (p. 95). Wiley-VCH Verlag GmbH & Co. KGaA.

Hussain, I., Jalil, A. A., Mamat, C. R., Siang, T. J., Rahman, A., Azami, M. S., & Adnan, R. H. (2019). New insights on the effect of the H2/CO ratio for enhancement of CO methanation over metal-free fibrous silica ZSM-5: Thermodynamic and mechanistic studies. *Energy Conversion and Management, 119*, 112056.

Itman, A., Cardoso, W. S., Gontijo, L. C., Silva, R. V., & Casteletti, L. C. (2013). Austenitic-ferritic stainless steel containing niobium. *REM: Revista Escola de Minas, 66*(4), 467–471.

Itman, A., Silva, R. V., Cardoso, W. S., & Casteletti, L. C. (2014). Effect of niobium in the phase transformation and corrosion resistance of one austenitic-ferritic stainless steel. *Materials Research, 17*(4), 01–08.

Kopyscinski, J., Schildhauer, T., & Biollaz, S. (2010). Production of synthetic natural gas (SNG) from coal and dry biomass—A technology review from 1950 to 2009. *Fuel, 89*(8), 1763–1783.

Koytsoumpa, E., & Karellas, S. (2018). Equilibrium and kinetic aspects for catalytic methanation focusing on CO2 derived substitute natural gas (SNG). *Renewable and Sustainable Energy Reviews, 94*, 536–550.

Kuznecova, I., & Gusca, J. (2017). Property based ranking of CO and CO_2 methanation catalysts. *Energy Procedia, 128*, 255–260.

Lee, W. J., Li, C., Prajitno, H., Yoo, J., Patel, J., Yang, Y., & Lim, S. (2020). Recent trend in thermal catalytic low temperature CO2 methanation: A critical review. *Catalysis Today*. https://doi.org/10.1016/j.cattod.2020.02.017

Lefebvre, J., Gotz, M., Bajohr, S., Reimert, R., & Kolb, T. (2015). Improvement of three-phase methanation reactor performance for steady-state and transient operation. *Fuel Processing Technology, 132*, 83–90.

Li, Y., Zhang, Q., Chai, R., Zhao, G., Liu, Y., & Lu, Y. (2015). Structured Ni-CeO_2-Al_2O_3/Ni-foam catalyst with enhanced heat transfer for substitute natural gas production by syngas methanation. *CheCatChem, 7*(9), 1427–1431.

Liu, J., Yu, J., Su, F., & Xu, G. (2014b). Intercorrelation of structure and performance of Ni-Mg/Al2O3 catalysts prepared with different methods for syngas methanation. *Catalysis Science & Technology, 4*(2), 472–481.

Liu, Q., Gu, F., Lu, X., Liu, Y., Li, H., Zhong, Z., Xu, G., & Su, F. (2014a). Enhanced catalytic performances of Ni/Al_2O_3 catalyst via addition of V2O3 for CO methanation. *Applied Catalysis, 488*, 37–47.

Liu, Z., Grande, C. A., Li, P., Yu, J., & Rodrigues, A. E. (2011). Multi-bed vacuum pressure swing adsorption for carbon dioxide capture from flue gas. *Separation and Purification Technology, 81*(3), 307–317.

Moller, B. (2014). Status of Bio2G and gasification testing at GTI. In *SGC international seminar on gasification, Malmo.*

Nahar, G. A., & Madhani, S. S. (2010). Thermodynamics of hydrogen production by the steam reforming of butanol: Analysis of inorganic gases and light hydrocarbons. *International Journal of Hydrogen Energy, 35*(1), 98–109.

Ngo, S. I., Lim, Y.-I., Lee, D., Go, K. S., & Seo, M. W. (2020). Flow behaviors, reaction kinetics, and optimal design of fixed-and fluidized-beds for CO2 methanation. *Fuel, 275.*

Quader, M. A., Ahmedb, S., Dawala, S. Z., & Nukman, Y. (2016). Present needs, recent progress and future trends of energy-efficient ultra-low carbon dioxide (CO_2) Steelmaking (ULCOS) program. *Renewable and Sustainable Energy Reviews, 55*, 537–549.

Rieke, S. (2015). Catalytic methanation—The Audi e-gas project as an example of industrialized technology for Power to gas. In *REGATEC 2015, Barcellona-Spain.*

Rönsch, S., Matthischke, S., Müller, M., & Eichler, P. (2014). Dynamische Simulation von Reaktoren zur Festbettmethanisierung. *Chemie Ingenieur Technik, 86*, 1198–1204.

Rönsch, S., Schneider, J., Matthischke, S., Schlüter, M., Götz, M., Lefebvre, J., Prabhakaran, P., & Bajohr, S. (2016). Review on methanation—From fundamentals to current projects. *Fuel, 166*, 276–296.

Rostrup-Nielsen, J., Pedersen, K., & Sehested, J. (2007). High temperature methanation: Sintering and structure sensitivity. *Applied Catalysis A: General, 330*, 134–138.

Schildhauer, T. J., & Biollaz, S. M. (2015). Reactors for catalytic methanation in the conversion of biomass to synthetic natural gas (SNG). *Chimia, 69*(10), 603–607.

Silva, R. R., Oliveira, H. A., Guarino, A. C., Toledo, B. B., Moura, B. T., & Pasos, F. B. (2016). Effect of support on methane decomposition for hydrogen production over cobalt catalysts. *International Journal of Hydrogen Energy, 41*(16), 6763–6772.

Su, X., Xu, J., Liang, B., Duan, H., Hou, B., & Huang, Y. (2016). Catalytic carbon dioxide hydrogenation to methane: A review of recent studies. *Journal of Energy Chemistry, 25*, 553–565.

Suopajärvi, H., Umeki, K., Mousa, E., Hedayati, A., Romar, H., Kemppainen, A., Wang, C., Phounglamcheik, A., Tuomikoski, S., Norberg, N., Andefors, A., Öhman, M., Lassi, U., & Fabritius, T. (2018). Use of biomass in integrated steelmaking—Status quo, future needs and comparison to other low-CO2 steel production technologies. *Applied Energy, 213*, 384–407.

Sushil, A., & Sandun, F. (2006). Hydrogen membrane separation techniques. *Industrial & Engineering Chemistry Research, 45*(3), 875–881.

Tada, S., Kikuchi, R., Takagaki, A., Sugawara, T., Oyama, S., & Satokawa, S. (2014). Effect of metal addition to Ru/TiO_2 catalyst on selective CO methanation. *Catalysis Today, 232*(1), 16–21.

Van Herwijnen, T., Van Doesburg, H., & De Jong, W. (1973). Kinetics of the methanation of CO and CO_2 on a nickel catalyst. *Journal od Catalysis, 28*(3), 391–402.

Wang, B., Yao, Y., Jiang, M., Li, Z., Ma, X., Qin, S., & Sun, Q. (2014). Effect of cobalt and its adding sequence on the catalytic performance of MoO3/Al2O3 toward sulfur-resistant methanation. *Journal of Energy Chemistry, 23*(1), 35–42.

Wang, H., Pei, Y., Qiao, M., & Baoning, Z. (2017). Advances in methanation catalysis. *The Royal Society of Chemistry, 29*, 1–28.

Yang, C., & Jackson, R. (2013). China's synthetic natural gas revolution. *Nature Climate Change, 3*, 852–854.

Zhang, G., Sun, T., Peng, J., Wang, S., & Wang, S. (2013). A comparison of Ni/SiC and Ni/Al2O3 catalyzed total methanation for production of synthetic natural gas. *Applied Catalysis A: General, 462–463*, 75–81.

Zhu, L., Yin, S., Wang, X., & Wang, S. (2015). Biochar: a new promising catalyst support using methanation as a probe reaction. *Energy Science & Engineering, 3*(2), 126–134.

Exergy, Anergy, and Sustainability

Hans E. Wettstein

Abstract Exergy and Anergy are important terms for understanding sustainability in the context of energy consumption. Well established in the scientific world, these terms are nearly unknown in public perception. This is astonishing regarding the man-driven climate change and the related energy politics. Simply spoken, Exergy is the usable feature of any form of energy and Anergy is its not usable feature. In this sense Exergy is the necessary driving feature for all living species and for all economic activities of mankind. Our most important sustainable Exergy source is the sun radiation.

This chapter explains the essence of Exergy and Anergy with its significance in a wide scope of applications like in daily routine, in education, in statistics as well as in science and technology development. A well-adapted understanding of these notations supports sustainable behavior within all education levels. A generalist reader of this chapter may omit reading the (later) subchapters addressed to thermodynamic specialists. However, the last chapter on statistics is addressed to generalists as well.

Keywords Sustainability · Exergy · Anergy

Nomenclature

e	Specific Exergy (kJ/kg)
h	Specific enthalpy (kJ/kg)
HHV	Higher Heating Value of a fuel = LHV + Steam Condensation heat in combustion products
ISO	International Organization for Standardization
J	Metric energy or exergy or heat unit "Joule"

In this chapter most acronyms and formulas are explained where they are used. The frequently used ones are explained in Nomenclature.

H. E. Wettstein (✉)
HEW, Zürich, Switzerland

F. P. García Márquez, B. Lev (eds.), *Sustainability*, International Series in Operations Research & Management Science 333,
https://doi.org/10.1007/978-3-031-16620-4_18

kWh	Energy unit (kilo watt hour) = 3.6 kJ (kilo joule =1000 J)
LHV	Lower Heating Value of a fuel (kJ/kg)
p	Pressure (bar)
s	Specific entropy (kJ/kg/K)
Subscript $_1$	Indicates "before a change of state"
Subscript $_2$	Indicates "after a change of state"
Subscript $_a$	Indicates ambient state
T	Temperature (K or °C)
T_a, T_h	Temperatures of an ambient and of a hot heat source (K)
W	Power unit Watt= J/s ; also multiples: kW, MW, GW, TW, PW (10^3, 10^6, 10^9, 10^{12}, 10^{14})

1 Introduction

The Industrial Revolution as transition to new manufacturing processes was mainly driven by the growing availability of mechanical power sources other than animal and human muscles, water wheels, and wind. This additional energy supply started with the development of steam engines in the eighteenth century and continued later with many other methods to generate mechanical power out of heat from burning wood and/or fossil fuels. The growing dominance of the use of fossil fuels has caused the current concerns of climate change. The public awareness on this has grown in the past few decades and launched the urgent call for sustainable behavior regarding energy consumption. In this chapter we discuss sustainability in the context of our energy sources.

Our current consumption of energy may be classified as electricity, mechanical power (i.e., for transport), heat, light, fuels, and food. Let us call such things simply different forms of energy. After consumption these things are spent. However, most of us had to learn at school, that energy cannot be spent because of the "energy conservation law." It can only be converted from one form of energy to a sum of other forms of energy. The contradiction to the generally used expression "energy consumption" can only be understood by the knowledge that any energy conversion from one form to another desired form is mostly combined with a formation of another form of energy, which is not the main intention.

Evidences for this may a light bulb, which transforms electricity into light and heat as an undesired side effect or the combustion motor of a car, which transforms fuel energy into the desired power and (undesired) heat. In the latter case the undesired heat represents typically even much more energy than the desired power due to a physical law called "second law of thermodynamics".

The notations Exergy and Anergy offer here an easily understandable quantification of such effects. Any defined quantity of the (indestructible) Energy (in one or more forms) is the sum of its Exergy and its Anergy. The Exergy can be destroyed, which means consumed. The Exergy part of an energy form can in principle be converted into other forms of energy containing together equal or less Exergy. Thus Exergy can only remain or decline while Anergy can only remain or grow.

Simplified Exergy is the consumable part of energy and Anergy is the not consumable part. That means that the Exergy part of energy can be converted into any other forms of energy without theoretical restrictions, while Anergy can never be converted into Exergy. The above-mentioned so-called second law of thermodynamics quantifies this natural law. Unfortunately this is in education only communicated in complex formulas, which form a nightmare even for mechanical engineering students. The trivial hidden commonplace in this has not yet arrived in the general public perception as it should.

This commonplace means simply that **mechanical and electrical energies are pure Exergy**. These energy forms can be converted into any other energy forms without theoretical restrictions. This makes them to the most valuable forms of energy. Energy form conversion in electric motors or generators is associated with some losses (formation of Anergy) due to practical imperfections such as friction. However, these losses remain practically in the order of less than 20%. Technical improvements reducing these losses are possible and ongoing.

The above-mentioned second law of thermodynamics has its big impact for all energy conversions involving heat. Heat is always a mixture of Exergy and Anergy with the Exergy share growing with the temperature of the hot substance, which carries the heat. Quantification of this requires more explanations given in the subchapters below, which are addressed to higher education levels. Historically the notations Exergy and Anergy were "invented" in the context of heat to power conversion. However, the above described generalization for electrical and mechanical energy is a trivial commonplace for everybody.

Statistics of the world's energy consumption is ruled in (many!) details by UN (United Nations) and IEA (International Energy Agency), in order to allow comparability of regions and countries. However, in these rules the notations Exergy and Anergy are still ignored. This can lead unfortunately to the physically correct conclusion that environmental heat (which is pure Anergy) is counted as a primary energy source, which is a practical nonsense. Another paradoxical effect on statistics is that the primary energy consumption per capita is reduced by the replacement of fossil fuels by a factor of around 3 while the per capita consumption of Exergy (which represents real convenience) does not change at all. This is often declared as an efficiency gain, a statement which is certainly not applicable in this case. However, it is the result of valid regulations as indicated in the last subchapter below.

2 Exergy and Anergy Explanation for Low Education Level

The experience of kids at the age of around 10 is sufficient to dock some knowledge on Exergy and Anergy. They know the use of buttons and switches to switch on light, electronic devices or to start a ride in an elevator. They also have felt the mechanical power of accelerating vehicles of different kind. They may also know the feeling of resistance when they want to move their bike. A teacher can start the lesson with such or similar everyday experiences. It is nearby to communicate that electrical power is behind the function of most button-operated appliances while in vehicles or bikes something called mechanical power is needed to move.

A next step is to discuss where electricity is coming from. Some kids may know that there are cables with wires inside of devices and overhead lines connecting to power plants. A visit of a power plant, if nearby, is helpful to explain that most electrical power is generated from mechanical power in a power plant, which may be a wind turbine, a water turbine, or a thermal plant generating mechanical power as an intermediate step for the generation of electrical power.

The notations "hot" and "cold" are well established in the mind of our kids at this age. They have personal experience with the heat of the sun and according to weather and climate with the effect of heating or air conditioning with a cooling effect. In the kid's first perception mechanical power, electricity and heat as a feeling of too hot or being too cold are three different things. Typically the teacher connects these three things with the abstract term "energy" in his (sometimes lengthy) lessons. This is scientifically not wrong, but it is not adequate for the state of the kid's minds and it becomes in a later age the cause of the misunderstandings addressed in the chapter on statistics and energy politics. Additionally this does not help the kids to understand sustainable attitudes regarding energy consumption in their everyday life.

As of here the teaching needs another focus: The kid's experience of the three different things (mechanical power, electricity, and heat) must be maintained. However, mechanical power and electricity have a common property, which is called "Exergy." This is the stuff which drives everything that moves, lives, glows, or radiates. In the kid's experience this can be the power of muscles driving a bike as well as the electric motor help of an e-bike or a running horse or an electric light bulb or many, many other examples, which a teacher could address. Exergy can be consumed. A biker is becoming tired and he needs food to replace the spent Exergy. The battery of an e-bike needs to be charged with the electrical Exergy coming from a power plant. Consumed Exergy is gone, but where? This can be used as a bridge to explain the features of heat.

An everyday experience is that friction generates heat. As an example this is known from the brakes of a bike. After climbing a hill by using the Exergy of the muscles the brakes are becoming hot during the downhill ride. At the end the brakes are hot and the biker is tired. The Exergy has been transformed into heat, which is stored in the brake. It is well known that converting this heat back into Exergy is not possible. The best engineers could do this only to a very limited share of the initial Exergy. Thus this initial Exergy has been transformed into heat. Heat is only partly Exergy. And if the heat is available at *e* low temperature the Exergy part of heat becomes small. This Exergy part disappears completely when the brakes have cooled down to ambient temperature.

It is a simple convention that heat consists of a part Exergy and another part Anergy. How much of each applies is a matter of calculation, but with the plausible statement that the Exergy part becomes higher if the heat is hotter, what means it is available at a higher temperature, like from a fire. However, if the hot gases from a fire in a stove cool down, its Exergy part drops. Thus a stove used for heating a house and fired with gas or oil or wood is a perfect Exergy distraction machine, because all heat produced is finally cooled down to ambient temperature. In other words the stove has transformed the Exergy contained in the fire into Anergy.

Now the term Energy (and its conservation law) can be introduced on a sound base, also in connection with the required measurement units. Again the kids have the experience that their parents pay for electricity. The typical unit is the kWh (kilo-Watt-hour) on this bill and this is the measure for the consumed electricity, which can be called as well electrical energy or Exergy. Thus the most used unit for Exergy is the kWh and indeed energy is measured with the same unit and also Anergy. Thc unit Joule may be introduced here or in connection with another lesson as the unit for mechanical work with the simple equivalence:

$$1\ \mathrm{kWh} = 3{,}600{,}000\ \mathrm{Joule} = 3{,}600{,}000\ \mathrm{Watt\ \ sec}$$

At a higher age of may be 12, the kids may have heard the expression "energy consumption" in the colloquial language. In many countries the energy conservation law is addressed at this level of basic education. The contradiction between these two statements can only be solved by explaining the following take-home messages:

- Electrical and mechanical energy are a form of energy, which is called Exergy.
- A quantity of heat is another form of energy which is always a sum of Exergy and Anergy.
- Electrical and mechanical energy forms are the best usable energy forms represented by its common feature called "exergy." From these we can produce light, power, heat, drive a car, etc. This can be done in very efficient but also in inefficient ways. In all these cases more or less Exergy is additionally lost.

As a note for the expert the third statement includes both the first and the second law of thermodynamics, but without the usual quantification in formulas as mostly addressed in the average or higher education levels. In this form it is understandable for kids and it has the following practical aspects for addressing sustainability in the daily life:

1. We consume Exergy, not energy. And we have to pay for it.
2. Exergy is the free usable form of energy.
3. Anergy is the not usable form of Energy (unless an Exergy input is consumed additionally. Example: Heat pump).
4. Electricity is pure Exergy. Consuming it for light or heat production with resistors of all kind converts most of it immediately into heat with a temperature-dependent share of Anergy.
5. Electricity and mechanical energy can be converted into each other without theoretical restrictions. However, some friction or similar losses occur always in the corresponding machines.

Fig. 1 A stove with a flame converts exergy into anergy (example 8 in the text)

6. Heat from firing fossil fuels is used in many countries for heating, transport, and electricity production. The corresponding CO_2 emission drives the undesired climate change.
7. The brake of a car or a bike is a perfect Exergy destruction machine.
8. Burning a fuel generates a very hot flame with a high share of Exergy. But simply warming up some water with it destroys most Exergy, which is in the flame (Fig. 1).

3 Exergy and Anergy for Average Education Level (Ranging from High School to College)

As known from didactics, young students have to be received with the expectable practical experience of their age, which may not include much about energy. However, they know that we need light, heat, cooling, mechanical power, electricity. They may also be acquainted with computers, gas for the car, heating oil, and more such things. We suggest to use such experiences to illustrate our proposed content below.

A first look at existing syllabuses in different countries reveals that the "energy conservation law" and some notions of thermodynamics like "heat engine" or "heat pump" are frequently addressed. In one or another form also the second law of thermodynamics is addressed in a simplified version. Here the following three chapters are recommended for a syllabus aimed at this level:

3.1 History of the Energy Conservation Law

Cutting back to the years before 1842, there was no relation between heat (measured in calories) and mechanical energy measured at that time in m × kp (meters times kilopond) or ft × lbf (foot times pound force). Mechanical energy could be consumed or used up. There was no conservation law. Unfortunately this interpretation persists in the current colloquial language (as "energy consumption") in spite of the latter changed meaning of the word energy as explained below.

However, in 1842 **Julius Robert von Mayer** published that water, if falling down 365 m (under isolated and atmospheric conditions) warms up by exactly 1 °C. This statement connected heat and mechanical energy. Based on this Hermann von Helmholtz formulated the **energy conservation law**. In 1842 **James Prescot Joule** summarized this in the way that heat and mechanical energy are proportional with a constant factor called "**Heat Equivalent**."

The heat equivalent persisted because of the continued use of the different units for heat and mechanical power until 1948. In this year the "**Ninth General Conference on Weights and Measures**" decided the "**Joule**" to become the sole unit of energy, valid for any form of energy such as heat, mechanical energy, electrical power, and other forms of energy. Of course this did not exclude the use of other Joule-derived units such as kWh. This decision made the previous explicit use of a heat equivalent factor obsolete in many scientific and technical formulas.

The most used energy units are nowadays:

1 Joule = 1 Watt second (Ws) = kg m^2/s^2 (the last expression is the definition in metric units)

1 kWh = 1 kilo-Watt-hour = 1000 W × 3600 s = 3.6 Million Joules = 3.6 MJ

The kWh is now probably the most used energy unit with the k standing for a factor of 1000. Its equivalent is also 3.6 MJ with the M standing for a factor of a million.

The old fashioned units for heat "calorie" (cal) and "kcal" survive still in tables of nutrition facts. The size of the numbers and the historic distinction of heat and mechanical energy have led to the formation of many other energy units, of which Fig. 2 shows a selection. Even in current official energy statistics the unit "toe" (ton of oil equivalent) is still frequently used. As another strange example the unit BTU (British Thermal Unit) is still used for natural gas trade.

3.2 *Energy Conversion Machines*

Different energy forms can be transformed into other energy forms. However, this is mostly subject to complex machines, which have been developed with hard efforts in history. The following chapters show important examples in historical sequence.

Energy Units			**Power Output Units**		
Joule		J	Watt W	=	J/sec
KiloJoule kJ	kJ	10^3 J	kW	=	10^3 W
MegaJoule	MJ	10^6 J	MW	=	10^6 W
GigaJoule	GJ	10^9 J	GW	=	10^9 W
TeraJoule	TJ	10^{12} J	TW	=	10^{12} W
PetaJoule	PJ	10^{15}J	PW	=	10^{15}W
kWh = 3600kJ					
MWh = 3600MJ					

Fig. 2 These are the most used units for energy and power output. However, there was a long history, which has generated an uncontrolled growth of the number of units besides the mentioned ones. Some unit names trace its initial purpose or origin, such as: toe = 41.8 GJ = 11.63 MWh (ton of oil equivalent), BTU = 1.05506 kJ (British Thermal Unit), MBTU = 293.07 kWh (still used in natural gas trade), cal = 4.1868 Joule (calorie, still used in nutrition)

3.3 Hydropower and Wind Power

Hydro and wind were in the preindustrial time the most important sources of mechanical energy other than animal and human muscles.

The transformation of the energy from moving water or from elevated water basins into mechanical energy has already been developed by the Romans. The **Hierapolis sawmill** from the third century AC is one of the first documented water wheels driving a crank shaft and applied for moving a saw. Water wheels have been used in Europe since the fourteenth century. As an example they were used in Zürich for mills and for pumping water to a higher elevation. As of the nineteenth century the wheels were gradually improved and in the twentieth century replaced by water turbines of the modern types "**Kaplan**," "**Francis**," "**Pelton**" for which efficiencies up to above 90% are reported. The transformation of the energy from the moving water to a rotating shaft does only suffer from friction and similar effects. There are no theoretical restrictions like discussed below for all energy transformations involving heat. In other words, the energy of pressurized or moving water, the energy of the rotating shaft of a turbine, and the electric energy produced by a coupled generator are pure Exergy.

Nowadays (2021) modern hydropower plants account for around 15% of the worldwide electricity production and this is still the largest source of renewable and sustainable Exergy still ahead of wind and solar power sources. In this case the real primary (and sustainable) Exergy source is the sun radiation, which drives wind, weather, evaporation, and rain.

Wind power was used since preindustrial time for driving mills and more importantly for sailing ships. Based on recent technologies it has become one of the sustainable supplies of electrical energy. It has the advantage over photo voltaic resources that it is not directly dependent on sun light and less dependent on seasonal fluctuations. It may further grow to become the dominant electric power source in the future.

3.4 Heat to Power Conversion

Machines for power generation out of heat are totally different from hydro power engines. This is known from Carnot's theory (as explained below). Such typical heat to power cycles use generally a fluid. They have typically the following four steps:

1. A cold fluid is **compressed** while consuming mechanical energy.
   (a) The fluid can be water or any gas
2. The compressed fluid is **heated up**
   (a) This can happen by hot surfaces (heat exchange) or by internal combustion like in gas turbines (GTs), gasoline engines, or diesel engines
3. The heated fluid is **expanded** while yielding mechanical power output.
   (a) A part of the power output is used for the initial compression. The remaining part is the usable power output
4. The low pressure fluid is cooled down restoring the initial state
   (a) This closes the circle, which led to the name "Cycle," for such processes.

Both compressing and expanding can be made with reciprocating piston engines or with rotating turbomachinery called "Compressor" and "Turbine." The machines described below are often called "prime movers." These are engines that convert one or more forms of energy (mostly heat) into mechanical energy, in most cases by turning a shaft. In the latter case the power output is often called "shaft power."

3.5 Steam Engines and Steam Turbines, the Historically First Prime Movers

The historically first conversion of heat to power was made with reciprocating steam engines. The first attempts from 1690 until after 1802 were driven by empiric considerations of the engineering pioneers. An overview is in Table 1.

The fundamental understanding of such conversion processes was first written down by Sadi Carnot in 1824 (Carnot, 1824). This fundamental detection was brought into general science and engineering only several decades later, after his death (see Sect. 3.9 below).

Reciprocating steam engines were a nearby idea to expand hot pressurized steam in an efficient manner in order to generate mechanical power. Such engines

Table 1 The most important first steam engines and turbines in history

Year	Inventor	Innovation steps
1690	Denis Papin	First piston engine, first pressure cooker with safety valve (Fig. 3)
1712	Thomas Newcomen	Steam engine with Balancier and Feed water pump
1776	James Watt	Steam Engine with Condenser, shaft drive, Speed control. The thermal efficiency was indicated with 3%
1802	Richard Trevithick	High pressure (145 psi/10 bar patented), first steam locomotive
1841	John Penn	Paddle steamer engine. (Diesbar live steam pressure 2.5 bar, ASME Landmark #245, earlier steamers are not preserved)
1880–1914	ASME	Formation and growth of the ASME Boiler and Pressure Vessel code
1878	Corliss	Harris-Corliss Steam Engine, Randall Brothers, Inc. ASME Landmark #110
1894	Parsons	First ship with steam turbine (Turbinia, ASME Landmark #73)

dominated until the first decade of the twentieth century. They had an impressive impact on industrial development by making mechanical power available not only along rivers with hydro power but also where coal could be delivered as the main fuel in use.

As of around 1900 the steam turbines were developed. Its first benefits over reciprocating steam engines were a lower specific weight, resulting in smaller size for a given power output and a smoother operation with less vibration. Until 2030 steam turbines dominated the world's markets for driving ships and electric power supply (see below). Reciprocating steam engines remained in use until the nineteen-fifties for railway locomotives in certain countries.

Nuclear reactors have become since 1956 a widely used heat source for power plants, which generate the mechanical energy with steam turbines and produce electrical energy in the driven generator. This technology emits only very little CO_2. Thus it is considered climate-neutral. However, many environmentalists consider it as not sustainable because of safety concerns with its operation as well as with the nuclear waste handling.

Today steam turbines dominate the world's electricity production and reciprocating engines are used only in niche cases, sometimes also motivated by the need for preservation of the cultural and historical heritage (Fig. 3).

3.6 Other "Prime Movers"

Since the beginning of the twentieth century the following most important prime movers have been developed:

Fig. 3 Left: First pressure cooker with weight loaded safety valve from Denis Papin in 1690. This was an early spin-off from the first heat to power engine. Right: first steam locomotive of Richard Trevithick in 1802

- The gasoline engine: This is a reciprocating engine, which has become the most used car drive. It converts the chemical energy of gasoline into the mechanical energy, which drives the car and its other systems including the electric devices.
- The diesel engine: The diesel engine is a reciprocating engine like the gasoline engine but with typically a higher compression, which causes self-ignition. Diesel engines are more efficient than gasoline engines and they are built for cars as well as in much larger size for driving large ships. In this role they have replaced most of the formerly used steam turbines due to its better thermal efficiency.
- The gas turbine (GT) is the latest developed prime mover, which has become the exclusive drive for the current air traffic. The current engines are called turbofans because its prime mover is a gas turbine driving a larger fan, which generated most of the thrust needed for the flight. The other important GT application is driving electric generators. GT-based power plants are mostly combined with steam turbines, which use the exhaust heat of the GT. Such "combined cycle power plants" are nowadays the most efficient of the existing plants and this plant type accounts for around 30% of the current world's electricity production. Such plants normally burn natural gas and emit per produced electric kWh between half and a third of the carbon dioxide compared to a coal fired steam plant.

3.7 *Electric Transmission*

Electric energy transmission has been introduced since around 1890 in the industrialized countries. It has changed the economic and social system drastically. The conversion of mechanical energy from hydro power or from any prime mover has first been introduced to supply light and this was soon followed by electric motors.

The electric grids were gradually built up and made both light and mechanical power available in larger spaces.

A reliable electric supply became a key driver for industrial and economic development in the twentieth century and it is up to now indispensable for the development of robotics, digitalization, and modern communication. Regarding sustainability, an electrification of all applications for all energy supply is the current trend because of the Exergy feature of the electric energy. The current growth of the numbers of electric battery driven cars underlines this statement.

3.8 Heat Pumps

Heat pumps are devices, which provide either heating or cooling in an Exergy efficient manner. Heat pumps are driven by mechanical power of an electric or another motor. The most used applications last from room heating to air conditioning including cooling in all kind of refrigerators. There is a large variety of thermal cycles using different fluids which cannot be shown here. A descriptive explanation is that a heat pump transports heat from a lower to a higher temperature level by consuming Exergy.

In industrialized countries electric driven heat pumps for room heating replace gradually the fossil fired heating systems. A new application of heat pumps is now mandatory in battery-driven electric road vehicles. Due to the limited capacity of the current battery technology it is simply not affordable to install an electric resistor heating because this is a perfect Exergy destruction device eating up the driving range of such vehicles in cold ambient. However, the automobile industry has already experience with air conditioners using a corresponding heat pump technology.

3.9 Carnot Factor. The Historic Root of Exergy and Anergy

Nicholas Léonard Sadi Carnot (1796–1832) launched in 1824 the first scientific approach to understand the conversion of heat into power. His book with 43 pages has the French title "Reflexions sur la puissance motrice du feu et sur les machines propres à développer cette puissance" (Reflections on the Motive Power of Fire and on Machines Fitted to Develop that Power). This remained his only publication. I cite here from the ASME Historical Landmark Brochure (2021) of the American Society of Mechanical Engineering: "The book's author, Nicholas Léonard Sadi Carnot (1796–1832), a French engineer and physicist, published this book when he was only 28 years old. Two of the critical questions he sought to answer were whether using a substance	

(continued)

besides steam might improve the performance of heat engines and whether heat engines could be 100% efficient, that is, convert all of the heat they receive into useful work. Carnot concluded that theoretically the working substance did not matter (but that for practical reasons some might be better than others) because the efficiency of a heat engine depended on the temperatures between which it works, not on the working substance. Higher efficiency is achieved when there is a greater difference between the temperature at which heat is supplied to a heat engine and the temperature at which heat is rejected or discharged from that engine."

Sadi Carnot in a painting from Louis-Léopold Boilly (1761–1845)

The contemporary scientific community realized only later how pioneering this was. And the most surprising is that he and his contemporaries still believed that heat and mechanical energy was something different. His brother published only in 1872 a second edition combined with Sadi's heritage, from which we learn that he later changed his mind and considered heat as another form of energy. He even had determined the heat equivalent with a realistic value around 10 years before Julius Robert von Mayer. This was not public, but it demonstrates that he was close to the energy conservation law before he died too early from cholera in 1832.

Later scientists, first Clapeyron, later William Thomson (the later Lord Kelvin) and Clausius interpreted and confirmed Carnot's theory. The key of this in today's language is the "Carnot Heat Machine," which is a theoretically ideal device capable of converting heat to power as well as reverse. This device operates in the following two phases:

1. A gas with ambient temperature is **compressed** while consuming mechanical energy in both steps:
   (a) In the first step compression is ideally isothermal at ambient temperature T_a while heat is rejected to ambient.
   (b) In the second step compression is ideal adiabatic (under ideal thermal insulation).
2. The fluid is **expanded** *while yielding energy output in both steps:*
   (a) In the first step expansion is ideally isothermal at the temperature T_h while the fluid absorbs heat from the hot reservoir.
   (b) In the second step expansion is ideally adiabatic while temperature of the fluid drops to ambient.

(c) A part of the energy output from the two expansion steps is used to cover the mechanical energy needed for the two compression steps. The remaining part is the usable mechanical Energy output.

In other words, this "Carnot Cycle" transports heat from a reservoir with a higher temperature T_h down to a reservoir with ambient temperature T_a while mechanical power output is generated.

The calculation for this ideal cycle, using the property data of an ideal gas, reveals for the heat to power application a very simple formula for the efficiency, which is called the "Carnot Factor"

$$\mathrm{CF} = \frac{\text{Mechanical Energy Output}}{\text{Consumed heat from the hot reservoir}} = 1 - \frac{T_a}{T_h} \quad (1)$$

The Carnot Cycle, defined this way, is reversible, in the sense that its operation can be reversed. In the reversed operation mode it transports heat from a reservoir with ambient temperature T_a up to a reservoir with a higher temperature T_h by consuming exactly the before delivered Mechanical Energy Output.

We omit showing the derivation here. But this Carnot Factor has become the historic root of the notation Exergy. Zoran Rant (1956) has suggested the word "Exergy" E for the several expressions, that were used before for the maximum available work from a quantity of heat Q coming from e reservoir with given Temperature T_h in an ambient with temperature T_a.

$$E = Q\,(1 - T_a/T_h) \quad (2)$$

Indeed "Exergy" was a very lucky and successful neologism, which was based on Rant's excellent knowledge of different languages. Shortly after he suggested also the word "Anergy" for any energy part, which cannot be converted into Exergy. These two words have been well adopted by the thermodynamic community in all languages of industrialized countries. It was a nearby generalization to declare Exergy to be the part of the energy of any system, which can be converted into mechanical energy, which itself can be converted into any other form of energy, whether it is electricity or heat or also into any kind of Anergy. However, this is still ignored in the general public as well as in the official energy statistics.

The Carnot factor is one of the basic influences on the energy consumption. Thus Fig. 4 is a visualization of this factor.

Now some words on the use of the notation Exergy in industrial technologies. There is a well-known saying about all kind of thermodynamic cycles that:

Exergy loss is a lost opportunity to generate useful power

This indicates that analysis of Exergy losses for thermodynamic cycles is a well-established method to recognize opportunities and threads. There are thousands of publications with Exergy loss analysis and most probably much more unpublished ones under the cover of the industrial know-how secrecy. As an example the author

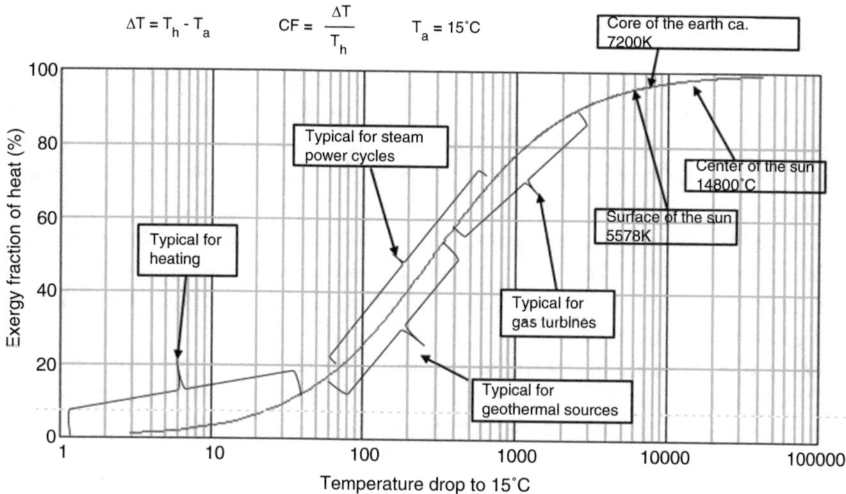

Fig. 4 Red is the Carnot Factor as a function of the temperature drop in Kelvin. This curve shows it for an ambient temperature of assumed 15 °C. Some important temperature ranges for technical energy form conversion and temperature levels are indicated in this graph. Note: T and T_a are in the Carnot formula the absolute temperatures in K (Kelvin). Conversion: T in °C + 273.15 = T in K

has published "Exergy Loss Considerations in Education for a Turbofan Power Cycle" (Wettstein, 2018a).

However, the methods to analyze such specific losses are typically subject to the bachelor or higher education levels. For this reason they are commented in the chapters below. Remembering the role of Sadi Carnot, he might be happy to hear that his ideal reversible machine has become the benchmark for all such things. The neologism Exergy was formed 124 years after his death. However, it was just a new name for his achievement. Summarizing again in other words:

The Carnot Heat Engine is the theoretically best possible sustainable device for its purpose.

The expression "Carnot Engine" can even be used in a proverbial sense for any best possible engine for its purpose.

4 Exergy and Anergy in Thermodynamic Education (Bachelor Level in Mechanical Engineering)

In a bachelor level course on basic thermodynamics the following topics are typically available:

- Introduction of the notations heat and temperature. Related measurements are defined and explained based on the notation "thermal equilibrium." The absolute temperature scale may be explained here or in connection with the second law of thermodynamics.

- First law of Thermodynamics with the mechanical equivalent of heat and with the explanation of energy units for different forms of energy such as mechanical, electrical, heat, and chemical energies. The mechanical energy explanation includes its different appearances as potential, kinetic, movement against a force and elastic energies.
- The definitions of a thermodynamic state of a body including solids and fluids in liquid or gaseous state as well as mixtures are given by thermal and calorific equations of state. The notations external work, internal energy, and enthalpy are explained in this chapter.
- For single phase fluids the thermal equations of state are explained at least for ideal gases but preferably for real gases as well. In all these cases two variables out of temperature, pressure, enthalpy, and entropy (if explained before) define the thermal state.
- The second law of thermodynamics is typically explained based on the Carnot cycle and its reversal. The notation of a state variable for a fluid is mostly explained again in this context and applied to variables such as specific entropy and specific enthalpy.
- More or less examples for thermal cycles for conversion of heat into power or reverse are given at this state. This mostly includes cycles for generation of heat and refrigeration.
- Describing the quantity of a material with its mass with conversion into molar quantities is used wherever chemical reactions are involved.

The reader of the following chapter should know these topics as well as the knowledge of the previous chapter for a beneficial digestion. At this education level a quantification of thermodynamic effects becomes important. Figure 5 is another representation of the second law of thermodynamics based on the notations Exergy and Anergy.

However, there are other forms or consequences of the second law such as:

- Heat can only be exchanged from hot to cold
- The efficiency of a Carnot Heat Engine cannot be exceeded
- All Processes running spontaneously in one direction are irreversible
- All Processes, in which friction takes place, are irreversible
- In a closed system the total entropy can stay constant or grow. It never decreases.

The specific entropy of a substance is a thermodynamic property of state, which is typically explained and discussed in Bachelor courses. Also the second law is typically taught based on the notation entropy as in the last case of above list. This is not deepened here.

Students of this level have to resolve the contradiction between the frequently heard notation "Energy consumption" and the energy conservation law. This contradiction is best removed with the statement that only the Exergy part of energy is used up by "consuming energy." However, in my experience many teachers for this

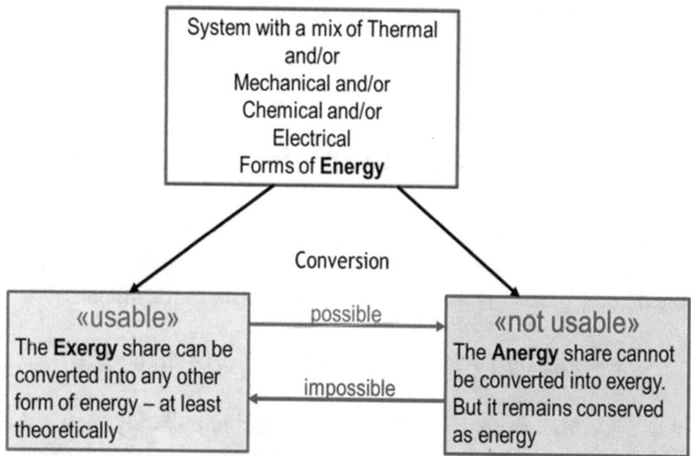

Fig. 5 A simple explanation of Exergy and of the second law of thermodynamics. This figure expresses the most demonstrative form of the second law of thermodynamics: "Anergy cannot be converted into Exergy while the reverse conversion is possible"

Table 2 Wrong words used in colloquial language

Wrong	Correct
Energy consumption	Exergy consumption
Energy loss	Strict distinction of "Exergy loss" and "Anergy discharge"
Using up energy	Using up Exergy or convert Exergy to Anergy

level said: "Exergy is something for advanced students only". This statement is backed by the fact that in such courses Exergy is only treated in combination with entropy and with heat to power cycles. In many courses the statement that both electricity and mechanical energy is pure Exergy is missing. In Table 2 the following corrections are suggested:

A student of this level should also learn what this means regarding sustainability:

An Exergy loss is per se not sustainable. But it cannot be completely avoided by any human activity. Therefore in a **modern understanding of sustainable behavior** any primary source of Exergy should come from a sustainable source such as sun radiation, hydro or wind energies. Of course in an environment of limited or restricted sustainable Exergy supply as well as with consumption from non-sustainable Exergy sources (such as fossil fuels) any reduction of Exergy losses is a step to more sustainable behavior.

Declaring the use of fossil fuels to be a non-sustainable activity has its reason in the facts that such fuels are typically hydrocarbons, which have been formed from oxygen and carbon dioxide by photosynthesis in plants using Exergy of the sunlight over millions of years. Burning fossil hydrocarbons in the current extent disturbs the current equilibrium in the earth's biosphere and it is considered as the main cause of the current global warming.

The following subchapters show how Exergy is quantified and used in frequent situations.

4.1 Exergy of a Pressurized Hot Fluid

The classic approach to teach about Exergy is to calculate the specific Exergy of a hot pressurized fuel with temperature T and pressure p referring to an ambient with temperature T_a and pressure p_a. The well-known formula for this is:

$$e=[h(T,\ p)-h(T_a,\ p_a)]-T_a\,[s(T,\ p)-s(T_a,\ p_a)] \qquad (3)$$

Here e is a specific Exergy in kJ/kg, h the specific enthalpy, and s the specific entropy. The latter two are state variables while the specific Exergy is not because it is additionally dependent on the environment conditions.

This formula is derived from the heat input (for heating the fluid up from ambient to T) multiplied with the CF and integrated over temperature from T_a to T. The state variables enthalpy h and entropy s represent in this sense a hint to write this expression without the integration sign. At Carnot's time enthalpy was called "heat content" (from which we have inherited the still used formula sign "h") and the first definition of the state variable entropy comes from Rudolf Clausius in 1850 based on Carnot's theory.

Note: The specific Exergy in mechanical engineering is mostly indicated in kJ/kg while in chemical context mostly the unit kJ/mol is applied.

4.2 Exergy of Fuels

If the substance in consideration is used as a fuel, we have to include additionally in Eq. (3) the stoichiometric chemical combustion reaction with oxygen of initially ambient condition. Doing this is a bulky exercise not shown here. But there are simple approximations or detours. Zoran Rant showed already in 1961 (Rant, 1961) that the Exergy of a selection of liquid hydrocarbons from C5 to C12 (chemical compounds containing oxygen and between 5 and 12 atoms of carbon) is between the Lower Heating Value (LHV) and the Higher Heating Value (HHV), typically around 3–4.5% above the LHV. The internet portal http://www.exergoecology.com/ offers an Exergy calculator for those interested in more accurate numbers for pure substances (The Exergoecology Portal, 2011).

For natural gas and liquid hydrocarbon fuels the Exergy content is not (yet) used in trade and values are difficult to obtain because these fuels are typically composed of many different hydrocarbons dependent on the origin. However, the heating value is relatively easy to measure and it forms the base for fuel trade. Thus in many cases the LHV is used as reference for the Exergy of a fuel. For Methane (the main constituent of natural gas) the mentioned portal gives a specific Exergy of 51.8934 MJ/kg, while LHV = 50 MJ/kg and HHV = 55.5 MJ/kg.

Cutting back to history, the fossil fuels (coal, oil, natural gas) were the key drivers for economic development. The classic heat to power cycles as indicated above operated all with a combustion step before the conversion of heat into mechanical power. With the introduction of the Exergy of heat also the Exergy of a fuel caught the interest of research.

4.3 Exergy in the Context of Efficiency Definitions

Efficiency is in general understanding a ratio describing the quality of a process by dividing a result with the needed effort. In the case of a thermal cycle we talk about a "**thermal efficiency**." This is the ratio of power output and heat input. Both measured in Joule or related to time in Watt. The above discussed Carnot factor is a thermal efficiency in this sense. However, in Carnot's time in 1824 the notation of efficiency for a thermal process could not exist, because heat and mechanical energy were considered to be totally different things. This underlines the brilliance of his consideration. Considerably later, with the detection of the mechanical heat equivalent and the follow-up development both quantities were recognized as different forms of energy measurable now in Joule or if related to time in Watt. Such an efficiency is always a number between 0 and 1.

An "**Exergetic efficiency**" could be defined by dividing an Exergy output with an Exergy input. Also here both output and input are measured in energy or power units. However, such efficiencies have not yet found its way into public perception, although this would be the best key number for the technical quality of a process.

In the case that output and input cannot be measured with the same units the word efficiency is sometimes abused. Better designations for this case are intensity (such as energy intensity of a production value, i.e. in kWh/$).

Also a frequently observed misleading attitude is the designation of an efficiency for a combined heat and power plant. Such plants have power and heat (mostly for district heating) as output and fuel heat as input, both measured with energy units. In order not to confuse with a (real) thermal efficiency this should rather be called "**use Factor of heat**."

An old fashioned, but still frequently used way of indicating thermal efficiencies is the so-called **Heat Rate**. This is a reciprocal efficiency. It is often indicated with different units such as MBTU/MWh.

Table 3 gives the magnitude of exergetic efficiencies for frequent applications:

Table 3 Exergetic efficiencies of frequent technical energy conversion processes

Device	Input energy form >> output energy form	Influenced by second law	Exergetic efficiency
Turbofan at start, zero speed (Wettstein, 2014)	Kerosene >> thrust power output	Yes	0%
Turbofan in cruising condition (Wettstein, 2014)	Kerosene >> thrust power output	Yes	Up to 36%
Room heating 22 °C with fire	Fuel >> heat	Yes	10%
Electrical resistor heating 22 °C	Electrical power >> heat	Yes	10%
Hot water 80 °C using fired or electric boiler	Fuel or elect.>> heat	Yes	ca. 20%
Room heating 22 °C with heat pump	electric power >> heat	Yes	20–40%
Gasoline engine (i.e. in a car)	gas >> shaft power	Yes	10–20%
Photovoltaics	Sun light >> electric power	Partly	10–20%
Nuclear power plant	reactor heat>> electric power	Yes	32–35%
Large Diesel engine	liquid fuel>> shaft power	Yes	45–55%
Gas Turbine Combined Cycle (GTPP) Plant	Natural gas >> Electric power output	Yes	50–64%
Hydro Power Plant (only dissipation and friction losses)	Potential energy of water >> shaft power	No	87–91%
Wind Turbine (limited by Betz law)	Kinetic energy >> shaft power	No	<59%
Electric generator (size dependent)	Shaft power >> Electric power	No	80–99%

4.4 Exergy Losses and Sankey Diagrams

Analysis of Exergy losses in any technical device can be enlightening for detection of opportunities and weaknesses. Scientific literature gives many examples. Among them we mention here an analysis of large contemporary Gas Turbine Combined Cycle (GTCC) plants (Wettstein, 2020a), which supply some 30% of the World's electricity. This is a successful and rather complex technology, for which this publication indicates the Exergy losses within the key components.

A both descriptive and historic visualization of such results are the so-called Sankey diagrams of Exergy losses. Already in 1961 Peter Grassmann (a pioneer of the application of the Exergy notation also in education) suggested this in his textbook (Grassmann, 1961). It is informative to study the differences of the energy and the Exergy considerations in Fig. 6. The Exergy loss V_k of combustion is indeed blown up in this case as discussed above. Exergy considerations play an important role in the chemical engineering technology. As shown in Fig. 6 Sankey diagrams are also used for indicating energy flows. More about this is mentioned in the statistic chapter below.

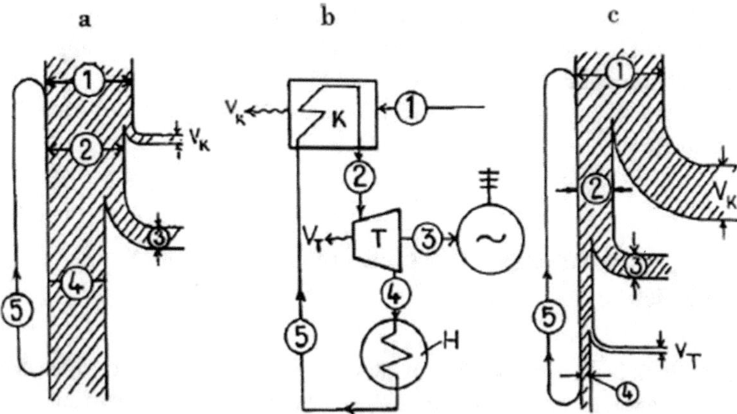

Fig. 6 Example of an early Sankey diagram from Grassmann (1961) for a 500 °C/80 bar steam power plant of that time: K, Boiler; T, Turbine; H, condensation heat receiving "condenser"; (**a**) Sankey diagram for energy; (**b**) flow scheme of the cycle; (**c**) Sankey diagram for Exergy; V_k, Energy/Exergy loss in Boiler; 1, Fuel Energy/Exergy to boiler; 2, steam flow from boiler to turbine; 3, Mechanical power to generator; 4, exhaust flow to condensor H; 5, Feedwater to Boiler

4.5 Exergy Penetration in the Society

A Google search for "Exergy" and its translations in other languages such as Exergie (Ge and Fr), Exergia (Sp), Essergia or Exergia (It), эксергия (Ru) gives more than two million hits with the majority for the English word Exergy. These hits involve not only Exergy loss analysis issues, but also education issues, society, and company names as well as general explanation documents and videos. On the other hand, the 2021 still limited penetration of the colloquial language space is illustrated by the fact that one of the most used online dictionaries https://dict.leo.org only gives the English-German translation while it ignores this word for all translations into its other supported languages by suggesting energy instead. Even the Merriam-Webster online thesaurus ignores still the word Exergy. Personal contacts of the author with scientists from China, Japan, and Korea revealed that they are well familiar with the notations Exergy and Anergy also in their native languages.

The Wikipedia's in the western languages all explain the notations Exergy and Anergy.

Finally a bachelor of any field should at least know that Exergy is the usable quality part of energy, which drives all life, all movements, and all economy while the rest of the energy is Anergy, which is characterized by its total inability to move anything.

A bizarreness related to Anergy happened as of 1867: The well-established scientist Rudolf Clausius suggested for the final thermal equilibrium of the universe—seen as a closed system—the apocalyptic expression "heat death." This alarmed some society circles. However, the detection followed that the universe had plenty of Exergy sources driven by nuclear fusion processes supplying sufficient hot radiation for billions of years. Thus, the apocalyptic expectation disappeared from public perception. More about this can be found under "Heat death of the universe." In our now learned wording the heat death would be simply the transformation of all Exergy into Anergy.

5 Exergy and Anergy in Technology Development and Science

Processes which include conversion of energy forms are used in most industries and in most production processes. And in such processes friction, heat exchange, mixing of substances, chemical reactions, electric resistors, and more effects cause Exergy losses. Understanding and quantifying these losses is a frequent task with the aim of finding opportunities for improvements and ways to reduce the Exergy losses. In this chapter some basic tasks for quantifying Exergy losses in thermal cycles are described following partly the paper (Wettstein, 2018a) of the author.

5.1 Exergy Loss of Heat Exchange

Any definition of Exergy is related to reference ambient conditions (pressure, temperature, and composition). We suggest to relate the specific Exergy always to the standard ambient conditions according to the international standard atmosphere, which corresponds to ISO 2314 at altitude zero (air with 15 °C, 1.01315 bar and 60% relative humidity).

A quantity of heat Q with the temperature T has an Exergy E according to the Carnot factor where T_a is the applicable ambient temperature:

Exergy E of a quantity of Heat Q with temperature T:

$$E = Q \times \frac{T - T_a}{T} \tag{4}$$

The corresponding Anergy A is:

Anergy of a quantity of Heat Q with temperature T:

$$A = Q \times \frac{T_a}{T} \tag{5}$$

Figure 7 visualizes the Exergy loss if a quantity of heat is exchanged from a hotter to a colder heat reservoir in an ambient with temperature T_a.

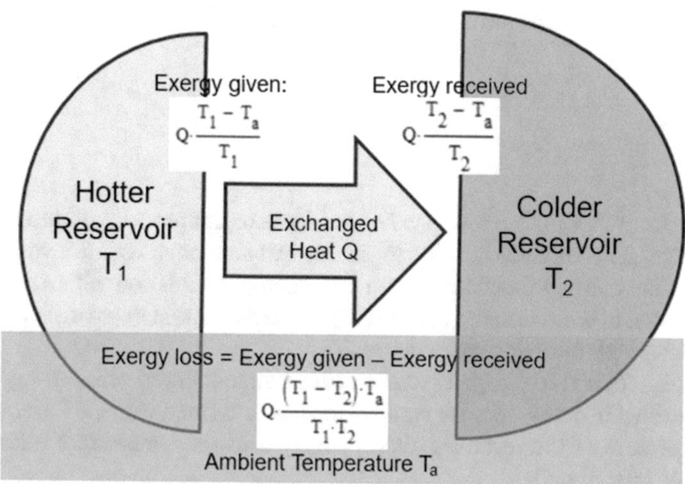

Fig. 7 Visualization of a heat exchange between two reservoirs of different temperature T_1 and T_2 in an ambient of temperature T_a. This is simply derived from the Carnot Factors as shown here. **The message is: The loss of Exergy grows with the difference of the temperature and it drops with a higher temperature level**

5.2 *Exergy Loss by Combustion*

Burning a fuel is the most used way to introduce the fuel Exergy into the cycle of a prime mover. Typical fuels like liquid hydrocarbons or natural gas supplied with ambient temperature and burnt with ambient air in a stoichiometric condition can generate heat with a very high temperature (above 2000 °C) and therefore with a high Carnot factor. But the existing thermal cycles cannot receive such high temperature because of the material limitations. Processes with the highest temperature capabilities are using the so-called internal combustion in compressed air, which is directly expanded after combustion. Such engines are either reciprocating engines or continuous like gas turbines (GTs). We limit the consideration to the latter for simplicity.

Currently the highest hot gas temperatures that can be managed continuously in GTs are in the 1500–1700 °C range with combustion occurring at the nearly constant pressure p.

Using formula (3) with p representing the combustion pressure and T_{in}, T_{out} the average temperatures of the cycle fluid at the combustor inlet and outlet we can calculate the Exergy received by the fluid:

Specific Exergy received in a combustor by the combustion gas mixture:

$$e = h(p, T_{out}) - h(p, T_{in}) - T_a[s(p, T_{out}) - s(p, T_{in})] \tag{6}$$

Alternatively we can use the Carnot factor directly resulting in formula (6a).

Specific Exergy received by the combustion gas mixture:

$$e = \int_{T_{in}}^{T_{out}} \mathrm{Cp}(T) \times \left(1 - \frac{T_a}{T}\right) dT \tag{6a}$$

In this formula Cp(T) is the specific heat for constant pressure of the combustion gas mixture. It can replace formula (6) in this and similar cases thus avoiding either the use of the entropy function or even motivating its omission within a thermodynamic course. In both formulas (6) and (6a) it is assumed that the oxidant and the fuel are received with the same temperature T_{in}.

Formulas (6) and (6a) expressed in words mean that the specific Exergy (per unit mass) received in the cycle is the combustion heat weighted with the Carnot factor of the temperature of the receiving fluid integrated in the temperature interval from combustor inlet to outlet.

The Exergy loss in a GT combustor is often approximated with the fuel LHV instead of using the fuel Exergy (if applicable including the Exergy used for heating and compressing the fuel to its combustor inlet condition) minus the Exergy received by the combustion gas according to formulas (6) or (6a).

If there is a temperature difference of the oxidant and the fuel, an additional mixing consideration is needed. We suggest solving this for both the energy and the Exergy balance of the combustor by declaring the reaction to occur at the lower temperature of oxidant or fuel reception. Then both heat and Exergy contents of the hotter reactant above this reaction temperature are considered on the input side (together with the LHV). The heating up of the combustion gas by the heating value is then considered to start at the declared reaction temperature. For GT combustors more details on this issue are given in Wettstein (2020b).

It is important to note that in most cases the Exergy loss by combustion is the largest single Exergy loss in GT cycles. The best contemporary GT Combined Cycles achieve up to 64% thermal efficiency. This implies around 36% Exergy losses of which the combustion Exergy loss amounts to around 25% (Wettstein, 2020b). A minor contribution to the received Exergy comes with the fuel into the cycle caused by the fuel pressure. It is mostly neglected, even in catalogue data for gas fired gas turbines. For a liquid fuel this part is indeed negligible. The dominance of the Exergy loss by combustion forms a strong motivation for the original engine manufacturers to increase the firing temperature. This is the main trend for further reduction of the combustion loss in GTs.

5.3 *Heat Loss to Ambient*

Any heat loss to ambient of a process fluid is associated with a specific Exergy loss according to an expression like (3) or (4). Heat losses occur from all hot parts, all coolers, and from the bottoming heat losses with the exhaust gas. In large gas turbines the (external) heat losses from the hot components are typically small fractions of a % (related to the heat received with the fuel heating value).

Estimations for the Exergy loss are easy to make by using the Carnot Factor weighted heat determined either from a measured heat balance or from heat transfer considerations.

5.4 Exergy Loss by Mixing

Mixing fluids with the same pressure and different temperature and/or different composition generates an Exergy loss. This case can generally be treated by applying formula (3) added for the two mixed fluids minus the result of the same formula for the mixed fluid with temperature and composition resulting from heat and mass balance across the mixing range.

In the case of gas turbines (GT) with the so-called open cooling systems compressed air is bypassing the combustor. It mixes into the hot gas in different locations after having carried out its cooling duty in combustor and turbine. In GT technology a simplified method to consider this is generally used, which is described in the standards ISO 2314 (Rant, 1961) or ISO 3977-3 and (Wettstein, 2020b). According to this a "mixed turbine inlet temperature" T_{mix} is defined by adding the fuel heat to the fluid with flow rate and composition as it leaves the turbine blading cold end (but of course considering the (mixed) oxidant's combustor inlet pressure and temperature). This allows to calculate the turbine exit temperature based on a polytropic efficiency, pressure ratio, and the heat balance of the combustor without knowledge about the cooling air share, which is never published by the OEMs.

A frequently applied method for assessing Exergy losses for mixing of different substances with the same temperature considers reversible and isothermal expansion or compression from the partial pressure of each substance to a partial pressure of the same substance in the reference ambient. Such losses are typically small in relative terms and technical exploitation is rare so far. However, such exploitation, for example, of natural salt concentration gradients in the sea (combined with temperature differences) has a potential for sustainable Exergy supply. However, the challenge is huge due to the large mass of water involved for a reasonable power output.

More details of such calculations can be found in the referenced publications of the author, especially (Wettstein, 2020b)

5.5 *Exergy Loss by Friction and Dissipation in Turbomachinery*

Turbomachinery is used either to compress or to expand fluids. This is always combined with friction at solid walls and dissipation within the fluid.

- If the fluid is compressed these effects convert mechanical energy (Exergy) from the rotor into heat received with the local temperature of the fluid, thus reducing the intended Exergy growth of the fluid.
- If the fluid is expanded these effects convert Exergy from the fluid into heat with the local temperature of the fluid, thus reducing the intended mechanical energy output of the shaft and increasing the specific exergy of the exiting fluid.

Compressors or turbines change the state of the fluid from p_1, T_1 to p_2, T_2. Assuming that the dissipation share remains constant along the change path leads to the so-called polytropic efficiency, a number between 0 and 1. A polytropic efficiency of 1 indicates a reversible process. 1 minus the polytropic efficiency is the Exergy converted into heat with the local fluid temperature. This heat contains still Exergy which must be considered for evaluating the Exergy loss by using formula (2).

If averages from states 1 and 2 can be measured for an existing turbomachine the (averaged) polytropic efficiency is determined. However, its calculation depends on the fluid too. For an ideal gas the calculation can be made with simple formulas.

The challenge is only that ideal gases do not exist in nature. Nevertheless air and combustion gas as typically used for gas turbines (Brayton cycle) behave rather close to an ideal gas. The mentioned formulas and with them the polytropic efficiency have therefore been widely adopted in the gas turbine industry. The industry holds internal measurement data from real engines. This allows them to compensate the remaining small mistakes caused by the real gas nature of the involved gas mixtures by calibration. The magnitude of such small mistakes for air gases was indicated in a paper of the author in 2014 (Wettstein, 2014).

However, steam is far away from ideal gas behavior. It even includes condensation to water (droplets) within expansion. For steam engines and turbines (Rankine cycle) the polytropic efficiency calculation required bulky and complicated approximations. Historically the equation of state for water and steam was listed in tables and for calculation of the expansion process represented in temperature-entropy or in enthalpy-entropy diagrams. This favored the use of the so-called isentropic efficiency:

$$\eta_{\mathrm{is}} = \frac{\text{real specific enthalpy drop}}{\text{isentropic (ideal)specific enthalpy drop}}$$

In the diagrams this could easily be handled. However, nowadays all the required thermodynamic properties of steam are available in computer readable applications.

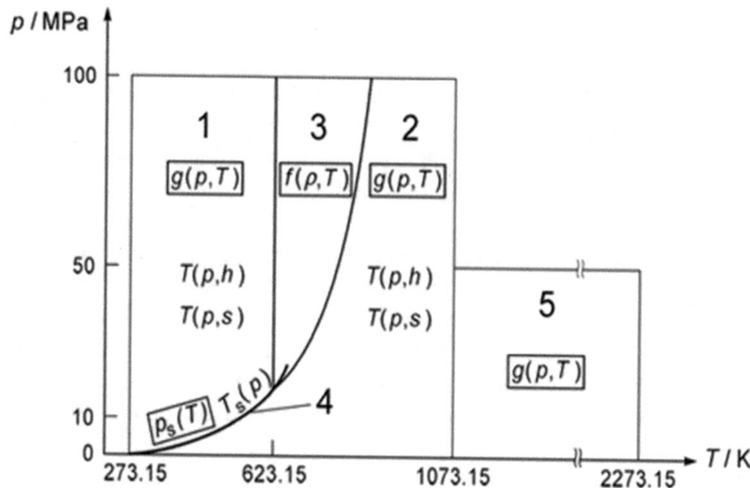

Fig. 8 Current definition range of the IAWPS thermodynamic properties of water and steam in a pressure temperature diagram. This is from IAPWS (2007)

The most used current water-steam database is maintained by IAWPS, an international organization (Fig. 8 and IAPWS, 2007).

For small pressure ratio p_1/p_2 the isentropic efficiency is identical with the polytropic efficiency. This is the reason why the polytropic efficiency sometimes is named "small stage efficiency," especially in communities, who favor the application of the isentropic efficiency.

The calculation of a polytropic efficiency for a larger pressure ratio requires a recursive integration, which was not possible in the time before computers were available. For a larger expansion pressure ratio the isentropic efficiency is larger than the polytropic efficiency. This is explained in more detail in Wettstein (2021). This paper also explains the direct incremental and recursive calculation of the polytropic efficiency for steam expansion. The favored method for this under adiabatic contitions was called CDRA (Constant Dissipation Rate Algorithm).

However, the simple but important message from these learnings is that the isentropic efficiency depends additionally from the pressure ratio and not only from energy conversion rate in the turbine. The benchmark measure for the aerodynamic quality of steam turbines remains the polytropic efficiency. This key number allows comparison also for turbines with different pressure ratios in contrast to the isentropic efficiency.

Indeed there is a recent trend to apply the polytropic efficiency to steam expansion too. The motivation is that this efficiency definition is the only one, which clearly separates conversion of steam Exergy to work (still pure Exergy) from the conversion to heat with the local steam temperature. An overview of this with further references was published in 2021 (Wettstein, 2021).

5.6 Additional Remark About Large Pressure Ratio

The above-mentioned CDRA allows calculation for steam of the exit condition of any blading assuming a constant polytropic efficiency for any pressure ratio. The assumption of constant polytropic efficiency along a blading makes sense in case of a blading with repeat stages. This was applied for most early axial compressor developments and also for the high pressure part of steam turbines.

In case of a cooled (gas) turbine blading the first stages suffer more from the cooling fluid discharges than the rear stages. In very conical low pressure steam turbines the steam flow field has a three-dimensional pattern, which requires a three-dimensional aerodynamic design. In such cases it can be more realistic to distinguish stages with different polytropic efficiencies. This applies especially if more detailed information on the blading is available and/or if the pressure ratio is large.

5.7 Exergy Losses by Cycle-Internal Pressure Losses "Throttling"

We suggest to characterize generally internal pressure losses with its relative pressure drop Δp related to the upstream pressure. In thermal cycles they are typically relatively small and without a temperature change. Therefore we can use for this case the ideal gas approximation with specific work of an ideal isothermal expansion being equal to an expression like $RT \times \ln(p_1/p_2)$ where the heat input must be equal to the specific work. In case of a pure pressure drop this work is all lost and an equal amount of heat remains (correctly spoken as "internal energy") in the discharged fluid. Using the Carnot factor for assessing this remaining heat the specific Exergy loss of the pressure drop gives us the ideal gas approximation for the specific Exergy loss e_{pd} by a relative pressure drop deltap

Ideal gas approximation for the specific Exergy loss e_{pd} by a relative pressure drop Δp:

$$e_{\mathrm{pd}} = R \times T_{\mathrm{amb}} \times \ln\left(\frac{1}{1 - \Delta p}\right) \tag{7}$$

Here R is the gas constant of the fluid mixture and T_{amb} is the (locally applicable) ambient temperature. Astonishingly this Exergy loss does not depend on the fluid temperature, which allows application of this formula also for heat exchangers with its variable temperature. This method neglects the Joule-Thomson effect, which occurs for real gases but not for ideal gases.

In principle the CDRA could also be applied for pressure drops. This requires a limit value consideration for the polytropic efficiency approaching zero because in the limit case both enthalpy difference and efficiency are zero (causing a division 0/0). However, the CDRA considers the Joule-Thomson effect dependent on the quality of the applied real gas formulation, which is in most cases insufficient in the desired data range. Thus this method cannot be recommended. It remains an issue for further research.

5.8 *Exergy Loss with the Exhaust Gas*

It is a general intention of an efficient design of thermal cycles to exhaust only gases with a low temperature. Its Exergy can easily be calculated with Eq. (3). However, there are limitations for reducing this temperature, such as:

- In boilers the condensation of water in the exhaust is mostly prevented by sufficient temperature in the order of around 90 °C because of an often observed content of sulfur dioxide, which forms with condensed water a corrosive acid.
- In turbofans the hot core cannot be expanded to a near ambient temperature. The reasons and the corresponding Exergy losses are explained in detail in a paper of the author (United Nations, 2018).

5.9 *Mechanical Power Losses and Conversion Systems*

Mechanical power losses can generally be assumed for the shaft friction in bearings and windage losses. It may include the parasitic power for the lube oil systems and other auxiliaries including all mechanical power transmissions. In large prime mover engines this is in the order of up to 1.5% of a thermodynamic shaft power. Details depend on the specific engine. This loss is like friction a direct Exergy loss if the generated heat is finally discharged to ambient temperature.

Mechanical power exploitation from sea-wave energy, for which already demonstrators have been operated is a sustainable technology. Waves are driven by wind, which is a result of the abundant sun radiation. Such engines suffer only from its internal friction as a mechanical power loss.

The use of tide caused level differences for generating hydropower exists in a few plants already. However, this may not be considered sustainable in a long-term perspective due to its (very small) contribution to the slowing down of the earth's rotation. It appears that both wave and tide exploitation will remain a small niche of the future Exergy supply.

6 Exergy in Statistics and in Energy Politics

Exergy statistics is regulated rigorously regarding its content definitions. The relevant documents are issued by the United Nations (UN) and by the International Energy Agency (IEA). The purpose of such regulation is to allow supranational consolidation of all kind of relevant categories of energy consumption. We have the following key documents:

- Issued by UN: "International Recommendations for Energy Statistics" (IRES) (United Nations, Department of Economic and Social Affairs, Statistics Division, 2018) and a final draft under the name "Energy Statistics Compilers Manual" issued in 2016 (United Nations, 2016).
- Issued by IEA: "Key World Energy Statistics,"(IEA, 2021) "Energy Efficiency Indicators: Fundamentals on Statistics," (IEA, 2014) and "Energy Efficiency Indicators, Highlights" (IEA, 2016).

These five documents are too detailed, to be referred here in depth. The Compilers Manual recommends types and entities of the energy consumers to consider. It classifies numerous energy products and distinguishes between renewable and non-renewable. The IRES declared the key principle that the (quote) "primary energy form should be the first energy form downstream in the production process for which multiple energy uses are practical." This means that "the physical energy value of the primary energy form is used for the production figure. For primary electricity, this is simply the gross generation figure for the source" (unquote). The International Energy Agency (IEA) uses the acronym TPES for "Total Primary Energy Supply."

The primary energy from fossil fuels is declared by IRES as gross (or net-) calorific value (GCV/NCV). They recommend to use preferably the NCV also for the so-called energy products. In the technical literature GCV/NCV is mostly referred to as HHV/LHV for Higher or Lower Heating Value. The difference is the water condensation heat which is in the combustion gas. Thus the two values are only different for fuels containing hydrogen. In all here mentioned references of UN and IEA the notations Exergy and Anergy do not occur.

The achievements of the international consolidation of energy statistics are impressive. There is an excellent overview of it given directly in Sankey diagrams for countries, regions, and the world composed by IEA. This is a website with this link: https://www.iea.org/sankey/#?c=World&s=Balance (IEA, n.d.).

However, the conservatism of the IEA community is expressed in the energy units coming as a standard in the above-mentioned energy Sankey diagrams: "Millions of tons of oil equivalent." The energy units can be switched to Petajoules like used in Fig. 9. As a step to understand the progress of electrification the website contains a view "Final consumption" which contains electricity. Calling this website on 2021-11-14 delivered the consolidated data from every year from 1973 to 2019. The live diagram indicates some more information but no Exergy consumption. Summarizing, this is a huge database, which is easily accessible.

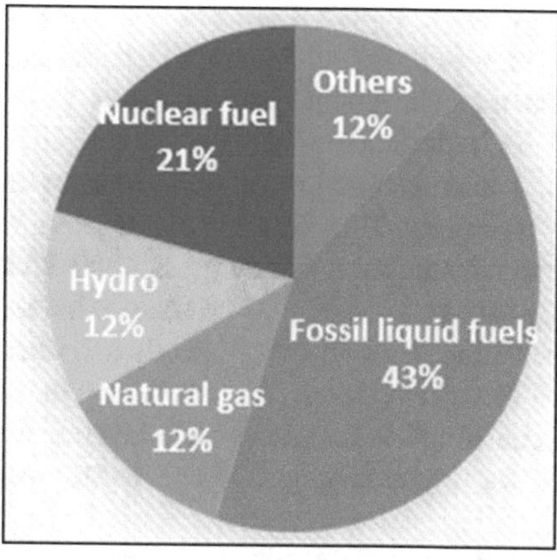

Total 1058PJ/yr or 4090W/capita

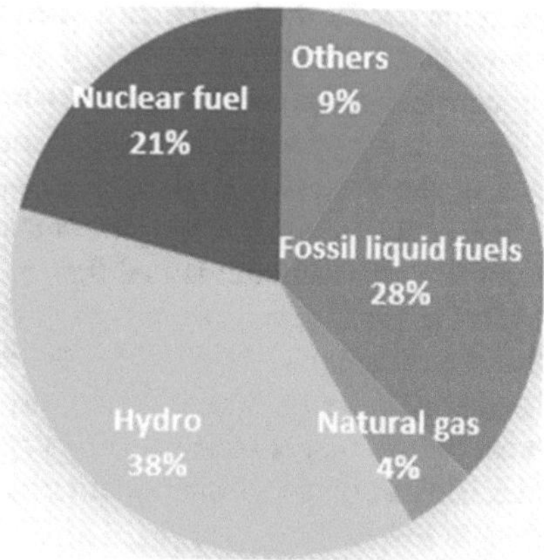

Total 348 PJ/yr or 1346W/capita

Fig. 9 Comparison of energy (left) and Exergy (right) based statistics of Switzerland in 2016. PJ, PetaJoule = 10^{15} Joule; W, watt

Energy politics is strongly supported by energy statistics, which is used for indicating both targets and achievements. In spite of the fact that only the Exergy share of energy is usable in the human society, the official statistics deals with energy only. Indeed political arguing with Exergy does not help a politician to attract more followers or voters as long as this notation is not established in the colloquial language. It is therefore a lengthy process to establish Exergy and Anergy within a general public space in spite of the trivial fact that nobody can survive without Exergy supply and nobody can consume energy in the true sense of the word. The outputs of the international energy statistic community reflect this situation. The author has described this in a 2018 paper (Wettstein, 2018b) from which some explaining texts are used here.

The fact that we need Exergy and not Anergy has a surprising consequence for the development of the energy statistics: In the ongoing energy transition the consumption of fossil and nuclear fuels is reduced in favor of renewable energy sources such as PV, Wind, and Hydro. The simple sense of the above-mentioned regulation of the statistics is that fossil and nuclear fuels are counted as heating value of the fuel and as primary nuclear heat while the mentioned renewables are counted as primary energy in the form of the produced electricity. The first are in all typical consumption cases converted into an Exergy form of energy while electricity is already Exergy. Indeed the society does not reduce its Exergy consumption but it increases the share of the renewable energies. This reduces the primary energy consumption in statistics drastically, even without any efficiency improvements, which reduce the Exergy losses.

The author has shown this paradoxical influence on the energy statistics of Switzerland in the above-mentioned paper (Wettstein, 2018b). This paper shows the consumed Exergy from the different primary sources based on plausible assumptions for the finally consumed Exergy share. Figure 9 is from this paper. It shows pie diagrams for the energy consumption in the official energy statistics and its conversion into an Exergy consumption pie. The comparison indicates in the energy pie a systematic overestimation of the importance of the fossil liquid fuels and of the natural gas and a corresponding underestimation of the hydropower, which supplied some 60% of the country's electricity. The other important learning from this consideration is that the per capita consumption of energy is 4090 Watt while the per Capita consumption of Exergy is only 1346 Watt. The issue is indeed that the decarbonization leads to an electrification of most energy needs including heating with heat pumps and mobility with electric vehicles. The electrification of most energy supply means that finally the consumption remains on the level of the needed Exergy plus a surplus of up to 20% for the remaining inherent resistor and friction losses.

This causes that in the view of the existing energy statistics the consumption drops from above 4000 Watt per capita to less than 2000 Watt per capita without any savings, reduction of comfort or efficiency improvements referring to this database of 2016 for Switzerland. In the total per capita consumption figures this statistics is typical for any industrialized country with a low share of coal-based energy consumption.

As a first step into the electrified future the reader may look into another current open source website. It indicates the current cross border electrical power flows as well as the country internal specific CO_2 emission per electrical kWh: https://app.electricitymap.org/zone/BR-N?solar=false&remote=true&wind=false. This website showed the European countries fairly completely at end of 2021. It is expanding worldwide based on volunteer contributions. In this case we remember again that the indicated electricity flows are Exergy flows.

Primary sources based on geothermal systems are shown as heat in the current statistics, while the usable Exergy is typically around 20% or less, due to the limited temperature. The significance of such sources is therefore drastically overestimated in the current energy-based statistics view.

The primary source "environmental heat" also shows up in some current energy statistics. But such sources have no Exergy at all. They can only be used by sacrificing Exergy, for example with heat pumps. It makes no sense to consider environmental heat as a primary energy source. It is indeed an Anergy source necessarily needed for heat pumps. However, there may be some exceptions in this respect for systems where solar heat is used directly for seasonal heat storage or for systems using natural thermal reservoirs like sea water or others. Such heat indeed saves Exergy during the discharge of the stored heat and therefore in an Exergy based statistics it could be considered with a Carnot factor related to the ambient conditions of the discharge period (in winter for heat discharge and in summer for cooling service). We may remember the historic time before the heat pump invention, in which ice from winter was stored for cooling services in summer.

7 Conclusions

The above Sects. 2–6 lead to the following summarized conclusions:

- The typical current low education level teaching about energy does not support a sustainable behavior. It must be completed with the adequate Exergy and Anergy notations and examples.
- In the average education level (College) the "must know knowledge about sustainability" is that electric and mechanical energies are the most valuable energy forms called "Exergy."
- In the Bachelor level of education the understanding of sustainability is fostered by the knowledge of typical energy conversion devices with its effect on the balance of Exergy losses.
- In the scientific and research level of education a deep understanding of exergy loss mechanisms in the technical and economic applications is needed for sustainable solutions.
- In the energy statistic field of education misinterpretations in the context of the current energy transition are nearly unavoidable if exergy considerations are ignored.

References

Carnot, N. L. S. (1824). *Réflexions sur la puissance motrice du feu et sur les machines propres à développer cette puissance*. 43 pages, 2nd edition in French: Annales scientifiques de l'École Normale Supérieure Sér. 2, 1, 1872, pp. 393–457. This booklet is subject to ASME Landmark #676 of Mechanical Engineering.

Grassmann, P. (1961). *Physikalische Grundlagen der Chemie-ingenieur-Technik*. Edition Sauerländer Aarau and Frankfurt am Main.

IAPWS. (2007). Revised Release on the IAPWS industrial formulation 1997 for the thermodynamic properties of water and steam (The revision only relates to the extension of region 5 to 50 MPa).

International Energy Agency (IEA). (2014). Energy efficiency indicators: fundamentals on statistics. 387 pages, online accessible.

International Energy Agency (IEA). (2016). Energy efficiency indicators, highlights. 154 pages, internet accessible.

International Energy Agency (IEA). (2021). Key World energy statistics. Annual publication summarizing the on line available "comprehensive view on energy production, transformation and final use". IEA consists in Oct 2021 of 30 member states, the European Commission and 8 associated states.

International Energy Agency (IEA). (n.d.). Web based statistics, http://www.iea.org/statistics/ ; direct link to Energy Sankey diagrams of world, regions and countries: https://www.iea.org/Sankey/

Rant, Z. (1956). Exergie, ein neues Wort für technische Arbeitsfähigkeit. *Forschung auf dem Gebiete des Ingenieurwesens, 22*, 36–37. ZDB-ID 212959-0.

Rant, Z. (1961). Zur Bestimmung der spezifischen Exergie von Brennstoffen. *Allg. Wärmetechnik, 10*(9), 172–176.

The Exergoecology Portal. (2011). http://www.exergoecology.com/

United Nations. (2016). Energy statistics compilers manual [Final draft subject to official editing], Series F No. 118, (186 pages)

United Nations, Department of Economic and Social Affairs, Statistics Division. (2018). International Recommendations for Energy Statistics (IRES), Series M No. 93; ISBN: 978-92-1-161584-5; 168 pages

Wettstein, H. E. (2014). Polytropic change of state calculations. In *Proceedings of the ASME international mechanical engineering congress & exposition IMECE2014-36202, Montreal.*

Wettstein, H. E. (2018a). Exergy loss considerations in education for a turbofan power cycle. In *Proceedings of ASME Turbo Expo 2018 turbomachinery technical conference and exposition GT2018-75048.*

Wettstein, H. E. (2018b). Exergy based energy statistics. GPPS-2018-23.

Wettstein, H. E. (2020a). Quality key numbers of gas turbine combined cycles. In *Proceedings of ASME Turbo Expo 2020 turbomachinery technical conference and exposition GT2020-14508.*

Wettstein, H. E. (2020b). How is a correct GT combustor heat balance established? In *Proceedings of ASME Turbo Expo 2020 turbomachinery technical conference and exposition, GT2020-14235*

Wettstein, H. E. (2021, June 7–11). Survey of calculation methods for polytropic efficiencies. In *Proceedings of ASME Turbo Expo 2021 turbomachinery technical conference and exposition GT2021, virtual, online, GT2021-59967.*

Simulating Complex Relationships Between Pollutants and the Environment Using Regression Splines: A Case Study for Landfill Leachate

Arpita H. Bhatt, Richa V. Karanjekar, Said Altouqi, Melanie L. Sattler, Victoria C. P. Chen, and M. D. Sahadat Hossain

Abstract Pollutant concentrations in the environment are functions of source variables, environmental variables, and time. Relationships among these variables are complex and unknown, therefore multiple linear regression may not be applicable. One solution is use of Multivariate Adaptive Regression Splines (MARS). MARS is a flexible statistical method, used for machine learning, but it is not typically appropriate for sustainability applications due to limited data. While most machine learning algorithms require a large amount of data and may not yield interpretable results, by contrast, MARS in conjunction with design of experiments has the ability to find important interpretable relationships in limited data. Laboratory data were collected under a controlled environment to discover how leachate composition changes as functions of ambient temperature, rainfall intensities, and waste composition. The MARS application to landfill leachate demonstrates a promise in capturing complex relationships, illustrated via MARS 3D interaction plots.

Supplementary Information The online version contains supplementary material available at https://doi.org/10.1007/978-3-031-16620-4_19.

A. H. Bhatt (✉) · M. L. Sattler · M. D. S. Hossain
Department of Civil Engineering, University of Texas at Arlington (UTA), Arlington, TX, USA
e-mail: arpita.bhatt@uta.edu; sattler@uta.edu; hossain@uta.edu

R. V. Karanjekar
Department of Civil Engineering, University of Texas at Arlington (UTA), Arlington, TX, USA

ERM, Houston, TX, USA

S. Altouqi
Department of Civil Engineering, University of Texas at Arlington (UTA), Arlington, TX, USA

Oman Environmental Services Holding Company, Al-Azaiba, Oman
e-mail: said.altouqi@beah.om

V. C. P. Chen
Department of Industrial, Manufacturing, and Systems Engineering, University of Texas at Arlington, Arlington, TX, USA
e-mail: chen@uta.edu

F. P. García Márquez, B. Lev (eds.), *Sustainability*, International Series in Operations Research & Management Science 333,
https://doi.org/10.1007/978-3-031-16620-4_19

Keywords Landfill · Solid waste · Leachate composition · Waste management · Multivariate adaptive regression splines (MARS) · Statistical models

1 Introduction

Modeling fate and transport of pollutants in the environment is critical to environmental management. Models or simulations can be used to predict human exposure to air and water pollutants and to design treatment systems to reduce exposure. Models are particularly important for sources of pollution which are not built yet and can also considerably reduce monitoring costs for existing sources. Concentrations of pollutants in the environment are functions of variables related to sources and the environment, as well as time. Relationships among these variables are often complicated and undetermined, which makes multiple linear regression difficult to apply. This is particularly true when concentrations are not steady state, meaning that time must be included as a predictor variable as well. One solution is use of Multivariate Adaptive Regression Splines (MARS, Friedman, 1991).

MARS is a flexible statistical modeling method that has been employed in a wide variety of applications from finance (Lee et al. 2006) to airlines (Pilla et al., 2012) to human physiology (Serrano et al., 2020). MARS builds models by fitting piecewise-linear regression, meaning, the nonlinearity of a model is approximated through the use of separate regression slopes that are connected by "knots." There are two stages of modeling. In the first stage, MARS constructs an overly large model by adding truncated linear basis functions for different knot locations. In the second stage, basis functions are selected for deletion based on improvement in model fit until the model fit cannot be improved. In sustainability applications, MARS has been used to model air pollution (García Nieto & Álvarez Antón, 2014; Gocheva-Ilieva et al., 2020; Kisi et al., 2017) and water pollution in lakes, streams, and rivers (García Nieto et al., 2011; 2016; Kisi & Parmar, 2016; Nacar et al., 2020; Zahiri & Nezaratian, 2020). It has been used to forecast medical and municipal solid waste generation (Fan et al., 2021; Karpušenkaitė et al., 2016; Szul & Knaga, 2019). However, it has not been applied to the problem of predicting water pollutant concentrations in landfill leachate.

Leachate management poses one of the most critical challenges for landfills (Johannessen & Boyer, 1999; UNEP (United Nations Environment Program) and CalRecovery, 2005). Landfill leachate contains a variety of pollutants, which if not monitored and controlled can leach into surrounding surface and ground waters (Johannessen & Boyer, 1999; Rapti-Caputo & Vaccaro, 2006). A variety of studies have measured landfill leachate quality (Gettinby et al.; 1996; Kouzeli-Katsiri et al., 1993; Kulikowska & Klimiuk, 2008; Mishra et al., 2017; Mohammad-pajooh et al., 2017), as well as variables that impact leachate quality, including temperature, moisture content, and waste composition (Al-Yaqout & Hamoda, 2003; Kulikowska & Klimiuk, 2008; Sourmunen et al., 2008; Tatsi & Zouboulis, 2002). Designing a leachate treatment system for a new landfill, however, requires predicting leachate quality for a system not yet built. A general model to forecast landfill leachate composition would be beneficial. Such a model could also reduce the requirements for monitoring leachate quality, both during the landfill's active life (50–100 years) and during post-closure (30 years); such monitoring can be very costly and time consuming.

Thus far, the few studies that have attempted to forecast leachate chemical composition using statistical techniques have concentrated on an individual landfill, or several local landfills (Eusuf et al., 2007; Gómez Martín et al., 1995a, 1995b; Kylefors, 2003). Zoughi & Ghavidel (2009) did develop an artificial neural network model from lab-scale landfill reactors to predict leachate concentrations, but the study was limited to ammonia-nitrogen and chemical oxygen demand (COD) and did not predict these as functions of ambient temperature, rainfall, or waste components, which have been documented in a variety of studies to impact leachate composition (El-Fadel et al., 2002; Kjeldsen et al., 2002; Kulikowska & Klimiuk, 2008). It would be beneficial to develop a global model rather than a region-specific model to estimate leachate composition from any landfill. Ideally, the global model would require only basic input information, such as ambient temperature, rainfall rate, and waste components.

Thus, the goal of this study was to investigate MARS as an approach to developing a global model to predict landfill leachate quality. Specific study objectives were:

1. To monitor leachate parameters as functions of time from simulated landfill reactors with various waste compositions maintained at differing rainfall intensities and temperatures.
2. Based on data from Obj. 1, to develop MARS statistical models for estimating leachate parameter concentrations as functions of temperature, rainfall intensities, time and waste components.

The novelty of our work is to apply MARS in conjunction with design of experiments to find important interpretable relationships in limited data and enable prediction of leachate composition over time, for any landfill in the world, based on the inputs for temperature, rainfall, and waste components. MARS models for predicting chemical and biochemical oxygen demand (COD and BOD) have been published (Bhatt et al., 2017). This chapter presents MARS models for forecasting pH, total alkalinity, TDS, TSS, VSS, Cl^-, NH_3-N as functions of rainfall intensities, time, temperature, and waste components.

2 Materials and Methods

2.1 Experimental Design

Waste components chosen were organic constituents of paper, food, yard & wood, and textile waste, along with inorganics. Experiments were conducted at 70, 85, and 100 °F, and average rainfall intensities of 2, 6, and 12 mm/day. Therefore, the two factors of main interest—temperature and rainfall—were each included at three levels and in all combinations; thus, $3^2 = 9$ combinations of rainfall and temperature were incorporated into the experimental design.

A cyclic incomplete block design was used (Dean & Voss, 1999). Due to the constrained area in constant temperature rooms, as well as the considerable time required to monitor reactors, it is prohibitive to conduct runs for every waste composition with each temperature-rainfall combination. The incomplete block design provided a minimal number of reactors such that all temperature-rainfall combinations were represented, but only a subset of combinations with waste compositions were included. The chosen combinations spanned the extents of rainfall intensities, temperatures, and waste compositions that were realistic and enabled quality MARS models. Since MARS is a regression-based method, replication of duplicate conditions in the reactors was not necessary.

The incomplete block design had the following settings listed below:

- No. of treatments (v) = temperature & rainfall combinations, $3^2 = 9$
- No. of blocks (b) = waste compositions = 9
- Block size (k) = # of times a block occurs in the experimental design = 3
- # of times a treatment occurs in the design (r) = 3

The total number of simulated landfill reactors can be calculated as $vr = bk = 27$. In the design, each degradable waste (food, paper, yard, and textile) was allowed to vary from 0 to 100%. Inorganic waste ranged only from 0 to 40% (Mason et al., 2003). The waste compositions are combinations of the waste components, and the nine selected waste compositions were selected to satisfy the following:

(a) Percentages for reactor components summed equal to one hundred.
(b) The correlation among waste component percents was minimalized to avoid inducing correlation between degradation rates.
(c) Euclidean distance in the five-dimensional space of waste combinations (yard, paper, textile, food, and inorganic waste percents) was maximized.

Various combinations were identified in the space representing the dimensions of the waste components. Each waste composition is a point in this space, and the Euclidean distances were calculated between the points. The set of nine waste compositions was chosen such that they satisfied the above three conditions and maximized the Euclidean distances separating the points (Chen et al., 2006). Table 1 shows the nine combinations of waste included in the experimental design.

Table 1 Waste component % by weight for nine combinations

Component	Component % by weight for each combination								
	a	b	c	d	e	f	g	h	i
Food	100	0	0	0	0	60	30	10	20
Paper	0	100	0	0	60	0	10	30	20
Textile	0	0	100	0	0	30	0	60	20
Yard	0	0	0	100	0	10	60	0	20
Inorganic	0	0	0	0	40	0	0	0	20

Table 2 Temperature, rainfall, and waste composition grouping for landfill reactors

Rainfall Intensity (mm/day)	Temperature (°F)	Waste Component Grouping								
		a	b	c	d	e	f	g	h	i
2	70		1					2		3
2	85	4		5					6	
2	100		7		8					9
6	70	10*		11*		12				
6	85		13		14		15			
6	100			16		17		18		
12	70				19		20		21	
12	85					22		23		24
12	100	25					26		27	

Note: Blue indicates a lab-scale reactor number
[a]R10 and R11 failed due to extreme acid accumulation and washout, respectively

Table 2 summarizes the treatments and block combinations used to assemble and operate the lab-scale landfill reactors according to the experimental design. For instance, Reactor 16 received 6 mm/day rainfall intensity at 100 °F and has c waste component combination, which is 100% textile waste (Table 1).

2.2 Waste Collection and Preparation

Rather than using landfill waste, fresh waste components were collected separately, so that they would not have degraded or been contaminated by unknown materials. Food waste was acquired from UTA's dining hall, and from local Thai and Indian restaurants. Food waste from these developing countries was used because developing nations usually have greater percentages of food waste in their landfills. Mixed yard waste (50% branches/leaves and 50% grass) was collected from the university grounds. To reduce decomposition, yard and food wastes were stored at 4 °C until utilized. Paper wastes (office paper, newspapers, mail, magazines, tissues, towels, diapers, corrugated boxes, and cartons) were gathered from recycling bins, restaurants, and grocery stores. Large paper pieces were cut to 10 cm × 15 cm size to fit in the reactors, but not shredded to better resemble real landfilled waste. Paper wastes were then blended in proportions equal to the national average for paper waste percentages in the USA (EPA, 1996). Cotton and synthetic textile scraps were acquired from alteration shops. Larger scraps were cut into 10 cm × 15 cm pieces to fit into the reactors. Inorganics (aluminum cans, plastic bottles, and concrete chunks) were gathered from recycling bins at UTA and the structural testing laboratory. Finally, waste components were mixed by their weight percent according to the experimental design (Table 2).

2.3 Reactor Setup and Filling

The set of 27 simulated landfills were constructed using 6 gallon (16 l) HDPE reactors. The choice of 6 gallons was used to avoid finely shredding waste, so that the waste size would more closely represent that of an actual landfill. Waste components were placed in the HDPE reactors. A U-tube manometer was used to leak-check the reactors prior to filling.

Since the waste used was collected prior to landfilling and had not yet degraded, anaerobic microorganisms were not initially present; hence, sludge from an anaerobic digester was used to seed each reactor, as has been done in previous studies (Di Addario et al., 2017; He et al., 2007; Ogata et al., 2016; Sang et al., 2008), to achieve 10–12% of the weight of the waste. Anaerobic digester microorganisms are often used to seed lab-scale landfill reactors, because obtaining a representative sample of microorganisms from the waste itself is difficult due to the heterogeneous composition of landfilled waste.

After filling and seeding, reactors were weighed, sealed, and placed in constant temperature rooms, as shown in Fig. 1. Reactors were maintained at temperatures and rainfall rates, as described in the experimental design, from 139 to 370 day. Leachate was not recirculated, as in typical landfill operations.

2.4 Leachate Parameter Measurements

Leachate samples from the 27 lab-scale landfill reactors were collected regularly, and parameters were measured using methods summarized in Table 3. Leachate volume and water added were measured daily.

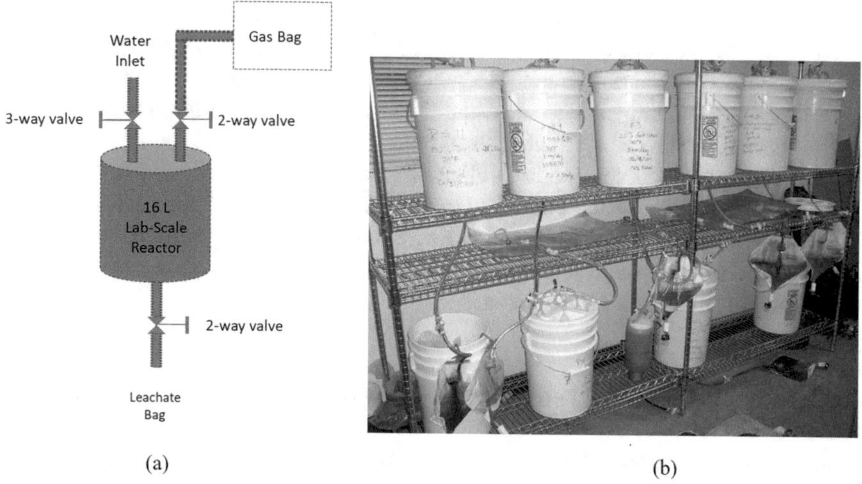

Fig. 1 (**a**) Schematic of reactor setup and (**b**) reactors in constant temperature room

Table 3 Methods for leachate parameter measurements

Leachate parameters	Methods
Total Alkalinity as $CaCO_3$	Standard Method 2320 B
Total Suspended Solids (TSS)	Standard Method 2540 D
Volatile Suspended Solids (VSS)	Standard Method 2540 E
pH, Conductivity, and Total Dissolved Solids (TDS), Ammonia-Nitrogen (NH3-N)	IntelliCAL probes from Hach
Chloride Ions (Cl^-)	SensION+ ISE probe from Hach

2.4.1 pH

The pH for waste degradation is one of the most vital environmental factors. It expresses the hydrogen-ion concentration. Sample pH was measured daily using a pH probe. In the anaerobic acid phase of waste decomposition, carboxylic acids accumulate, decreasing the pH. However, after methanogenesis starts, carboxylic acids are expended, and pH rises. In the last phase, pH keeps rising and then stabilizes.

2.4.2 Total Alkalinity as Calcium Carbonate ($CaCO_3$)

Alkalinity measures the buffer capacity for resisting a drop in pH resulting from acid addition. In order to obtain the acid-neutralizing capacity of leachate samples, methyl orange end points could not be used for titration due to the dark leachate color. Since the existing buffer system was mainly due to weak acids with ionization constants around 1×10^{-5} and their concentrations varied in each leachate sample, as well as due to the dark color of leachate samples, it was necessary to prepare acid titration curves to determine the inflection points for each sample being analyzed.

Alkalinity of samples was measured volumetrically by titration with 0.2 N sulfuric acid (H_2SO_4) as a titrant. The amount of leachate sample used for a titration varied, as certain samples had high alkalinity values, which was used in the beginning as an initial estimate. The titration curves were prepared for each sample to determine the inflection points for the selection of stoichiometric end points. Differences in the stoichiometric end points occurred in all samples. This may be due to different concentrations of fatty acids.

2.4.3 Conductivity and Total Dissolved Solids (TDS)

Conductivity measures a liquid's ability to transmit an electric current and relies upon the total concentration of ions. Conductivity is directly proportional to total dissolved solids (TDS). Conductivity (micromhos per centimeter) can be multiplied by an empirical factor (ranging from 0.55 to 0.9, depending on soluble components and temperature) to obtain TDS (in mg/l) (Eaton & Franson, 2005). The conductivity and TDS were measured using a Hach IntelliCAL probe.

2.4.4 Total Suspended Solids (TSS) and Volatile Suspended Solids (VSS)

For TSS analysis, filters were prepared following the standard method (2540 D) and weighed before analysis. Well-mixed, measured leachate was filtered on a filter disk by applying suction and then washing with distilled water. Filters were dried in an oven at 103–105 °C for one hour and burned in a furnace at 550 °C. TSS measures the portion of solids retained by a filter. The increase in weight of the filter after drying at 103–105 °C for one hour represents TSS, while the weight lost on ignition at 550 °C is VSS.

2.4.5 Chloride and Ammonia-Nitrogen

Chloride is a non-degradable, inorganic macro component. Change in chloride concentration is often used to evaluate leachate dilution. Chloride ions in leachate samples were measured using a chloride ion-specific electrode (ISE, Hach). Most nitrogen in municipal solid waste (MSW) is in the form of ammonia, created when proteins and amino acids degrade. Ammonia-nitrogen in leachate samples was measured using an ammonia-nitrogen ISE (Hach).

2.5 Data Evaluation and Model Development

This study, along with our companion work on BOD/COD (Bhatt et al., 2017), utilized a MARS statistical modeling approach in forecasting leachate parameters. Initially, a standard multiple linear regression was attempted, but inadequate. MARS is a flexible statistical modeling method with assumptions that are less rigid than for linear regression. MARS is commonly used for machine learning, but machine learning is not typically appropriate for sustainability applications due to the limited availability of data. While most machine learning algorithms require a large amount of data and may not yield interpretable results, by contrast, MARS in conjunction with design of experiments has the ability to find important interpretable relationships in limited data. MARS models provide implementable prediction models that can be used to simulate relationships in practice and can guide directions for future research. MARS runs were performed via Salford Predictive Modeler (SPM) Builder, Version 6.6 in our study.

2.6 Model Fitting and Selection

To form basic functions, truncated linear or "hinge" functions (linear functions that apply over part of the range of an independent variable) bend at knots to allow a piecewise-linear structure that can approximate a curved relationship. In MARS, the piecewise-linear structure is eventually "smoothed" by replacing the hinge function with cubic or quintic spline forms that satisy a continuous derivative. In our study, hinge functions were used to approximate the relationship between the response variables (leachate parameters: pH, total alkalinity, conductivity, TDS, TSS, VSS, Cl^{-}, and NH_3-N) and predictor factors (time, temperature, rainfall, waste composition). For instance, a univariate basis function (BF_1) on the variable rainfall intensity might be defined as:

$$b_m(\text{temperature}) = \max(0, \text{rainfall}-10) \tag{1}$$

where the hinge function bends at a rainfall rate of 10, which is the knot value. The model fitting initializes with only a constant term, and then basis functions are added (first univariate terms and then interaction terms). For the first stage, the user specifies a maximum number of basis functions to build the "overly large" MARS model. The final MARS function has the form shown in Eq. (2) (for observations $i = 1, \ldots n$):

$$Y_i = \beta_0 + \sum_{m=1}^{M} \beta_m b_m(\boldsymbol{x}_i) + \varepsilon_i \qquad (2)$$

where:

Y_i = response variable
$\boldsymbol{x}_i$ = vector of predictor factors
β_0 = constant
β_m = coefficient for the m-th basis function, denoted as $b_m(\boldsymbol{x}_i)$ and
ε_i = random error term

The maximum number of basis functions was varied from 15 (default) up to 84, in increments of 1 or 2. Other input parameters were also varied until the predictive squared error (PSE) and adjusted-R^2 values leveled off at low and high values, respectively. The model at this point was designated as the best model.

3 Results and Discussion

3.1 Reactor Data

This section discusses the variation in leachate parameters (pH, total alkalinity, conductivity, TDS, TSS, VSS, Cl^-, and NH_3-N) for the pure cases with 100% of one waste component (food, paper, yard, or textile). While these pure cases are not realistic in practice, they provide comparison among these different types of waste. Moreover, the results for the mixtures of waste yielded results that were a blending of the pure cases. The leachate BOD/COD trends and MARS models results are provided elsewhere (Bhatt et al., 2017).

3.1.1 Leachate Parameter Trends for 100% Waste Reactors

Reactor #s 8, 14, and 19 contained 100% yard and were operated at temperatures of 100 °F, 85 °F, and 70 °F; and rainfall intensities of 2, 6, and 12 mm/day, respectively. Leachate parameter trends for 100% yard waste reactors are provided as an example in Fig. 2. Graphs of trends of leachate parameters with time for the remaining 100% waste reactors (food, paper, textile) are provided in supplemental material. R10 (100% food) and R11 (100% textile) failed because of acid accumulation and microorganism washout, respectively. The 100% food waste reactors (R4, R25) had the greatest concentrations for all leachate parameters due to most organic matter present, as food is the most easily degraded waste component.

Figure 2a shows an example pH pattern, where pH acts as a good indicator of microbial activity in the reactors. For most reactors, pH trends were neutral to alkaline after 20–30 days of operation, except for reactors carrying 100% food

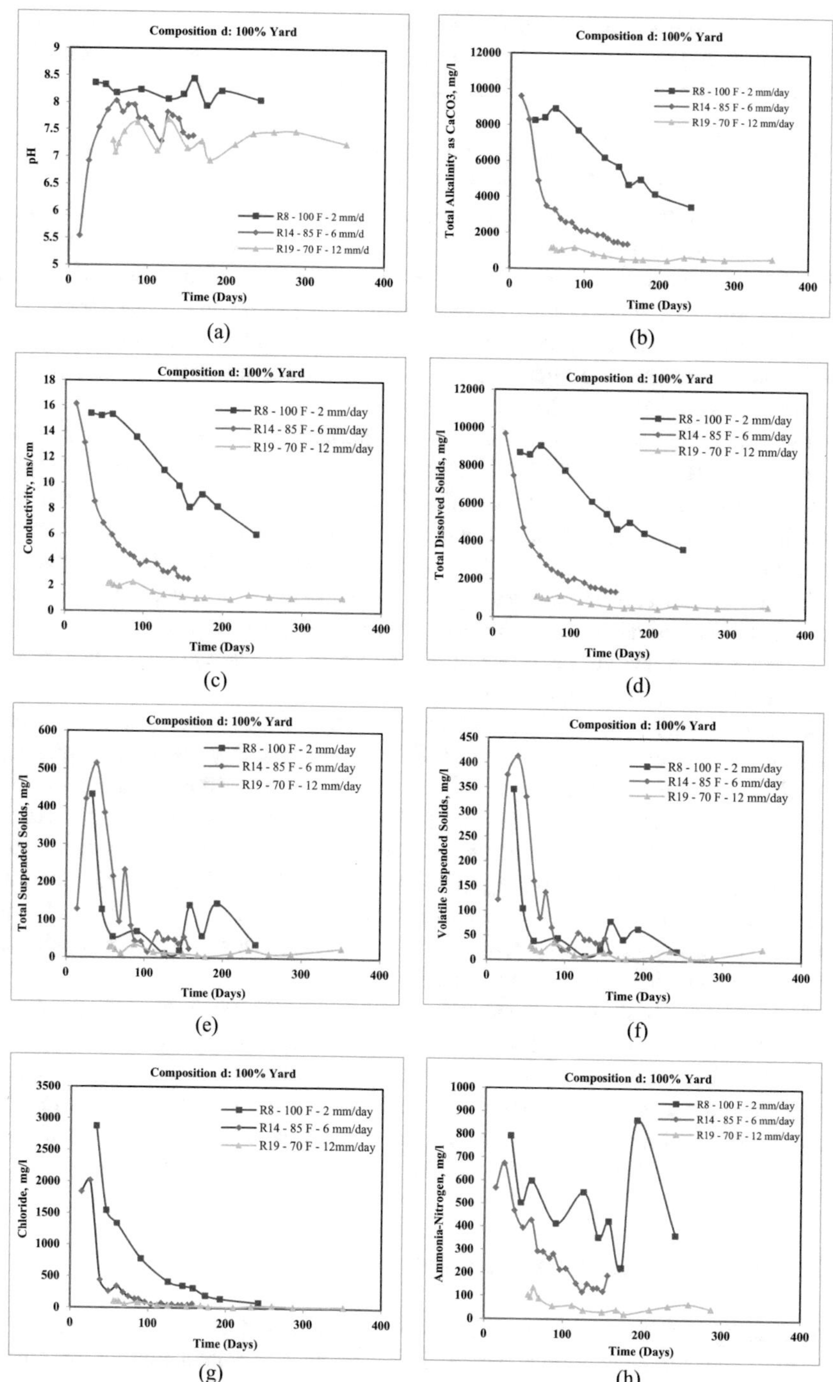

Fig. 2 Leachate parameters for 100% yard reactors: (**a**) pH; (**b**) total alkalinity; (**c**) conductivity; (**d**) TDS; (**e**) TSS; (**f**) VSS; (**g**) Cl^-; (**h**) NH_3-N

waste. The food waste reactors had a lag time of above 100 days for pH values to be stabilized. In the initial phase of reactor life, the pH is acidic due to carboxylic acid accretion. In the later stage, methanogens transform acids to methane and carbon dioxide and pH stabilizes.

Figure 2b illustrates alkalinity decreasing with time, with the exception of some small peaks. The alkalinity values for food waste reactors were higher than other reactors. This is due to organic waste decomposition generating bicarbonate, which is dissolved carbon dioxide and constitutes one of the main components of alkalinity. The trends for conductivity and TDS for the reactors are similar, as seen in Fig. 2c, d, demonstrating the relationship between these two parameters. Pure 100% food reactors (R# 4, 25) had higher values for conductivity and TDS compared to other 100% waste reactors. The higher values of TDS indicate the presence of inorganic materials (Nagarajan et al., 2012).

TSS and VSS, seen in Fig. 2e, f, are important parameters to determine the strength of a liquid. TSS and VSS values for 100% food waste reactors are significantly greater than other reactors in first 50 days of operation. Later, there was a significant drop in both parameters' values. This might be due to a faster decomposition rate of organic matter in early stages. The VSS/TSS for 100% food reactors were between 0.5 and 0.9, suggesting high biomass content in leachate from food waste reactors (Oliveira, 2012). There were multiple peaks observed for all 100% waste reactors, indicating the effect of combinations of different waste decomposition rates, rainfall, and temperature.

Chloride, seen in Fig. 2g, was present in high concentrations in leachate samples and decreased over time. Chlorine in MSW has two primary sources: plastics and household kitchen and garden waste (Zhou et al. 2019). The 100% food reactors had the highest chloride concentrations, while 100% yard reactors were second. Sizirici and Tansel (2010) and Reinhart and Grosh (1998) reported decreasing chloride trends over time as well. According to El-Fadel et al. (2002), chloride concentrations rise or fall incrementally or remain stable during monitoring. Chloride is minimally impacted by biological, physical, or chemical removal processes, but can be released into leachate under high moisture conditions.

Finally, Fig. 2h shows an example of decreasing ammonia trends with a few peaks, which was observed for all 100% waste reactors. Kjeldsen et al. (2002) reported that protein decomposition did not change ammonia concentrations over time. Since there is no mechanism for break down ammonia under methanogenic conditions, leaching is the only way that can lower ammonia concentration in leachate over time (Burton & Watson-Craik, 1998; Robinson, 1995).

3.1.2 Summary Statistics for Leachate Parameters

The decomposition of solid waste in landfills is largely due to microbial activities and thus, leachate generation is directly impacted by microbial action. In this study, wide variations in the values of leachate quality parameters illustrate the importance of waste component composition, temperature, and rainfall intensities in determining

Table 4 Summary of leachate parameters statistics for lab-scale reactors

Leachate parameters	Range	Mean	Median	Standard deviation
	(values in mg/l unless otherwise mentioned)			
pH (no unit)	5.42–8.45	7.27	7.35	0.62
Total Alkalinity	150–18,500	2053.66	850.00	3014.39
Conductivity, ms/cm	0.365–28.3	3.44	1.42	4.78
Total Dissolved Solids (TDS)	104–17,360	1924.70	746.50	2841.05
Total Suspended Solids (TSS)	3–1542.5	96.45	48.00	150.98
Volatile Suspended Solids (VSS)	2–1345	73.13	37.50	111.31
Biochemical Oxygen Demand (BOD)	6.5–46,134.17	1885.66	83.76	5274.53
Chemical Oxygen Demand (COD)	61.75–64,032.16	3787.46	694.04	9033.34
Ammonia-Nitrogen	0.01–3163.33	112.86	20.45	242.42
Chloride	1.04–4266.4	188.61	40.31	441.02

leachate composition. Table 4 summarizes the statistics of leachate quality parameters for the simulated reactors.

3.1.3 Comparison of Leachate Parameters with Previous Studies

During the acid formation phase of a landfill, pH values are low due to the presence of volatile organic acids (VOA). Kjeldsen et al. (2002) observed a pH range of 4.5–9, which is comparable with the range observed in this study (5.4–8.5). El Khatib (2010) stated that temperature has no statistical significance on pH.

In the previous studies, the alkalinity range for actual landfills is 240–8965 mg/l and 4250–8250 mg/l for USA and Italy, respectively (Al-Yaqout & Hamoda, 2003). The observed upper value in this study (18,500 mg/l) is higher, probably because of concentrated leachate from various reactors due to low rainfall intensities.

Conductivity is expected to increase during the acidic phase, due to mobilization of metals and decreases in the methanogenic phase due to the complexation of metals with sulfides (Pohland et al., 1992). However, conductivity decreased over time for almost all reactors. In a previous study, conductivity values ranged 1–26 ms/cm (El Khatib, 2010). These values are comparable with this study (0.4–28.3 ms/cm).

Kylefors and Lagerkvist (1997) found total solids concentration to decline as the leachate transitioned from the acidogenic to methanogenic stage. In this study, TDS decreased with time for all reactors except few peaks. El Khatib (2010) concluded based on a *t*-test that temperature significantly impacts TDS.

The observed trend of TSS and VSS with time in this study is similar to each other. In a previous study, the TSS range was 191–740 mg/l and VSS range was 72–329 mg/l (Kulikowska & Klimiuk, 2008). The ranges observed here are wider than previously reported. The reason might be the faster and better degradation of wastes in a shortened time in the reactors compared to actual landfills.

NH_3-N concentrations are higher in 100% food, 100% yard, 60% food, and 60% yard reactors. El Khatib (2010) found via a *t*-test that temperature had a significant impact on higher ammonium concentration. This was also confirmed in this study. Many researchers (Barlaz et al., 2002; Burton & Watson-Craik, 1998; Robinson, 1995) stated that NH_3-N builds up in landfills because it does not degrade under anaerobic conditions. However, in this study ammonia concentrations tend to decrease with time for all reactors except for few fluctuations, likely due to washout.

Chloride concentrations in this study decrease over time due to the washout effect, although Tesseme and Chakma (2019) found that chloride did not decrease even after many years of leaching. The range observed in this study is comparable to that found by Kjeldsen et al. (2002).

3.1.4 Effect of Temperature

Reactor #s 20 and 26 had the same waste composition of 60% food, 30% textile, 10% yard, and same rainfall of 12 mm/day and were chosen to compare the impact of temperature. Figure 3 displays the leachate parameters trend for these two reactors. Reactor # 26 (100 °F) had higher concentrations of all leachate parameters than R# 20 (70 °F). This indicated that higher temperature increases the waste decomposition initially and in turn generates higher concentrations of contaminants. Approximately after 100 days, the concentration of almost all leachate parameters decreased faster for Reactor # 26 and was more comparable with R# 20.

3.2 *MARS Modeling for Leachate Parameters*

Table 5 displays PSE and adjusted R^2 value for each leachate parameter. Lower PSE and higher adjusted R^2 indicate better fitting models, where the adjusted R^2 is a penalized version of the coefficient of determination measuring the proportion of variability in the response variable that is explained by the MARS model. The MARS output for all leachate parameters (Bhatt, 2013) incorporates the selected model, regression information, ANOVA decomposition table, and relative variable importance table. The MARS tool is useful in capturing and uncovering some complex relationships between factors and leachate parameters as discussed below with few example plots. The MARS tool also has been applied to leachate Biochemical Oxygen Demand (BOD) and Chemical Oxygen Demand (COD) to illustrate the complex relationships uncovered by MARS (Bhatt et al., 2017).

3.2.1 Comparison of Importance of Factors in the MARS Models

The MARS relative variable importance scores are summarized in Table 6 for all eight leachate parameters. The variable importance scores are calculated within the

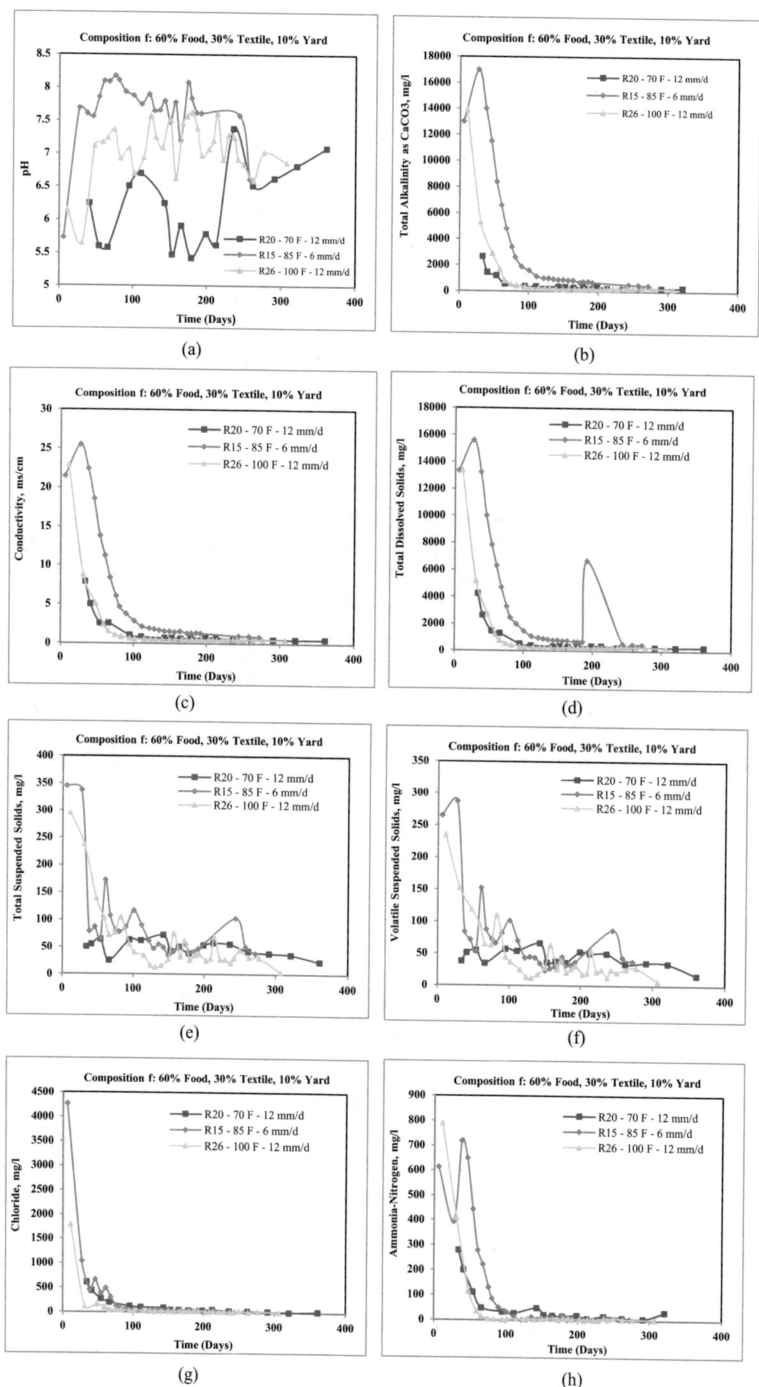

Fig. 3 Leachate parameters for 60% food, 30% textile, 10% yard reactors: (**a**) pH; (**b**) total alkalinity; (**c**) conductivity; (**d**) TDS; (**e**) TSS; (**f**) VSS; (**g**) Cl^-; (**h**) NH_3-N

Table 5 Leachate parameters predictive squared errors and adjusted-R^2 values for MARS final models

Leachate parameters	Max basis functions	Optimal model nr (# of terms in the final model)	Predictive squared error (PSE) for best model option	Adj. R^2
Total alkalinity	50	37	0.020	0.96
pH	34	26	0.159	0.76
Conductivity	50	31	0.015	0.96
TDS	50	33	0.022	0.94
TSS	28	20	0.069	0.72
VSS	28	19	0.058	0.78
Cl^-	34	18	0.065	0.86
NH_3-N	26	26	0.199	0.75

Table 6 MARS relative variable importance scores

Factors	Leachate Parameters							
	Alkalinity	pH	Conductivity	TDS	TSS	VSS	Chloride	NH_3-N
Time	100	100	100	100	91.16	93.35	100	87.32
Rain	76.33	62.27	89.75	94.23	100	100	65.14	44.7
Food	22.16	71.2	41.87	29.72	60.93	71.07	21.93	100
Yard	40.12	36.67	58.12	42.56	26.09	22.42	24.4	98.34
Paper	42.35		44.66	43.35	36.53	31.69	18.5	
Temp	24.53	57.95	18.19	20.97		41.14	16.55	25.6
Textile	31.74	29.54	24.86	29.6				
Leachate Volume	6.69	21.39	8.55	6.13	24.03		6.75	

	Most Important		>50		10-50		<10

MARS algorithm and are primarily based on the contribution of a factor variable in improving the fit of the MARS model, but also consider the number of model terms that include each factor. The relative importance value scores are scaled from 0 to 100 with the most important factor set to 100. Each column of Table 6 consists of the relative variable importance information for each MARS model, and the factors in the rows have been sorted to represent the overall relative importance across the eight MARS models. The table is color-coded to indicate the importance of various factors: dark blue for high importance (factors with top two values), light blue for medium importance (>50), lightest blue with low importance, and the lighter blue for the remaining numbers (10–50). It is seen that time and rain are the most important factors among all leachate parameters except pH and ammonia-nitrogen (NH_3-N). Higher NH_3-N levels occur with food and yard waste because of the presence of a high amount of nitrogen, and MARS has captured this with the high relative importance for the food and yard factors for NH_3-N. The finding that leachate parameter concentrations are most influenced by rain and time is sensible since rain dilutes the leachate concentrations, and the waste decomposition process continues over the course of landfill life (50–100 years). Overall, food, yard, paper, and temperature follow time and rain in the order of importance. Further, it is observed that time, rain, food, and yard factors appear in all eight leachate parameters. Finally, the least important is leachate volume.

Table 7 summarizes the breakdown of MARS model basis functions (bfs) from the analysis of variance across the eight leachate parameters. The table is divided into high (dark green), medium (light green), and low (lighter green) importance categories. Two types of basis functions occur in the MARS model: main effects that involve only one factor and two-factor interactions that model the nonlinear combined effect of two factors. Within each importance category of the table, the main effect model terms are listed first and sorted based on the frequency of appearance across all the leachate parameter MARS models, followed by the two-factor interactions, also sorted based on the frequency of appearance. Model terms, time and rain appear in the MARS models for all leachate parameters, while time-rain, time-food, time-leachvol appear in most. Model terms listed under the medium (light green) importance category appear in the MARS models for 4–6 leachate parameters. Factors listed under low importance category appear in the MARS models for three or fewer leachate parameters. The inclusion of many two-factor interaction model terms demonstrates the complexity of the relationships in leachate data. There is overall agreement between Tables 6 and 7, where Table 7 shows Time and Rain factors appearing in many model terms, followed by Food and Yard, and then the remaining factors.

3.2.2 MARS 3D Interaction Plots

Further understanding of the two-factor interaction effects can be examined via 3D interaction plots provided in the MARS output, which show the combined effect of two factors on a leachate parameter. Example graphs are illustrated below; detailed results are presented in Bhatt (2013). The effect on the specified leachate parameter is plotted in the z-axis, with the two factors varying in the plane of the x-axis and y-axis. Although MARS does not provide axis units, the units are °F for temperature, mm/day for rainfall, and mg/l for leachate parameters except pH.

Figure 4 provides Food-Time MARS 3D interaction plots for Conductivity and TDS, showing some interesting trends. In theory, there is a strong correlation between these two variables, which is influenced by the total amount of dissolved organic and inorganic matters in leachate. Conductivity is an indicator of dissolved inorganic ions. The time graphs (Figs. 2 and 3) for these variables show a similar downward trend with time. However, the Food-Time interaction 3D plots uncover a sharp dive in TDS concentration in the beginning, regardless of the food%. The Food-Time interaction 3D plots for TSS and VSS display a similar downward trend with time with food, which is expected as VSS is measured from the loss on ignition of the mass of measured TSS.

Table 7 MARS basis functions for leachate parameter

Factors	Leachate Parameters							
	Alkalinity #bfs	pH #bfs	Conductivity #bfs	TDS #bfs	TSS #bfs	VSS #bfs	Chloride #bfs	NH_3-N #bfs
Time	3	2	2	4	2	2	2	2
Rain	1	2	1	1	2	2	2	1
Time-Rain	6	1	5	5	1	2	5	
Time-Food	2	3	4	5	1	4		2
Time-LeachVol	1	2	1	1	3		1	2
Food		2	2			1		2
Yard	1		2	1				2
Time-Yard	1	5			1	1	2	2
Time-Temp	2	3	2	4				2
Paper-Rain	2		4	1	2	1		
Food-Rain		2			3	1	2	
Food-Temp		1			1	1	1	1
Textile	2							2
Temp		1						1
Paper				1				
Time-Textile	2		1	1				2
Time-Paper	2			3			1	
Yard-Rain	4					1	1	
Yard-Temp	2		2					1
Textile-Rain			2	3				
Rain-Temp					1	2	1	
LeachVol		1			2			
Textile-Temp	2	1						
Paper-Yard	2		1					
LeachVol-Textile	2							
Textile-Yard		1		1				
Paper-Temp				2				
LeachVol-Food					1			1
LeachVol-Yard								2
Food-Paper						1		
LeachVol-Rain								1

	High importance		Medium importance		Low Importance

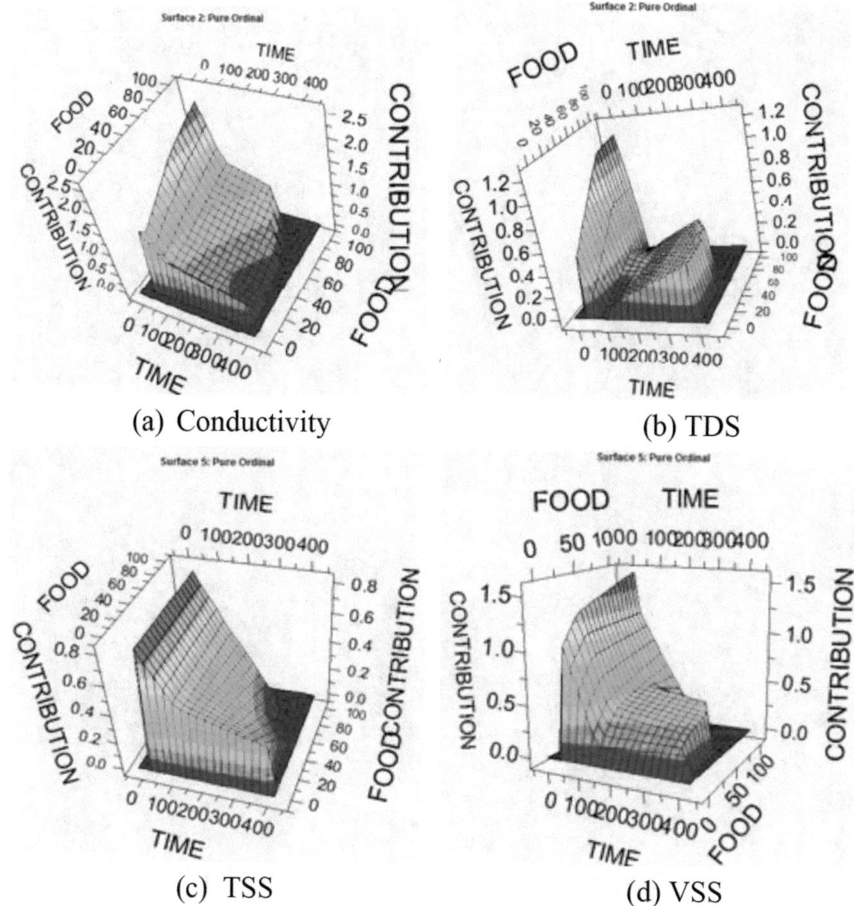

(a) Conductivity

(b) TDS

(c) TSS

(d) VSS

Fig. 4 MARS 3D interaction plots for food-time. (**a**) Conductivity, (**b**) TDS, (**c**) TSS, (**d**) VSS

Figure 5 displays Rain-Time MARS 3D interaction plots for alkalinity, pH, conductivity, and TDS. In general, with time and as rain increases, it is expected to have a downward trend of leachate quality parameters due to waste decomposition and dilution due to rainwater. This trend is observed in the Rain-Time 3D interaction plots for alkalinity. There is a strong chemical tie between alkalinity and pH. Alkalinity is a measure of acid-neutralizing capacity, also called the buffering capacity, while pH (acid ions) is a measure of how acidic or basic the sample is. The alkalinity resists the drop in pH, if acids are added. The Rain-Time 3D interaction plots for alkalinity and pH show similar trends, where in the beginning alkalinity concentration increases and reaches a peak when rain increases to 6 mm/day, and then it decreases with time. Whereas pH is low in the beginning then reaches a peak around 6 mm/day of rain and stabilizes with time. The Rain-Time 3D interaction plots for conductivity and TDS are also showing interesting trends. As discussed

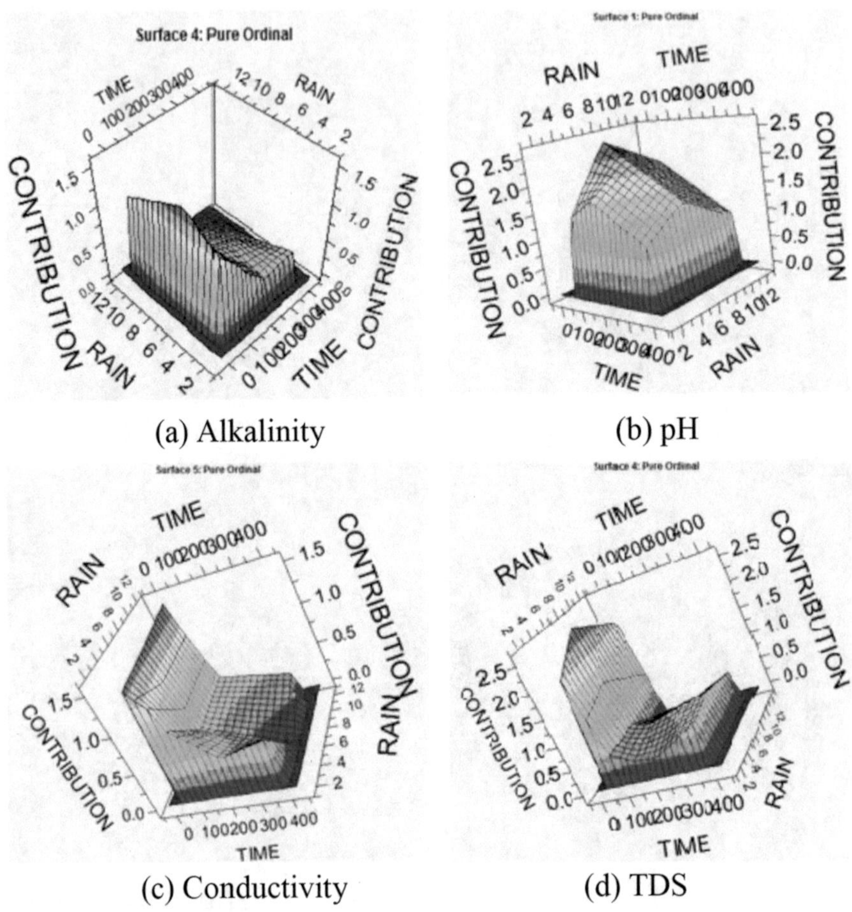

Fig. 5 MARS 3D interaction plots for rain-time. (**a**) Alkalinity, (**b**) pH, (**c**) conductivity, (**d**) TDS

earlier, there is a strong correlation. Conductivity is a measure of the ability of the liquid to carry an electric current, and this ability depends on the total concentration of ions. This is an indirect measurement of TDS. The MARS 3D interaction plots reveal that TDS is at a peak when rain is 6 mm/day. However, such a peak is not observed in conductivity, instead there is a gradual increase.

Figure 6 shows Time-Leachate Volume MARS 3D interaction plots for TSS and NH_3-N. It uncovers an unexplained trend in TSS concentration. However, the time plots of TSS for various reactors (Bhatt, 2013) indicate that there are a lot of fluctuations and that is what MARS appears to capture for the Time-Leachate Volume interaction. The NH_3-N 3D plot displays a high concentration of NH3-N initially with time, and when leachate volume is less, it then decreases over time with an increase in leachate volume. This might be because most of the waste would be degraded at later times, and high leachate volume creates a dilution effect.

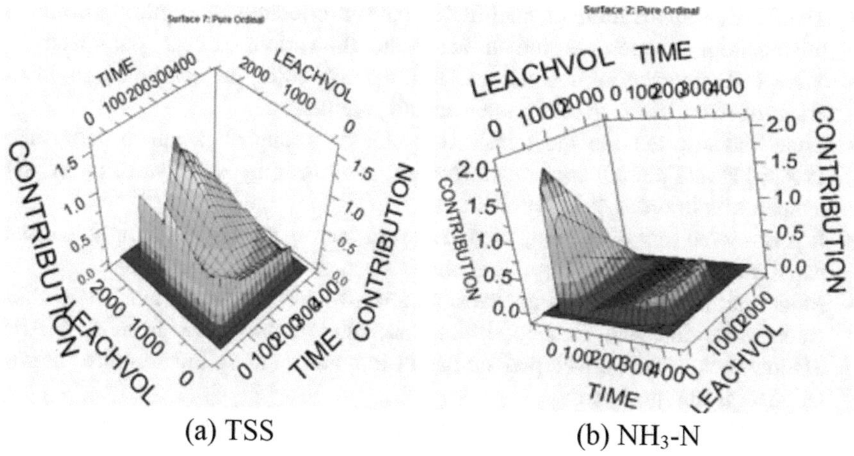

Fig. 6 MARS 3D interaction plots for time-leachate volume. (**a**) TSS, (**b**) NH_3-N

4 Future Work

Because simulated conditions are more idealized (e.g., uniform moisture, small waste pieces), waste decomposes more rapidly in the lab than in the field (Cruz & Barlaz, 2010). Hence, a "scale-up factor" needs to be developed from field leachate data to adjust the leachate models developed from laboratory data to field conditions. The MARS models were not originally developed using field data because it would have been difficult to obtain field data spanning the extensive range of rainfall intensities, temperatures, and waste component percentages studied in the lab. The controlled conditions of lab experiments were more suited to studying relationships among variables.

Once the scale-up factor is developed, more field data need to be obtained in order to validate the model. The effect of factors, such as age of waste and landfill operational parameters, should also be evaluated. A MARS approach can also be applied to predict heavy metals, natural organic matters (NOM), and per- and polyfluoroalkyl (PFAS) in landfill leachate.

5 Summary and Conclusions

A systematic, flexible, and novel approach using MARS to estimate leachate parameters from a landfill, given basic information—landfill's ambient temperature, annual rainfall intensities, and waste composition—is presented in this study. Laboratory data were collected using a cyclic incomplete block design to develop MARS models without any prior assumptions on the model equation form, and best fit models were selected.

- The MARS application in landfill leachate is effective in capturing complex relationships between the studied factors and the various leachate parameters
- MARS 3D interaction plots are useful in understanding the combined nonlinear effect of two factors affecting the leachate parameters.
- Time and rain are the most important factors among all leachate parameters except pH and ammonia-nitrogen (NH_3-N), followed by food, yard, paper, and temperature in order of importance.
- It is observed that time, rain, food, and yard factors appear in all eight leachate parameters and the least important one is leachate volume.
- Model terms time and rain appear in the MARS models for all leachate parameters; while time-rain, time-food, time-leachate volume appear in most. MARS 3D interaction plots developed are useful in understanding the complex interrelationships among factors.
- The MARS models for all leachate parameters (response variables) had the R^2 values of above 0.70, with three models over 0.90. The MARS provide implementable prediction models that can be used to simulate relationships in practice and can guide directions for future research.

References

Al-Yaqout, A., & Hamoda, M. (2003). Evaluation of landfill leachate in arid climate—A case study. *Environment International, 29*, 593–600.

Barlaz, M. A., Rooker, A. P., Kjeldsen, P., Gabr, M. A., & Borden, R. C. (2002). Critical evaluation of factors required to terminate the postclosure monitoring period at solid waste landfills. *Environmental Science & Technology, 36*(16), 3457–3464.

Bhatt, A. H. (2013). *Development of statistical models for predicting leachate parameters from simulated landfills*. Ph.D. Dissertation, University of Texas at Arlington.

Bhatt, A. H., Karanjekar, R. V., Altouqi, S., Sattler, M. L., Hossain, M. S., & Chen, V. P. (2017). Estimating landfill leachate BOD and COD based on rainfall, ambient temperature, and waste composition: Exploration of a MARS statistical approach. *Environmental Technology & Innovation, 8*, 1–16.

Burton, S. A., & Watson-Craik, I. A. (1998). Ammonia and nitrogen fluxes in landfill sites: applicability to sustainable landfilling. *Waste Management & Research, 16*(1), 41–53.

Chen, V. C., Tsui, K. L., Barton, R. R., & Meckesheimer, M. (2006). A review on design, modeling and applications of computer experiments. *IIE Transactions, 38*(4), 273–291.

Cruz, F. B. D. L., & Barlaz, M. A. (2010). Estimation of waste component-specific landfill decay rates using laboratory-scale decomposition data. *Environmental Science & Technology, 44*(12), 4722–4728.

Dean, A. M., & Voss, D. (1999). *Design and analysis of experiments* (Vol. 1). Springer.

Di Addario, M., Temizel, I., Edes, N., Onay, T. T., Demirel, B., Copty, N. K., & Ruggeri, B. (2017). Development of Fuzzylogic model to predict the effects of ZnO nanoparticles on methane production from simulated landfill. *Journal of Environmental Chemical Engineering, 5*(6), 5944–5953.

Eaton, A. D., & Franson, M. A. (2005). *Water Environment Federation. Standard methods for the examination of water & wastewater*. American Public Health Association.

El-Fadel, M., Bou-Zeid, E., Chahine, W., & Alayli, B. (2002). Temporal variation of leachate quality from pre-sorted and baled municipal solid waste with high organic and moisture content. *Waste Management, 22*(3), 269–282.

El Khatib, D. (2010). *Municipal solid waste in bioreactor landfills: A large scale study*. Ph.D. dissertation, University of Cincinnati.

Eusuf, M. A., Hossain, I., Noorbatcha, I. A., Zen, I. H. (2007). The effects of climate and waste composition on leachate and emissions of gas: A case study in Malaysian context. In *Proceedings of the international conference on sustainable solid waste management* (pp. 437–443).

Fan, L., Abbasi, M., Salehi, K., Band, S. S., Chau, K. W., & Mosavi, A. (2021). Introducing an evolutionary-decomposition model for prediction of municipal solid waste flow: application of intrinsic time-scale decomposition algorithm. *Engineering Applications of Computational Fluid Mechanics, 15*(1), 1159–1175. https://doi.org/10.1080/19942060.2021.1945496

Friedman, J. H. (1991). Multivariate adaptive regression splines. *The Annals of Statistics, 19*(1), 1–67.

García Nieto, P. J., & Álvarez Antón, J. C. (2014). Nonlinear air quality modeling using multivariate adaptive regression splines in Gijón urban area (Northern Spain) at local scale. *Applied Mathematics and Computation, 235*, 50–65. https://doi.org/10.1016/j.amc.2014.02.096. ISSN 0096-3003.

García Nieto, P. J., García-Gonzalo, E., Alonso Fernández, J. R., & Díaz Muñiz, C. (2016). Using evolutionary multivariate adaptive regression splines approach to evaluate the eutrophication in the Pozón de la Dolores lake (Northern Spain). *Ecological Engineering, 94*, 136–151. https://doi.org/10.1016/j.ecoleng.2016.05.047. ISSN 0925-8574.

García Nieto, P. J., Sánchez Lasheras, F., de Cos Juez, F. J., & Alonso Fernández, J. R. (2011). Study of cyanotoxins presence from experimental cyanobacteria concentrations using a new data mining methodology based on multivariate adaptive regression splines in Trasona reservoir (Northern Spain). *Journal of Hazardous Materials, 195*, 414–421. https://doi.org/10.1016/j.jhazmat.2011.08.061. ISSN 0304-3894.

Gettinby, J. H., Sarsby, R. W., & Nedwell, J. C. (1996). The composition of leachate from landfilled refuse. In *Proceedings of the Institution of Civil Engineers-Municipal Engineer* (Vol. 115, No. 1, pp. 47–59). Thomas Telford-ICE Virtual Library.

Gocheva-Ilieva, S. G., Ivanov, A. V., Voynikova, D. S., Stoimenova, M. P. (2020). Modeling of PM10 air pollution in urban environment using MARS. In I. Lirkov & S. Margenov (Eds.), *Large-scale scientific computing. LSSC 2019*. Lecture notes in computer science (Vol. 11958). Springer. https://doi-org.ezproxy.uta.edu/10.1007/978-3-030-41032-2_27.

Gómez Martín, M. A., Antigüedad Auzmendi, I., & Pérez Olozaga, C. (1995a). Landfill leachate: variation of quality with quantity. In T. H. Christensen, R. Cossu, & R. Stegmann (Eds.), *Sardinia'95, Fifth international landfill symposium, S. Margheritha di Pula, Cagliari, Italy, 2–6 October* (Vol. I(III), pp. 345–354). CISA (Environmental Sanitary Engineering Centre).

Gómez Martín, M. A., Antigüedad Auzmendi, I., & Pérez Olozaga, C. (1995b). Multivariate analysis of leachate analytical data from different landfills in the same area. In T. H. Christensen, R. Cossu, & R. Stegmann (Eds.), *Sardinia'95, fifth international landfill symposium, S. Margheritha di Pula, Cagliari, Italy, 2–6 October* (Vol. I(III), pp. 365–376). CISA (Environmental Sanitary Engineering Centre).

He, R., Liu, X., Zhang, Z., & Shen, D. (2007). Characteristics of the bioreactor landfill system using an anaerobic–aerobic process for nitrogen removal. *Bioresource Technology, 98*, 2526–2532.

Johannessen, L. M., & Boyer, G. (1999). *Observations of solid waste landfills in developing countries: Africa, Asia, and Latin America*. Urban Development Division, Waste Management Anchor Team, The World Bank.

Karpušenkaitė, A., Ruzgas, T., & Denafas, G. (2016). Forecasting medical waste generation using short and extra short datasets: Case study of Lithuania. *Waste Management & Research, 34*(4), 378.

Kisi, O., & Parmar, K. S. (2016). Application of least square support vector machine and multivariate adaptive regression spline models in long term prediction of river water pollution.

Journal of Hydrology, 534, 104–112. https://doi.org/10.1016/j.jhydrol.2015.12.014. ISSN 0022-1694.

Kisi, O., Parmar, K. S., Soni, K., & Demir, V. (2017). Modeling of air pollutants using least square support vector regression, multivariate adaptive regression spline, and M5 model tree models. *Air Quality, Atmosphere, & Health, 10*(7), 873–883. https://doi.org/10.1007/s11869-017-0477-9

Kjeldsen, P., Barlaz, M. A., Rooker, A. P., Baun, A., Ledin, A., & Christensen, T. H. (2002). Present and long-term composition of MSW landfill leachate: a review. *Critical Reviews in Environmental Science and Technology, 32*(4), 297–336.

Kouzeli-Katsiri, A., Christioulas, D., & Bosdogianni, A. (1993). Leachate degradation after recirculation. In S. M. di Pula (Ed.), *Proc., Sardinia 93—4th int. landfill symposium* (pp. 1007–1018).

Kulikowska, D., & Klimiuk, E. (2008). The effect of landfill age on municipal leachate composition. *Bioresource Technology, 99*(13), 5981–5985.

Kylefors, K. (2003). Evaluation of leachate composition by multivariate data analysis (MVDA). *Journal of Environmental Management, 68*(4), 367–376.

Kylefors, K., & Lagerkvist, A. (1997, October). Changes of leachate quality with degradation phases and time. In *Sixth international landfill symposium* (pp. 13–17).

Lee, T. S., Chiu, C. C., Chou, Y. C., & Lu, C. J. (2006). Mining the customer credit using classification and regression tree and multivariate adaptive regression splines. *Computational Statistics & Data Analysis, 50*(4), 1113–1130.

Mason, R. L., Gunst, R. F., & Hess, J. L. (2003). *Statistical design and analysis of experiments: With applications to engineering and science*. Wiley.

Mishra, H., Karmakar, S., Kumar, R., & Singh, J. (2017). A framework for assessing uncertainty associated with human health risks from MSW landfill leachate contamination. *Risk Analysis, 37*(7), 1237–1255.

Mohammad-pajooh, E., Weichgrebe, D., & Cuff, G. (2017). Municipal landfill leachate characteristics and feasibility of retrofitting existing treatment systems with deammonification—A full scale survey. *Journal of Environmental Management, 187*, 354–364.

Nacar, S., Mete, B., & Bayram, A. (2020). Estimation of daily dissolved oxygen concentration for river water quality using conventional regression analysis, multivariate adaptive regression splines, and TreeNet techniques. *Environmental Monitoring and Assessment, 192*, 752. https://doi-org.ezproxy.uta.edu/10.1007/s10661-020-08649-9

Nagarajan, R., Thirumalaisamy, S., & Lakshumanan, E. (2012). Impact of leachate on groundwater pollution due to non-engineered municipal solid waste landfill sites of erode city, Tamil Nadu, India. *Iranian Journal of Environmental Health Science & Engineering, 9*(1), 35.

Ogata, Y., Ishigaki, T., Nakagawa, M., & Yamada, M. (2016). Effect of increasing salinity on biogas production in waste landfills with leachate recirculation: A lab-scale model study. *Biotechnology Reports, 10*, 111–116.

Oliveira, P. M. T. D. (2012). *Evaluation of different biological landfill leachate treatment systems for facilities in Portugal*. Masters thesis, University of Porto.

Pilla, V. L., Rosenberger, J. M., Chen, V., Engsuwan, N., & Siddappa, S. (2012). A multivariate adaptive regression splines cutting plane approach for solving a two-stage stochastic programming fleet assignment model. *European Journal of Operational Research, 216*(1), 162–171.

Pohland, F. G., Cross, W. H., Gould, J. P., & Reinhart, D. R. (1992). *The behavior and assimilation of organic priority pollutants codisposed with municipal refuse*. EPA Cooperation Agreement CR-812158, 1.

Rapti-Caputo, D., & Vaccaro, C. (2006). Geochemical evidences of landfill leachate in groundwater. *Engineering Geology, 85*(1–2), 111–121.

Reinhart, D. R., & Grosh, C. J. (1998). Analysis of Florida MSW landfill leachate quality. In *Analysis of Florida MSW landfill leachate quality*. University of Central Florida/CEED.

Robinson, H. D. (1995). The technical aspects of controlled waste management. *A review of the composition of leachates from domestic wastes in landfill sites.* Report for the UK Department of the Environment. Waste Science and Research, Aspinwall & Company, Ltd.

Sang, N. N., Soda, S., Sei, K., & Ike, M. (2008). Effect of aeration on stabilization of organic solid waste and microbial population dynamics in lab-scale landfill bioreactors. *Journal of Bioscience and Bioengineering, 106*(5), 425–432. https://doi.org/10.1263/jbb.106.425

Serrano, N. B., Sánchez, A. S., Lasheras, F. S., Iglesias-Rodríguez, F. J., & Valverde, G. F. (2020). Identification of gender differences in the factors influencing shoulders, neck and upper limb MSD by means of multivariate adaptive regression splines (MARS). *Applied Ergonomics, 82*, 102981.

Sizirici, B., & Tansel, B. (2010). Projection of landfill stabilization period by time series analysis of leachate quality and transformation trends of VOCs. *Waste Management, 30*(1), 82–91.

Sourmunen, K., Ettala, M., & Rintala, J. (2008). Internal leachate quality in a municipal solid waste landfill: Vertical, horizontal and temporal variation and impacts of leachate recirculation. *Journal of Hazardous Materials, 160*(2), 601–607.

Szul, T., & Knaga, J. (2019). Identification and analysis of sets variables for of municipal waste management modelling. *Geosciences, 9*(11), 458. https://doi.org/10.3390/geosciences9110458

Tatsi, A., & Zouboulis, A. (2002). A field investigation of the quantity and quality of leachate from a municipal solid waste landfill in a Mediterranean climate (Thessaloniki, Greece). *Advances in Environmental Research, 6*(3), 207–219.

Tesseme, T., & Chakma, S. (2019). Trend analysis of long-term MSW leachate characteristics. In *Advances in waste management* (pp. 143–153). Springer.

UNEP (United Nations Environment Program) and CalRecovery. (2005). *Solid waste management: Regional overviews and information sources* (Vol. II). International Environmental Technology Centre, UNEP.

U.S. Environmental Protection Agency (EPA). (1996). *Characterization of municipal solid waste in the United States: EPA/530-R-96–001*. Office of Solid Waste.

Zahiri, J., & Nezaratian, H. (2020). Estimation of transverse mixing coefficient in streams using M5, MARS, GA, and PSO approaches. *Environmental Science and Pollution Research, 27*, 14553–14566. https://doi.org/10.1007/s11356-020-07802-8

Zhou, S., Liu, C., & Zhang, L. (2019). Critical review on the chemical reaction pathways underpinning the primary decomposition behavior of chlorine-bearing compounds under simulated municipal solid waste incineration conditions. *Energy & Fuels, 34*(1), 1–15.

Zoughi, M., & Ghavidel, A. (2009). Neural network modeling and prediction of methane fraction in biogas from landfill bioreactors.

Printed in the United States
by Baker & Taylor Publisher Services